Topics in
Current Physics

10

Topics in Current Physics   Founded by Helmut K. V. Lotsch

Synchrotron Radiation

Techniques and Applications

Edited by C. Kunz

With Contributions by
K. Codling   W. Gudat   E. E. Koch
A. Kotani   C. Kunz   D. W. Lynch
E. M. Rowe   B. F. Sonntag   Y. Toyozawa

With 162 Figures

Springer-Verlag Berlin Heidelberg New York 1979

Professor Dr. Christof Kunz

II. Institut für Experimentalphysik, Universität Hamburg,
Luruper Chaussee 149, 2000 Hamburg 50, Fed. Rep. of Germany

ISBN 3-540-09149-1 Springer-Verlag Berlin Heidelberg New York
ISBN 0-387-09149-1 Springer-Verlag New York Heidelberg Berlin

Library of Congress Cataloging in Publication Data. Main entry under title: Synchrotron radiation.
(Topics in current physics ; v. 10) Includes bibliographical references and index. 1. Synchrotron
radiation. I. Kunz, Christof, 1936– II. Codling, K. III. Series. QC787.S9S95 539.7′35 78-24275

Offset printing and bookbinding: Konrad Triltsch, Graphischer Betrieb, Würzburg
2153/3130-543210

Preface

Synchrotron radiation as a spectroscopic research tool has undergone a most interesting and astonishing historical development and has now come to the stage of an exciting boom. The machines which produce synchrotron radiation were built and developed exclusively for other purposes in the past, namely high-energy physics. At the same time, however, they involuntarily became better and better light sources for the spectral range from the visible to the hard x-ray region. Now we are at the point that the first few storage rings have gone into operation as machines dedicated to synchrotron radiation and several more are in the stage of construction and planning. All this was brought about by the successful research performed during the past fifteen years in which several groups all over the world have participated at different accelerator centers mostly symbiotic with high-energy physics. As it happens with a young and rapidly developing field, the number of reviews and monographs is still minute. The objective of this book is to fill an apparent gap and to provide a sound basis for those who are interested in synchrotron radiation and its applications.

The first three chapters are concerned with synchrotron radiation and the instrumentation involved. The history and the fundamental quantitative properties of synchrotron radiation are introduced in Chap. 1. Chapter 2 deals with the machines producing synchrotron radiation in order to provide the necessary basic understanding of the functioning of these light sources, while the set-up of synchrotron radiation laboratories and the often fairly special instrumentation, both for the vacuum ultraviolet (VUV) and the X-rays, is described in Chap. 3.

The next four chapters are mainly concerned with spectroscopy of electronic states with binding energy of less than 500 eV. It was the VUV range which attracted the first great interest in synchrotron radiation, since before its advent no systematic investigation of deep electronic states was possible. Chapter 4 gives the theoretical background, while the following three chapters deal with the experiments performed in atomic physics (Chap.5), molecular physics (Chap.6) and solid-state physics (Chap.7). The central parts of these chapters are concerned with the primary processes, namely absorption and reflection. Secondary processes are frequently incorporated in Chaps. 5 and 6, but they are not exhaustively reviewed here.

VI

The organization of the present Topics volume was guided by the knowledge that other
books are arriving on the horizon. Two volumes of the series Topics in Applied
Physics (namely 26 and 27) are devoted to Photoemission in Solids under the editorship
of M. Cardona and L. Ley, and X-Ray Lithography was already treated in volume 22 of
the same series on X-Ray Optics edited by H.J. Queisser. These treatments together
provide a good overview of current research effort, but they do not exhaustively
cover the subject of synchrotron radiation. This field is even broader. It has found
numerous applications to chemistry and has tremendous importance for the functioning
of the complicated high-energy physics acceletors and storage rings themselves.
Emission of synchrotron radiation from astrophysical objects is also of considerable
interest.

Hamburg, November 1978   *C. Kunz*

Contents

List of Contributors

CODLING, KEITH

   Reading University, J.J. Thomson Physical Laboratory, White Knights,
   Reading RG 62 AF, Great Britain

GUDAT, WOLFGANG

   Institut für Festkörperforschung, Kernforschungsanlage Jülich GmbH,
   Postfach 1913, 5170 Jülich, Fed. Rep. of Germany

KOCH, ERNST-ECKHARD

   Deutsches Elektronen-Synchrotron DESY, 2000 Hamburg 52, Notkestrasse 85
   Fed. Rep. of Germany

KOTANI, AKIO

   The Research Institute for Iron, Steel and Other Metals,
   Tohoku University, Sendai, Japan

KUNZ, CHRISTOF

   Deutsches Elektronen-Synchrotron DESY, 2000 Hamburg 52, Notkestrasse 85
   Fed. Rep. of Germany

LYNCH, DAVID W.

   Ames Laboratory - USDOE and Department of Physics, Iowa State University, Ames,
   Iowa 50010, USA

ROWE, EDNOR M.

   Physical Sciences Laboratory, University of Wisconsin-Madison, P.O. Box
   6, Stoughton, Wisc. 53589, USA

SONNTAG, BERND F.

    II.Institut für Experimentalphysik, Universität Hamburg, Luruper Chaussee
149, 2000 Hamburg 50, Fed. Rep. of Germany

TOYOZAWA, YUTAKA

    The Institute for Solid State Physics, The University of Tokyo,
Roppongi, Minato-Ku, Tokyo 106, Japan

1. Introduction – Properties of Synchrotron Radiation

C. Kunz

With 9 Figures

Synchrotron radiation (SR) is electromagnetic radiation emitted by charged particles moving on circular orbits with highly relativistic velocities. In present day's accelerators and storage rings only the lightest charged particles, electrons and positrons, have a velocity sufficiently near to the velocity of light, c, to give off SR with an intensity of practical importance. Not only synchrotrons but also betatrons, storage rings and any bending magnet in a particle beam line can be a source of SR, the name originates from the fact that its first observation was made at the General Electric synchrotron in Schenectady (USA) in 1946. It is sometimes also called "magnetic bremsstrahlung".

Synchrotron radiation originally was just one of the by-products of high-energy particle acceleration. When the accelerators became larger, SR turned out to be the main mechanism for energy loss and an appreciable fraction of the microwave power fed to the accelerator cavities is needed to compensate this loss. On the other side, the damping mechanism of beam oscillations brought about by SR in storage rings is a positive aspect for accelerator technology, e.g. for the injection procedure.

The mechanism of SR emission is comparable to that of an oscillating dipole, which is the picture one obtains when projecting the circular orbit sideways. The intensity, however, is not confined to the fundamental frequency of revolution which is in the MHz range but due to the δ-function like concentration of charge at the particle and due to the relativistic velocities harmonics contribute up to very high order. The spectrum emitted extends from the visible through the vacuum-ultraviolet, soft X-ray range far into the X-ray range proper (see Fig.1.1). The individual harmonic lines are smeared out thus leading to a continuous spectrum. The following outstanding properties make SR one of the most useful sources for spectroscopy at photon energies above the visible:

1) Continuous spectrum from the infrared to the X-ray region,

2) collimation of the emitted radiation in the instantaneous direction of flight of the emitting particles (angular spread is in the order of 1 mrad),

2

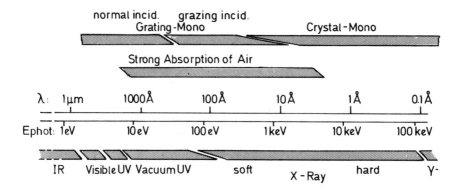

SPECTRAL REGIONS FOR USING SYNCHROTRON RADIATION

Fig. 1.1 Nomenclature and basic facts about electromagnetic radiation from $1 - 10^5$ eV

3) linear polarization with the electric vector parallel to the plane of the orbit,

4) circular polarization above and below the plane of the orbit,

5) time structure with pulse lengths down to 100 ps,

6) absolute calculability of all the properties of the source,

7) cleanliness of the source since light emission takes place in ultrahigh vacuum in contrast to gas discharge or spark lamps.

It is, however, necessary to point out that any high-power laser in its regime surpasses SR in spectral brightness. While a laser has all its power concentrated in a narrow spectral range the large power of synchrotron radiation is spread out over an enormous spectral range in most of which no laser sources yet exist. On the other hand, the recent ideas about SR emission from beams with sinusoidally oscillating paths, the so-called periodic wigglers, can lead to a spectral concentration of SR, if not to a free electron laser in the vacuum ultraviolet or X-ray regions.

Before synchrotron radiation can profitably be applied monochromatization is necessary in most of the cases. A thorough quantitative knowledge of all its properties is needed for the construction of a good instrumentation. After a short historical review in Sect. 1.1 a quantitative treatment of the SR properties is given in Sect. 1.2. While the advantages of SR over classical sources in practically all respects is quite clear in the vacuum ultraviolet region (6 eV - 6 keV), a detailed comparison has to be made for each experiment in the X-ray region. These considerations will be discussed in Sect. 1.3. The references of this chapter are mainly concerned with the basic facts about SR while reviews and papers on instrumentation and applications are cited in the corresponding chapters.

1.1 Historical Development

While IVANENKO and POMERANCHUK (1944) [1.1] and independently somewhat later SCHWINGER (1946) [1.2,3] were the first who worked out the theory of SR for circular particle accelerators, its origin can be traced back to the end of the nineteenth century. LIENARD [1.4] and SCHOTT [1.5-7] showed that an electron moving on a circular orbit is a strong source of electromagnetic radiation. SCHOTT's work was initially an attempt to create a classical model of the stable atom. Quantization of angular momentum, which solved the problems of the nonradiating atom, is a negligible ingredience for the huge "quantum orbits" with 1 - 1000 m diameter in present day's synchrotrons and storage rings. It was shown that quantum mechanical corrections are usually not larger than 10^{-5} [1.3,8-11]. Subsequently to the pioneering theoretical papers practically all aspects of SR emission were clarified [1.12-31]. The only aspect of SR which may need some further considerations is that of coherence, although there have been several treatments concerned with special questions [1.13,29-31].

The first experimental observation of SR is very illustratively discussed in a letter in Physics Today in 1975 by BALDWIN [1.32]. BLEWETT [1.33] was the first experimentalist who became interested in observing SR. In 1946 he measured [1.33] the energy loss due to SR in a betatron as a contraction of the orbit. Since this was a 100 MeV betatron he would have been able also to "see" SR if the vacuum system had not been made out of black ceramics. BLEWETT looked for an emission in the microwave region without success. One year later a technical assistant (Floyd Haber) [1.32] working at one of the first synchrotrons, the 70 MeV General Electric machine, was the first man to "see" SR with the help of a mirror. This observation was correctly explained and published by ELDER et al. [1.34].

Thereafter the properties of SR were studied systematically at several places [1.35-55] in addition to the 70 MeV synchrotron in Schenectady, namely the 250 MeV synchrotron in Moscow, the 300 MeV synchrotron at the Cornell University, the 180 MeV synchrotron at the NBS in Washington, and the 6 GeV synchrotron DESY in Hamburg. The increasing size of these machines is depicted in Fig. 1.2.

Development of Synchrotrons (SY) and Storage Rings (ST)

• < 0.1 GeV	GE BETATRON	1945		
	GE SYNCHROTRON	1946		
○ < 0.5 GeV	CORNELL I SY	1953	TANTALUS ST	1968
	MOSCOW I SY	1956	ACO ST	1973*
	NBS SY	1961	SURF II ST	1975
	BONN I SY	1962	INS-SOR ST	1976
	GLASGOW SY	1971*		
○ < 2 GeV	MOSCOW II SY	1961	VEPP 2M ST	1976
	CORNELL II SY	1961	DCI ST	1977
	FRASCATI SY	1962	ADONE ST	1978+
	TOKYO SY	1963	SRS ST	1979
	LUSY SY	1976*	ALADDIN ST	1980
			BROOKHAVEN I ST	1981
			BERLIN ST	1981
			PAMPUS ST	1982
○ < 8 GeV	CEA SY	1964	SPEAR ST	1973
	DESY SY	1964	DORIS ST	1974
	NINA SY	1966	VEPP 3 ST	1977
	BONN II SY	1968	VEPP 4 ST	1978
	ARUS SY	1970	BROOKHAVEN II ST	1981
			PHOTON FACTORY ST	1981
> 8 GeV			PETRA ST	1978
			PEP ST	1979
			CESR ST	1979

*Approximate start of SR activities

Fig. 1.2 Growth of sources for SR since 1945. The circles drawn are approximately on scale

TOMBOULIAN and co-workers were the first to demonstrate the feasibility to do useful spectroscopy in the vacuum ultraviolet with SR [1.44] at Cornell, while this was exploited systematically for the first time from 1961 on by MADDEN and CODLING [1.47-50] at the NBS predominantly for clarifying the absorption behaviour of rare gases in the vacuum ultraviolet.

It is interesting to ask the question why it took a fairly long time from the first observation of SR to its useful and systematic application. There are probably two reasons, the first is the relative complexity of this light source compared to what spectroscopists were used before, and the second reason has its origin in the fact that up to very recently practically all the accelerators and storage rings producing SR were in the hands of high-energy particle physicists and were optimized and scheduled according to their needs. Long-term prospectives looked reliable only at a few places in order to attract spectroscopists to make a long-term investment. It took another 15 years after 1960 before the first storage ring designed and built specifically as a light source went into operation [1.56]. Although our present day's storage ring sources would not have been developed to such high standards without the enormous potential of high-energy physics and although the quality of their machines is not so different from what is ideal for spectroscopy the large expansion of SR came only with the advent of the idea of storage rings dedicated to SR work.

Table 1.1 Synchrotrons used as light sources. [E: particle energy; R: magnet radius; I: maximum current (during the acceleration period); ε_c: characteristic photon energy; SR: synchrotron radiation]

Name	Location	E[GeV]	R[m]	I[mA]	ε_c[eV]	Remarks
Group I, ε_c = 1 - 60 eV						
PTB	Braunschweig	0.14	0.46		13	
SURF I	Washington	0.24	0.83	10	37	closed down
MOSCOW I	Moscow	0.25			~40	closed down
Group II, ε_c = 60 - 2000 eV						
GLASGOW	Glasgow	0.33	1.25	0.1	64	closed down
CORNELL I	Ithaca	0.32	1		73	closed down
BONN I	Bonn	0.5	1.7	30	163	SR Lab
C-60	Moscow	0.68	2	10	349	SR Lab
CORNELL II	Ithaca	0.7	1.5		508	closed down
FRASCATI	Frascati	1.1	3.6	15	821	closed down
LUSY	Lund	1.2	3.65	10	1050	SR station
INS-SOR I	Tokyo	1.3	4.0	60	1220	SR Lab
SIRIUS	Tomsk	1.3	4.2	20	1160	SR Lab
PACHRA	Moscow	1.3	4.0	100	1220	dedicated, SR Lab
Group III, ε_c = 2 - 30 keV						
BONN II	Bonn	2.5	7.65	30	4530	SR Lab
NINA	Daresbury	5.0	20.8	50	13,300	closed down
CEA	Cambridge/Mass.	6.0	26.0	30	18,400	closed down
ARUS	Yerevan	6.0	24.65	20	19,500	SR Lab
DESY	Hamburg	7.5	31.7	30	25,500	2 SR Labs
Group IV, $\varepsilon_c \geq$ 30 keV						
CORNELL III	Ithaca	12	120	2	32,000	closed down

Table 1.2 Storage rings (most of them dedicated as light sources). [E: particle
energy; R: magnetic radius; I: maximum current; ϵ_c: characteristic photon energy;
SR: synchrotron radiation]. The numbers on the proposed storage rings usually are
subject to changes; in the wiggler sections ϵ_c will be much larger than the values
quoted for bending magnets.

Name	Location	E[GeV]	R[m]	I[mA]	ϵ_c[eV]	State
Group I, $\epsilon_c \leq 60$ eV						
TANTALUS I	Stoughton	0.24	0.64	100	48	dedicated
SURF II	Washington	0.24	0.83	30	37	dedicated
INS-SOR II	Tokyo	0.30	1.1	200	54	dedicated
Group II, ϵ_c = 60 - 2000 eV						
ACO	Orsay	0.55	1.11	100	333	dedicated
VEPP-2M	Novosibirsk	0.67	∿2	∿100	∿350	SR Lab
BROOKHAVEN I	Upton	0.70	∿2	1000	∿400	dedicated, plan. stage add. wigglers
ALADDIN	Stoughton	0.75	2.08	1000	450	dedicated, plan. stage add. wigglers
BESSY	Berlin	0.8	1.8	300	630	planning stage
SILVA	California	0.8	∿1.8	300	630	proposed
ADONE	Frascati	1.5	5.0	60	1500	SR Lab
PAMPUS	Amsterdam	1.5	4.17	500	1800	dedicated, proposed add. wigglers
Group III, ϵ_c = 2 - 30 keV						
SRS (Nina II)	Daresbury	2.0	5.55	1000	3200	dedicated, add. wiggles under con-struction
BROOKHAVEN II	Upton	2.5	∿8.0	1000	4300	dedicated, plan. stage add. wigglers
DCI	Orsay	1.8	3.82	400	3390	SR Lab
MOSCOW	Moscow	2	5	1000	3500	dedicated, proposed
VEPP 3	Novosibirsk	2.2	6.15	80-500	3800	SR Lab, wiggler, under construction
PHOTON FACTORY	Japan	2.5	8.0	500	∿4300	dedicated, plan. stage
DORIS	Hamburg	5	12.12	100	22900	2 SR Labs
SPEAR	Stanford	4	12.7	60	11200	SR Labs
VEPP 4	Novosibirsk	6	33	100	14500	under constr., SR Labs
GROUP IV, $\epsilon_c \geq 30$ keV						
CESR	Ithaca	8	32	100	35000	under construction, SR Lab
PEP	Stanford	15	170	100	44000	under construction
PETRA	Hamburg	19	200	90	75000	

From 1960-1970 several larger centers developed at places where favorable conditions were found: NBS, Washington [1.50], INS, Tokyo [1.57,58], DESY, Hamburg
[1.55], this was augmented by a few smaller activities like the ones in Bonn and
Frascati. Only after about 1970 TANTALUS in Stoughton, Wisconsin [1.59-61], as
the first storage ring, not built, but operated exclusively for SR work, came into

full operation. After 1974 the same conditions became true for ACO at Orsay [1.62].
1976 the 300 MeV INS-SOR storage ring in Tokyo [1.56] went into operation, the first
ring designed as a light source. At about the same time SURF II at the NBS,
Washington, came into action [1.63]. Of the large storage rings presently available
we want to mention especially VEPP 2M and VEPP 3 at Novosibirsk [1.64], SPEAR at
Stanford [1.65] and DORIS at Hamburg [1.66]. The latter two have expansion programs
underway which are described for DORIS [1.67,68] and for SPEAR [1.65]. Further expansions of the laboratory are planned for ADONE at Frascati. New dedicated storage
rings are under preparation with SRS at Daresbury [1.69], ALADDIN at Stoughton
[1.70], two storage rings at the National Synchrotron Light Source Center at
Brookhaven [1.71], PAMPUS in Amsterdam [1.72], the Photon Factory project in Japan
and a small storage ring in Berlin. Several more dedicated storage rings are proposed
all over the world, especially for industrial applications, and several studies on
future needs have been prepared [1.72]. Tables 1.1 and 1.2 give a compilation of all
the machines which are operated or planned as SR sources, see also [1.73]. The SRS
and BROOKHAVEN II machines will install wiggles at high magnetic fields to shift
ε_c far beyond the values given in Table 1.2. Many further references from the early
days of SR can be found in a bibliography by MARR et al. [1.74].

1.2 Quantitative Properties

1.2.1 Equations for Ideal Orbits

The derivation of the fundamental equations describing the emission of SR can be
found nowadays in modern textbooks of electrodynamics [1.75-77]. There are also
several reviews on this topic [1.44,55,78-82]. It starts from the equation for I,
the radiated power per unit time interval of an electron moving on an arbitrary path
$\underline{r}(t)$ with velocity $\underline{v}$, momentum $\underline{p} = \gamma m\underline{v}$ and energy $E = \gamma mc^2$ [1.77]

$$I = \frac{2e^2}{3m^2c^3}\left[\left(\frac{d\underline{p}}{d\tau}\right)^2 - \frac{1}{c^2}\left(\frac{dE}{d\tau}\right)^2\right] \tag{1.1}$$

where e is the electron charge, m its mass, c the velocity of light and $d\tau = dt/\gamma$
the element of proper time.

In specializing this in a fairly straightforward manner for a single electron
moving with constant velocity v on a circular orbit (the second term in the square
bracket vanishes) one obtains the power $I(\lambda,\psi)$ radiated off around the whole orbit
in cgs units in erg/(s · electron) into a wavelength interval $d\lambda$ and an interval $d\psi$
of the azimutal angle ψ (see Fig.1.3):

$$I(\lambda,\psi) = \frac{27}{32\pi^3} \frac{e^2 c}{R^2} \left(\frac{\lambda_c}{\lambda}\right)^4 \gamma^8 \left[1 + (\gamma\psi)^2\right]^2 \left\{K^2_{2/3}(\xi) + \frac{(\gamma\psi)^2}{1 + (\gamma\psi)^2} K^2_{1/3}(\xi)\right\} \quad , \quad (1.2)$$

with

$\lambda_c = \frac{4\pi R}{3} \gamma^{-3}$ the "characteristic wavelength"   ,

$$\xi = \frac{\lambda_c}{2\lambda} \left[1 + (\gamma\psi)^2\right]^{3/2} \quad , \qquad\qquad (1.3)$$

$\gamma = \frac{E}{mc^2}$,

where R denotes the radius of curvature, $K_{1/3}$ and $K_{2/3}$ the modified Bessel functions of the second kind [1.83].

The two terms in the curly brackets of (1.2) are associated with the intensities in the two directions of polarization, $I_\parallel$ and $I_\perp$, having the electric vector parallel and perpendicular to the plane of the orbit. If we define the linear degree of polarization P_L as usual we obtain

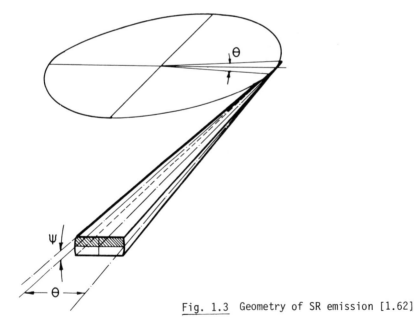

Fig. 1.3 Geometry of SR emission [1.62]

$$P_L = \frac{I_{\shortparallel} - I_{\perp}}{I_{\shortparallel} + I_{\perp}} = \frac{K^2_{2/3}(\xi) - \dfrac{(\gamma\psi)^2}{1 + (\gamma\psi)^2} K^2_{1/3}(\xi)}{K^2_{2/3}(\xi) + \dfrac{(\gamma\psi)^2}{1 + (\gamma\psi)^2} K^2_{1/3}(\xi)} \tag{1.4}$$

In Fig. 1.4 $I_{\shortparallel}$, $I_{\perp}$ and $P_{\perp}$ are plotted for a specific storage ring DORIS [1.84]. Since the two components of the electric vector have a well defined phase relation with respect to each other, namely $+ \pi/2$ or $- \pi/2$ above or below the plane of the orbit, respectively, one can also calculate a degree of circular polarization P_c, with the decomposition of the elliptically polarized wave into right- and left-hand waves with intensities I_R and $I_{\perp}$.

$$P_c = \frac{I_R - I_L}{I_R + I_L} = \pm \frac{\sqrt{I_{\shortparallel} I_L}}{I_{\shortparallel} + I_{\perp}} \tag{1.5}$$

where the positive and negative signs correspond to $\psi > 0$ or $\psi < 0$, respectively.

Equation (1.2) can be integrated over all azimutal angles ψ yielding the total power radiated by a single electron per second and unit wavelength interval, namely

$$I(\lambda) = \frac{3^{5/2}}{16\pi^2} \frac{e^2 c}{R^3} \gamma^7 (\frac{\lambda_c}{\lambda})^3 \int_{\lambda_c/\lambda}^{\infty} dn \, K_{5/3}(n) \tag{1.6}$$

DORIS 3.5 GeV

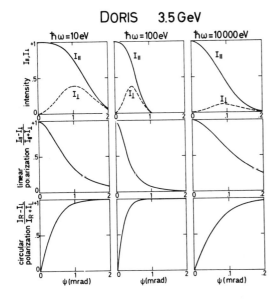

Fig. 1.4 Angular distribution of the two components $I_{\shortparallel}$ and $I_{\perp}$ (parallel and perpendicular to the orbit). Also the linear degree of polarization according to (1.4) and the circular degree of polarization according to (1.5) are given. [1.84]

where $K_{5/3}$ is another Bessel function of the second kind [1.83]. Since (1.6) depends only on the ratio λ/λ_c and the machine parameters R and E this function can be obtained from a universal function $F(\lambda/\lambda_c)$ with the appropriate prefactors. Appropriate functions of the type $F(\lambda/\lambda_c)$ have been graphically displayed by several authors. Nowadays most of the SR centers have developed computer programs which allow for a rapid calculation of the intensities for different geometries.

When plotting (1.6) as a function of λ, the intensity goes through a maximum value at $\lambda_m = 0.42 \lambda_c$. Experiments have been performed at wavelengths as short as $\lambda_m/4$. When looking at Fig. 1.5 one realizes that the intensity drops very rapidly below λ_c. It is also noted from this figure that other useful plots as, e.g., the number of photons per second and photon energy interval must not go through a maximum at all. Thus λ_m is a fairly arbitrary quantity.

For practical cases the power distribution in ψ can be of interest. In this case (1.2) can be integrated over all wavelengths yielding

$$I(\psi) = \frac{e^2 c}{R^2} \gamma^5 [1 + (\gamma\psi)^2]^{-5/2} \left[\frac{7}{16} + \frac{5}{16} \frac{(\gamma\psi)^2}{1 + (\gamma\psi)^2} \right] \quad . \tag{1.7}$$

Further integration finally yields the total power emitted by one single electron

$$I_{tot} = \frac{2}{3} \frac{e^2 c}{R^2} \gamma^4 \tag{1.8}$$

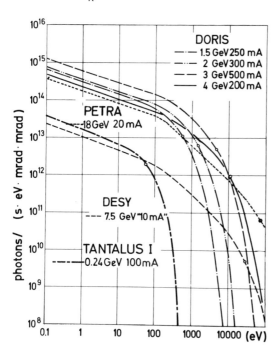

Fig. 1.5 Spectral distribution of intensity in an aperture 1 mrad wide and 1 mrad high centered at a tangential direction. The open circles indicate the characteristic energy ε_c according to (1.13). This aperture is well filled at low-photon energies while only the part near the orbit is illuminated with hard X-rays. While 1 mrad horizontally is typically accepted by an experiment at a large storage ring, 10 mrad can easily be accepted at a small storage ring like, e.g. TANTALUS I.

All these expressions for intensities, (1.2), (1.6), (1.7), and (1.8), can be transformed into intensities per unit current by multiplying the right-hand sides by $2\pi R/(ec)$. Then the current has to be measured in electrostatic units. If the current is measured in Ampere, the above equations must be multiplied by $3.10^9\ 2\pi R/(ec)$. In transforming the intensities per electron into intensities per unit current all those problems with actual machines which deviate from ideal circles due to the insertion of straight sections are eliminated. It is quite obvious that the only quantity which determines the observed intensity coming from the curved sections of a machine is the current which passes through these sections. For questions concerning the linear dependence of SR intensity with particle number see Sect. 1.2.2 below.

In the following we note in short a few useful expressions. Several more can be found in the report by GREEN [1.85].

$$B[k\ gauss] \cdot R[m] = 33.35\ E[GeV] \tag{1.9}$$

where B is the magnetic field.

$$\delta E[keV] = 88.5\ E^4[GeV]/R[m] \tag{1.10}$$

is the energy loss of one particle per turn.

$$\gamma = 1957\ E[GeV], \tag{1.11}$$

$$\lambda_c[\text{Å}] = 5.59\ R[m]/E^3[GeV] = 186.4/(B[k\ gauss]E^2[GeV]), \tag{1.12}$$

$$\epsilon_c[eV] = 2218\ E^3[GeV]/R[m] = 66.51\ B[k\ gauss]E^2[GeV], \tag{1.13}$$

with $\epsilon_c = hc/\lambda_c$, h: Planck's constant, $\epsilon_c[eV] = 12400/\lambda_c[\text{Å}]$.

$$I[phot/(s \cdot eV \cdot mrad)] \approx 4.5 \cdot 10^{12}\ R^{1/3}[m]\epsilon^{-2/3}[eV]j[mA]$$
$$\text{for } \epsilon << \epsilon_c, \tag{1.14}$$

where I is the intensity integrated over all angles ψ per horizontal angular intervall $\Theta = 1$ mrad (see Fig.1.3) and j is the current.

The total power radiated is

$$I[Watt] = 88.5\ E^4[GeV]j[mA]/R[m]. \tag{1.15}$$

The angular spread $\Delta\psi$ (FWHM) of intensity can be approximated [1.77]as

$$\Delta\psi \approx \frac{2}{\gamma}\left(\frac{\epsilon_c}{\epsilon}\right)^{1/3} \quad \text{for } \epsilon << \epsilon_c$$

$$\Delta\psi \approx \frac{2}{\gamma}\left(\frac{\epsilon_c}{3\epsilon}\right)^{1/2} \quad \text{for } \epsilon >> \epsilon_c \tag{1.16}$$

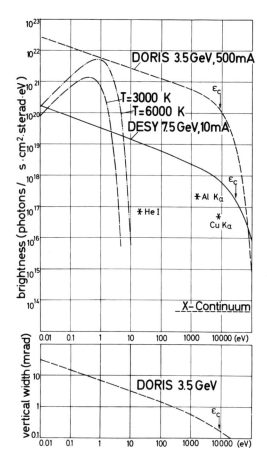

Fig. 1.6 Brightness of SR of DORIS (Based on source size 1 × 10 mm²) DESY (source size 3 × 10 mm²), Cu Kα characteristic radiation and Bremsstrahlung from a 60 kW X-ray tube (estimated from an effective source size of 1 × 1 mm²), Al Kα characteristic radiation from a 5 kW X-ray tube (2 mm diameter spot size) and He I resonance line [1.111] (estimated source size 10 mm diameter, 20 meV line width and a guessed collimation of the 10^{13} emitted photons quoted into 0.01 sterad). Brightness of black-body radiation is also given. [1.86]

The luminosity $\eta(\psi,\epsilon)$ is defined as the number of photons per unit area of the source, A, (which is the cross section of the beam) and per unit solid angle. In contrast to many classical sources $\eta(\psi,\epsilon)$ is a very anisotropic quantity. From (1.16) and (1.14) we obtain for $\epsilon \ll \epsilon_c$ as the maximum luminosity in the plane of the orbit

$$\eta(0,\epsilon) \; \alpha \; jR^{2/3} \; A^{-1}\epsilon^{-1/3}, \tag{1.17}$$

where A is the area of the source. This is plotted for a practical case in Fig.1.6 together with the $\Delta\psi$ which was not obtained from the approximations (1.16) but from an actual computer calculation [1.86,87].

1.2.2 Considerations for Real Orbits

a) Coherence

In several equations of the previous section it is implicitly assumed that SR emission depends linearly on the number of electrons (the current). This holds only true as long as the positions of the electrons on the orbit are distributed statistically on the scale of the wavelength of light under consideration. Definitely no such correlation can be expected for optical and X-ray wavelengths. On the other hand, at very long wavelengths which are much longer than the separation of the individual bunches of electrons in the machine there is no emission, a direct current does not emit radiation. This is a consequence of coherence. We further mention the very strong radiation at the wavelength which belongs to the regular separation of bunches on the orbit. This radiation is coherent with the particle current and can be quadratic in intensity with the number of electrons per bunch. Interactions with the microwave cavities and the vacuum pipe occurs through this emission [1.88]. The radiation can lead to an indirect coupling of the oscillation of different bunches along the orbit. In this case one deals with storage-ring machine physics. Only the thorough investigation of these processes can lead to a better understanding and mastering of storage rings with very high currents (see also Chap. 2). Theoretical treatments of the coherence problems can be found in [1.29-31].

WAVELENGTH SHIFTER

MULTIPOLE WIGGLER MAGNET

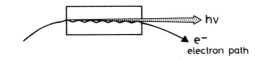

HELICAL WIGGLER

Fig. 1.7 Different types of wiggles: A wavelength shifter is just a sharp bend which serves to save energy in dedicated storage rings as SRS, BROOK-HAVEN II, PHOTON FACTORY (see Table 1.2) by localizing the X-ray emission at a few points along the orbit. Multipole wigglers can be of planar or helical shape. Both serve to obtain high intensities in narrow spectral bands. [1.68]

b) Periodic Wigglers

A special case of coherence is achieved with periodic wigglers also called undulators [1.89-93]. The device generates a periodic deflection of a beam by small angles in a straight section of the storage ring (see Fig.1.7). Up to one hundred wiggles are planned for such devices. The technical part is described in Chap. 2. In this case coherent emission of a single particle from the different equivalent points along its path is obtained. The coherence condition requires that the time difference for an electron and the light at the distance of two wiggles λ_o is equal to the period of the light wave. This leads to a peak in the spectrum at a wavelength

$$\lambda_p \approx \lambda_o/\gamma^2 \; . \tag{1.18}$$

The intensity at the peak position is proportional to n^2, where n is the number of wiggles. As for an example, we take E = 5 GeV, λ_o = 10 cm obtaining $\lambda_p \approx 10 \; \overset{\circ}{\text{A}}$. Linear and helical wigglers have been proposed. Their emission differs with respect to the angular distribution and the polarization, predominantly linear and predominantly circular respectively [1.89]. Visible radiation from a linear wiggler in a synchrotron was recently measured successfully at Tomsk [1.93a]. In this context stimulated emission within a wiggler is also discussed. After this effect was demonstrated [1.94-96] in the infrared, hopes have been raised for the production of a free electron X-ray laser [1.97]. Most probably this goal is not achievable without the production of a high efficiency resonator which appears to be very difficult from present knowledge.

c) Synchrotron Accelerators

Another modification of the equations in the previous section is needed to describe synchrotrons in which electrons are accelerated according to $\sin^2\omega_o t$, $\sin\omega_o t$ or other more complicated laws like e.g. in the so called "flat top" operation [1.98]. Some of the necessary integrations can be carried out analytically (see e.g. [1.55]) but it is usually more convenient to do a numerical computer integration for actual cases.

d) Beam Cross Section and Divergency

We now want to consider the modifications which are due to the actual cross section area and the divergency of the particle beam. Usually the linear beam dimensions are considerably less than 10 mm and the divergency in the vertical direction is less than 1 mrad. The particle dimensions σ_x, σ_y at any position of the orbit are obtained from the β-functions and the emittances ϵ [(1.85,99,100] as e.g. $\sigma_y = \sqrt{\epsilon\beta}$ (see Chap.2). σ_y is one standard deviation of a gaussian distribution so that the full width at half maximum is given by

$$\text{FWHM}_y = 2.35 \; \sigma_y = 2.35 \; \sqrt{\epsilon\beta} \; . \tag{1.19}$$

The angular divergency of the particle beam is usually negligible compared to the natural divergency of SR emission in the vacuum ultraviolet. It can be very important, however, in the X-ray region. The angular divergency is given by

$$\text{FWHM}_{y'} = 2.35 \, \sigma_{y'} = 2.35 \, \sqrt{\frac{\epsilon}{\beta}} \sqrt{1 + \frac{\beta'^2}{4}} \,. \tag{1.20}$$

Note the importance of the second square root factor which is due to the oblique orientation of the phase space ellipse [1.85]. Usually the angular distribution of SR is approximated by a gaussian distribution and then the total angular width is obtained by quadratic addition [1.85,101] (see also Fig.3.7). As a consequence the maximum brightness as described by (1.17) and shown in Fig.1.6 can be reduced in practical cases considerably in the X-ray regime especially for the highest energies. Further, the degree of linear polarization in the plane of the orbit and the degree of circular polarization above and below this plane are reduced. Thus, design goals for dedicated X-ray machines should rather aim at a reduction of beam divergency (with the exception of a few ports for special experiments) while the main goal for vacuum ultraviolet machines is a reduction of the beam size at all beam ports in order to increase the brightness.

It should be mentioned that (1.19) and (1.20) are calculated usually for a beam optics with an emittance which depends on the beam energy E like $\epsilon \alpha E^2$. Electrostatic and electromagnetic interactions of the particles within the beam at higher currents will usually lead to an increase of ϵ with j [1.88,101].

1.2.3 Time Structure

Time structure is impregnated onto the light beam from different origins.:
1) In a sychrotron there is a natural repetition rate due to the acceleration cycle, e.g. 50 Hz at DESY. The electrons are accelerated usually for 10 ms from low energies up to a maximum energy of e.g. 7.5 GeV. Depending on the wavelength interval under inspection, radiation occurs during the last 2-7 ms. At the 10th ms electrons are extracted and a pause occurs till the 20th ms when the cycle is repeated.

2) In both synchrotrons and storage rings any structure in the charge distribution around the orbit is repeated with the frequency of revolution. This has, e.g., at DESY and DORIS a periodicity of 1μs. In synchrotrons the bunch pockets are usually filled only for 3/4 of the circumference. Storage rings are frequently operated with only one bunch filled. In this case extremely short flashes (see below) are repeated at long intervals.

3) Only discrete stable positions occur around the orbit with a separation determined by the microwave frequency. If all these pockets are filled this leads to a pulse structure with 2 ns intervals at DORIS. The pulse length is in the order of 200 ps.

16

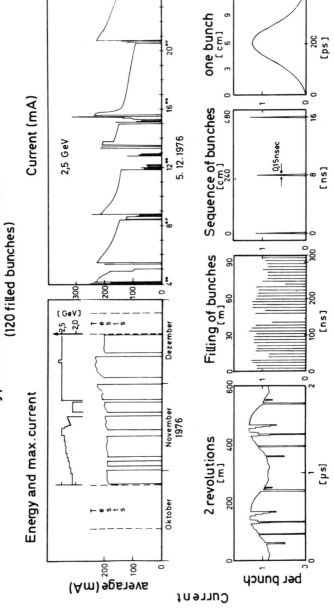

Fig. 1.8 Typical time structure of the SR emission from a storage ring (DORIS) at different expansions of the time scale. [1.102]

Figure 1.8 gives a survey of the time structure of Doris on all levels of time
scale, even the very long term operational aspects are shown as accumulated by
KOSUCH [1.102]. Time structure has several aspects for the experimentalist:

a) The duty cycle has consequences for dead-time and count-loss considerations. It
   must be carefully considered when planning and evaluating an experiment.

b) Pulsed excitation of decaying processes like luminescence is a favourable tool
   for obtaining decay times. In addition transient phenomena can be observed with
   time resolution.

c) The luminosity of a SR source, as shown in Fig.1.6, is a time averaged luminosity.
   The peak luminosity during the pulse can be higher by three orders of magnitude.
   This property could be applied to the investitation of nonlinear processes and
   photon mixing.

1.3 Comparison with Other Sources

The final decision of an experimentalist to use SR instead of a conventional light
source will usually be based on many aspects and involves all the details of an experiment. Intensity, nevertheless, in many cases is the dominant criterion and therefore we shall give detailed discussions here of this property of SR and other
comparable light sources. Fig.1.6, showing the brightness, is a good starting point
for such an analysis [1.85,103]. For actual experiments it is always necessary to
remember that the differences in brightness come into play only if the acceptance
a' of the experiment is matched to the emittance e' of the SR source. In many cases
an experiment will at least in the vertical direction, accept more than what is
emitted from a SR source. In this case the classical source gains by roughly a factor a'/e' if a' and e' are steric acceptance and emittance, respectively. Although
a SR source emits horizontally into 2π it is fairly difficult to extract more than
50 mrad from a magnet chamber in spite of mirror concepts [1.104] showing that this
is principally possible. The intensity ratio of an experiment with SR compared to a
classical source is therefore given by

$$\frac{I_{SR}}{I_{CL}} = \frac{\eta_{SR}}{\eta_{CL}} \cdot \frac{e'}{a'} \quad , \tag{1.21}$$

where η_{SR} and η_{CL} are the respective brightnesses.

1.3.1 Infrared and Visible Range

The brightness of a very intense SR source, like DORIS, at a wavelength of 100 μm (0.01 eV) is by about two orders of magnitude higher than that of a high pressure mercury vapor lamp. The development of tunable lasers, however, has discouraged prospective users. On the other hand, there are still gaps in the spectrum not covered by lasers and Fourier spectroscopy should be an appropriate technique. Two detailed papers are devoted to the infrared [1.105,106]. The visible and the near ultraviolet is covered by a sufficient number of other sources and therefore not of real interest for SR work. Nevertheless the pulsed time structure was used in the UV for an experiment to measure luminescence decays [1.107]. This is quite simply achieved since SR in this region can be extracted through a window.

1.3.2 Vacuum Ultraviolet Range

Table 1.3 gives a compilation of the most important continuum and resonance-line sources [1.108,109]. It is easily recognized that none of these sources covers a very large spectral range. The rare gas continuum sources, nevertheless, are fairly important. They give an intensity behind a monochromator which usually is by one order of magnitude less than that from a synchrotron and further reduced in comparison to a storage ring. The peak intensity for the He continuum at 800 Å was reported to be in the order of 10^9 Photons/(s·Å) [1.109]. There exist, however, some versions of these lamps which are optimized to a point that they can almost compete

Table 1.3 VUV and soft X-ray radiation sources. [1.108]

Source	Photon energy range [eV]	Linewidth	Remarks
He continuum	$12 \leq \hbar\omega \leq 21$	-	
Ne continuum	$12.4 \leq \hbar\omega \leq 16.8$	-	pressures of 50-200 torr
Ar continuum	$8 \leq \hbar\omega \leq 11.8$	-	
Kr continuum	$6.9 \leq \hbar\omega \leq 9.9$	-	
Xe continuum	$6.2 \leq \hbar\omega \leq 8.4$	-	
Hinteregger lamp	$4 \leq \hbar\omega \leq 14$	multiline	
He I resonance lamp	21.2	∿1 meV ?	
He II resonance lamp	40.8	<10 meV	pressures of 0.1 - 0.5 torr
Ne I resonance lamp	16.8	∿1 meV	
Ne II resonance lamp	26.9	<10 meV?	
BRV Source	$4 \leq \hbar\omega \leq 250$	continuum and multiline	10^{-4} torr
YM X-ray	132.3	0.5 eV	
ZrM X-rays	151.4	0.8 eV	
NbM X-rays	171.4	1.2 eV	
RhM X-rays	260.4	4.0 eV	
TiM X-rays	452	-	
MgK X-rays	1254	∿0.7 eV	
Al K_α X-rays	1487	∿0.8 eV	
Cu K_α X-rays	8055	∿2.5 eV	

with a SR source at one point in the spectrum [1.110]. Definitely the HeI resonance lamp used for photoemission experiments at 21.2 eV is superior to a SR source if it is used without a monochromator. In the order of $5 \cdot 10^{11}$ Photons/s at the sample have been reported [1.111]. This intensity might be achieved one day also with very favourably matched monochromators at dedicated storage rings. In this context also the recent development of excimer lasers in the energy region below 10 eV has to be mentioned [1.111a]. In summary, SR definitely is superior to any other source in the vacuum ultraviolet if a source is needed which is tunable over a wide spectral range.

1.3.3 X-Rays

As mentioned already before, the intensity of SR experiments in the X-ray region needs a more careful analysis than in the vacuum ultraviolet [1.86,103,112]. In the X-ray region powerful X-ray generators exist [1.113] which consume up to 100 kW of power for rotating copper anodes and in the order of 5 kW for rotating aluminum anodes. As a rule of thumb 10^{-3} of the power is emitted as X-rays, one half as the bremsstrahlen continuum and one half as characteristic lines. From Fig.1.6 it is quite evident that the bremsstrahlen continuum cannot really compete with SR. Thus all the experiments which need tunability profit considerably from the very high intensity of SR. In addition, the good collimation makes simple plane crystal mono-chromators very efficient.

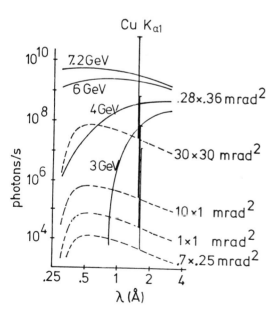

Fig. 1.9 Quantitative comparison of the intensity of SR from DESY at different energies and of typical Cu anode X-ray tube (45 keV, 50 mA) in different apertures. [1.112]

For experiments, which need only discrete wavelengths, a comparison with the characteristic X-ray lines is appropriate unless a well defined wavelength is needed which might not be available. If good collimation is necessary as for X-ray topography then again the high brightness of SR dominates. In cases, like photoelectric emission, monochromators can be constructed which accept 100×100 mrad^2. In this case the ratio a'/e' is $5 \cdot 10^3$ for a 10×0.2 mrad^2 emission and equal dimensions of the source. In this case other criteria may be involved when making a decision whether or not it might be worthwhile to use SR (see Sect. 3.6.2).

RABE [1.112] has given a detailed comparison not for the optimum case but for a more practical case which is summarized in Fig.1.9. He compared SR from DESY with X-rays from a 45 kV, 50 mA (about 2.2 kW) X-ray tube. Advantage or disadvantage of SR depend very much on the angular acceptance of the experiments.

1.4 Acknowledgements

The author wishes to thank numerous colleagues at DESY and at the other SR laboratories who have helped in many discussions to clarify the matter presented in this chapter. Special thanks go to E. Thumann for typing and preparing the manuscript and to V. Fischer, W. Knaut, J. Schmidt and M. Sörensen for producing the figures.

References

1.1 D.D. Ivanenko, I. Pommeranchuk: Phys. Rev. 65, 343 (1944)
1.2 J. Schwinger: Phys. Rev. 70, 798 (1946)
1.3 J. Schwinger: Phys. Rev. 75, 1912 (1949)
1.4 A. Lienard: L'Eclairage Electr. 16, 5 (1898)
1.5 G.A. Schott: Ann. d. Physik 24, 641 (1907)
1.6 G.A. Schott: Phil. Mag. 13, 189 (1907)
1.7 G.A. Schott: *Electromagnetic Radiation* (Cambridge University Press, Cambridge, 1912)
1.8 A.A. Sokolov, N.P. Klepikov, I.M. Ternov: J. Exper. Theoret. Phys. (USSR) 23, 632 (1952)
1.9 J. Schwinger: Proc. Nat. Acad. Sci. (USA) 40, 132 (1954)
1.10 A.A. Sokolov: Nuovo Cim. Suppl. 3.X, 743 (1956)
1.11 F. Gutbrod: Z. Physik 168,177 (1962)
1.12 L. Artsimovitch, I. Pommeranchuk: J. Exper. Theoret. Phys. (USSR) 16, 379 (1946)
1.13 L.I. Schiff: Rev. Sci. Instr. 17, 6 (1946)
1.14 D.D. Ivanenko, A.A. Sokolov: Doklady Akad. Nauk SSSR 59, 1551 (1948)
1.15 G. Parzen: Phys. Rev. 84, 235 (1951)
1.16 D.L. Judd, J.V. Lepore, M. Ruderman, P. Wolff: Phys. Rev. 86, 123 (1952)
1.17 H. Olsen, H. Wergeland: Phys. Rev. 86, 123 (1952)
1.18 L.I. Schiff: Am. J. Phys. 20, 474 (1952)
1.19 A.A. Sokolov, N.P. Klepikov, I.M. Ternov: J. Exper. Theoret. Phys. (USSR) 24, 249 (1953)
1.20 A.A. Sokolov, N.P. Klepikov, I.M. Ternov: Doklady Akad. Nauk SSSR 89, 665 (1953)
1.21 A.A. Sokolov, I.M. Ternov: J. Exper. Theor. Phys. (USSR) 25, 698 (1953)
1.22 A.A. Sokolov, I.M. Ternov: Doklady Akad. Nauk SSSR 92, 537 (1953)

21

1.23 M. Neumann: Phys. Rev. 90, 682 (1953)
1.24 A.A. Sokolov, I.M. Ternov: Sov. Phys. JETP 1, 227 (1955)
1.25 A.A. Sokolov: Sov. Phys. Doklady 2, 573 (1957)
1.26 A.A. Sokolov, I.M. Ternov: Sov. Phys. JETP 4, 396 (1957)
1.27 A.A. Sokolov, I.M. Ternov: Sov. Phys. Doklady 8, 1203 (1964)
1.28 A.A. Sokolov: Wiss. Z. Fr.-Schiller-Univ. Jena, Math.-Nat. Reihe 15, 101 (1966)
1.29 E.M. McMillan: Phys. Rev. 68, 144 (1945)
1.30 J.S. Nodvick, D.S. Saxon: Phys. Rev. 96, 180 (1954)
1.31 C. Bénard, M. Rousseau: J. Opt. Soc. Am. 64, 1433 (1974)
1.32 G.C. Baldwin: Physics Today 28, 9 (1975)
1.33 J.P. Blewett: Phys. Rev. 69, 87 (1946)
1.34 F.R. Elder, A.M. Gurewitsch, R.V. Langmuir, H.C. Pollock: Phys. Rev. 71, 829 (1947)
1.35 F.R. Elder, A.M. Gurewitsch, R.V. Langmuir, H.C. Pollock: J. Appl. Phys. 18, 810 (1947)
1.36 F.R. Elder, R.V. Langmuir. H.C. Pollock: Phys, Rev. 74, 52 (1948)
1.37 I.M. Ado, P.A. Cherenkov: Sov. Phys. Doklady 1, 517 (1956)
1.38 F.A. Korolev, O.F. Kulikov: Opt. Spectr. 8, 3 (1960)
1.39 F.A. Korolev, O.F. Kulikov, A.S. Yarov: J. Exper. Theor. Phys. (USSR) 43, 1653 (1962)
1.40 F.A. Korolev, A.G. Ershov, O.F. Kulikov: Sov.Phys. Doklady 5, 1011 (1961)
1.41 F.A. Korolev, V.S. Markov, E.M. Akimov, O.F. Kulikov: Sov. Phys. Doklady 1, 568 (1956)
1.42 P.L. Hartman, D.H. Tomboulian: Phys. Rev. 87, 233 (1952)
1.43 P.L. Hartman, D.H. Tomboulian: Phys. Rev. 91, 1577 (1953)
1.44 D.H. Tomboulian, P.L. Hartman: Phys. Rev. 102, 1423 (1956)
1.45 D.H. Tomboulian, D.E. Bedo: J. Appl. Phys. 29, 804 (1958)
1.46 P. Joos: Phys. Rev. Lett. 4, 558 (1960)
1.47 R.P. Madden, K. Codling: Phys. Rev. Lett. 10, 516 (1963)
1.48 R.P. Madden, K. Codling: Phys. Rev. Lett. 12, 106 (1964)
1.49 R.P. Madden, K. Codling: J. Opt. Soc. Am. 54, 268 (1964)
1.50 K. Codling, R.P. Madden: J. Appl. Phys. 36, 380 (1965)
1.51 R. Haensel, G. Bathow, E. Freytag: Proc. 5th Intern. Conf. High Energy Accelerators (Frascati 1965) p. 257
1.52 G. Bathow, E. Freytag, R. Haensel: J. Appl. Phys. 37, 3449 (1966)
1.53 R. Haensel, B. Sonntag: J. Appl. Phys. 38, 3031 (1967)
1.54 R. Haensel: Dissertation University of Hamburg (1966)
1.55 R. Haensel, C. Kunz: Z. Angew. Physik 23, 276 (1967)
1.56 T. Miyahara et al., SOR Ring, an electron storage ring dedicated to spectroscopy, I. Design; II. Construction and Performance; Institute for Nuclear Study University of Tokyo, Reports INS TH-107, 108 (June 1976)
1.57 T. Sagawa, Y. Iguchi, M. Sasanuma, T.Nasu, S. Yamaguchi, S. Fujiwara, M. Nakamura, A. Ejiri, T. Masuoka, T. Sasaki, T. Oshio: J. Phys. Soc. Japan 21, 2587 (1966)
1.58 T. Sagawa, Y. Iguchi, M. Sasanuma, A. Ejiri, S. Fujiwara, M. Yokota, S. Yamaguchi, M. Nakamura, T. Sasaki, T. Oshio: J. Phys. Soc. Japan 21, 2602 (1966)
1.59 F.C. Brown, P.L. Hartman, P.G. Kruger, B. Lax, R.A. Smith, G. M. Vineyard: Synchrotron Radiation as a Source for the Spectroscopy of Solids, N.R.C. Solid State Panel Subcommittee Rep. (March 1966)
1.60 E.M. Rowe, F.E. Mills: Particle Accel. 4, 221 (1973)
1.61 C. Gähwiller, F.C. Brown, H. Fujita: Rev. Sci. Instrum. 41, 1275 (1970)
1.62 P. Dagneaux, C. Depautex, P. Dhez, J. Durup, Y. Farge, R. Fourme, P.-M. Guyon, P. Jaeglé, S. Leach, R. Lopez-Delgado, G. Morel, R. Pinchaux, P. Thiry, C. Vermeil, F. Wuilleumier: Ann. Phys. 9, 9 (1975) and P.M. Guyon, C. Depautex, G. Morel, Rev. Sci. Instrum. 47, 1347 (1976)
1.63 D.L. Ederer, S.C. Ebner: A user Guide to SURF II, (NBS, Washington 1976)
1.64 V.V. Anashin, A.N. Skrinsky et al., "VEPP-2M and VEEP 3M storage rings as SR sources"; 5th All-Union Conf. on Particle Accelerators, Dubna (Oct. 1976)
1.65 K.O. Hodgson, H. Winick, G. Chu (eds.): Synchrotron Radiation Research and the Stanford Synchrotron Radiation Project, SSRP Report N. 76/100, Aug. 1976
1.66 E.E. Koch, C. Kunz, E.W. Weiner: Optik 45, 395 (1976)

1.67 H.-J. Behrend, E.E. Koch, C. Kunz, G. Mühlhaupt: In *Synchrotron Radiation and New Developments*, ed. by F. Wuilleumier and Y. Farge, Special Issue Nuclear Instrum. and Methods Vol. 152 (North-Holland, Amsterdam 1978) p. 37

1.68 E.E. Koch, C. Kunz (eds.): *Synchrotronstrahlung bei DESY, Ein Handbuch für Benutzer* (DESY, Hamburg, July 1977)

1.69 D.J. Thompson: "Design and progress with the SRS storage ring" Proc. Vth All Union Conf. on Charged Particle Accelerators, Dubna (Oct. 1976) and DL/SRF/P54 (Nov. 1976)
Daresbury Laboratory, Design Study for a dedicated Source of Synchrotron Radiation, DL/SRF/R2 1975

1.70 E.M. Rowe et al., A proposal to the National Science Foundation for the expansion of the synchrotron radiation center of the University of Wisconsin, Madison (Dec. 1976)

1.71 Synchrotron Radiation Research Facility, Conceptual Design Report, June 1976, Brookhaven Nat. Lab. BNL Report 21589, Proposal for a National Synchrotron Light Source, Vol. I, Vol. II, BNL Report 50595, (Febr. 1977)

1.72 PAMPUS, A proposal for a 1.5 GeV Electron Storage Ring as a dedicated Synchrotron Radiation Source, FOM-Institute Amsterdam, Report NK 235 (Oct. 1976)

1.72'An Assessment of the National Need for Facilities Dedicated to the Production of Synchrotron Radiation. Report to the National Academy of Sciences, Washington, D.C. (1976) Projekt Synchrotronstrahlungs-Speicherring, Bericht einer Studiengruppe Gutachterausschuß "Synchrotronstrahlung" bei der Koordinierungsstelle Hochenergiephysik, Jan. (1977); Synchrotron Radiation a Perspective for Europe, prepared by a working group on synchrotron radiation at the European Science Foundation, ESF, Strasbourg (1977)

1.73 J.Wm. McGowan, E.M. Rowe (eds.): Proc. of Synchrotron Radiation Facilities, Quebec Summer Workshop, Report University of Western Ontario, London, Ontario (1976)

1.74 G.V. Marr, I.H. Munro, J.C.C. Sharp: *Synchrotron radiation: A bibliography*, Daresbury, Nuclear Physics Laboratory, Report DNPL/R 24 (1972) and DL/TM 127 (1974)

1.75 D.D. Ivanenko, A.A. Sokolov: *Klassische Feldtheorie* (Akademie-Verlag, Berlin 1953)

1.76 A. Sommerfeld: *Vorlesungen über Theoretische Physik III* (Akademische Verlagsgesellschaft, Geest & Portig, Leipzig 1961)

1.77 J.D. Jackson: *Classical Electrodynamics* (John Wiley, New York, London 1962)

1.78 A.A. Sokolov, J.M. Ternov: *Synchrotron Radiation* Pergamon Press, London 1968)

1.79 R.P. Godwin: In: Springer Tracts in Modern Physics, Vol. 51 ed. by G. Höhler (Springer: Berlin, Heidelberg, New York 1969) p. 1

1.80 R.P. Madden: In *X-ray Spectroscopy*, ed. by L.V. Azaroff (McGraw Hill, New York 1974) p. 338

1.81 R. Haensel, G. Zimmerer: Proc. Intern. Summer School on Synchrotron Radiation Research (Alghero 1976) Vol. 1, ed. by A.N. Mancini and I.F. Quercia, Intern. College on Applied Physics and Istituto Nazionale di Fisica, p. 409

1.82 E. Tamm (ed.): *Synchrotronoje islutschenie* (russ.) Akademie der Wissenschaften der UdSSR, Lebedev Institut Vol. 80 (Nauka, Moscow 1975)

1.83 G.N. Watson: *Bessel Functions* (The McMillan Company, New York 1945) p. 188

1.84 C. Kunz: Physikal. Blätter $\underline{32}$, 9 (1976)

1.85 G.K. Green: "Spectra and Optics of Synchrotron Radiation", BNL Report 50522 (April 1976)

1.86 C. Kunz: In: *Vacuum Ultraviolet Radiation Physics*, ed. by E.E. Koch, R. Haensel and C. Kunz (Pergamon-Vieweg, Braunschweig 1974) p. 753

1.87 R. Klucker: unpublished

1.88 A. Piwinski, A. Wrulich: Excitation of Betatron-Synchrotron Resonances by a Dispersion in the Cavities, Report DESY 76/07 (1976)
R.D. Kohaupt: "Parametric excitation of longitudinal quadrupole instabilities for short particle bunches", DESY internal report H3-76/2

1.89 H. Winick, T. Knight (eds.): *Wiggler Magnets*, A Collection of Material presented at the Wiggler Workshop held at SLAC March 1977, SSRP Report No 77/05 (May 1977)

1.90 D.F. Alferov, Yu.A. Bashmakov, E.G. Bessanov: Sov. Phys.-Techn. Phys. $\underline{17}$, 1540 (1973) and $\underline{18}$, 1336 (1974)

1.91 B. Kincaid: J. Appl. Phys. $\underline{48}$, 2684 (1977)
1.92 J.P. Blewett, R. Chasman: J. Appl. Phys. $\underline{48}$, 2692 (1977)
1.93 G. Chu: Report SLAC 1782 (1976) and [Ref. 1.65 p. 195]
1.93a A.N. Didenko, A.V. Kozhevnikov. A.F. Medvedev, M.M. Nikitin: Pisma v. Zhurnal.
      Techn. Phys. $\underline{4}$, 689 (1978)
1.94 J.M. Madey: J. Appl. Phys. $\underline{42}$, 1906 (1971)
1.95 L.R. Elias, W.M. Fairbank, J.M. Madey, H.A. Schwettman, T.I. Smith: Phys. Rev.
      Lett. $\underline{36}$, 717 (1976)
1.96 D.A.G. Deacon, C.R. Elias, J.M.J. Madey, G.J. Ramian, H.A. Schwettman, T.I.
      Smith: Phys. Rev. Lett. $\underline{38}$, 892 (1977)
1.97 P. Csonka: Particle Accel. $\underline{7}$, 9 and 255 (1975/76)
1.98 G. Hemmie: internal report DESY S1-73/2
1.99 M. Sands: "The Physics of Electron Storage Rings - An Introduction" SLAC-Re-
      port 121, Stanford (Nov. 1970)
1.100 K.G. Steffen: *High Energy Beam Optics* (Interscience Publishers, Wiley and
      Sons, New York 1965)
1.101 P. Pianetta, I. Lindau: J. Electron Spectr. $\underline{11}$, 13 (1977)
1.102 N. Kosuch: [Ref. 1.68, p. 40]
      R. Fischer, R. Rossmanith: IEE-Trans. Nucl. Science $\underline{20}$, 549 (1973)
1.103 P. Eisenberger: In Proc. Study Symposium on Research Applications of Synchro-
      tron Radiation, Brookhaven National Laboratory (Sept. 1972) ed. by R.E. Watson,
      M.L. Perlman: BNL Report 50 381 (1973) p. 83
1.104 R.López-Delgado, H. Szwarc: Opt. Commun. $\underline{19}$, 286 (1976)
1.105 J.R. Stevenson, H. Ellis, R. Bartlett: Appl. Opt. $\underline{12}$, 2884 (1973)
1.106 P.Meyer, P. Lagarde: J. Physique $\underline{37}$, 1387 (1976)
1.107 R. López-Delgado, A. Tramer, I.H. Munro: Chem. Phys. $\underline{5}$, 72 (1974)
1.108 E.E. Koch: In *Interaction of radiation with condensed matter*, Vol. II, ed by
      L.A. Self (publication of the Trieste Center for Theoretical Physics Int.
      Atomic Energy Agency, Wien 1976) Chap. 19 p. 225
1.109 J.A.R. Samson:"*Techniques of Vacuum Ultraviolet Spectroscopy* (Wiley and Sons,
      New York 1967)
1.110 K. Radler: private communication
1.111 see, e.g., J.A. Kinsinger, W.L. Stebbings, R.A. Valenzi, J.W. Taylor: Anal.
      Chem. $\underline{44}$, 773 (1972)
1.111a C.K. Rhodes (ed.): *Excimer Lasers*, Topics in Appl. Phys., Vol 30 (Springer:
      Berlin, Heidelberg, New York 1979
1.112 P. Rabe: [Ref. 1.68, p. 59]
1.113 B.K. Agarwal: *X-Ray Spectroscopy, An Introduction*, Springer Series in Optical
      Sciences (Springer: Berlin, Heidelberg, New York 1979)

2. The Synchrotron Radiation Source

E. M. Rowe

With 8 Figures

Historically, the first laboratory observations [2.1,2] of the phenomenon that has become known as synchrotron radiation were carried out with electron accelerators that were built for quite a different purpose. So, also, were the early efforts to utilize synchrotron radiation for spectroscopic researches. TOMBOULIAN and HARTMANN [2.3] used the Cornell 0.32 GeV synchrotron to demonstrate the technique in 1956. CODLING and MADDEN [2.4] used the 0.180 GeV synchrotron at the National Bureau of Standards in their investigations of photo excitation spectra of the noble gases beginning in 1962, and HAENSEL and his co-workers worked originally with the 6 then 7.5 GeV synchrotron DESY and later with the 3.5 GeV electron-positron storage ring DORIS to estabish their research programs. All of these machines were constructed for high-energy nuclear research and the investigations carried out by these workers were quite secondary to the main purposes of the laboratories that hosted these and most of the later efforts.

As interest in the use of synchrotron radiation has developed, some machines have been made available for the exclusive use of vacuum ultraviolet and soft X-ray spectroscopists. Notable here were the commitment of the NBS synchrotron SURF I exclusively to the use of Madden and his group in 1968, the commissioning of TANTALUS I [2.5,6], the 0.24 GeV storage ring at the University of Wisconsin, as a dedicated synchrotron radiation source in 1968 and more recently the assignment to the LURE group at Orsay of the 0.54 GeV electron-positron storage ring ACO for dedicated operation (see also Chap. 1, especially Fig. 1.2 and Tables 1.1 and 1.2 for more details).

However, none of these machines, regardless of their present status, was designed to be a synchrotron radiation source. Further, at the time of this writing, the dedicated machines in use form a minority of the machines now used worldwide as synchrotron radiation sources.

In 1965, a group under the leadership of SASAKI began the study of a 0.30 GeV storage ring designed especially to be a synchrotron radiation source at the Uni--versity of Tokyo and in 1971 ROWE [2.7] described a 0.96 GeV ring based on the TANTALUS I design. Shortly thereafter, the design of this machine was modified and the energy raised to 1.76 GeV reflecting the needs of the spectroscopists for higher brightness and access to the hard X-ray range [2.8]. Thereafter, other designs were developed, but to date only the machine designed by SASAKI and his co-workers has come into operation, thereby becoming the first purposely built electron accelerator for synchrotron radiation research [2.9].

All of the machines used originally were synchrotrons; machines designed to accelerate and deliver on target, repetitively, electrons at as high an energy as the state of accelerator technology at the time made possible. Since in synchrotrons electrons are usually injected at a relatively low energy and then raised to a higher energy, the spectrum of the synchrotron radiation produced by these machines varies cyclically during the acceleration process. In addition, for at least half of the time, no radiation is produced at all as the magnetic fields of the accelerator are reset to the injection level. The electron current and, therefore, the radiation intensity varies from cycle to cycle as well as the electron orbit position and beam cross section. Thus, these machines are not ideal as radiation sources. However, advances in high energy nuclear physics, the field of research responsible for the existence of usable synchrotron radiation sources, provided the remedy for the shortcomings of the synchrotron when the interests of high energy physicists turned to achieving extremely high center of mass energies by colliding counter rotating beams of charged particles. This resulted in the development of what is probably the most remarkable and sophisticated instrument in the history of scientific research: the electron-position storage ring. It was necessary for the successful development of these machines that accelerator technologists address and solve complex problems of beam cross section, position and stability in order to achieve the high luminosities required for adequate interaction rates between the colliding beams.

Once these problems were solved, the electron-positron storage ring became the premier tool of the high-energy physicist. But also, by the same achievement, the accelerator technologists have provided the spectroscopist with a radiation source of immense power and flexibility. Thus, while much work has been done and will be done yet with synchrotrons, in the future synchrotron radiation research will be dominated by investigators having access to storage rings and it is with these machines that we will now concern ourselves. In Sect. 2.1 the fundamental concepts of the theory of orbit dynamics in storage rings are introduced at a level which should help the spectroscopist to obtain a certain basic understanding of the machines he is using. Sect. 2.2 is concerned with the design considerations of a spectroscopic light source while two specific examples of a small and a large dedicated storage ring are described in Sect. 2.3.

2.1 Fundamental Concepts

In this section we will give a brief overview of the more important aspects of the
theory of particle accelerators. Accelerator science and technology has reached
a very high level of sophistication during the three decades since the end of World
War II, thus to attempt to do more would far exceed the scope of this chapter. Those
wishing to extend their understanding of this interesting and important application
of classical physics are referred to any of several standard texts [2.10,11,12].

2.1.1 Orbit Dynamics

In essence, an electron storage ring consists of an array of bending magnets ar-
ranged so as to cause a group of electrons at some energy to follow a path, hereafter called the orbit, which closes upon itself in such a way that at the point
of closure, s_0, the two ends of the orbit match not only in radius but also in r'
where the prime denotes the derivative with respect to s, the distance measured
along the orbit. While the orbit will usually be in a plane, this is not a necessity
nor must the total angle swept by the orbit be limited to 2π radians.

a) Betatron Oscillations

These requirements are relatively easy to meet since the radius of curvature of
the orbit at any point is determined exactly by the energy of the electron and
the magnetic field that it moves in. Thus, the design of the central orbit is a
matter of rather simple geometry. However, the number of electrons on the central
orbit of any real machine amounts to at most a very small number. Therefore, provisions must be made to accommodate electrons with a spread in transverse and longitudinal momenta in any practical machine. To contain particles with transverse
momenta components, magnetic fields with transverse gradients are incorporated,
either in the bending magnets or in separate magnetic elements called quadrupoles.
The existence of such gradient fields in the guide field structure (often called
the lattice) leads to the result that particles with transverse momenta experience
restoring forces proportional to their displacement from the central orbit, and,
therefore, oscillate about the central orbit. These oscillations are called betatron
oscillations. In contrast to photon optical systems, in general, magnetic elements
that exert restoring forces in one dimension have quite the opposite effect on
motions in the direction in quadrature with that dimension. Thus, it is the case in
most storage rings (and synchrotrons for that matter) that the lattice of the guide
field will contain at least two kinds of elements: magnets that bend and focus horizontally and magnets that bend horizontally and focus vertically. The more modern
separated function designs used in storage rings contain three types of elements:
bending, vertically focusing, and horizontally focusing.

With the inclusion of focal elements the periodicity of the storage ring changes from that of the bending magnets to one defined by the repetition of a basic unit cell consisting of one or more bending magnets and some number of focal elements. The machine will be made up of a number, generally divisible by two, of these unit cells.

It is common practice to speak of the tune of the machine, i.e. the number of betatron oscillations per revolution, as a whole number plus a fractional part. The value of the tune, called the ν or the q value, will depend on the details of the lattice, and as a general rule, small beam cross sections require large ν values. The number of oscillations in either transverse mode per revolution must not, obviously, be a rational number lest imperfections in the magnet lattice drive the transverse motions of the electrons.

Further, as is predicted by the equations of motion of the electrons which are non-linear, the tune values must not satisfy

$$\pm\, p\,\nu_x \pm q\nu_z = m \tag{2.1}$$

where p, q, and m are integers, and the subscripts x and z refer to the radial and vertical motions, respectively.

Since the lattice will consist of a series of focusing and defocusing elements as well as magnets, the cross section of the beam will vary as a function of circumferential position. The beam will have its maximum width at horizontally focusing elements and its maximum height at vertically focusing elements. Thus the circulating beam can be characterized by two functions, called β functions, which describe these variations in the horizontal and vertical beam dimensions through relationships of the form

$$\Delta x(s) = a\sqrt{\beta_x(s)}\,\cos(\phi_x - \psi) \quad , \tag{2.2}$$

where $\Delta x(s)$ is the distance of the electron from the equilibrium orbit, a and ψ are initial conditions peculiar to the particular electron trajectory under consideration and ϕ is the phase advance per revolution of the betatron oscillation. If one views the trajectory of an electron at a fixed location on the orbit, s_0, the electron displacement from the central orbit will vary sinusoidally over a number of revolutions, always remaining within the limiting values $\pm x(s_0)$ or $\pm y(s_0)$ which occur for $\phi = \psi$. Within the limiting envelopes, x(s) and y(s), the electrons execute oscillations of all phases and all amplitudes up to a limit set by the maximum transverse momenta that can be contained by the lattice. The β functions are fixed in space relative to the machine lattice and must meet the condition that $\beta(s_0) = \beta(s_0 + L)$ where L is the length of the orbit.

b) Betatron Oscillations of Off-energy Particles

The focusing elements that allow electrons with transverse momenta components to circulate on stable trajectories also provide the means by which electrons with energy errors relative to the median, or reference electron, energy also find stable orbits. The mechanism here is that an electron with an energy error will undergo a greater or lesser angular deviation than the reference electron on passing through a bending magnet. The radially focusing quadrupoles will act so as to correct this deviation through forces directed towards the central orbit. However, the trajectory followed by the off energy electron will not lie on the central orbit but will be displaced by a distance which varies as a function of s. This displacement is described by a function of the form

$$\Delta x(s,\Delta E) = \eta(s) \frac{\Delta E}{E} , \qquad (2.3)$$

where $\Delta x(s,\Delta E)$ is the departure of the off energy electron from the median energy orbit and ΔE is the energy error. The η function is fixed in space relative to the lattice as are the β functions and obeys the condition $\eta(\Delta E,s) = \eta(\Delta E,s + L)$, where L is the orbit circumference. The shape of the η function and therefore of the off energy trajectories will resemble the β_x function qualitatively and off energy particles with transverse momenta components will execute betatron oscillations about their displaced orbits.

The electrons of higher energy are affected less by the focusing elements than the reference particle while the electrons of lower energy are affected more. Hence, their betatron tunes differ from the nominal tune of the machine. This tune shift with energy is characteristic of strong focusing machines, and can reduce the usable aperture of the machine to a value far smaller than the physical aperture of the vacuum chamber. The tune shift as a function of energy relative to the reference orbit is called the chromaticity and is defined as

$$\xi = \frac{\Delta \nu}{\nu} \frac{p}{\Delta p} . \qquad (2.4)$$

In all storage rings, additional focusing elements producing sextupolar fields are incorporated in the lattice to control the chromaticity, both to increase the usable aperture and to control certain instabilities in the circulating electron beam. These capabilities are particularly important during injection when the maximum aperture is required.

c) Phase Focusing and Synchrotron Oscillations

Since the off energy electrons move on trajectories either of smaller or larger average radii than the orbit of the reference particle, it is clear that after a number of revolutions, the off energy particle must accumulate a phase error with respect to the radio frequency acceleration wave because the acceleration wave is synchronous only with the reference particle. However, these phase errors result in a differential accelerating process which in turn results in phase focusing. All that is required is that the accelerating wave be of an amplitude greater than necessary to restore the mean energy loss per revolution of the reference particle. Then the reference particle, if it is to remain synchronous, must cross the accelerating gap at a phase ϕ_0 relative to the accelerating field such that

$$V_s = V_0 \cos(h\omega_0 t \pm \phi_0) \tag{2.5}$$

where V_s is the voltage necessary to restore the mean energy loss, V_0 is the peak gap potential, ω_0 is the revolution frequency of the reference particle, h is any whole number and it is assumed that when $\cos(h\omega_0 t) = 1$, the accelerating field is in the correct polarity to impart energy to the electrons. Now consider a particle with energy greater than that of the reference particle. Then, because for any electron energy of interest here $v \approx c$ and therefore the circumference of its trajectory will be somewhat greater than that of the reference orbit, the off energy particle will arrive at the accelerating gap somewhat later than the reference particle. If ϕ_0 is positive the off energy particle will, as a consequence, receive less energy than the mean energy loss per revolution and will thus tend to be returned to the synchronous phase. A similar argument may be made for the particle with less energy than the reference particle. Both signs for ϕ_0 are given in the expression because at very low energies ($v \ll c$) ϕ_0 must be negative for phase focusing. The energy at which ϕ_0 changes signs is known as the transition energy. Normally, injection into electron storage rings takes place at energies very much higher than the transition energy, therefore we will not concern ourselves further with it.

Since the off energy particles experience restoring forces which tend to return them to the mean energy, it is reasonable to expect them to oscillate about the synchronous phase. This is indeed the case, and as a result the electrons oscillate between energies somewhat higher and lower than the reference energy. The number of these oscillations, called synchrotron oscillations, per revolution is generally denoted by the symbol ν_s. However, in contrast to the betatron oscillations, which are at frequencies higher than ω_0, synchrotron oscillation frequencies are generally very much lower than ω_0. The maximum particle energy error that may be accommodated by the storage ring will be determined by the over voltage factor, that is the ratio between V_0 and V_s, and the maximum value of the η function permitted by the vacuum chamber aperture.

2.1.2 Radiation Damping

At first thought, it would seem that quite aside from the question of the mean energy loss to synchrotron radiation, the electron beam circulating in a storage ring must eventually be lost because its cross section should continually grow larger due to the emission process: an electron emits a photon, the energy of the photon must be subtracted from the energy of the electron, and as a result the electron now finds itself on an orbit for which its energy is wrong and it must, therefore, begin both betatron and synchrotron oscillations about a new central orbit displaced radially from the reference orbit. The combined amplitudes of these oscillations, after a sufficiently large number of emission events, would, without the intervention of some other process, grow larger than the vacuum chamber cross section and the electron should be lost horizontally. Vertically, a similar argument can be made. The expectation value of the photon emission angle, no matter how small, is still finite, and, therefore, conservation of momentum assures us that the vertical motion must also grow, although not so rapidly as the horizontal motion.

Paradoxically, the process which gives rise to this growth in beam size, synchrotron radiation, also provides a strong damping mechanism in a properly designed lattice. Consider an electron with a total momentum $\underline{p}$ which is at some angle δ relative to the central orbit. At this point in its orbit, the electron emits a photon and its momentum becomes $\underline{p} - \Delta\underline{p}$ still at an angle of very nearly δ to the central orbit. Upon passing through the accelerating structure the electron will be given an increment of momentum $\Delta\underline{p}_0$ equal to the average momentum loss per turn, but the direction of $\Delta\underline{p}_0$ will be parallel with the central orbit. The increment of momentum loss $\Delta\underline{p}$ had a transverse as well as a longitudinal component, but the increment of momentum gain has only a longitudinal component. Thus the transverse components of the electron momentum must diminish and, therefore, so must the amplitude of the transverse or betatron motions.

As presented, this discussion only applies rigorously to the vertical motion of the electrons because this motion is essentially energy independent. However, both the horizontal betatron motion and the synchrotron motion of the electrons manifest themselves in displacements of the radial position of the electrons. So long as the rate of radiation loss does not depend on the displacement of the electron trajectory from the reference particle orbit horizontally, as is the case in separated function lattices, the argument just presented also holds for the radial betatron motion. However, if the bending magnet fields have gradients, the rate of energy loss will depend on radial position since $dE/dt \propto E^4/\rho$, where α is the local radius of curvature of the trajectory. In particular, if the bending magnets have gradients that produce horizontal focusing, the radial betatron motions driven by the emission process may actually be enhanced beyond the capabilities of the radio frequency acceleration process to provide damping.

The synchrotron motions will also damp but because of a somewhat more direct mechanism. As the electrons move on synchrotron oscillations, they depart from the central energy. In contrast to betatron oscillations which in general have ν values of one or more, synchrotron oscillation ν values are generally very much less than unity. Therefore, during a synchrotron oscillation an electron will execute many revolutions as its trajectory moves between energies somewhat higher and somewhat lower than the reference orbit energy while still receiving, on the average, an energy increment nearly equal to the median energy loss per revolution. Since the energy loss per revolution varies as the fourth power of the electron energy, particles at either higher or lower energies than the median will tend to be returned to the median energy. The beam dimensions will reach equilibrium when the emission effects which drive synchrotron and betatron motions are just balanced by the radiation damping. However, the equilibrium size will depend on the details of the storage ring lattice.

2.1.3 Beam Lifetime

Since in a properly designed storage ring all motions damp, the mechanisms leading to beam loss are only scattering on residual gas molecules in the vacuum chamber, quantum fluctuations and electron-electron scattering (Touschek effect). The effects of the first of these can be minimized by utilizing a high conductance distributed pumping system, that is by incorporating sputter ion pumping elements which utilize the fringe field of the bending magnets in the vacuum chamber. This approach puts the pumps where they are needed most: close to the electron orbit. Quantum fluctuation is a term which describes the fact that the maximum energy loss per turn of an electron can be orders of magnitude greater than the mean. Thus, if the accelerating voltage is not great enough, an appreciable number of particles will lose enough energy each revolution to lose synchronism with the accelerating system and be lost. Fortunately, the degree of over-voltage required to achieve acceptable beam lifetime against this effect is not great. Generally, a factor of two or three is sufficient.

The Touschek effect is somewhat more troublesome. In this process, electrons which are, as we have discussed, always executing betatron oscillations of some amplitude about the equilibrium orbit scatter off of each other. During such a scattering event, some transverse momentum is transferred into longitudinal momentum and the resulting longitudinal momentum error can be sufficiently large so as to cause both electrons to lose synchronism with the accelerating field and be lost. In general, the higher the energy the larger the transverse momentum and, therefore, the greater the probability for loss following a scattering event. Furthermore, since a small beam cross section and, therefore, a high charge density is a design goal in synchrotron radiation sources, these machines are particularly susceptible to this loss mechanism. Once again, since the loss is through the longitudinal dimension, over-voltage in the accelerating system must be supplied, but the over-voltage factor has a rather interesting dependence upon the energy of the electrons. At low energies (E < 0.1 GeV) the over-voltage factor is rather modest; of the order 2 to 3. In the intermediate range (0.15 GeV < E < 1 GeV), and depending on the nature of the storage ring lattice, the over-voltage factor required for reasonable beam life-time against the Touschek effect may be as large as 10^2. As the electron energy becomes greater (E > 1.5 GeV) the over-voltage factor decreases and approaches unity. The reason for this behavior can be understood when it is remembered that the equilibrium transverse momenta of the stored beam increase approximately as the square of the electron energy while the synchrotron radiation energy loss, and therefore the accelerating voltage required, increases as the fourth power of the energy. Therefore, at high energies an over-voltage factor of two implies a voltage excess of perhaps 10^6 volts which provides very large energy acceptance for the electrons indeed.

2.1.4 Beam Cross Section

From the discussion thus far, it is reasonable to expect that equilibrium cross section of the circulating electron beam will be considerably smaller vertically than it is horizontally. However, the beam's vertical dimension is never as small as is predicted theoretically. This is because of coupling of the horizontal and vertical motions of the electrons brought about by imperfections in the guide field lattice. For example, consider the effects of one horizontally focusing quadrupole with its magnetic axis not parallel to the median plane of the orbit. A radially displaced electron entering this quadrupole will experience a transverse impulse with a vertical component which will manifest itself in a vertical displacement of the electron further along the orbit. This displacement will result in the electron receiving a vertical impulse back towards the median plane in the next vertically focusing quadrupole and thus some of its radial momentum will have been transformed into a vertical betatron oscillation.

Other effects can increase the equilibrium beam cross-sectional dimensions both horizontally, vertically, and longitudinally. For example, the radio frequency structure through which energy is restored to the beam to make up for the energy lost to synchrotron radiation, is, generally, a cavity. Such structures are capable of supporting a large number of modes, some of which are characterized by magnetic fields transverse to the electron path. While it is most unlikely that such modes would be harmonically related to the fundamental resonance frequency of the cavity, still they will present some impedance at these harmonics and, since the electron beam is bunched quite tightly, the beam will drive these higher modes and rather large transverse radio frequency deflecting fields can be set up. In cases such that

$$\pm p\nu_x \pm q\nu_z \pm r\nu_s = m \qquad (2.6)$$

where p. q, r and m are integers, the radial and/or vertical motions of the electrons will be considerably enhanced through coupling effects. These are called synchro-betatron resonances.

Other causes of enhanced beam cross-sectional dimensions are collective effects
resulting from the large current densities in the electron beam itself and the interaction of the electrons with their own image currents and wake fields in the vacuum
chamber. In another class of instabilities, the electron motions are driven by the
presence of ions trapped in the potential well of the electron beam. Fortunately,
almost all of the effects mentioned here manifest themselves as coherent motions of
the beam and therefore may be suppressed by the application of inverse feedback to
the electron beam.

In the foregoing, we have made an effort to discuss some of the fundamental
characteristics of electron accelerators in general and electron storage rings in
particular as they affect the design of such machines in their application as
synchrotron radiation sources. In Sect. 2.3 we will describe two machines and discuss
their designs in light of the material just developed.

2.2 Design Considerations

Because of the growing demand for access to synchrotron radiation sources and the
increasing level of sophistication of synchrotron radiation users it is reasonable
that a number of electron accelerators have been designed specifically to be intense sources of synchrotron radiation. These machines are, of course, all storage
rings and, in general, are quite similar to the electron-positron storage rings now
in use by high-energy physicists. However, because of their application, these
specially designed "light bulb" machines differ significantly in detail from their
precursors. For example, it is generally a design objective of the electron-positron
machine designer to minimize the energy loss to synchrotron radiation while reaching
the highest particle energy possible. In contrast, the designer of the synchrotron
radiation source seeks to maximize the energy loss to synchrotron radiation for a
given electron energy. Thus the designers of colliding beam machines will employ the
lowest magnetic fields feasible while the radiation source designer will employ the
highest magnetic fields practicable. Interestingly, both designers have the same
ultimate aim: the achievement of their design goals, particle energy in one case
and spectral range in the other, with the least expenditure of radio frequency
energy.

2.2.1 Magnetic Field and Energy

In the case of the synchrotron radiation source, the choice of the highest magnetic fields practicable pays an additional dividend since the orbit radius will be
reduced and, therefore, the first optical elements in the photon beam lines may be

placed closer to the electron orbit. This allows a larger solid angle of the radiation pattern to be collected with mirrors of manageable dimensions. This is a matter of considerable importance in VUV and soft X-ray range where the use of grazing incidence focusing optics has reached a rather high degree of development (for details see Chap. 3).

The choice of electron energy for the synchrotron radiation source will depend on the wavelength range that the machine is designed to produce. It can be shown easily that at wavelengths such that $\lambda \gg \lambda_c$, the spectral intensities of all synchrotron radiation sources are very nearly the same, exhibiting only a weak dependency on the electron path radius in the bending magnets. Therefore, there is little improvement in this part of spectrum to be gained by building a machine of higher energy than necessary to produce the desired spectral range. Conversely, if the machine is to support a substantial research program requiring an intense, hard X-ray source, there can really be no substitute for electron energy. Obviously, a high energy machine can be used for VUV and soft X-ray researches. However, this can lead to serious problems of radiation damage to the VUV optical systems caused by hard X-ray contamination of the photon beams, and it does increase the higher-order output of the monochromators used in the longer wavelength ranges. But of more importance, the large radius of curvature of the electron trajectory characteristic of high energy machines makes it difficult to install low f value grazing incidence optics to gather the radiation from the electrons. This leads to difficulties in achieving large solid-angle acceptances in the photon optical systems. It has, in fact, been determined that bending magnet fields of between 1.2 and 1.5 Tesla [2.13] are optimum for synchrotron radiation sources, thus the energy of a machine designed to be VUV soft X-ray source with spectral capabilities up to perhaps 1 keV should be approximately 700-800 MeV.

2.2.2 Lattice

The energy of the machine having been determined by the machine's ultimate use, the lattice must now be considered. Here, many compromises must be effected. The geometry of the central orbit is relatively easy to lay out but the determination of the number, placement, and order of the focal elements is a somewhat more complicated problem. The shape of the envelope functions and the off energy function, both of which follow the distribution of the focal elements, is an important consideration here. The first aim of the synchrotron radiation source designer is to specify a distribution of focal elements which will give the electron beam desirable characteristics; one of the most important of which is generally considered to be minimum cross sectional dimensions in one or both of the transverse coordinates at the source points. This may be done by minimizing the β and η functions and their derivatives in the bending magnets. The dimensions achieved will depend only on the betatron tune of the machine since the emission processes which enlarge the beam and the action of the acceleration system which provides the damping mechanism are both prescribed entirely by the energy of the electrons and the radius of curvature in the bending magnets.

Large tune values are achieved through high strength focal elements, however there are limits to this process. The lattice consists of a series of identical cells and the phase advance of the betatron oscillations per unit cell normally must not exceed π. Further, the strength of the focal elements required to reach a given set of betatron tunes may not be technically feasible. It is, therefore, more common to reach high tune values through the use of many cells and moderately strong focal elements. Here again, there are limits. High tune machines tend to large chromaticities which in turn lead to small usable apertures for the electron beam. The chromaticity may in principle be corrected to acceptable values, but again, the sextupole strength required may not be possible.

The design of the lattice will also be affected by purely practical considerations such as the availability of sufficient unencumbered space between lattice elements for the accelerating system, the injection system and the electron beam monitoring system. In addition, care must be taken that elements of the lattice do not interfere with the photon beam lines themselves. Meeting all of these requirements may well result in a lattice of less than optimal properties.

The process by which the lattice is designed is thus seen to be an iterative process aimed at optimizing some properties of the machine at the expense of others. This procedure is generally carried out with the aid of powerful and sophisticated interactive computer programs [2.14,15,16].

2.2.3 Injector

The source of the electrons to be stored will have a considerable effect on the design of the storage ring. Ideally, the electrons should be injected into the ring at as high an energy as possible. In the past, this has placed constraints on the possible locations of electron-positron storage rings because the cost of an accelerator suitable for use as an injector for such machines is extremely high; generally an order of magnitude or more greater than that of the storage ring itself. Thus, all electron-positron storage rings built so far, save one, have been constructed at the sites of existing high energy electron accelerators. However, while the advantages of high injection energy cannot be denied, two rather recent developments have changed this situation, at least as far as VUV soft X-ray synchrotron radiation sources are concerned. These are the development of the modern microtron [2.17] as an efficient, inexpensive source of electrons at energies up to 100 MeV and the concept of the accelerating storage ring. By designing the storage ring so that injection may be carried out at an energy within reach of an accelerator of the microtron type and with sufficiently precise control capability so that acceleration to the operating energy without significant particle loss is possible, the possession of a high energy linear accelerator or synchrotron is no longer a necessary precondition for undertaking the construction of a storage ring of 500 to 1000 MeV.

Unfortunately, machines of the two to four GeV class, that is storage rings designed to produce hard X-rays, still require injection energies of the order of one GeV or more. There are three reasons for this. Firstly, to inject at, say, 0.1 GeV into such a machine implies that the magnetic fields be controllable with high precision over a range of excitations a factor of 20 to 40 in extent. Secondly, the damping times of the electron transverse motion and energy are proportional to the product of γ^{-3} ($\gamma=E/mc^2$) and the circumference of the trajectory. Hence, in a large circumference machine operating at low energy, the damping times become very large and stacking, that is accumulating the stored electron beam through multiple injections, becomes difficult. Finally, the circulating beam current thresholds for most of the beam instabilities mentioned in Sect. 2.1.4 exhibit strong dependencies on the ratio of γ to the machine circumference. Thus, the accumulation of a large circulating beam may be impossible at low energies in a large machine because of the destructive effects of beam instabilities.

However, it should be pointed out here that one electron storage ring makes an admirable injector for another electron storage ring [2.18]. Thus, an intermediate energy storage ring can serve both as an optimized VUV and soft X-ray source and as an injector for a high energy storage ring designed to be a hard X-ray source. Except for the short periods of time when the smaller ring is serving as an injector, there need be little interference between the research programs at the two sources.

2.2.4 Accelerating System

In the design of the radio frequency accelerating system for the synchrotron radiation source, a choice involving compromises must also be made. This is the operating frequency or, more properly, the harmonic number. Because of space limitations, only the very smallest storage rings utilize radio frequency systems which operate at the fundamental revolution frequency of the electrons. It is common practice to choose a harmonic number which results in a system frequency in the range of 50 to 500 MHz. It is quite easy to show that, for a given voltage gradient, the power loss in the accelerating structure decreases as the inverse square root of the harmonic number. Further, the length of electron bunches decreases with the harmonic numbers, an advantage in time-resolved fluorescence and photoemission studies. Also, the physical size of the accelerating structures decrease with the harmonic number.

On the other hand, at the higher harmonic numbers, where these advantages are most pronounced, the cost of radio frequency generating equipment is considerably greater and significantly larger overvoltage factors are required to achieve adequate Touschek lifetimes. Further, the very short electron bunches characteristic of high-harmonic operation are more likely to excite very high frequency electromagnetic fields within the vacuum chamber and accelerating cavities which in turn can react on the beam resulting in increased beam cross section and bunch length. These problems tend to make high harmonic number radio frequency systems less attractive than lower frequency systems even though the longer bunch structure is not so favorable for time resolved studies.

The vacuum chamber for the synchrotron radiation source presents all of the same technological problems as the vacuum chamber for the colliding beam storage ring as well as some uniquely its own. It must, of course, sustain a high vacuum environment for circulating electrons and be electrically clean so that the very short bunches of electrons, which may reach current densities far in excess of 100 amperes/mm^2, do not excite electromagnetic wave distributions in it which, through interaction with the electrons, could increase the beam dimensions both transversely and longitudinally or, in extreme cases, actually destroy the beam. In addition, means for removing the energy deposited on the outer wall of the vacuum chamber through synchrotron radiation must be provided. Indeed, in very high energy machines, the removal of this energy requires that the vacuum chamber be made of aluminum for high heat conductivity and include integral water cooling passages on its outer periphery. Since the outer periphery of the vacuum chamber must also be penetrated in many places for synchrotron radiation ports, the problems of maintaining electrical cleanliness as well as providing adequate cooling at these locations can be difficult.

2.2.5 Energy Shifter Wigglers

We have pointed out the advantages of the intermediate energy machine over the high energy machine as source for VUV and soft X-rays above. However, the choice of an intermediate energy for the VUV soft X-ray source does not necessarily preclude the use of such a machine as a hard X-ray source. It was pointed out in a previous section that in the first approximation the design of the central orbit of a storage ring is a geometrical problem; all that is required is that the orbit close upon itself. The details of the shape of the orbit are largely at the disposal of the designer, thus he may choose to include relatively long, field-free regions in the lattice. In these regions, called long straight sections, very high field (super-conducting) magnet arrays may be installed to create synchrotron radiation source points where the critical wavelength can be as much as a factor of five shorter than in a normal magnet [2.8]. To install such an array, called a wiggler, in a machine without destroying the stability of the electron orbits requires that electron optical properties of the insertion be carefully matched to the β and η functions in the rest of the lattice and, of course, the increased loss of energy per turn must be allowed for in the design of the radio frequency accelerating system. Again, as pointed out above, while for the hard X-ray source there is no satisfactory substitute for electron energy, the inclusion of one or more wigglers in a lower energy machine gives the machine an extra degree of utility for a rather modest expenditure.

2.2.6 Multipole Wigglers (Undulators)

Another device that can be installed in a long straight section is the low field
multipole wiggler [2.19]. The purpose of such a device is not to extend the spectral
range of the source, but rather to create a source point with special spectral properties. If the multipole wiggler has many periods (of the order of one hundred) and
if the field is comparatively low (0.1 - 0.2 T), the spectrum of the radiation it
produces will consist of a relatively low level synchrotron radiation continuum, with
λ_c appropriate for the electron energy and the strength of the multipole fields, but
with a quasi-monochromatic line several orders of magnitude more intense than the
continuum background superimposed on it. The wavelength of this line, λ, is simply
related to the length of the multipole period, λ_0, through a relativistic Doppler
shift, as discussed in Sect. 1.2.2, e.g. $\lambda \approx \lambda_0 \gamma^{-2}$ where γ is the relativistic energy
of the electron beam.

A variant of the transverse multipole wiggler is the helical wiggler. The wavelength of the "monochromatic" radiation produced by this device is related to the
helix period length in exactly the same way as in the transverse multipole, however
in this case the synchrotron radiation will be circularly polarized. It is of interest to note here that recently laser action has been observed in an electron beam tra-
versing a helical wiggler [2.20]. This development presents the possibility of
tunable cw lasers for the optical and near ultraviolet range with average powers
in the tens of kilowatts through the combination of a multipole wiggler and a
suitably designed electron storage ring.

2.3 Design Examples

In this section we will discuss the designs of two storage rings; one to be used
by VUV and soft X-ray spectroscopists, the other to be used primarily for the production of hard X-rays for biological, crystallographic, and other hard X-ray researches.
In spite of their differing purposes and spectral capabilities, these machines have
many common design features. However, their lattices are quite dissimilar reflecting both their differing applications and basic differences in the philosophies of
their designers.

The first machine, which has been named ALADDIN, [2.21] is of intermediate
energy (0.75 GeV) and has been designed to be an intense source for the wavelength
range extending from the infrared to 6.5 Å. It was designed by personnel of the
Synchrotron Radiation Center of the University of Wisconsin-Madison. The second
machine, named the "National Synchrotron Radiation Light Source" [2.22] was con-
ceived by members of the staff of the Brookhaven National Laboratories. It is

intended primarily to be an intense source for the range 10 to 0.1 Å although it can be used as a source for the VUV and soft X-ray range.

2.3.1 ALADDIN

An overall plan view of the machine is shown in Fig. 2.1. The injector machine will be mounted inside of the storage ring so that valuable experimental area outside of the ring will not be taken up by the electron beam transfer system, the injector machine and its radiation shielding, and its ancillaries. While mounting of as much equipment as possible within the ring is attractive because of the spa-

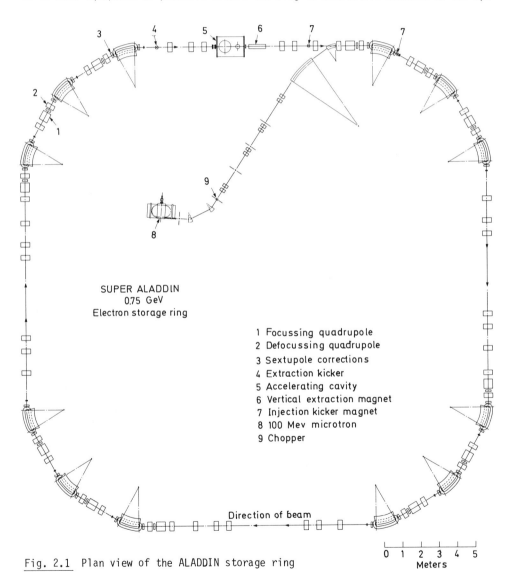

SUPER ALADDIN
0.75 GeV
Electron storage ring

1 Focussing quadrupole
2 Defocussing quadrupole
3 Sextupole corrections
4 Extraction kicker
5 Accelerating cavity
6 Vertical extraction magnet
7 Injection kicker magnet
8 100 Mev microtron
9 Chopper

Direction of beam

0 1 2 3 4 5
Meters

Fig. 2.1 Plan view of the ALADDIN storage ring

tial economies that can be realized, the creation of a low noise environment for the investigators using the machine is, in our judgment, far more important. Therefore, the storage ring power supplies, coolant circulating pumps and similarly acoustically noisy items will not be mounted in this area.

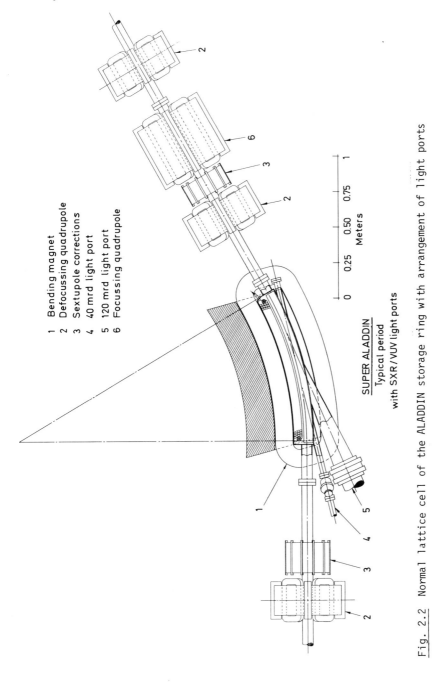

1 Bending magnet
2 Defocussing quadrupole
3 Sextupole corrections
4 40 mrd light port
5 120 mrd light port
6 Focussing quadrupole

SUPER ALADDIN
Typical period
with SXR/VUV light ports

0   0.25   0.50   0.75   1
Meters

Fig. 2.2 Normal lattice cell of the ALADDIN storage ring with arrangement of light ports

a) Lattice

The lattice for this machine conforms to the design principles which have been developed in previous sections. An overview of the machine is exhibited in Fig 2.1 and a unit cell of the lattice is shown in Fig 2.2. The lattice structure of ALADDIN consists of twelve unit cells and four long straight sections. The ALADDIN concept, as is quite typical for such machine designs, has undergone several changes. The arrangement described here was sometimes also called SUPER ALADDIN, since it contains more quadrupoles for reducing the beam size at the source points than the original concept. Each unit cell is comprised of a multi-pole correction magnet, a vertically focusing quadrupole, a radially focusing quadrupole, a vertically focusing quadrupole, a bending magnet, and a second multipole correction magnet, in that order. The unit cell is asymmetrical, that is the quadrupole triplett is not mounted halfway between bending magnets, so as to facilitate the installation of the photon beam lines of which there are two in each bending magnet. In this lattice, the major part of the focusing in the unit cells is provided by the quadrupole triplett with some additional vertical focusing taking place at the entrance and exit edges of the bending magnet. However, the bending magnet field has no gradient: it is a pure dipole, thus it provides neither horizontal nor vertical focusing. This is a design feature which is required if damping is to be achieved in the horizontal, vertical, and longitudinal modes of the particle motion for all possible excitations of the focusing elements.

The placement of the focusing elements assures us that the minima in vertical and horizontal beam cross sections will occur in the bending magnets, again for all possible excitations of the focal elements. This is a desirable property in a storage ring designed to be a synchrotron radiation source since it results in all source points being identical and of the highest possible brightness. The variation in shape of the electron beam cross section as it traverses the unit cell can be seen in Fig 2.3 where the β and η functions are displayed. As has been mentioned, the shape of these functions is largely independent of focal element strength and,

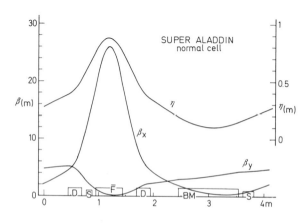

Fig. 2.3 Variation of the envelope functions β_x, β_y, and the off-energy function η in a normal cell of ALADDIN along the orbit. (F: focusing, D: defocusing quadrupoles, S: sextupole, BM: bending magnet)

therefore, the betatron tune. However, their amplitudes are tune-dependent; the lower the tune the larger the maximum and minimum amplitudes. In addition, lower tunes result in smaller variations in the β and η functions.

The lattice includes four long straight sections. The motivation for these inclusions is twofold. One of the long straight sections is required to provide space for the installation of the radio frequency cavity and the beam injection and ejection systems without interference with the photon beam lines. The other long straight sections can provide the space for any of the several special insertions mentioned in Sects. 2.25 and 2.26. One of the long straight sections might also be used as the source region for an inverse Compton scattered photon beam generated through the interaction of a cw laser beam with the circulating electron beam. Such a system utilizing a He-Ne laser of modest power (<10 watts) would produce a Γ ray beam with a peak energy of ∼ 40 MeV with a very desirable duty cycle and high average intensity.

As is to be expected, the electron beam envelope functions change in the long straight sections. However, as is shown in Fig 2.4, the major change is in the β_x (horizontal) function, thus the increase in beam cross section in the long straight sections will have mimimum effect on the output flux of a monochromator utilizing the synchrotron radiation from a transverse wiggler mounted in a long straight section. The β and η functions as shown indicate that the electron trajectories will be nearly parallel in the long straight sections as is required to obtain the highest degree of monochromaticity from a helical wiggler. If necessary, the excitation of the long straight section quadrupoles can be altered to produce very small values of β thus creating "low β insertions" for the installation of very high brightness three pole transverse wigglers. In this mode of operation the beam cross-sectional area in the long straight sections would be reduced by a factor of about 5 over that in the bending magnets.

The design and construction of the quadrupoles and bending magnets are essentially the same as was employed on TANTALUS I but with some modifications based on experience

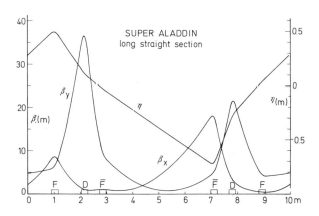

Fig. 2.4 Variation of the envelope and off-energy functions along a straight section of ALADDIN

gained during the construction and operation of this machine. As with TANTALUS I, the magnets and quadrupoles will be laminated, a construction practice which has been adopted in all recent electron storage ring designs. The "inside out" design of the TANTALUS I magnets has been retained but the use of weldments and through bolts to hold the magnet and quadrupole blocks together has been superseded by the more modern technique of gluing the laminations together [2.23]. Edge corrections in the bending magnets to maintain constant magnetic lengths, and therefore orbit position, over the full range of magnet excitation will be employed.

b) Vacuum System

The ALADDIN vacuum system will be constructed of stainless steel and will employ internal sputter ion pumps mounted in the bending magnet fringe field region. These elements are electrostatically screened from the circulating electron beam and are in an ideal location to deal with the increased gas load during stored beam conditions resulting from photoelectron induced gas desorption. The vacuum chamber will be of rectangular cross section in the bending magnets to accommodate the sputter ion pumps and circular in cross section in the short straight sections between normal cells. In the long straight sections, the cross section will also be circular but of

Table 2.1 Parameters of ALADDIN (F: focusing, D: defocusing quadrupols, B: bending magnets, 0: drift section)

General		Magnet	
Energy	750 MeV	Betatron Tunes	
Revolution Frequency	3.38 MHz	Radial	$\sim$7.15
Critical Wavelength	27.6 Å	Vertical	$\sim$7.15
Damping Times		Momentum Compaction	0.021
Radial	21.8 ms	Basic Unit	
Vertical	22.3 ms	Focusing Order	DFDBO
Energy	11.2 ms	Bending Magnet	
RF System		Radius	2.08 m
Harmonic	15	Field Index	0
Number of Cavities	1	Edge Angles	15°
Max. Volts Per Turn	250 kV	Quadrupoles	
RF Power to Beam (1 A)	13.4 kW	Max. Grad.	1.6 kG/cm
Vacuum System		Phase Advance	
Construction	Stainless Steel, Bakeable	Radial	154.5°
Pressure	10^{-9} to 10^{-10} Torr	Vertical	184.5°
Corrections		Long Straight	
Sextupoles and Higher Order		Focusing Order	FDFOFDF
Multipole Magnets	24	Central Drift	
		Space	4 m
Injection		Quadrupoles	
Type - Multiturn using shorted transmission		Max. Grad.	1.6 kG/cm
line inflector with two pulsed dipoles.		Phase Advance	
Source - 100 MeV Racetrack Microtron with		Radial	360°
5 cell accelerating and focusing		Vertical	180°
structures. Two 180° bending and			
focusing end magnets			

increased diameter to accommodate the injection and fast beam dump extraction equipment to be installed there.

To provide an electrically clean vacuum chamber in the synchrotron radiation source in spite of the many penetrations for photon beam ports, the apertures in the vacuum chamber walls will be circumferential slits of a height just adequate to pass the photon beams: Because these slits are thin and parallel to the beam image currents flowing in the chamber walls, they will cause minimal electrical disturbance and also serve as effective conductance limitations for gas flowing from the beam lines into the storage ring. In order to reduce the effects of the electrical discontinuities in the vacuum chamber at points where the chamber cross section changes, i.e. at the entrances and exits of bending magnets, the chamber will taper smoothly from one cross section to the next, much the same way in which such transitions are effected in radio frequency transmission lines. Mounted in every short straight section will be position sensitive electrodes and feedback electrodes for monitoring the operation of the storage ring and controlling coherent instabilities.

c) Accelerating System

For the radio frequency accelerating system, an operating frequency of approximately 50 MHz has been chosen as a best compromise in that it makes possible a reasonably high shunt impedance accelerating cavity with a large degree of freedom from the high-order deflecting modes that have been responsible for increased beam cross sections in other storage rings. This choice of frequency also allows us to use the very well developed gridded power tubes available now rather than the still somewhat experimental klystrons that would be necessary for frequencies in the 200 to 500 MHz range. The required energy gain per turn in ALADDIN operating at 750 MeV is 12.5 keV. With one ampere circulating, the radio frequency system must supply 12.5 kW to the electron beam plus power to make up for cavity losses. Operating with an over-voltage factor of twenty for an acceptable Touschek lifetime, the total radio frequency demand will be less than 40 kW at maximum beam current. The cavity will be constructed of copper and will be evacuated. Provisions for servo tuning the cavity to compensate for beam induced currents will be included.

d) Injector

The injector accelerator for ALADDIN will be a 100 MeV race track microtron. This machine is a development of the 44 MeV microtron designed and built by the Synchrotron Radiation Center operations group and which is now in service as the injector for TANTALUS I. This microtron has the distinction of not only being the highest energy machine of this type built to date but also of being the first microtron designed and built in the United States that operated successfully. The 100 MeV microtron differs from the 44 MeV machine in that it is of the split sector

Table 2.2   100 MeV Microtron Parameters

Type	Split sector
Number of orbits	20
Magnetic field	~1.0 T
Accelerating radio frequency	2.85 MHz
Radio frequency power	6 MW
Radio frequency source	Magnetron
Accelerating structure	6 Cell standing wave π mode slot coupled
Energy gain per orbit	5.0 MeV
Beam current (pulse)	30 mA
Beam pulse length	2.5 μs
Principal operating parameters:	
ν	1
Ω	10

design to allow the use of a multi-cell accelerating structure. As a consequence, the 100 MeV microtron operates at a much higher energy gain per turn than the 44 MeV microtron, and therefore requires only twenty orbits to achieve the design energy. This is to be compared with the thirty-four orbits required in the 44 MeV machine. Nevertheless, the principles of operation of the two machines are identical. A list of principle design parameters of the 100 MeV microtron is given in Table 2.2.

Upon extraction from the microtron, the electrons will pass through a beam transport system twelve meters long to the ALADDIN injection plane. The beam transport system is shown in plan view in Fig. 2.1 The first two bending magnets perform an achromatic bend and in conjunction with the first quadrupole doublet produce a double focus with zero energy dispersion at the beam aperture stop plane. The "chopper" will deflect the beam into the beam stop aperture in synchronism with the ALADDIN radio frequency system. Thus, the only electrons to arrive at the injection plane will be those in the proper phase for radio frequency capture. This technique will not only reduce the energy deposition in ALADDIN and, more importantly, the uncontrolled stray radiation during injection, but will also permit the control of the number and circumferential distribution of bunches filled. The second pair of quadrupole doublets produce a de-magnified image of the source which is in turn transformed in betatron phase space by the remaining beam transport system elements so as to match the admittance ellipse of ALADDIN at the injection plane.

Injection of the electrons into ALADDIN will follow conventional multi-turn techniques. The scheme is illustrated schematically in Fig 2.5 which gives a phase space representation of the ALADDIN injection plane on which are shown the ALADDIN admittance area, the microtron emittance area, and the boundaries imposed by the injection septum and the vacuum chamber. At injection time the ALADDIN admittance

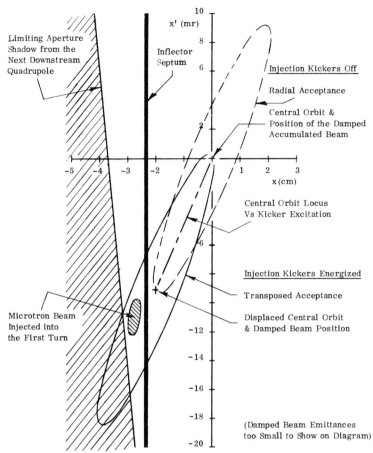

<u>Fig. 2.5</u> Radial phase space diagram of injection into ALADDIN-injector exit view (zero energy dispersion case). Shown is the first turn injection at full kicker excitation (see text). The ellipses mark apertures and beam positions, see also Sect. 3.1.3

area is deflected by the upstream kicker so that a portion of it overlaps the area behind the septum. This deflection is corrected by the downstream kicker. Thus, electrons injected into the overlap area follow the normal orbit around the machine after passing the downstream kicker until they reach the upstream kicker whence they again follow the deflected orbit. During the time required for one orbit (0.29 μs) the excitation of the two kickers is reduced sufficiently so that the electrons will now pass on the right side of the septum and will remain in the machine. Now, more electrons will be injected into the remaining overlap area, and, in fact, the process will continue for the time required for five orbits before the overlap area becomes zero with all the electrons injected executing stable betatron oscillations within the admittance area which will now be centered about the position of the cen-

tral (undeflected) orbit. With the 30 mA pulsed beam current expected from the microtron and taking account for the approximately 66% loss due to the chopper, the current accumulated in each injection event will be about 50 mA. Once captured in the
storage ring, the electron transverse momenta and energy spread will damp so as to
occupy a very small region at the center of the admittance ellipse within a few
seconds of injection and then the process may be carried out again. This injection
scheme, which is called stacking, is widely used on electron storage rings.

e) Computer Control

Computer monitoring and control of accelerators is a highly developed technology
and the high level of performance and reliability now required of synchrotron radiation sources makes it imperative that this technology be exploited. Fortunately,
the continued rapid development of microprocessors and standard I/O modules, such
as CAMAC, has made it possible to set up a special purpose computer network, such
as will be required to monitor and control the many subsystems of ALADDIN, at considerably less cost than previously. The philosophy chosen to institute computer
control of ALADDIN is as follows. The main control computer of the system will be a
fairly large minicomputer. This computer will act as an executive presiding over an
array of microprocessors, one or more at each subsystem of the ring. The microprocessors will communicate to the subsystems through standard I/O systems such as
CAMAC. In operation, the microprocessors would be responsible for the direct control
of the subsystems according to programs and commands fed to them by the executive
computer. One of the most important functions of the microprocessors, particularly
during the running-in period, will be that of monitoring the performance of the subsystems with as high an update rate as possible. To have an "instant playback" of
systems parameters to diagnose intermittent abnormal operation or systems failure in
any storage ring can be invaluable.

As the radiation facility develops, other tasks that the computer network will
be expected to take on are monitoring the status of the photon beam lines and the
responsibility for fast dump of the stored beam in the event of subsystems failure.

During stored beam time and machine downtime when the demands on the executive
computer are minimal, it will be made available to the users for program development, simple data reduction tasks and mass data storage and transfer.

2.3.2 The National Synchrotron Light Source (NSLS)

This machine was designed to meet the needs of investigators working in X-ray physics,
crystallography, and biology for an immensely intense source of extremely high
brightness in an energy range extending from two keV to 30 keV and beyond. The parameters of the machine are given in Table 2.3 and a plan view is shown in Fig. 2.6.
A complete lattice element is exhibited in Fig. 2.7.

Table 2.3 Parameters of the NSLS (Brookhaven) source

General	
Energy	2.5 GeV
Revolution Frequency	1.98 MHz
Critical Wavelength	2.9 Å in arc magnets
Damping Times	
Radial	4.3 ms
Vertical	4.3 ms
Energy	2.15 ms
RF System	
Harmonic	25
Number of Cavities	2
Max. Volts Per Turn	800 kV
RF Power to Beam (1 A)	584 kW without wigglers
Vacuum System	
Construction	Alumininum, Bakeable to 150°C
Pressure	10^{-9} to 10^{-10} Torr
Corrections	
Sextupoles and Higher Order	
Multipole Magnets	36
Injection	
Type	Multiturn using shorted transmission line inflector with two pulsed dipoles
Source	700 MeV separated function slow cycling synchrotron.
Lattice: FODO	
Preinjector: 50 - 100 MeV linac	
Magnet	
Betatron Tunes	
Radial	~10.8
Vertical	~6.8
Momentum Compaction	0.0056
Basic Unit	
Focusing Order	$FDBBFBBDFDO_{2\ 3\ 3}$
Bending Magnet	
Radius	8.17 m
Field Index	0
Edge Angles	5.62° at BB
Quadrupoles	
Max. Grad.	2.2 kG/cm
Phase Advance	
Radial	325.8°
Vertical	205.5°
Long Straight Central	
Drift Space	5 m

A study of these figures will reveal several fundamental differences between this machine and the previous example. In the case of the NSLS, the complete lattice element is considerably more complicated; involving eight bending magnets and eleven quadrupoles. Six such elements, known as super periods, make up the lattice of this ring. An examination of the β and η functions, shown in Fig. 2.8, discloses that

50

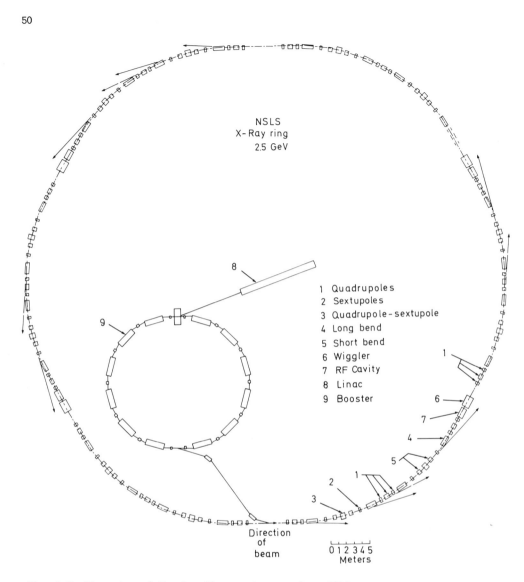

NSLS
X-Ray ring
2.5 GeV

1 Quadrupoles
2 Sextupoles
3 Quadrupole-sextupole
4 Long bend
5 Short bend
6 Wiggler
7 RF Cavity
8 Linac
9 Booster

Direction
of
beam

0 1 2 3 4 5
Meters

Fig. 2.6 Plan view of the Brookhaven storage ring, NSLS

there is a mirror symmetry between the two halves of the complete lattice element, thus even though the super periodicity is six, the fundamental periodicity of the structure is twelve.

This, plus the fact that there is a significant phase advance in both betatron motions in each of the six long straight sections explains why the phase advance per complete lattice element does not violate the limit on phase advance per lattice element mentioned in Sect. 2.1.1. But the behavior of the η and β functions in the long straight sections is the most important feature of this design. It will be noticed that at the centers of the straight sections, these functions become vaninshingly small. Therefore, at these points the electron beam cross sections will

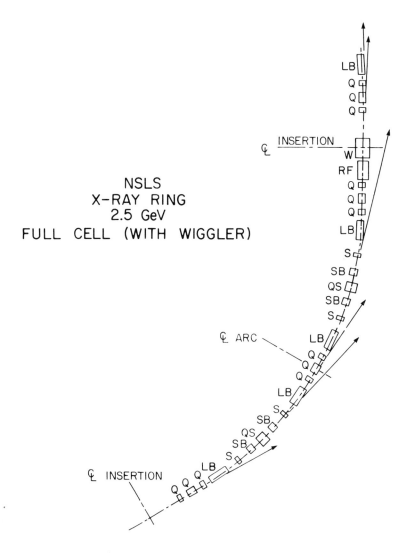

Fig. 2.7 Sextant of the 2.5 GeV NSLS-ring. (W: wiggler). The other symbols are explained in the text

also be very small indeed, and it is at these locations that the installation of three pole transverse wigglers is planned. At a field of 4.0 T and an electron energy of 2.5 GeV the radiation from these wigglers will have a value of λ_c of 0.7 Å (17.5 keV).

At the bending magnets, the electron beam cross sections are, of course, larger but not so large as to seriously reduce the performance of the machine as a source at lower energies. However, it must be kept in mind that this machine has been

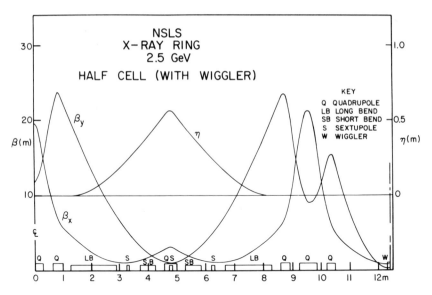

Fig. 2.8 Variation of the envelope and off-energy functions in a half cell of the 2.5 GeV NSLS-ring

designed as a very high performance hard X-ray source and it is expected that it will find its major application in this service.

In achieving these beam characteristics in the long straight sections a number of compromises have had to be made. For example, the betatron tune of the machine is quite high necessitating the installation of a number of rather strong multipole correction magnets which must be positioned at points of the lattice that would, ideally, be occupied by bending magnets and quadrupoles. This would obviously compli- cate the design of these elements. Alternatively, the position and length of the magnets could be changed, but this will degrade the performance of the machine some- what. However, the reduction is not expected to be serious, hence, this is the choice that was made. The effect of this approach on the lattice configuration may be clearly seen in Fig. 2.7 where it will be noticed that two of the bending magnets (denoted by SB) are one third the length of the normal bending magnets (denoted by LB). This was required to make space for the sextupole correction magnets (denoted by S). However, a third sextupole was required per lattice element and this will be combined with the horizontally focusing quadrupole in the center of the element. This combination element is denoted by QS. The S sextupoles correct the chromaticity in the vertical plane and the QS sextupole corrects the chromaticity in the horizon- tal plane.

Beyond the optical properties of lattice, the construction details of the NSLS differ in only two important aspects from the previous example. Firstly, because of the very large energy density of the synchrotron radiation at the vacuum chamber walls in the bending magnets, the vacuum chamber will be fabricated from aluminum extrusions with channels for circulating water through their outer periphery. The construction techniques that must be utilized with aluminum in order to achieve high vacua are somewhat more complicated than those employed with stainless steel, however, these techniques are now well developed.

Secondly, because the machine energy is high, the injection energy must also be high: 700 MeV. To reach this energy economically, a slow cycling synchrotron, called a "booster", will be used. Interestingly, this machine itself would have been considered a major high-energy accelerator as short a period of time as fifteen years ago. It too will be of the modern separated function design with its lattice designed to achieve damping of both the transverse and longitudinal electron motions. This feature leads to some additional complication in the injector machine but the advantages of injecting a well damped beam into a storage ring that were discussed in Sect. 2.2.3 far outweigh the added expense.

References

2.1 J.P. Blewett: Phys. Rev. 69, 87 (1946)
2.2 F.R. Elder, R.V. Langmuire, H.C. Pollock: Phys. Rev. 74, 52 (1948)
2.3 D.H. Tomboulian, P.L. Hartman: Phys. Rev. 102, 1423 (1956)
2.4 R. Madden, K. Codling: J. Appl. Phys. 36, 380 (1965)
2.5 E.M. Rowe, R.A. Otte, C.H. Pruett, J.D. Steben: IEEE Trans. NS-6, 159 (1969)
2.6 E.M. Rowe, F.E. Mills: Particle Accel. 4, 211 (1973)
2.7 E.M. Rowe, C.H. Pruett, J.D. Steben, W.R. Winter, Jr.: IEEE NS-18, 210 (1971)
2.8 E.M. Rowe: "Research Applications of Synchrotron Radiation, 1," Brookhaven National Laboratory's Report 50381 (unpublished)
2.9 T. Miyahara, H. Kitamura, T. Katayama, M. Watanabe, S. Sato, E. Ishiguro, M. Endo, Shigeo Yamaguchi, T. Yamakawa, Seitaro Yamaguchi, T. Sasaki: Particle Accel. 7, 163 (1976)
2.10 M. Sands: "The Physics of Electron Storage Rings," Stanford Linear Accelerator Center Report 121 (unpublished)
2.11 M.S. Livingston, J.P. Blewett: Particle Accelerators (McGraw-Hill New York/1962)
2.12 H. Bruck: Accelerateurs Circulaires de Particules (Presses Universitaires de France, Paris 1966), in French
2.13 N. Marks, M.W. Poole: "The Choice of Dipole Magnetic Field for the SRS," Daresbury Nuclear Physics Laboratory Report DL/TM129 (1974)
2.14 A.A. Garrin, A.S. Kenney: "SYNCH: A Computer System for Synchrotron Design and Orbit Analysis," University of California-Berkeley (1974)
2.15 K.L. Brown, D.C. Carey, Ch. Iselin, F. Rothacker: "TRANSPORT: A Computer Program for Designing Charged Particle Beam Transport Systems," Stanford Linear Accelerator Center Report 91 (1974)
2.16 E. Keil, Y. Marti, B.W. Montague, A. Sudboe: "AGS: The ISR Computer Program for Synchrotron Design, Orbit Analysis and Insertion Matching," CERN Report 75-13, Geneva (1975)
2.17 S.P. Kapitza, V.N. Melekhin: The Microtron (Moscow, 1969, in Russian); English translation to be published (Harwood Academic Publishers, London) as Vol. 1 of "Accelerators and Storage Rings"

2.18 This technique was first demonstrated by members of the late Academician Professor G.I. Budker's group at the Institute for Nuclear Physics of the Siberian SSR at Novosibirsk

2.19 H. Winick, T. Night (Eds.): "Wiggler Magnets", Stanford Synchrotron Radiation Project Report 77/05 (1977)

2.20 D.A.G. Deacon, L.R. Elias, J.M.J. Madey, G.J. Ramian, H.A. Schwettmann, T.I. Smith: Phys. Rev. Lett. 38, 892 (1977)

2.21 A proposal to the National Science Foundation for the expansion of the Synchrotron Radiation Center of the University of Wisconsin-Madison, The Synchrotron Radiation Center, University of Wisconsin-Madison (1976)

2.22 Proposal for a National Synchrotron Light Source, Brookhaven National Laboratory Report 50595, Vols. I, II (1977)

2.23 M.W. Poole, N. Marks, A.G. Wardle: "Magnets for the Booster Synchrotron and Storage Ring of the Daresbury Synchrotron Radiation Source", in Proceedings of the Fifth International Conference on Magnet Technology (Frascati 1975)

3. Instrumentation for Spectroscopy and other Applications

W. Gudat and C. Kunz

With 53 Figures

The peculiar nature of a synchrotron radiation (SR) source has given rise to a special instrumentation [3.1-9] which differs in many respects from that used with ordinary light sources [3.10-16]. In addition to new monochromator designs there are other new interesting instrumental developments which are due to the fact that with synchrotron radiation a strong light source became available in large spectral regions for the first time.

Those properties of a synchrotron radiation source (see also Chap.1) which have an important influence on the instruments (especially on monochromator design) are: 1) The immobility of the source, 2) the good vertical collimation of the SR emission, 3) the continuous white spectrum, 4) the polarization with a predominantly horizontal electric vector of polarization, 5) the fact that the source (namely the electron beam) is usually by far wider horizontally than vertically, 6) the large distance between the source and the experimental equipment, which is either determined by radiation safety considerations or at least by the finite size of the bending magnets and other installations near the source.

An initial attempt to review the instrumentation for SR was made in 1973 at the first international users meeting [3.2] at which one of the present authors also gave a contribution [3.17]. Even in these early days it was barely possible to give a complete survey on the activities in this field. Nowadays, with the rapid development of SR technologies, it has become even more difficult. We have therefore to apologize if recent published or unpublished developments have escaped our attention. New developments, which in many cases are still unpublished, are coming up with reflecting coatings, reduction of straylight, holographic gratings, zone plates, filters, soft X-ray lithography and microscopy, new monochromator designs, UHV monochromators, new X-ray monochromator designs and many other applications of the X-ray part of the spectrum. We refer those readers who are eager to learn more about details of the instrumentation than we could include in this chapter to handbooks [3.7,8,18-20] and the preprint series which are issued at DESY (Hamburg), SSRP (Stanford), SRC (Daresbury), LURE (Orsay). There is also available a survey on the present and future European [3.21] activities and those planned in the USA [3.22].

In accordance with the much longer development of instrumentation in the vacuum ultraviolet region (namely the region of grating monochromators) this type of instrumentation will occupy a large fraction of this chapter. The X-ray field is in a process of rapid expansion but the number of instruments uniquely designed for SR sources is still limited. Already the lay-out of the beam lines determine to a large extent the type of instruments which can be used efficiently for monochromatization. Of course, small (up to 1 GeV) and large (typically 3 GeV) SR sources require completely different types of laboratories. Examples of such laboratories are described in Sect. 3.1. The necessary optical components and crystals are described in Sect. 3.2. Sections 3.3 and 3.4 are concerned with the vacuum ultraviolet and X-ray monochromators, respectively. These are the most important units of instrumentation, since only a few experiments can do without them. Although there are probably no detectors and filters in use which could not be operated also with other light sources they deserve a special section in this chapter, namely Sect. 3.5. Synchrotron radiation covers a spectral range by far larger than any other source. Selective detectors and filters are therefore needed, in addition to monochromators, in order to avoid excessive stray-light and higher-order radiation. In Sect. 3.6, finally, we describe several typical experimental arrangements which exemplify the way experiments in the different subfields using SR are carried out. Several important techniques in a SR laboratory, like e.g. vacuum technology and computerization, had to be omitted almost completely since there are not so many aspects of these techniques which would differ from their applications elsewhere.

3.1 Lay-out and Operation of Laboratories

3.1.1 VUV Laboratory at a Small Storage Ring

A small storage ring for the VUV is usually a machine with a beam energy in the range 200-800 MeV and of a typical size up to 10 m in diameter (see Chap.2). Storage rings of such a size can be located in the center of a single hall and many experiments can be grouped around the machine. Usually each experiment can have its own beam line and more than ten beam lines can be installed quite easily. We take as a typical installation at a small storage ring the TANTALUS I laboratory [3.20, 23-29] in Stoughton, Wisconsin, which is shown in Figs. 3.1 a) and b). This is a 240 MeV storage ring with an average diameter of about 3 m (see also Tables 1.1 and 1.2).

From the point of view of space for installing experiments this size is already a minimum. It would be possible with present day superconducting magnets to miniaturize such a storage ring further. Ideas of this nature were around in the early days of SR work (e.g. building of a storage ring with 50 cm diameter) but have not been followed up any further since it became evident that the installation of many different experiments at small distances from the source point would not be possible

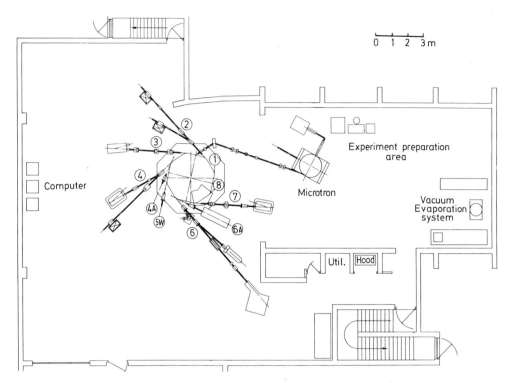

Fig. 3.1a Arrangement of beam ports and experiments at the 0.24 GeV storage ring
TANTALUS I at Stoughton/Wisconsin. (With permission by E.M. Rowe).

with such a mini-storage ring. Twice as large machines as TANTALUS I are probably
of the optimum size for offering sufficient space for the installation of equipment.
Unless very small storage rings become so simple that it will be affordable to use
them as a kind of laboratory source for only one or two experiments this idea appears to have no future prospectives.

One important aspect of experimental installations at a small storage ring is the
short distance between the source and the first optical element. While in principle
an unlimited use of optical lenses (focusing mirrors) renders such geometrical considerations irrelevant, practical considerations are usually in favor of short optical paths. There are several reasons why this holds:

1) The only useful optical elements with imaging properties in the VUV are reflecting mirrors, and optical gratings. Usually the beam line optics consists of one
   or several mirrors. The efficiency of such mirrors, however, is fairly low and
   even at grazing incidence a loss of intensity of at least 50% due to limited
   reflectivity and stray-light has to be tolerated. This limits the number of optical elements and consequently the flexibility of design.

Fig. 3.1b   Photograph of the laboratory at the TANTALUS I ring [3.28]

2) A focusing mirror may need a complicated shape, which causes its price to increase
   considerably with its size. A mirror intercepting a beam of the same angular width
   at a shorter distance is cheaper and easier available.

3) Beams of a certain angular spread are blowing up to considerable dimensions over
   large distances. This requires wide and expensive UHV beam lines, large valves
   and mirror boxes, etc. A further disadvantage of such wide beam-lines is their
   large conductivity and therefore the necessity for large vacuum pumps in cases
   where a pressure difference between an experiment and the storage ring has to be
   held up along the beam line by differential pumping.

   At TANTALUS I, which is the first dedicated storage ring used extensively in the
VUV region the vacuum system is not extremely sophisticated. Fairly small mirrors
(about 4 cm wide and 10 cm long) intercept 10 mrad horizontally. They fit into small
mirror boxes. Beam lines are mostly made of 35 mm inner diameter tubes sealed with
2 3/4" o.d. conflat flanges. Hand operated valves allow for the vacuum isolation between beam lines and the storage ring. A vacuum isolation chamber with a fairly
large volume and a high speed ion getter pump is located between the experiment and
the storage ring. It serves, together with the narrow beam lines, as a differential

pumping system and would reduce the effect of a slow increase of pressure in the
experimental arrangement to the storage ring. It is not of much help in the event
of a catastrophic vacuum break-down. The short distances between experiment and
storage ring are adverse to installation of fast closing valves. This is one of the
big problems and dangers at the TANTALUS storage ring. Vacuum break-down at one of
the experiments could ruin detectors and other equipment not only in the storage
ring itself, but also for all the experiments which are connected to it at the same
time.

Quite some effort probably has to go into vacuum safety considerations at future
small storage rings. There is a tendency to operate many experiments in the future
with more complicated detectors and other sensitive and expensive equipment simultaneously. One has just to imagine the damage which would result from a vacuum failure in one experiment to all the others if several of them have one or more of the
very expansive channel plate arrays (see Sect.3.5) in operation. For protection
accoustic delay lines [3.19,30,31] were developed at the ACO storage ring and are
introduced into the beam lines there (see below).

Alignment of instruments at TANTALUS I is done quite easily by making use of the
visible part of the spectrum. Since free access is possible all around the storage
ring after a beam is stored, optical elements can be adjusted in the same way as it
would be done with a laboratory source. If vacuum lines are temporarily closed off
with windows in order to obtain the visible beam out in the air, care has to be
taken to use windows with zero wedge angle in order to avoid beam deflections. The
vertical center of the beam can be found by using a polarizer foil set at 90 degrees to the horizontal polarization direction of synchrotron radiation. Extinction
will occur in the plane of the orbit according to Fig. 1.4.

While these small storage rings can be built in a way that radioactive radiation
due to particle loss during operation causes no danger (the lost particles are
buried under appropriate conditions in the yokes of the magnets) the experimental
area has to be cleared from people during injection into the storage ring. As a
consequence also those experiments which are in the process of installing equipment
have to interrupt their work at periodic intervals.

3.1.2 VUV and X-Ray Laboratory at a Large Storage Ring

Storage rings operating at beam energies in the region of 3 GeV have average diameters of 50 m or more (see Chap.2 and Tables 1.1 and 1.2). Further, radiation safety
is an important factor with such machines, because of their emission of hard X-rays
and also because of potential hazards due to particle losses. Accordingly, there is
usually heavy shielding set up between the experimental area and the storage ring.
It is customary to make the shielding so effective that the experimental area with
the exception of small regions around the direct beam lines does not need to be
cleared from people during the experiment. In this respect, synchrotrons are more

dangerous sources of radioactive radiation due to the continuous refilling of particles. They need a much heavier shielding all the time. One of the unwanted consequences of this radiation shielding is a larger distance between the source point
and the experiment than it would be necessary already from the size of the magnets
and other installations. It is proposed [3.32] to insert scrapers at two points along
the orbit after storage of beam is accomplished. Any particles getting out of step
would be lost there. Consequently all the other parts are accessible when a stored
beam is in orbit, if only care is taken for the much lighter X-ray shielding. This
concept which is proposed for the SRS storage ring under construction at Daresbury
[3.32], however, demands evacuation of people from all the experimental area during
the injection process.

Figures 3.2 a) and b) show the arrangement of the laboratory at the DORIS storage
rings [3.8,33]. This is quite typical also for the laboratory at the SPEAR storage
ring at Stanford [3.7] although the distances and specific arrangements of beams
differ. At DORIS a water cooled copper absorber, which is incorporated in the vacuum chambers within the magnets, absorbs most of the emitted radiation. A narrow fan
of SR penetrates a hole in the absorber and enters the beam line. A water cooled
removable beam stop protects closed valves from overheating by SR. A beam shutter
BS protects the experimental area during injection into DORIS and, when closed,

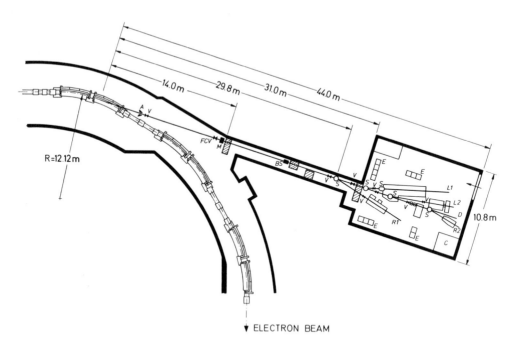

Fig. 3.2a Layout of the synchrotron radiation laboratory at DORIS. Shown is one
quadrant of the storage ring, the beam line to the laboratory and the arrangement
of the experiments. For details see text [3.33]

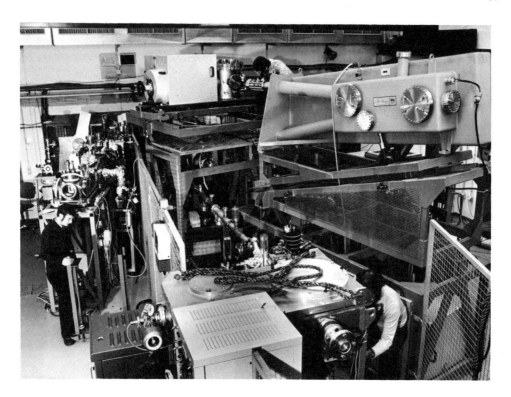

<u>Fig. 3.2b</u> Photograph of the experimental area in the DORIS laboratory viewed in the direction towards the source

allows access to experiments in the direct beam D. This direct beam is collimated to a very narrow region at the experimental area by several shielding blocks. Access to this region is blocked while the beam is on, mainly because of the potential danger of a loss of the stored beam into the direction of the laboratory. A permanent magnet M serves to deflect any charged particles in such a beam while the γ quanta would continue along the beam line. In the radiation safety considerations this magnet serves to reduce the total dosis in the laboratory by one order of magnitude. At the SPEAR laboratory individual beam shutters are available for each X-ray beam. At this laboratory, due to special conditions of the injection into the storage ring, beam shutters may be open during injection provided that a very narrow region around the direct beam is inaccessible. This is achieved by a so-called 'hutch system'. The shielding is so close to the beam that only small lead covered doors are necessary to provide sufficient access to the experiment. Beam shutters can only be opened if electrical switches confirm that all openings of the hutch are closed. The hutches are sufficiently small to prevent accidental enclosure of any person inside the hutch.

At DORIS an automatic fast closing valve with all-metal seals is located in the beam line near the storage ring (FCV). It is activated by vacuum sensors at the

experimental area located 18 m upstream from it. The closing time of ~30 ms is sufficiently short for stopping a shock wave from entering DORIS. This valve (FCV),
however, is not completely tight. Another slowly closing UHV valve nearer to DORIS
has to be activated simultaneously in order to prevent a slow venting of the machine.
Such a system might not be effective anymore if the distance between the experimental
area and the source is reduced to less than 15 m which is geometrically possible in
principle. For such reduced beam line lengths at the ACO storage ring at Orsay an
accoustic delay line [3.19,30,31] was constructed which will increase the transit
time of a shock wave by 190 ms. Such a device, however, is quite clumsy.

At the DORIS laboratory a beam of 3.8 mrad horizontal aperture is available.
Grazing incidence mirrors split off secondary beams so that altogether four simultaneously operational beams are formed (Fig.3.3). At SPEAR the first mirrors are
located much nearer to the source point. Figure 3.4 [3.7] shows a beam arrangement
in which altogether 18 mrad are utilized for four different beams. Similar arrangements are also planned for a new laboratory at DORIS [3.8] and other storage rings
of similar sizes. The beam splitting and the arrangement of the optical elements of
the different monochromators and experiments in the DORIS laboratory are visualized
in Fig. 3.3. Only one station is reserved for X-ray experiments. In a second special

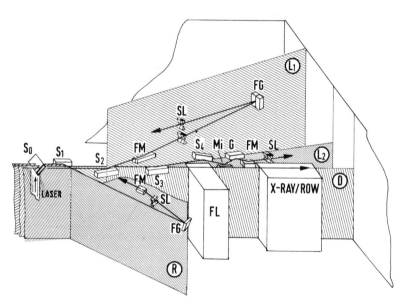

Fig. 3.3 Optical layout of the vacuum ultraviolet beam lines at the DORIS laboratory. The beam (from left) is split into four independent beams L_1, L_2, D and R,
S_1 to S_4: plane mirrors, FM: focusing mirrors, FG: concave gratings, G: plane grating, Mi: flat mirrors, SL: entrance and exit slits, respectively, FL: X-ray fluorescence experiment, X-RAY/ROW: space for an X-ray or Rowland monochromator, the monochromators HONORMI, FLIPPER and HIGITI are occupying the beams L_1, L_2 and R,
respectively [3.33]

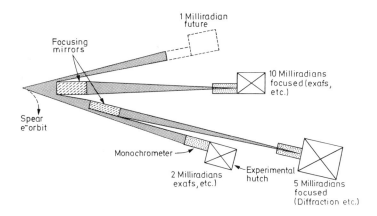

Fig. 3.4 X-ray beam layout at the SSRL storage ring laboratory at SPEAR for 'beam line II'. In two cases focusing torroidal mirrors serve to concentrate up to 10 mrad of the horizontal fan onto the sample [3.7]

X-ray laboratory devoted to molecular biology and administered by the European Molecular Biology Laboratory (EMBL) the X-ray radiation is used from 1978 on. In this laboratory [3.34] four X-ray beams are split off by very grazing incidence mirrors and by crystals in an arrangement quite similar to that at the SPEAR laboratory (see Fig.3.4).

The determination and continuous monitoring of the vertical beam position is of considerable importance for these very long beams especially in the X-ray region where, according to Chap. 1, the vertical angular spread of intensity is very narrow. If the beam at 10 keV is misaligned at a distance of 40 m from the source by only 10 mm, this could mean a loss in intensity of one to two orders of magnitude depending on the aperture of the experimental arrangement. This becomes even more serious, if the experiment needs a well defined degree of polarization. Misalignments then will not only affect the result quantitatively but also qualitatively.

There are several methods in use to determine the position of the beam. The center of the beam can be determined visually with a polarizer foil oriented perpendicular to the main direction of polarization (see Sect.3.1.1). At high powers it may, however, not be wise to extract the visible light through a window directly exposed to the beam (even if the observer is protected by lead glass) because the glass may break. If X-rays are involved a TV camera imaging a fluorescence screen can be used. Another method is exposure of a ZnS covered screen behind a Be-window. The beam shutter can be closed after an exposure of about 1 minute and the ZnS screen, if kept in the dark, will fluoresce for about 5 minutes with good visibility. Additional absorbers serve to isolate the hard radiation which gives a good indication of the center of the beam. At 40 m distance at DESY, for instance, a bright band of less than 5 mm vertical width will allow for a determination of the beam level with an accuracy of 0.5 mm. The same procedure could be carried out with microscope slides. After exposure a blackening due to radiation induced defects marks the beam position. In this case several of these glass monitors could be placed along the beam and if these glass pieces have no appreciable wedge angle a laser beam can be easily brought into position afterwards to simulate the beam for making adjustments. There is also a special type of X-ray paper in usage which shows different colors depending on beam intensity (Kodak, linagraph-paper).

After having determined the beam position the next problem is long term stability of this position. Two possible effects have to be expected: displacement of the electron beam at the source point and changes of the direction of the beam. The second effect is the dominating one with long beam lines. These parameters can be influenced by the activation of correction magnets. This requires a more or less continuous monitoring of the beam position. Figure 3.5 shows two monitor systems which have been realized. Figure 3.5a shows a system of 3 metal strips mounted in the DORIS Lab. [3.8] at the front end of the beam shutter BS. The photoemission current is measured. A similar system is used at SPEAR. Maximum current from the central strip and equal currents from the upper and lower strips are required. This system is operative only when the beam shutter is closed. With DORIS, however, beam positions are very reproducible for one setting of energy and beam optics. Only when these parameters are changed drastic deviations occur (by as much as 1.5 mrad) which need to be corrected for before experiments are resumed. Figure 3.5b shows the SPEAR monitor [3.7] consisting of a graded screen cut out for the part of the beam used by the experiments. This allows for a continuous supervision.

There is one point which can become important with an extensive usage of synchrotron radiation from one storage ring at several points. The electron beam cannot be

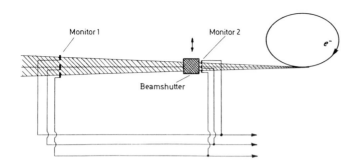

Fig. 3.5a Position monitors for the vertical beam position. Photoemission monitors at the DORIS laboratory monitoring the photocurrent from three insulated metal strips where the center of the arrangement is adjusted to the ideal plane of the orbit. Either the system at the front end of the beam shutter (Monitor 2) or another at the experimental area (Monitor 1) can be used

Fig. 3.5b Position monitors for the vertical beam position. Photograph of a fluorescence screen monitor used at SPEAR at 22 m from the source point. The divisions are
spaced by 1 cm. The black rectangle is cut out in the center to allow the used beam
to pass to the experiments [3.7]

moved around in a storage ring easily by correction coils in such a way that it meets
externally determined conditions in a series of several magnets. Even if these conditions are met at one energy and with one beam optics they might not be fulfillable
with other parameters. This might necessitate a complicated mechanical realignment
procedure of the storage ring magnets or a frequent realignment of experimental
equipment.

   With synchrotrons, e.g. like DESY, motion and deformation of the beam during the
acceleration period (10 ms) can take place [3.34a]. These can be especially large
at the end of the acceleration cycle when electrons are extracted by exciting orbit
deformations. Experiments can use gate signals to eliminate such time intervals. It
is also of great importance with synchrotrons to monitor the beam intensity which is
available at the sample. Figure 3.6 shows as an example three curves which were measured simultaneously [3.35]. Figure 3.6a is a spectrum without any monitoring exhibiting all the fluctuations of the source. For obtaining curve b) the signal was divided
by the signal from a monitor which was located in front of the experiment while curve
c) shows a normalization of the spectrum to a reference signal which monitors exactly the same light beam which is hitting the sample. Obviously the last method is
the best for suppressing fluctuations.

   Although synchrotrons are usually inferior to storage rings as SR light sources,
they have properties which can be of some value for special experiments. Thus, the

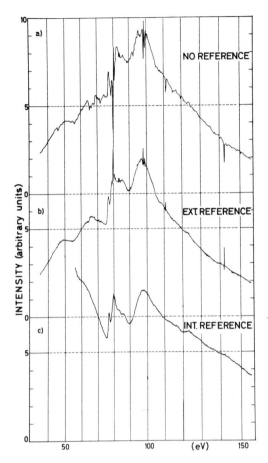

Fig. 3.6 Original spectra registered simultaneously by an open photomultiplier with an AL_2O_3 coated cathode measured at DESY: a) without reference monitor, b) the same spectrum divided by a reference signal obtained from a cathode surrounding the entrance aperture of the monochromator, c) the spectrum divided by a reference signal obtained from the photoelectrons of a gold coated mirror reflecting the monochromatized beam [3.35]

much simpler vacuum system (10^{-6} Torr) allows for the investigation of 'dirty' gasses or solids in a much easier way. Any equipment for which there is no inherent need for ultrahigh vacuum conditions can be installed much simpler and cheaper. Often the very high photon energies are used from synchrotrons since they are not produced by the present generation of storage rings (although PETRA, PEP, CESR, see Tables 1.1 and 1.2 will be available soon).

3.1.3 Beam Line Optics

a) General Considerations

Practically any optical instrument has a certain maximum acceptance defined as the area of a real or virtual diaphragm times the maximum solid angle tolerable for the beams passing through this diaphragm. The goal of a beam line optics is to image the source in such a way onto this diaphragm that the acceptance is filled. If this is achieved the intensity throughput with a given brightness of the source is maximized. Imaging is governed by Abbe's law (Liouvilles' theorem). For these considerations we have ignored intensity losses in the optical elements. No further geometrical gain in intensity is then possible. There are, however, three points: 1) the goal defined above can be achieved in an elegant (say simple, efficient, low-cost) manner or in a more complicated way tolerating reflection losses of many mirrors or other components, 2) when constructing new instruments their acceptance can be increased and 3) they can be matched from the outset to the emittance of a SR source rendering additional optical coupling elements unnecessary.

Unlike with radiation in the visible and near vacuum ultraviolet focusing elements in the soft X-ray region and even more in the hard X-ray region are restricted considerably. It is fairly common practice to avoid too many focusing mirrors at small grazing angles and bent gratings or crystals if possible.

There are several techniques which are helpful in designing optical arrangements in the vacuum ultraviolet, soft and hard X-ray regions. First-order imaging properties of normal and grazing incidence focusing mirrors are described by a focal length applying the classical laws of optics. Spherical mirrors at grazing incidence usually have considerable aberrations already at fairly small apertures (see Sect.3.2). Usually the imaging properties including the aberrations can be obtained in a fairly straight-forward way by applying geometrical optics either analytically by taking into account higher order terms of the characteristic optical functions or by ray tracing which can be achieved with the help of modern computers. Ruled gratings are treated quite similarily with a few modifications. It should be noted that even a plane grating has imaging properties (see e.g. [3.36]). Crystals, as they are used for X-ray monochromators, are treated in first order again by ray optics. Imaging properties of crystals with surfaces cut at an angle to the lattice planes [3.37] come into play quite similarily to the just mentioned case of optical gratings (see Sect.3.4).

b)   The Phase Space Method

A method which is especially useful in treating X-ray optics at a storage ring came up recently [3.38-40] and was borrowed from the theory of charged particle beam transport systems with which it has an intimate relationship at the source point anyhow. It is a phase space technique as illustrated in Fig. 3.7. We demonstrate it here for the vertical extension of the SR source and for its vertical divergency described in a y and y' coordinate system where y is a vertical coordinate and y' is its derivative. SR be emitted at $Z = 0$ from an electron beam whose distribution in space and angle be described by gaussian functions. Then the contour for, e.g., one standard deviation of beam intensity in the y, y' plane is an ellipse (see also

Chaps.1 and 2). The emission of SR adds to the angular width not to the vertical extension of the source. In order to be able to handle the convolution of beam and SR distributions mathematically the SR angular distribution is also approximated by a gaussian distribution. Then the convolution results in ellipses in phase space again. These are also given in Fig. 3.7a and describe now the properties of the SR source. Figure 3.7b shows how the ellipses transform when we let the beam expand along its path. The area of the ellipse is an invariant (Abbe's or Liouville's theorem!). We also show the part of the ellipse which is cut out by a horizontal slit at distance $Z = \ell$. This slit can be transformed back to the origin $Z = 0$ and this is shown in Fig. 3.7b. In Figs. 3.7c and d we show the equivalent construction for a source point where the electron optics gives an oblique ellipse. For special applications it could be useful to consider separately ellipses for the two directions of polarization in such diagrams.

In general the transformation of a coordinate y_0, y_0' at the origin $Z = 0$ of the source to a point y, y' at a certain position $Z = \ell$ of the beam line is described mathematically by a series of matrix multiplications where different elements of the beam line correspond to a unique matrix each. There are matrices for drift sections, focusing elements and dispersive elements which can be worked out for each case [3.38-40]. The corresponding problem of finding the source coordinate y_0, y_0' corresponding to a coordinate y, y' at an instrumental aperture is solved by the inverse transformations.

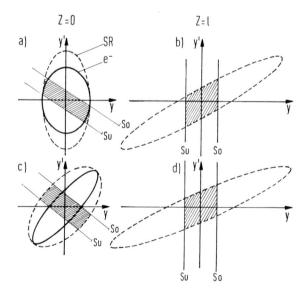

Z=0   Z=l

Fig. 3.7 Phase space ellipses for the vertical coordinate of the electron beam alone (solid lines) and including the SR divergency (dashed lines). The source is located at $Z = 0$, the light beam expands in the positive Z direction. The upright ellipse for SR a) is simply sheared while the beam moves to $Z = \ell$ (b). The invariants are the intersections with the y-axis, the projection on the y' axis and the area. A slit (with boundaries S_u and S_o) inserted at $Z = \ell$ can be projected back to the origin (a). The shaded area is an invariant also. Figures c and d show the same transformation for an already tilted electron beam ellipse (divergent electron beam!)

c) Magic Mirrors

One of the important properties, especially of large storage rings, is time structure. The pulse structure of the electron current can be as short as 150 picoseconds
(see Chap.1). The question arises, how an optical system collecting light from the
same bunch of electrons emitted at different times at different parts of the orbit
will influence this pulse length. LOPEZ-DELGADO and SZWARC [3.41] have shown that
there exist specially shaped mirrors winding around an idealized circular storage
ring (Fig.3.8) which could collect all the light into a focus without distortion of
the pulse structure. The method to treat the problem is to consider both, the time
an electron travels around the orbit (with practically the velocity of light) to a
certain point at which emission takes place and the time, the light needs from there
on its optical path to the focus. Apart from the time structure problems, from these
considerations follows the interesting result that a radiating circle with well collimated emission can be imaged to one point.

   For practical purposes one can show that any other focusing optics collecting
radiation from practically achievable horizontal apertures will produce only minor
distortions of the pulses. As for an example we calculate for DORIS that a mirror
collecting 100 mrad of horizontal radiation would increase the pulse length by
3 ps. This is only a 2% increase of a 150 ps pulse. We therefore come to the conclusion that the magic mirror concept is most probably without practical importance
for ordinary beam line concepts. It appears to be mainly of conceptual interest.

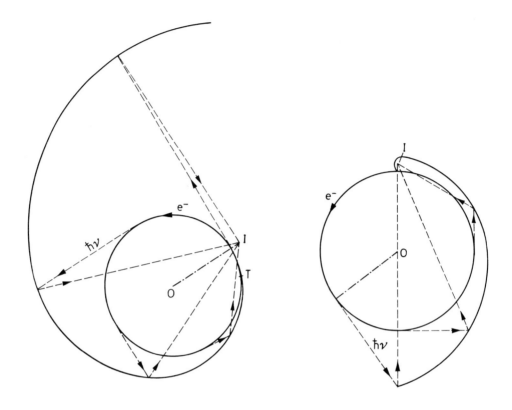

Fig. 3.8 Two classes of the 'magic mirror' arrangement which is capable of focusing SR from a circular orbit onto one point I without introducing any time distortion of the pulses [3.41]

3.2 Optical Components

Every monochromator or spectrograph must have an optical element to disperse the radiation of the light source. Since several decades mechanically ruled reflection gratings have been used with various degrees of success in the spectral range of the VUV down to wavelengths of about 20 Å. At wavelengths below about 10 Å crystals are the most efficient dispersers. The newly developed transmission gratings give promise of covering the transition region. Gratings made on special blanks and bent crystals already have certain focusing properties, but in general especially shaped mirrors have to be used for a stigmatic imaging of the light source. Holographically made gratings offer the possibility of directly compensating some aberrations. In order to maximize the throughput of an optical system each component should have the highest attainable reflectivity in the desired wavelength range. In the near VUV multilayer reflective coatings may help to solve this problem.

In the following paragraphs we want to summarize some of the more important properties of the various optical components mentioned above. No in-depth discussion is intended since adequate treatments have been given in several monographs and review articles [3.10-17,42-45].

3.2.1 Mirrors and Reflective Coatings

a) General Remarks

The discussion of Sect. 3.1 has already stressed some of the severe constraints which are imposed on mirror design for SR beamlines at high current and high energy storage rings. For a stigmatic imaging an excellent optical figure is necessary, especially due to the long beamlines and the required grazing incidence optics. Mirrors should have very low stray light which implies extremely small rms roughness in a wavelength range of soft X-rays. Synchrotron radiation covers an enormous wavelength range which also requires high reflectivity of mirrors and reflective coatings in a broad spectral range. Furthermore, at grazing angles of a few degrees but with wide acceptance angles, large sizes are necessary which are difficult to make. In addition, the material must be suitable for UHV requirements of the beamlines and the optical instruments. And last but not least, mirrors, which are close to the tangential point have to withstand power densities up to hundred watts/cm^2 or more depending on the angle of incidence of the light. In view of these stringent requirements it is not surprising that there is no unique answer of what one might consider as the optimum mirror. At present, extensive investigations are being carried out [3.46,47] in order to clarify these problems.

We begin our discussion on mirrors and reflective coatings with a brief summary of the optical response of a solid to electromagnetic radiation.

b) Reflectivity in the Vacuum Ultraviolet

The optical response of a solid to electromagnetic radiation is, in principle, described by either of two frequency dependent complex quantities, the well-known complex dielectric function $\hat{\varepsilon} = \varepsilon_1(\omega) + i\varepsilon_2(\omega)$ and the complex index of refraction $\tilde{n} = n + ik$. Both complex quantities are related through $\tilde{n}^2 = \tilde{\varepsilon}$ as a direct result of travelling wave solutions to Maxwell's equations for an absorbing optical medium. At normal incidence the reflectivity R of a surface with negligible surface roughness can be calculated from Fresnel's equation

$$R = |\tilde{r}|^2 = \left|\frac{\tilde{n}-1}{\tilde{n}+1}\right|^2 = \frac{(n-1)^2 + k^2}{(n+1)^2 + k^2} \qquad\qquad (3.1)$$

with $\tilde{r}$ being the complex reflexion coefficient. When n approaches one, the reflectivity can become very small. This behavior is seen from Fig. 3.9 which displays the normal incidence reflectivity of aluminium, gold and carbon over a wide energy range [3.48]. These materials as well as others [3.48-53] exhibit a steep decrease in reflectivity beyond 30 to 40 eV which is approximately proportional to the fourth power of the wavelength. This simply means that normal incidence mirrors are no

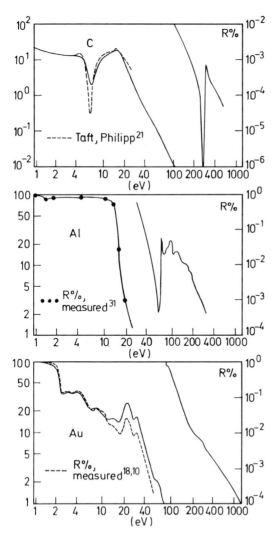

Fig. 3.9 Normal incidence reflectivity of C, Al, and Au as calculated from a Kramers-Kronig consistent set of optical constants. For comparison experimental data are included [3.48]. For references of the experimental data see also [3.48]

longer practical. However, at high angles of incidence Θ the reflectivity increases considerably according to Fresnel's equations which in this case depend on the polarization of the light. For an absorbing medium with the E-vector of the light perpendicular to the plane of incidence the complex reflection coefficient $\tilde{r}_s$ is [3.54]

$$\tilde{r}_s = \frac{\cos\Theta - (\tilde{\varepsilon} - \sin^2\Theta)^{1/2}}{\cos\Theta + (\tilde{\varepsilon} - \sin^2\Theta)^{1/2}} . \tag{3.2}$$

and with E parallel to the plane of incidence* one finds

$$\tilde{r}_p = \frac{\tilde{\varepsilon}\cos\Theta - (\tilde{\varepsilon} - \sin^2\Theta)^{1/2}}{\tilde{\varepsilon}\cos\Theta + (\tilde{\varepsilon} - \sin^2\Theta)^{1/2}} . \tag{3.3}$$

As usual, the square of $|\tilde{r}_s|$ and $|\tilde{r}_p|$ give the reflectivities R_s and R_p respectively. For partly polarized radiation the reflectivity is obtained from

$$R = 0.5 \, [R_p(1 + P) + R_s(1 - P)] \tag{3.4}$$

where P is the degree of polarization defined as

$$P = (I_p - I_s)/(I_p + I_s) \quad , \tag{3.5}$$

I_p and I_s being the intensities of p and s polarized light, respectively.

   Figure 3.10 shows the reflectivity of Pt versus photon energy for various angles of incidence as determined experimentally with s-polarized SR light [3.56]. The data demonstrate the dependence of the reflected intensity on the angle of incidence and convincingly illustrate the principle of using a mirror at non-normal incidence as a high energy cut-off filter. As we shall see below, this constitutes an important element in the designs of monochromators and beamlines. Since the synchrotron radiation has a continuous wide-band spectrum, harmonics can seriously contaminate the radiation that have been monochromatized by a grating or a crystal.

   Another property of synchrotron radiation is of importance when using mirror optics, namely the horizontal polarization of the electric vector E of the light. In Fig. 3.11 we show the calculated reflectivity R_s and R_p for Pt as a function of the angle of incidence. As already stated the reflectivity is quite small at normal incidence, but for R_p it is even smaller at intermediate angles, namely at the well-known Brewster angle. For little absorption, i.e. small k, R_p is considerably smaller

---

*For p-polarization the validity range of Fresnel's equations has to be carefully considered. According to KLIEWER [3.55] important deviations occur for simple metals.

74

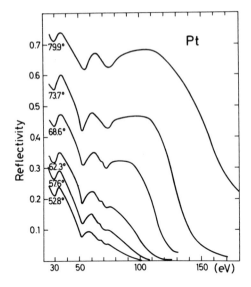

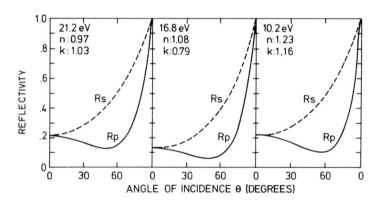

Fig. 3.10 The experimentally determined reflectivity of Pt is plotted
for various angles of incidence
versus photon energy [3.56]

Fig. 3.11 The calculated reflectivity of Pt for perpendicular R_s and parallel R_p
polarization is plotted for three different photon energies [3.57]

than R_s. Obviously, this has to be considered in the design of beam lines and optical instruments in order to enhance the intensity and the degree of polarization.

Equations (3.2) and (3.3) are the exact Fresnel equations and thus are also valid
for the soft X-ray region. But they are often simplified by treating the unit decre-
ments to the complex refractive index δ and β in $\tilde{n} = 1 - \delta + i\beta$ as small quantities
(typically of the order 10^{-3} to 10^{-6}) which implies that their squares and products
can be neglected. Furthermore, for small grazing angles one can replace $\sin\theta$ by the
angle θ and can neglect the polarization dependence, thus obtaining a single expression for the reflectivity of the X-ray region [3.13,58,59]. This rather lengthy expression which we do not want to write down, is valid for the wavelength region
below 10 Å. However, at longer wavelength the approximate theory can introduce

significant errors [3.60], since the reflectivity is no longer the same for both states of polarization. δ is roughly given by [3.58]

$$\delta = 2.74 \times 10^{-6} \frac{Z\rho}{A} \lambda^2 \quad , \tag{3.6}$$

where Z is the atomic number, A the atomic weight (in g) and ρ the density (in g/cm^3) of the material. λ is expressed in Ångström units. The parameter δ determines that glancing angle Θ_c where total external reflection sets in a particular wavelength λ_c. With decreasing ϑ the real part of $\tilde{n}$ gets closer to 1 for a given β and thus the refraction becomes smaller. Therefore total reflection occurs with increasing δ at increasingly larger glancing angles Θ_c.

The free electron theory of simple metals can be utilized to obtain a relation between the wavelength and the critical glancing angle of total reflection, namely,

$$\sin\Theta_c = \lambda \cdot \left(\frac{N \cdot r_0}{\pi}\right)^{1/2} \tag{3.7}$$

where N is the number of electrons per cm^3 and r_0 is the classical radius of an electron ($r_0 = 0.28 \times 10^{-12}$ cm). Combining (3.6) and (3.7) gives the proportionality $\Theta_c \sim \delta^{1/2}$, as discussed above. It is also clear from (3.7) that a material with high electron density reflects a shorter wavelength at a given glancing angle than does a low density material. This is the reason why one uses for instance gold and platinum as coating materials for mirrors and grating. It also explains why a surface which has become contaminated by cracked hydrocarbons of pump oil and vacuum grease shows a strong decrease in reflectivity at the short wavelength limit. Equation (3.7) can only give a guide line in order to determine the short wavelength limit and the cut-off angle, respectively. This expression does not depend on the complete dielectric constant as it should. Moreover such effects as surface roughness are by no means included. As we shall see below this is of great importance. It is our experience that (3.7) results in a too optimistic value for Θ_c when the total number of electrons is counted. It is more reasonable to use an effective electron density which is determined by the number of electrons which can be excited with light of wavelength λ. The effective electron density can be estimated from sum rule plots as given by HAGEMANN et al. [3.48]. Although considerable efforts in utilizing the X-ray part of SR have been undertaken in the various SR laboratories, no recent systematic investigation of the cut-off angle as a function of wavelength and material appears to be available. The agreement between theory and the experimental data up to about 1965 has been thoroughly discussed by SAMSON [3.10].

The β-parameter in the expression for $\tilde{n}$ corresponds to the absorptive part of the complex index of refraction. Therefore β determines the shape of the onset of total reflection. It is a steep step-like threshold for small absorption ($\beta \sim 10^{-3}$). This

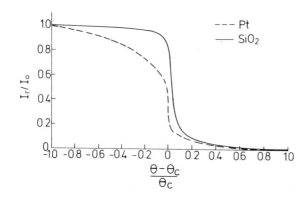

Fig. 3.12 The calculated reflectivity I_R/I_0 versus the
dimensionless quantity $(\theta-\theta_c)/\theta_c$
being the angle of total external
reflection demonstrates that
the more strongly absorbing Pt
does not show as steep an edge
as fused quartz [3.40]

is illustrated in Fig. 3.12 which depicts the calculated reflectivity of fused quartz
and of platinum versus the dimensionless quantity $(\theta - \theta_c)/\theta_c$ [3.40,59].At θ_c the
more strongly absorbing platinum (which is also the high electron density material,
see discussion above) has a reflectivity only about 0.4 times that of quartz. For
mirror design this should be kept in mind.

c) Coating Materials and Multilayer Coatings

Very often the optical surfaces of mirrors are overcoated with a thin layer of a
different material in order to enhance the reflectivity. The choice of the material
depends on various criteria: Obviously the wavelength range to be covered and the
angle of light incidence have to be considered, but also the ease of reproducible
production of thin layers is of importance as well as the strength of adhesion to
the substrate. Further constraints are due to the UHV requirements imposed by the
vacuum of storage rings. Finally, the coating must be resistant to contamination and
oxidation.

From the discussion of the previous paragraphs we know that mirrors at near normal incidence can be used efficiently to photon energies up to 30 to 40 eV. The energy range of the near UV up to about 12 eV is the domain of Al mirrors which are
overcoated with 100 to 250 Å of MgF_2 or LiF for oxidation protection. Such mirrors
can now be obtained commercially with a reflectivity up to 90% at energies below
10 eV. At higher energies the reflectivity deviates considerably from that shown
for pure Al in Fig. 3.9 due to the dielectric coating [3.44,10,61]. It should be
noted that the performance of Al mirrors is also influenced by surface roughness
and losses due to surface plasmons [3.62,63].

At higher photon energies between 10 eV and 40 eV one is better off with coatings
of noble metals such as gold and platinum (see Figs. 3.9-11) which offer maximum
reflectivities for Au and Pt of 17% and 23%, respectively [3.48,49,57]. These materials have proven to show reproducible reflectivities even after various bake-out
cycles.

It is well known that the maximum reflectivity of Au increases with increasing film thickness up to about 200 Å and then decreases again for thicker films [3.56, 65,66]. The explanation for this effect is that thin gold films deposited in high vacuum have an island structure which grows as the average film thickness increases to become a continuous film at about 100-300 Å. Thicker films are thought to become less smooth due to the formation of conglomerates. However, this strongly depends on the preparation conditions. It is believed that evaporation under ultrahigh vacuum conditions results in smoother films [3.46] if the substrate material is smooth. Polishing defects, for instance, with fairly broad structures (10-20 μm) and rms heights below 100 Å are accurately replicated by thin coatings [3.67,68]. The structure of the coating itself has typical dimensions of a few atomic layers in height.

Other materials which have been used as thin film coatings include Rh, Pd, Ir, Os and others. The reflectivity of these materials are given in [3.10,49-53,56,57]. General information on the preparation of thin film coatings is presented in [3.10, 12,69,70].

Recently an interesting new material has been suggested [3.71]. At least up to 25 eV SiC, prepared by chemical-vapor deposition, has a significantly higher reflectivity than any other material investigated to date (see Fig.3.13). According to REHN et al. [3.72] this is very likely due to the extremely smooth surfaces and the high volume plasmon energy of 22 eV of SiC. We want to stress that the data of Fig. 3.13 are experimentally determined and not the result of a Kramers-Kronig conversion of transmittance data to reflectivity data as has been done by HAGEMANN et al. [3.48]. That data, (see Fig.3.9 for some of those results) shows for Au and Cu a somewhat higher reflectivity than determined experimentally.

For the design of the Flipper monochromator at the DESY laboratory an extensive test program on reflectivities of various materials have been carried out. Evaluations are underway [3.73]. Preliminary statements are that carbon and tungsten are other high reflectivity coatings in the energy range 10 to about 40 eV.

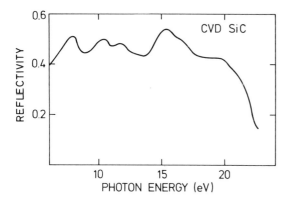

Fig. 3.13 The experimentally determined normal incidence reflectivity of chemical vapor deposited (CVD) SiC is about twice as high as any other material measured up to now in the energy range 5 to about 25 eV [3.14]

As previously described, we have to use grazing-incidence optics at energies above ~ 40 eV. The same materials, i.e. Au, Pt, Ir, W and C then also give good coatings. Depending on the angle of incidence, the reflectivity can be 70 to 80% for energies below about 150 eV (see Fig.3.10).

Quite recently it has been shown [3.74] that the reflectivity at near normal incidence can be enhanced substantially by interference effects in the reflectance of multilayer films using layered structures of Au/C and Cu/C. This technique is well known in the visible and of extremely wide spread application there [3.69].

The basic idea of using a sequence of alternating films of two materials with high and low absorption coefficients is to place the strong absorber in thin layers at the nodes of the desired wavefield. This implies that a fairly large penetration depth of the wavefield can be achieved with a participation of many layers in the reflection process [3.75]. The experimental values (Fig.3.14) obtained thus far were by a factor 7 higher than those of a single opaque gold film at energies around 70 eV. Further improvements seem to be possible [3.74,75,45]. But as with other interference systems there will be only a fairly limited wavelength interval which is efficiently reflected. This means that one would have to use a big set of differently coated mirrors in order to work in a wide spectral range. Since the thickness of the single layers have to be in the order of the desired wavelengths to be reflected, the practical limitation of the useful spectral range, at least in the near future,

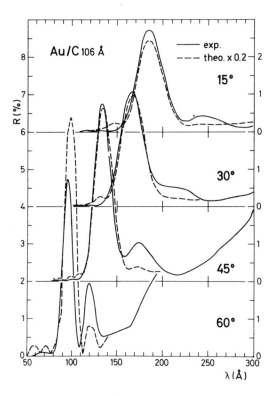

Fig. 3.14 Reflectivity R of a multi-layer interference mirror with 4.5 periods of Au/C on glass as a function of photon wavelength λ. The average periodicity is 106 Å. The angle of incidence α is varied in steps from 15° to 60°. Solid line: experimental values, dashed line: theoretical values based on optical constants of [3.48] (for the glass substrate $\tilde{\varepsilon} = 0.9 + 0.1$ i was used independent of wavelength). The theoretical curve is reduced by a factor of 0.2 [3.74]

will certainly be at wavelengths above about 50 Å due to the enormous problems of
making reproducibly thin smooth layers. Nevertheless, the technique of multilayer
interference coatings in the far VUV appears to be one of the most promising mirror
technologies.

Obviously, a knowledge of the optical constants and reflectivity of various materials is necessary for a proper design of optical systems for use in the extreme VUV.
But there is still only very little experimental information available at photon energies between 200 and 600 eV. It is noteworthy that a large fraction of the data
has been obtained with use of characteristic line sources, i.e. without SR. Once
again, most of the literature up to 1965 has been reviewed by SAMSON [3.10]. Since
that includes the relevant literature, we just want to add a few comments. The most
comprehensive study of the optical constants is due to LUKIRSKII and coworkers
[3.76-79]. In the wavelength range 7 to 190 Å they investigated the angular dependence of the reflectivity of more than 15 metals, semiconductors and insulators. In
Fig.3.15 we reproduce their results for Au and for C. As expected, the reflectivity
of C drops faster with increasing grazing angle than does Au. It should be mentioned
that the data of LUKIRSKII and coworkers [3.76-79] were obtained on air exposed
samples. But firstly, the penetration depth of the light is fairly large and secondly, we are interested in the behavior of materials useful as coating materials of
optical components in air-exposed systems, and thus we consider the data as very
useful.

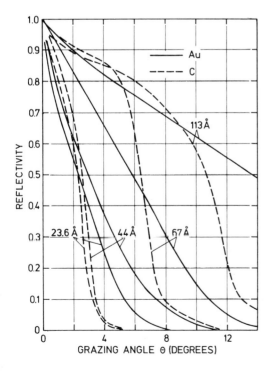

Fig. 3.15 The measured reflectivity
of Au and C is shown vs. the grazing
angle Θ of incidence for various
photon energies. Note the steep
decrease in the reflectivity of C
with increasing angle Θ [3.77]

The wavelength range around the carbon K edge at 44 Å is experimentally the most difficult range to work at. As already mentioned previously, this is due to the fact that after prolonged use of mirrors a contamination layer of cracked hydrocarbons builds up resulting in a decrease of reflectivity. The build-up is a product of light intensity, time, and residual gas composition. At the DESY synchrotron laboratory, for instance, the beam lines are operated at a relatively high pressure of $\sim 10^{-6}$ Torr and then it takes only a few days, until a sizeable intensity drop, i.e. reflectivity drop, is observed. Even at the DORIS laboratory with beamlines operating in the 10^{-9} Torr range this effect is seen to occur. However, much longer exposure times have to be accumulated. As an illustration for the contamination as well as on other radiation induced effects we show a beam-splitter mirror in Fig. 3.16 which has been used in the beam line of the DORIS laboratory. In addition to the surface effects, a discoloration is seen inside the mirror material. It is interesting to note that the visual inspection of a Cu beam-splitter mirror of the Stanford SR Laboratory did not show evidence for carbon build-up even after about 2 years of normal use under clean UHV conditions [3.80]. In contrast to the glass mirror at DORIS the Cu mirror is held at a constant temperature slightly above room temperature [3.81].

Quite generally, the soft X-ray region is served best with mirrors at grazing incidence. In particular, for the performance in the high-energy range the coating material and the substrate material are of importance. There is some experimental evidence that surface roughness can be an even more important parameter. In the next section this question shall be studied in somewhat more detail.

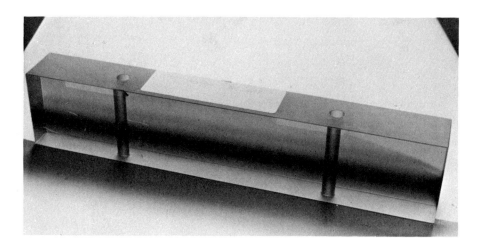

Fig. 3.16   The radiation damage of SR is demonstrated. A glass beam-splitter mirror which has been used for several months in the beam-line of the DORIS SR laboratory under normal operating conditions (E $\sim$ 1.7-2.5 GeV, I $\sim$ up to 300 mA) shows strong discoloration within the glass and cracks at the optical surface

d) Scattering and Stray Light

Any non-ideal surface shows diffuse reflectivity of VUV radiation, i.e. scattering [3.83]. Until recently mainly scattering measurements with visible light have been done showing a spectral dependence of the fraction of scattered light intensity proportional to λ^{-4} [3.82,83]. This wavelength dependence indicates a Rayleigh-type scattering with the characteristic fourth power law. From diffraction by random irregularities, which is a second kind of scattering mechanism, one expects a proportionality to λ^{-2}. Measurements on extremely smooth SiC surfaces with a rms roughness of 4 to 15 Å (Fig.3.17) show very little scattering in the VUV up to 25 eV. The data of Fig. 3.17, replotted versus λ^{-2}, do not show a Rayleigh-type scattering for the SiC mirror in contrast to the Mo coated quartz mirror, where the Rayleigh scattering is dominating. The total scattered intensity was found to be less by about a factor of 10 to 20 compared to the metal-coated highly polished quartz surfaces and about a factor of 100 less than from a Pt coated Cu mirror.[3.80].

As it appears, the available theories on the scattering of VUV light are not yet sufficiently sophisticated [3.72,83,84]. But the scarce experimental results indicate that the scattering strongly depends on the angle of incidence. Furthermore, for a given surface roughness the optical scattering rises monotonically with increasing photon energy and reduction of scattering becomes increasingly important for high photon energy applications. In fact, a Pt coated Cu mirror used for an extended period in the primary SR beam at SPEAR was found to have a rms roughness of about 186 Å and a cut-off energy of about 500 eV with a monochromatization ratio of approximately 10 : 1 at 300 eV, whereas the corresponding numbers for a new Cu mirror having a rms roughness of 30 Å were 60 : 1 at about 280 eV and a high energy cut-off at least 600 eV [3.80]; see also Fig. 3.31.

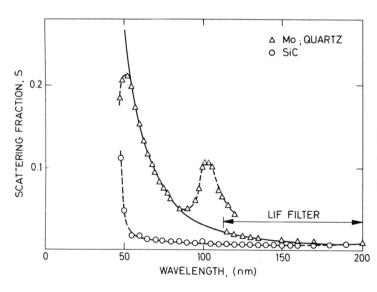

Fig. 3.17 Total scattering spectra for Mo coated fused-quartz and SiC mirrors show the superior performance of SiC as a VUV mirror material. The points joined by clashed lines are inaccurate due to stray light or second-order light. The solid curves are best fits according to the expression $S = A/\lambda^4 + B/\lambda^2$ [3.82]

All one can say at present about scattering in the VUV from the available data is that it appears to be of crucial importance for the soft X-ray range. But even in the range below 300 eV one should care for smallest available surface roughness.

e) Mirror Substrate Materials

From spectroscopy in the visible and infrared a highly developed technology of glass mirrors is available. But those mirrors do not have to fulfill all the requirements which are necessary for mirrors in storage ring SR beamlines. Rather the constraints on materials for VUV mirrors are closer to those in advanced laser technologies. The main reason is the high power density of SR on mirrors. Half of the total power of SR is emitted at shorter wavelengths than λ_c (see Chap.1). For high energy storage rings and synchrotrons with λ_c of ~ 10 Å this means that most mirrors working at not too grazing angles absorb about half the power which is emitted into that solid angle. Obviously, this causes problems with heat dissipation and additionally problems with distribution of charge due to photoeffect in insulators. At the DORIS storage ring, for instance, it was observed that glass mirrors (see Fig.3.16) become cracked at the surface even at a distance of about 20 m. Thus mirror cooling of metal substrates is necessary for mirrors close to the electron orbit. At the Stanford SR laboratory platinum-plated ultra-smooth copper mirrors have been installed in such positions and they have proven very successful [3.80,81]. The Pt coating did not show signs of peeling or other deterioration which has been observed on conventional glass mirrors with metal coatings.

In a recent comparison [3.71] of laser mirror substrate materials it turned out that SiC also has excellent VUV mirror characteristics. It has a relatively high electrical conductivity and, as already mentioned, it can be polished to a super-smooth surface with rms roughness of about 7 Å which is very stable and extremely hard. Moreover, the reflectivity at wavelengths above ~ 500 Å is higher than that of any other material measured so far [3.72].

For all mirrors in a SR beamline behind the very first one, glass ceramic appears to be the most favorable material[*]. These ultralow thermal expansion materials [3.85, 86] guarantee the highest precision optical figures and the smoothest surface of mirrors even under UHV conditions and moderate SR power densities. Table 3.1 gives the thermal expansion coefficients as well as the long time dimensional stability for a number of glass ceramics. For comparison also some metals have been included.

---

*According to most recent results by NIEMANN et al. [3.86a] there are indications that the optical figure of the surfaces of Zerodur mirrors undergoes irreversible changes after irradiation by hard X-rays.

Table 3.1 Thermal properties of mirror substrate materials

Material	Average daily length change $\Delta L/L \cdot 10^{-9}$	Thermal expansion, range $10° - 30°C$ $\alpha \cdot 10^7 \ °C^{-1}$	Thermal conductivity, at $27°C$ $W/m \cdot °C$
Corning 7940 fused silica	-0.51 ± 0.03[a]	$+4.4 - 4.7$[a]	1.4[c]
Owens-Illinois Cer-Vit C-101 glass ceramic	0.50 ± 0.03[a]	$(-0.3) - (-1.2)$[a]	1.7[c]
Heraeus-Schott Zerodur glass ceramic	$0. \pm 0.03$[a]	$(-0.4) - (0.5)$[a]	1.7[c]
Heraeus-Schott Homosil fused silica	-0.56 ± 0.03[a]	$+4.5 - 5.0$[a]	1.4[c]
Simonds Saw & Steel Super Invar	$0. \pm 0.03$[a]	$(-3.0) - (-1.7)$[a]	-
SiC reaction bonded	-	43[b]	180[b]
Cu	-	168[b]	400[b]
Mo	-	51[b]	140[b]

[a] [3.85,86]

[b] [3.71]

[c] [3.87] and [Ref. 3.139, vol. 3]

f) Imaging in VUV

Some general considerations on imaging problems in the VUV are given in Sect. 3.1.3. Also the properties of the so-called magic mirror are discussed which is an optical element capable of collecting all the light of a wide solid angle into a focus without the time distortion of the pulse structure of synchrotrons and storage rings [3.41]. Detailed discussions are presented in the literature for normal and grazing incidence optics [3.10,15,16,88,89]. Here we just want to touch on some aspects which are related to the properties of SR.

It is well known that focussing is best achieved with ellipsoidal mirrors or with off-axis paraboloids. However, due to the very grazing angles of incidence often used in the VUV very large ratios of the semi-axis of the ellipsoids (typically 10-100) occur. Therefore these mirrors are extremely difficult to make with the necessary high accuracy of the optical figure. If one is able to get them at all, then they can be quite expensive. In many cases toroidal mirrors are used which come closest to the more complex elliptical mirrors. Quite generally one has the possibility to use two mirrors instead of one, e.g. two cylindrical mirrors with curvatures at right angles with respect to each other which will fully correct for one of the major aberrations of a spherical mirror at grazing incidence, namely the astigmatism. The other important aberration, the spherical aberration has to be considered anyhow. In a wavelength region say above 50 Å one can afford two grazing incidence reflections without too much intensity loss (see above). The use of two mirrors may save cost and time (of delivery) in addition to obtaining a smoother surface which can be produced more easily on surfaces of simpler shapes. Various attempts have also been made to bend mirrors with a simpler optical figure to obtain a more complex figure, e.g. to bend a cylindrical mirror in order to approximate an ellipsoidal mirror [3.90] or to bend an appropriately shaped plane mirror to also obtain an elliptical mirror by simple adjustment of a single set-screw [3.66]. These kind of systems have proven to be extremely useful in illuminating the entrance slit of normal and grazing incidence monochromators [3.91]. Perhaps more often bent mirrors have been utilized in the X-ray region (see Sect. 3.4.3) of the SR spectrum in order to enhance the acceptance of crystal monochromators or to focus the radiation. It is also quite common to directly focus with bent crystals.

3.2.2 Dispersive Elements

a) Reflection Grating Dispersors

The most commonly used type of dispersor in the VUV is the reflection grating. The simplest system is the plane grating which consists of a number of equidistant lines or grooves ruled on a smooth mirror-like surface. The basic grating equation is obtained by considering the condition for constructive interference between two parallel rays diffracted by two consecutive lines of the grating. The result is [3.10,16]

$$\pm m \cdot \lambda = d(\sin\alpha + \sin\beta) \qquad (3.8)$$

where m is an integer specifying the diffraction order and d is the spacing between the lines. If W is the ruled width then d = W/N, where N is the total number of lines α and β are the angles of incidence of the incoming light and the diffracted light, respectively. The sign convention is such that the negative sign applies when the spectrum lies between the central image (where α = β) and the tangent to the grating ("outside order"). The positive sign must be used, if the spectrum lies between the incident beam and the central image ("inside order"). If one considers the path difference between rays originating from opposite sides of the grating one can work out the theoretical resolving power $R_0 = \lambda/\Delta\lambda$ of the grating by applying Rayleigh's criterion. This gives

$$R_0 = (W/\lambda)(\sin\alpha + \sin\beta) = \frac{W \cdot m}{d} \qquad (3.9)$$

which shows that a high-resolution grating must have a large width W for a given wavelength and that it should be operated at large angles of incidence. The resolving power of a grating also increases with the spectral order number m. This can be seen from (3.8) and (3.9). How close the theoretical limit is approached strongly depends on the quality of the grating. For a spherical concave grating which shall be discussed in detail in the next section (3.9) does not hold [3.10]. Here one has to introduce an optimum width W_{opt} [3.10] of the grating to get an optimum resolving power R_{opt} equal to $0.92 \cdot W_{opt}$ (m/d).

The angular dispersion of the grating is obtained by differentiation of (3.8) with respect to the diffraction angle β

$$\frac{d\beta}{d\lambda} = \frac{m}{d\,\cos\beta} \qquad (3.10)$$

Plane gratings already have certain imaging properties due to the dispersion of the light causing the beam cross-section to change after reflection. Simultaneously the virtual size and distance of the light source changes. This is of importance for plane grating grazing incidence monochromators working at SR sources without entrance slit as we shall see below (Sect.3.3.1)

b) Spherical Concave Grating

It was ROWLAND [3.92] who realized that the focusing properties of a mirror and the dispersive properties of a grating can be combined in a single optical component to give a monochromatized stigmatic image of a light source. In Fig. 3.18 ROWLAND's basic idea is illustrated for a concave spherical grating with radius R. A circle of radius R/2 which is tangent to the surface of the grating at its center and lying in a plane perpendicular to the direction of the grooves, is called the Rowland circle. A point light source lying on this circle will be focused on this circle. Since ROWLAND's comprehensive treatment the theory of the concave grating has been discussed extensively in the literature. In particular we want to mention the work of BEUTLER [3.93], and NAMIOKA [3.94,95] and, more recently, the work of WERNER [3.96] who was able to give a general focusing condition in a more complete formulation of the theory for all types of ruled surfaces. Also several reviews on the grating theory have been presented [3.10,12,16,43,84]. Here we want to briefly outline the general ideas and introduce the notation in order to have a basis for the later discussion on monochromators.

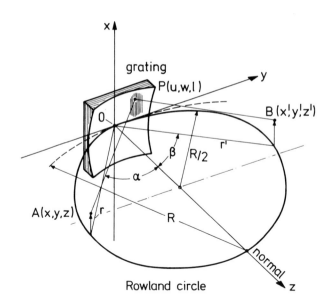

Fig. 3.18 Schematic of the Rowland Circle mounting for a concave spherical grating. Radiation from a point A is dispersed and focused by the grating at the point B with A and B lying on the Rowland Circle having a diameter R equal to the radius of curvature R of the spherical grating. α and β are the angles of incidence and diffraction, resp., measured in the meridional plane

In Fig. 3.18 we also show the conventional Cartesian coordinate system with the origin in the center of the diffraction grating. An incident ray from point A to point P on the grating surface is diffracted towards point B. Within the geometrical theory of the concave grating a characteristic light pass function F is defined [3.93-96] as

$$F = \overline{AP} + \overline{BP} + m \cdot w \cdot (\lambda/d) \qquad (3.11)$$

In addition to the optical path length AP + BP a third term is included which ensures that rays from different grooves of the grating reinforce one another, where w is a coordinate perpendicular to the grooves along the surface. The light pass function can be expanded as a power series in terms of the coordinates of the grating pupil and the spectral image information can be evaluated by applying Fermat's principle which says that the diffracted point B is located in such a way that the light pass function has a stationary value for any point P on the grating surface. Thus, in order to find the focusing conditions one has to take the partial derivatives of F with respect to the pupil coordinates which then can be related to the aberrations of the concave grating. WERNER [3.96] gives an extensive discussion on the choice of the coordinate system, in particular if one uses an off-plane mounting [3.93-95] the expression for the various terms of the light path function can be simplified due to approximations. The resulting equations are still very lengthy and we do not write them down. With increasing power of the series expansion of F the expressions contain successively higher inverse powers of R, r and r' with r and r' being the central rays in the meridional plane (see Fig.3.18). Since the grating dimensions are small compared to these parameters, the higher-order terms in the expansion rapidly decrease in magnitude. These higher-order terms can be associated with certain types of aberrations [3.10,93-96] i.e. with astigmatism, with coma, spectrum line curvature and with spherical aberration, respectively. The first-order term of the partial derivative of (3.11) with respect to the pupil coordinates gives [3.10]

$$\left(1 + \frac{x^2}{r^2}\right)^{1/2} (\sin\alpha + \sin\beta_0) = \frac{m\lambda}{d} \qquad (3.12)$$

and $\quad \frac{x}{r} = \frac{x_0'}{r_0'}$ \hfill (3.13)

with (r_0', β_0, x_0') being the coordinates of the image point for the central ray. Since $x^2/r^2 \ll 1$, Eq. (3.12) which represents the grating equation for the concave grating, is the same as (3.8) for the plane grating.

The second-order term of the series expansion gives the general image equation for the diffraction grating. In the dispersion plane which is usually called the horizontal focal plane, the focusing properties are described by

$$\frac{\cos^2\alpha}{r} - \frac{\cos\alpha}{R} + \frac{\cos^2\beta}{r_h'} - \frac{\cos\beta}{R} = 0, \qquad (3.14)$$

for the properties of the vertical focal plane, i.e. perpendicular to the dispersion plane, one finds

$$\frac{1}{r} - \frac{\cos\alpha}{R} + \frac{1}{r_v'} - \frac{\cos\beta}{R} = 0, \qquad (3.15)$$

where r_h' and r_v' are the distances to the horizontal and vertical foci, respectively. Eqs. (3.14) and (3.15) have the well-known Rowland circle and Wadsworth mountings as special solutions.
For the Rowland mounting the solution of (3.14) is

$$r = R \cos\alpha \text{ and } r_h' = R \cos\beta \qquad (3.16)$$

which is the equation of a circle (the Rowland circle) expressed in polar coordinates. For the Rowland mounting the entrance and exit slits and the grating lie on a circle with radius R/2. An appropriately bent photographic plate could be used, for instance, to record the full spectrum simultaneously. The solution of (3.15) allows to calculate the astigmatism i.e. the difference between the horizontal and vertical foci and is given by

$$r = \frac{R}{\cos\alpha} \; , \; r'_v = \frac{R}{\cos\beta} \; . \tag{3.17}$$

Rowland mountings offer optical advantages, since lower order aberration terms, for instance coma, become identically zero and spherical aberration remains as the major image distortion. Therefore, this mounting is prefered for high-resolution monochromators. At SR sources, where the light source direction is fixed, a scanning Rowland monochromator either generates a spatially varying exit beam which is obviously not desirable, or a fixed exit beam with a complex mechanical linkage of various optical components in addition to the grating. As already pointed out, there exists an optimum grating width for the concave grating W_{opt} [3.98] which is due to the rapid increase of the spherical aberration term with increasing grating width. W_{opt} depends on the angles α and β and on R, the radius of curvature but also on λ. Consider a 2 m grating with 1200 grooves per mm. The optimum width at 100 Å wavelength then is 104, 38 and 24 mm for angles of incidence of 7, 45 and 80 degrees, respectively. The maximum resolving power for a grating illuminated with a point source is [3.98]

$$R_{opt} = 0.92 \; W_{opt} \cdot \frac{m}{d} \; . \tag{3.18a}$$

For a grating much wider than the optimum width the resolution decreases to about 0.8 times R_{opt}. These theoretical limits are, however, usually not reached, since they require infinitly small slits or at least slit widths S smaller than $R \cdot \lambda/W_{opt}$. The examples for W_{opt} used above imply a slit width of 0.2, 0.5 and 0.8 µm, respectively, or less. The resolution for a spectrograph which is limited by the entrance slit widths is obtained from

$$R_{slit} = 0.92 \; \frac{R \cdot \lambda \cdot m}{S \cdot d} \tag{3.18b}$$

At λ = 100 Å the 2 m, 1200 gr/mm grating would give R_{slit} of about 500 with 50 µm slit width in contrast to R_{opt} of more than 2.6×10^4.

A second solution of (3.14) and (3.15) is described by

$$r = \infty \; \text{and} \; r'_h = \frac{R \cos^2\beta}{\cos\alpha + \cos\beta} \tag{3.19}$$

for the focus in the dispersion plane and

$$r = \infty \; \text{and} \; r'_v = \frac{R}{\cos\alpha + \cos\beta} \tag{3.20}$$

in the orthogonal plane. This is the Wadsworth mounting [3.93] which places the source at infinity. This type of mounting is the basis for a large number of monochromators which have been installed at SR laboratories at large accelerators or storage rings (see Sect.3.3), where one easily finds the ratio of $r/r' \gtrsim 30$. If one is working close to normal incidence (i.e. $\cos\beta \sim 1$), then $r'_h \sim r'_v$ and one obtains a point focus. At this point we want to note that Wadsworth mountings use the electron or positron beam width as entrance slit which therefore determines the attainable resolution. Approximate solutions of the equations obtained by setting partial derivatives of the light pass function zero have been given, e.g. the well known Seya-Namioka mounting which minimizes aberrations by the choice of the sum of the angles $\alpha + \beta$ to be 70.5^0 [3.10,99,100]. We also want to mention the asymmetric mounting due to POUEY [3.84,101]. Various monochromator mountings shall be described in Sect. 3.3, but there we do not want to elaboratè further on the complex mathematics of imaging properties but rather shall present their advantages and shortcomings.

c) Aspherical Concave Gratings

So far we have confined our discussion on spherical concave gratings. However, by various theoretical efforts [3.102,104] it has been shown that with aspherical gratings some aberrations occurring with spherical gratings vanish or are at least diminished by a large extent. With elliptical surfaces the ratio a/b of the half-axis can be determined in such a way that for a given angle of incidence the astigmatic aberration term is minimized for a range of diffraction angles. With a/b = 0.134 NAMIOKA [3.103] was able to reduce the astigmatism to about 25% of that of a spherical grating. However, it should be kept in mind that aspherical mirror surfaces are difficult to polish and it is even more difficult to rule on elliptical surfaces. Perhaps the future will bring progress with holographically produced gratings. This fairly new promising technique shall be described later on.

d) Efficiency and Blaze

It is of particular importance to know which fraction, s, of the light intensity incident on a grating is going into the various orders. While the groove spacing of a grating determines the angular separation of the spectral orders and wavelengths, it is the groove shape which controls thc partiLlon of the incident intensity into Lhe spectral orders. As discussed in Sect. 3.2.1 the total reflected intensity is determined by the coating material of the grating. It is this question of grating efficiency which we want to discuss in this paragraph.

Gratings having a sawtooth shape of the grooves are called blazed gratings with the blaze angle being the angle between the normal of the individual groove facet and the overall surface normal. The blaze angle helps to enhance the intensity in a certain wavelength region. This occurs, if the direction of the diffracted light-beam coincides with the direction of the beam specularly reflected from the groove facets. At near normal incidence the relevant blaze wavelength λ_B is obtained from

$$\lambda_B = 2d \cdot \sin\Theta_B \quad . \tag{3.21}$$

For a general angle of incidence α the blaze wavelength λ_B^+ is given by

$$\lambda_B^+ = \lambda_B \cdot \cos(\alpha - \Theta_B) \quad . \tag{3.22}$$

Quite often a grating blazed for the visible or ultraviolet is also blazed for an appropriate VUV wavelength at a grazing angle of incidence. The grating is also blazed for the wavelengths λ^+/m in m-th order. In the VUV region blazed gratings proved to be superior to lightly ruled gratings. Theoretical calculations on the grating efficiency as a function of groove shape have been performed to various degrees of sophistication. Here we just want to mention the comprehensive most recent work of LOEWEN et al. [3.105] and that of HUNTER and coworkers [3.106], which also provides more references. Since the actual groove shape is so important, however, an experimental determination of the efficiency is superior to any theoretical prediction. Experimental set-ups for this purpose are described [3.10,66,106,107]. Basically, one measures the diffracted intensity with a suitable reflectometer for a monochromatized beam of light impinging under fixed angle onto the grating in an in-plane mounting. In Fig. 3.19 we reproduce some of the results obtained by HAELBICH et al. [3.107] at 194 Å wavelength for various angles of incidence. The gratings were commercial Bausch and Lomb blazed plane replica gratings, i.e. an epoxy replica of an originally ruled grating, which were either new, used or used and recoated with Au to restore the reflectivity.

The new replica 9-2 and 10-1-1 of the same master grating with 1200 gr/mm show efficiencies with strong blaze maxima for the +1 (inside) order. The efficiency is between 12% and 14% for these conditions, the highest observed efficiencies were about 20%. One of the used gratings has been recoated with 600 Å of Au. Under the assumption that different replica from the same master are similar, one finds a deterioration by a factor of 25 from Fig. 3.19 and, this is important to note, after recoating with gold the original efficiency values are practically regained. As a general reference source for efficiency measurements we refer to the proceedings of the last VUV conferences [3.2,3,9,108,109]. We also reproduce a very recent compilation by JOHNSON [3.110] of the efficiencies of various (commercially or not) available gratings in Fig. 3.20 (see also Table 3.2). This data collection by JOHNSON [3.110] comprises both mechanically ruled gratings and holographically produced gratings. We shall talk about the latter grating in somewhat more detail later on. But here we see that both types have about the same efficiencies. This result was also obtained by HAELBICH et al. [3.107]. We also learn that the efficiency decreases

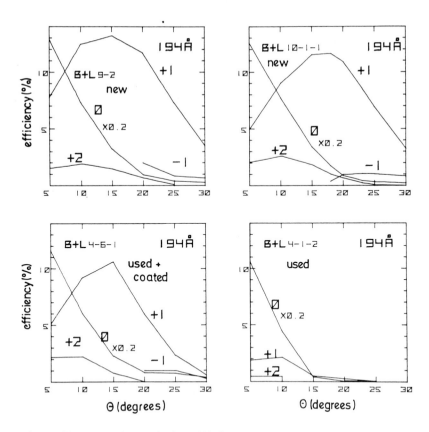

Fig. 3.19 Comparison of the efficiency of two new gratings (Bausch and Lomb No. 2588-9-2 and 2588-10-1) and two gratings used with synchrotron radiation (No. 2588-4-6-1 and 2588-4-1-2). Grating No. 3588-4-61 has been coated with 600 Å of gold after use. The O. order has been scaled down by a factor of 0.2 [3.107]

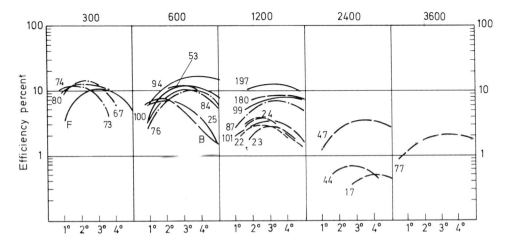

Fig. 3.20 Grating efficiencies for ruled and holographic gratings are plotted vs. grazing angle of incidence. The test wavelength is 44 Å. Further explanations are given in Table 3.2 [3.110]

Table 3.2 Grating efficiency $\lambda = 44$ Å

Radius	Origin	Catalogue No.	Serial No.	Type
17	2 m	B & L	35-52-37-800	2553-6-1-1	Shallow Blazed. Ruled. Replica.
22	plane	J.Y.	-	-	Holographic
23	2 m	B & L	35-52-40-700	2517-2-4-5	Shallow Blazed. Ruled. Pt. Replica.
24	2 m	B & L	35-52-40-700	2617-2-5-8	Shallow Blazed. Ruled. Pt. Replica.
25	2 m	B & L	35-52-40-400	2278-32-1-3	Shallow Blazed. Ruled. Pt. Replica.
44	2 m	B & L	35-62-41-800	1194-6-3	Shallow Blazed. Ruled. Pt. Replica.
47	2 m	B & L	35-52-41-800	1194-5-2	Shallow Blazed. Ruled. Au. Replica.
53	5 m	N.P.L.	-	156	Laminar. Ruled. Au.
67	plane	Göttingen/IC.	-	18	Laminar. Holographic. Au.
73	2 m	Göttingen/IC.	-	25	Laminar. Holographic. Au.
74	2 m	Göttingen/IC.	-	26	Laminar. Holographic. Au.
76	2 m	Göttingen/IC.	-	34	Laminar. Holographic. Au.
77	1 m	B & L	55-72-36-900	1068-1-1-1-4	Shallow Blazed. Ruled. Au. Replica.
80	5 m	Göttingen/IC.	-	21	Laminar. Holographic. Au.
84	5 m	Göttingen/IC.	-	37	Laminar. Holographic. Au.
87	5 m	Göttingen/IC.	-	43	Laminar. Holographic. Au.
94	plane	N.P.L.	-	194	Shallow Blaz. Holographic. Au.
99	plane	N.P.L.	-	177	Shallow Blaze. Holographic. Au.
100	5 m	N.P.L.	-	173	Laminar. Ruled. Original. Au.
101	2 m	B & L	35.52.40.700	2517-2-6-3	Shallow Blazed. Ruled. Pt. Replica.
180	1 m	Hyperfine Inc.	-	320	Shallow Blazed. Au. Ruled. Original.
197	1 m	Ion Tech. Ltd.	-	-	Shallow Blazed. Holographic. Au.
B	2 m	B & L	35-52-40-400	2278-30-2-6	Shallow Blazed. Ruled. Pt. Replica.
F	plane	N.P.L.	-	83	Laminar. Ruled.

with increasing line density. The data of Figs. 3.19 and 20 were obtained with in-plane mountings, i.e. incoming and diffracted rays are in the meridional plane of the grating. Very interesting off-plane mountings for grazing incidence have also been considered [3.16,96,97]. In extreme off-plane mountings the incoming light strikes the grating grooves almost parallel to their direction and not perpendicular as usual. This is called conical diffraction [3.111] and it gives rise to a reflection/diffraction contribution of the whole groove even at grazing incidence which is not the case for in-plane mountings. By these techniques WERNER [3.97] was able to get absolute efficiencies of 30% in the +1 order at 44 Å wavelength at grazing angles of 0.5 to 6°. This is certainly promising and warrants further investigations. However, only a fairly moderate resolution ($\sim$ 100) can be obtained this way. For certain kinds of secondary spectroscopies as for instance fluorescence this might be sufficient.

e) Holographic Gratings

The basic idea of present day advanced holographic grating technology is the relationship between interference fringes and the rulings of a grating. It was already in the last century that CORNU [3.112] realized that a grating with a systematic variation of the grating constant d has certain focusing properties and he was already able to demonstrate it experimentally. In the following decades various attempts have been made to produce such gratings, but in general those gratings were not useful for spectroscopic purposes [3.43]. In 1966 LABEYRIE [3.113] suggested the method of recording the interference fringes of two coherent laser light beams. In 1967 RUDOLPH and SCHMAHL [3.114] were able to produce a grating by recording the interference fringes of a laser beam in a positive photoresist material which was a thin layer on a flat optical substrate. By a special dissolution technique the exposed resist was converted into a laminar type grating structure which was useful for spectroscopy. In the same year the French company Jobin-Yvon designed the first commercial holographic grating [3.43]. This company which gave considerable impetus on the design of the modern holographic gratings which now actually have imaging properties. Of course, various other laboratories were also involved in the development of the theory and technology of the holographic grating. For example, SHANKOFF [3.115] and SHERIDON [3.116] succeeded in producing blazed holographic gratings. Because of lack of space we are not able to review all these valuable efforts. A recent review is available [3.117]. The main advantages of holographic gratings are as follows [3.43,106,117]:

- The spectra are completely free of ghosts which are known to occur to various degrees of magnitude on ruled gratings due to periodic irregularities in the groove spacing.

- Spectra have much lower stray light level and better signal to noise ratio.

- The size of gratings only depends on the optics used to produce the interference pattern (up to ~ 60 cm diameter is available).

- Steeply curved substrates (aspherical surfaces) can be used to produce the gratings (at least in principle).

- Wider choice of groove spacing is available (up to 10000 gr/mm).

- Finally and perhaps most importantly there is the possibility of introducing focusing properties when designing a grating.

Obviously this is a long list of advantages. It is legitimate, however, to ask for disadvantages or, in other words, for which application is the regular grating still the better choice? There is one important point to mention, namely at present, it is not possible to produce all-metal holographic gratings which are as good as described above. However, for certain applications in UHV bakable systems all metal gratings might be a necessity. Recently RUDOLPH and SCHMAHL [3.117] developed a method for producing gratings with all-metal structures. First tests performed at DESY indicate that the efficiency along the grating surface is not constant [3.107]. This is, however, a technical problem which can be overcome.

To date Jobin Yvon is the main supplier of holographic gratings of various types. According to the grating's properties they define 4 different types of gratings [3.43]:

- Type I is identical in terms of aberrations to a conventionally ruled grating and is therefore exchangeable.

- Type II has already some aberration compensation built in for certain types of monochromators. For instance, astigmatism reducing gratings are available for Seya-Namioka monochromators.

- Type III and IV are practically complete monochromators in themselves. They can act as a collimator, a diffraction grating and a focusing element even with aberration corrections. Wavelength scanning is achieved by a simple rotation of the grating.

A monochromator for the spectral range 1500 Å to 150 Å based on a type IV grating is in operation at the LURE SR laboratory with great success [3.118] and at present various other systems of this kind are under construction [3.119,120] for wavelength between 2000 Å and 20 Å. For extreme VUV light holographic gratings now appear to have higher efficiencies than ruled gratings [3.66,110].

The theory of the holographic grating has been worked out in detail in the last few years [3.118-122]. The light pass function, see (3.11), of the regular concave grating is extended in order to include also such terms which contain the coordinate of the optical arrangement used to produce the gratings. Then the procedure described above is applied: i.e. power series expansion of the light path function and application of Fermat's principle which then yields additional terms which are due to the grating lines. For a detailed description of the theory the reader is referred to the original literature [3.43,117,120,123].

f) Zone Plates and Transmission Gratings

Besides the reflection type dispersers which we discussed above in the VUV also transmission type dispersers exist. Until recently [3.10,124] Fresnel zone plates have not been used for practical purposes in the VUV. A zone plate is a "thin-film-optical-element" which consists of an array of concentric rings which alternatively are opaque and transparent to the incident radiation. The widths of the zones or equivalently the radii of circles r_n are choosen in such a way that the distance from the circumference of a given circle to an image point P on the symmetry axis of the zone plate differs from that of adjacent circles by $\lambda/2$. For this geometry and an illumination with parallel light of wavelength λ an enhancement of light intensity at the image point P will occur. The zone plate acts as a lens and the imaging properties are described by the well known focal equation for a lens. The focal length f of a zone plate for a wavelength λ is given by [3.124]

$$f = \frac{r_1^2}{m \cdot \lambda} \quad \text{with } m = 1,3,5,\dots \tag{3.23}$$

if $\lambda/16 << f$. This is a direct result of the equation describing the radii r_n of the n th ring

$$r_n^2 = f \cdot n \cdot \lambda + \frac{n^2\lambda^2}{16} \quad n = 1,2,3 \tag{3.24}$$

To a good approximation, the attainable resolution Δ is given by

$$\Delta = \frac{1.22 \cdot \lambda}{D} \cdot \frac{f}{m} = 1.22 \frac{\lambda}{2 \cdot r_n} \tag{3.25}$$

with D the diameter of the zone plate. For an amplitude zone plate with equal width of opaque and transparent zones the maximum efficiency is $1/(\pi^2 m^2)$. SCHMAHL and collaborators [3.125] used holographically made zone plates at wavelengths of 46 Å to 23 Å to focus SR onto a sample for microscopy purposes. The condensor zone plate for instance was made up of 2600 zones with $r_1 \sim 50$ μm (see also Sect.3.6.1). Recently it was suggested [3.126] to prepare fairly thick zone plates by the technique of X-ray lithography in order to use the channels between adjacent zones as reflectors with blaze properties. Also the possibilities of using phase zone plates [3.127] and zone plates with curved surfaces [3.128] have been discussed.

Transmission gratings are closely related to zone plates which disperse and focus the transmitted light. In fact, one can think of the zone plate as a radial transmission grating. The convention more or less is to name laminar amplitued transmission gratings simply as transmission gratings, i.e. an array of wires with

approximately rectangular cross-section. In the last two to three years the technology of producing such gratings for use in the X-ray and VUV region has been developed by DIJKSTRA and LANTWAARD [3.129]. Basically it is the same kind of holographic process as is used in the production of zone plates. Typical dimensions of
state of the art gratings [3.42,130] are a thickness of about 0.2 μm and 1000 lines
per mm. In order to be rigid enough over a large area there is a coarse support
structure of about 2 μm thickness. It is believed that the gratings can be constructed
to achieve first order transmission up to 25% in the region of anomalous dispersion
[3.130]. The efficiency of a transmission grating can be evaluated as follows. The
count rate of a detector which is received in a given order m as a function of photon energy or wave vector q is calculated from the N-slit diffraction pattern taking
into account the modulation by the simple slit diffraction pattern related to the
particular fractional slit opening a/d [3.42,130]. Since the "opaque" thin lines
also have a certain transmission in the VUV also the interference between waves
passing through transparent parts and those waves attenuated and phase shifted
coming through the wire have to be included. With known optical constants the efficiency versus photon energy has been calculated and compared with experimental data.
Fig. 3.21 a shows the ratio of efficiencies of m-th to 1-st order, part b) displays
the first-order to zero-order ratio values and c) gives the absolute efficiency of
the 1st order. The existing discrepancies are believed to be due [3.42,129] to the
difficulties in obtaining absolute normalization to the incident beam. One of the
potential advantages of transmission gratings is to obtain an optimum efficiency
performance for a given energy region by a suitable choice of the grating material
and its thickness. In Sect. 3.6.1 we shall discuss monochromator arrangements which
have been proposed or already been used. The properties of a transmission grating
behind a grazing incidence telescope for cosmic ray spectroscopy are discussed at
some detail [3.131].

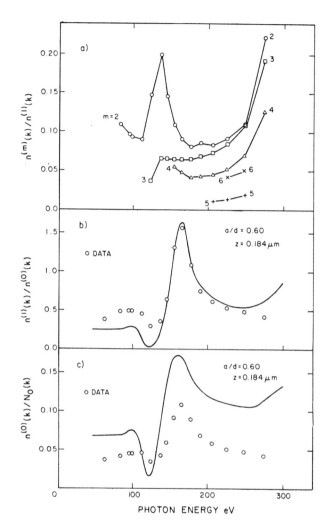

Fig. 3.21 The efficiency of a transmission grating is plotted. a) shows the relative efficieny of orders two through six compared to first order. b) gives the value of the first order to the zeroth order which has been used to determine the grating thickness. c) displays the absolute efficiency for the first order [3.130]

g) Crystals for Monochromators

The wavelength region where crystal diffraction can be used to monochromatize SR extends in principle up to wavelengths above 100 Å. If d_∞ is the geometrical distance between atomic planes then $\lambda_{max} = 2d_\infty$ is the maximum usable wavelength. Table 3.3 gives d_∞ for a few useful materials. The anorganic crystals extending to gypsum are usually quite resistive to the primary SR beam. High resolution can be achieved quite easily if the angular divergence of the incoming radiation is small (see Sect.3.4). For a perfectly parallel beam the resolution depends only on the single crystal reflection curve, as shown in Fig. 3.22. The relative photon energy resolution $\Delta\epsilon/\epsilon$ for zero crystal absorption depends thus only on the choice of the

98

Table 3.3 Useful values of lattice constants in X.U. ($\approx 10^{-3}$ Å). We give here the geometrical distance between atomic planes, d_∞, not corrected for the index of refraction

crystal	d_∞[X.U.]
silicon (220)	1920
germanium (220)	1997
quartz (11$\bar{2}$0)	2451
NaCl (100)	2814
calcite (100)	3029
silicon (111)	3135
germanium (111)	3262
quartz (10$\bar{1}$1)	3336
quartz (10$\bar{1}$0)	4246
gypsum (010)	7585
lead stearate	50300
cerotic acid	72500

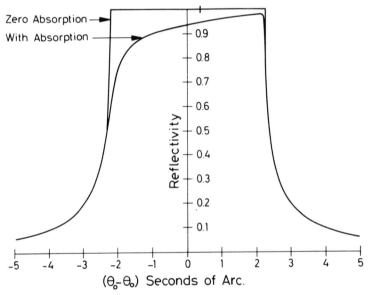

Fig. 3.22 Single crystal reflection profile for the 220 Bragg reflection from Si at λ = 1.38 Å, calculated with and without absorption [3.132]

proper Bragg reflection [3.132] and is·independent of the photon energy

$$\frac{\Delta\varepsilon}{\varepsilon} = \frac{4r_e}{\pi a^3} F_h \, d_\infty^2 \qquad (3.26)$$

where F_h is the structure factor, $r_e = e^2/mc^2$ and a is the lattice constant. Table 3.4 gives calculated resolutions [3.41,132] for a wide range of Bragg reflections (F_h is set constant here independent of the X-ray wavelength).

Table 3.4 Theoretical energy resolution for various Bragg reflections
[3.41,130]

crystal	hhl	$\Delta\epsilon/\epsilon$
germanium	111	$3.2 \cdot 10^{-4}$
	220	$1.5 \cdot 10^{-4}$
silicon	111	$1.33 \cdot 10^{-4}$
	220	$5.6 \cdot 10^{-5}$
	400	$2.3 \cdot 10^{-5}$
	440	$8.9 \cdot 10^{-6}$
	330	$8.1 \cdot 10^{-6}$
	444	$4.6 \cdot 10^{-6}$
	660	$2.2 \cdot 10^{-6}$
	555	$1.4 \cdot 10^{-6}$
	880	$6.6 \cdot 10^{-7}$

In some cases the very high resolution of such perfect crystals is not necessary and even not welcome, since it leads to a considerable loss of intensity. In some cases crystals with a mosaic spread [3.58], namely a crystal consisting of individual subcrystal slabs which are thin compared to the absorption length and having a certain variation in orientations about a mean orientation, can serve to pass a wider wavelength band. Crystals of pyrolythic graphite [3.133] with a mosaic spread of about 10 mrad can serve such a purpose. The disadvantage of using such crystals is an accompanying increase in the divergence of the beam after monochromatization. This could be avoided by producing a strong gradient in the lattice constant of a crystal in a depth smaller than the absorption length [3.134]. Alloying of a single crystal with a gradient in composition appears to be an appropriate method [3.135].

For use with very long wavelength we mention the so-called soap crystals listed at the end of Table 3.3 [3.136-139]. These crystals have not found any wide spread applications with SR yet due to their low resolution which is reported to be 0.03-0.016 [3.138]. Reflectivities are in the order of 1-5% for typically 100 layers in the 20-120 Å region [3.138]. For some applications, however, the large band-pass associated with the low resolution could be quite desirable. These crystals are on the boarder-line to the evaporated multilayer structures mentioned above (Sect.3.2.1).

Focusing with crystals can be achieved by bending them either cylindrically or even spherically [3.140,140b]. Usually the bending introduces defects which leads to a somewhat enlarged single crystal reflection curve compared to the ideal values, as shown in Fig. 3.22. A reduction in the size of a bundle (with an associated increase in its angular spread according to Liouville's theorem) can be accomplished by asymmetric cuts of a crystal [3.37], as shown in Fig. 3.23.

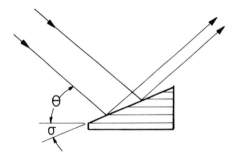

Fig. 3.23 Change of beam cross secttion when Bragg reflected from a crystal with its surface cut at an angle σ relative to the lattice planes

3.2.3 Filters and Polarizers

a) Filters and Higher Order Problems

Synchrotron radiation continuously covers the wavelength range from the infrared to the X-ray region. Therefore the spectrum of an optical disperser will contain an admixture of higher harmonics in addition to the fundamental wavelengths. Sometimes higher orders are used to improve the resolution of a monochromator, but generally the higher order contribution is an unwanted effect. In Sect. 3.2.1 we described the method of utilizing a mirror at a grazing angle of incidence as a high energy cut-off filter. Recently this technique has been extended to a system of two plane parallel mirrors. With a simple rotation of the mirrors the cut-off wavelength can be changed to a certain extent maintaining the direction of a slightly displaced exit beam [3.141]. However, often such a system does not provide the necessary spectral purity in particular in the longer wavelength region say above 100 Å due to the missing of a pronounced short wavelength cut-off. On the other hand, the transmittance of thin films of some materials does show the desired window-type spectral characteristic in limited wavelength intervals. For example, thin films of Al with a thickness of a few thousand Å (see Fig.3.24) have the property of transmitting a portion of the VUV spectrum from 72 eV (~170 Å) to about 18 eV (~680 Å), while blocking the longer wavelengths up to the visible and the short wavelengths down to about 50 Å. This transmission characteristic makes Al one of the most useful transmission filter materials in the VUV. Practically no higher order radiation is present behind an Al filter in the energy range 36 eV to 72 eV.

The transmittance curves $T(\hbar\omega)$ of Fig. 3.24 have been calculated for film thicknesses d = 1000 Å and 3000 Å according to the equation $T = I/I_0 = \exp(-\mu d)$, where I and I_0 are the transmitted and incident intensities, respectively, and μ is the linear absorption coefficient of the material which is related to the optical constants by

$$\mu = \frac{4\pi k}{\lambda}.$$ (3.27)

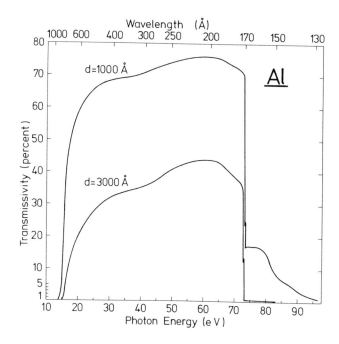

Fig. 3.24   The transmissivity of Al thin films vs. wavelength demonstrates the
excellent performance of Al as a VUV filter having window-type characteristic.
Curves calculated with (3.27) and optical absorption data of [3.48]

$T = \exp(-\mu d)$ is only valid, if the reflectance R<<1 and $k^2 << n^2$. This is the case
for Al for wavelengths up to about 620 Å [3.48]. At longer wavelengths a full treatment has to include possible interference effects which cause T and R to exhibit
oscillatory behavior as a function of d/λ [3.48,69,142]. In fact, in the near VUV
interference effects in multi-layer systems are used to produce transmission filters
with small bandwidth and high transmission [3.61,69,143].

For an exact determination of the transmittance of thin films the optical constants of the materials in question have to be known. At present only very little
data sets are available in the VUV [3.48,144]. Recently, a comprehensive collection
and tabulation of available absorption data for alkali halides has been published
[3.145]. Compilations of this kind can well serve as starting points for an evaluation of optical constants and sum rules [3.48,146]. In the X-ray region extensive
data compilations [3.147] for the mass absorption coefficient are available. However,
one should realize that the data are often incorrect at photon energies below about
1000 eV [3.68]. The reason is simply due to the fact that in most cases extrapolated
data had to be used rather than measured ones.

In the near VUV data have been accumulated on the transmissity of thin metal
films useful as transmission filters. SAMSON [3.10] presents an overview of the
available data up to 1965. Virtually none of these data have been obtained with SR.

In subsequent experimental studies with monochromatized SR often considerable differences have been found in the fine structure of the spectra [3.27,148-150]. Nevertheless, the compilation of SAMSON [3.10] still provides a very useful guide for the
selection of transmission filters. In Table 3.5 we give the useful wavelength ranges
of several often used materials along with the respective references. Some of the
materials are affected by contamination and oxidation resulting in change of transmittance, in formation of pinholes and ultimately in a destruction of the filter.
In such cases (e.g. Pr.) a protective overcoating of a very thin ($\sim$50 Å) carbon
layer has proven to be extremely useful [3.48]. The standard way of preparing VUV
filters is by means of various evaporation techniques [3.69,70]. HUNTER [3.153]
also gives a thorough discussion on the influence of pinholes, substrate material,
supporting mesh, contamination, aging effects, etc. on the transmission characteristic of filters. With Rowland monochromators and some other types of instruments it
is hardly possible to keep the spectra free of higher order contributions over large
wavelength regions. In such cases the ratio of first to higher orders can be estimated by measuring pronounced absorption edges. [3.17,154]. Another method has been
proposed by HOWELLS et al. [3.155] based on photoelectric measurements with retarding
potentials in order to stop electrons arising from high energy photons from higher-
order harmonics. It should also be mentioned that under certain circumstances predis-
persers of any kind can be utilized to select the various orders [3.36,156].

Table 3.5 Useful spectral range for several filter materials

material	range	reference
lime soda glass	throughout visible - 4 eV	3.10
quartz glass	- 7,5 eV	3.10
MgF_2	- 10.5 eV	3.10
LiF	- 11.5 eV	3.10
In	11 - 16.5 eV	3.10
Al	17 - 72 eV	3.48,27
Te	20 - 40 eV	
Mg	20 - 49 eV	3.48
Si	20 - 100 eV	3.27
C	45 - 270 eV	3.48
hydrocarbon films	45 - 270 eV	3.144,27
Pr	60 - 110 eV	3.48,152

b) Polarizers

The SR sources are the only ones in the VUV emitting highly polarized radiation
(compare Chap.1). The polarization state depends on the elevation angle ψ between
the electron orbit and the observer. For vanishing angle ψ the radiation is linearly
polarized with the electric field vector in the orbital plane. With increasing elevation angle the perpendicular field component increases with a relative phase

difference of plus or minus $\pi/2$ above or below the orbital plane, respectively. The absolute values of the electric field components are determined by the absolute value of the angle ψ (see Chap.1).

There is an increasing number of experiments using polarized synchrotron radiation in a variety of applications; for instance, the determination of optical constants with the reflectivity versus angle of incidence technique [3.56,150] (see also Chap.7), the study of selection rule effects on optically excited electronic transitions [3.157,158] and especially in the field of photoelectron spectroscopy [3.149,159]. All these experiments make use of a known degree of the polarization of the monochromatized light. There is another class of experiments in which the degree of polarization and the phase difference between the mutually perpendicular electric field components is determined, i.e. the ellipsometry technique. This technique is a standard method of determining optical constants in the visible and near UV [3.160]. However, in the VUV at wavelengths below the LiF cut-off (1050 Å) there appears to be only one experimental investigation of this kind [3.161]. This is certainly due to the fact that it is necessary to use reflection polarizers/ analyzers in order to analyze the polarization of the reflected light. [3.10,162]. Reflection polarizers are not as efficient as transmission polarizers which are in common use in the visible and in the near VUV. However, they are not limited to a certain wavelength range. Furthermore, they can have a high degree of linear polarization, if more than one reflection occurs at an appropriate angle.

The principle of reflection polarizers is evident from Fig. 3.11 where the reflectivity of Pt is plotted versus angle of incidence. If unpolarized light is reflected at an angle near to the so-called principal angle of incidence Θ_B, the degree of polarization of the reflected light acquires a maximum value depending on the optical constants of the reflecting material as a function of wavelength. If the reflecting material is a non-absorbing (k=0) dielectric this occurs at the Brewster angle Θ_B which is related to the index of refraction by Brewster's law. Then the reflected light is completely plane polarized, since R_p equals zero. This is not the case for the optical constants of a material as shown in Fig. 3.11 (k≠0). Here one has to include the absorptive part of the complex index of refraction and one defines a principal angle of incidence Θ_B for which the phase change is 90^0 in analogy to the ordinary Brewster law (k=0, [3.163]). DAMANY [3.163] gave an analytical equation for the angle Θ_p for which the polarization becomes a maximum and HUMPHREYS-OWEN [3.164] obtained a general solution for the angle Θ_m where the minimum in R_p occurs. For example, for the 16.8 eV curves of Fig.3.11 these three angles span a range of about 10^0.

Various types of reflection polarizers have been described in the literature including three (Fig.3.25a) and four-mirror configurations. Experimenters have used metal (mostly Au) coated mirrors [3.161-165] and combined metal and dielectric mirrors [3.162,166]. The latter configuration is particularly useful in a wavelength

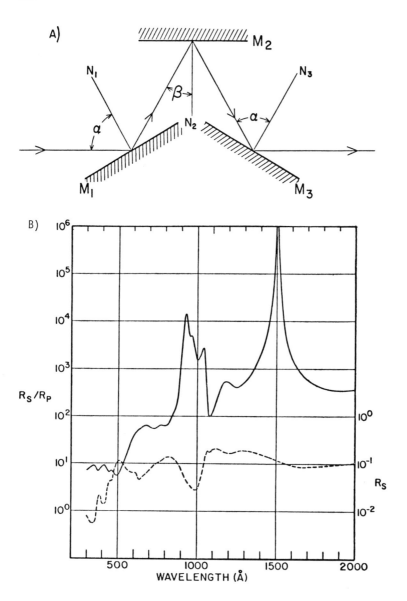

Fig. 3.25 Part A) shows a schematic diagram of a three mirror polarizer. The relation between the two angles of incidence is $2\alpha - \beta = 90^0$ for an undeviated light beam. B) The calculated polarization R_s/R_p (solid line) and the throughput R_s (dashed line) is plotted for a three mirror polarizer. The angle of incidence on the first and third mirror (both Al) is $\alpha = 73.5^0$. The second angle of incidence is $\beta = 57^0$ that is the Brewster angle at $\lambda = 1500$ Å for the second reflecting surface of MgF_2 [3.162]

range where for the dielectric a true Brewster reflection occurs (k=0), since it implies that the R_p component is zero for the appropriate angles of incidence. Three and four mirror polarizers are built in such a way that the beam is transmitted without deviations (see Fig.3.25a). For the three mirror system this implies that $2\alpha - \beta = 90^\circ$ holds with α and β approximating the Brewster angles in the desired wavelength interval. For the four mirror system all angles are equal.

HASS and HUNTER [3.162] have designed a three mirror system applying two Al and a MgF_2 reflection at angles $\alpha = 73.5^\circ$ and $\beta = 57^\circ$ which is approximately the Brewster angle for MgF_2 at 1500 Å. They calculated the ratio R_s/R_p in the wavelength range 2000 Å to 300 Å and the throughput R_s of the system. As expected their results (Fig.25b) for R_s/R_p show a pronounced spike at 1500 Å. But even for wavelengths below 900 Å there is a useful polarization.

Transmission polarizers make use of birefringence and their degree of polarization and throughput are much larger than for reflection type polarizers. However, they are limited in wavelength to about 1200 Å or 10 eV. Various types of transmission polarizers have been described in the literature including Rochon [3.167], Wollaston [3.168] and Sénarmont prisms [3.169] and others [3.10,162].

It should also be noted here that Bragg reflection at the appropriate angles is a good method to obtain polarizers and analyzers for the X-ray region.

3.3 VUV Monochromators

3.3.1 General Considerations

In the preceeding section we have described the most important properties of various optical components, e.g. mirrors, dispersers, filters etc. In this section we want to discuss the design characteristics of monochromators and spectrographs which are in use or have been used at the various SR laboratories.

It is quite common to make a distinction between a monochromator, a spectrometer and a spectrograph (although these terms are sometimes also used equivalently nowadays). All three systems generally work with an entrance slit and a grating in one or the other stigmatic mounting as discussed in Sect. 3.2.2. Both, a monochromator and a spectrometer also have an exit slit to separate out the desired band pass to be used in a subsequent experiment in case of a monochromator or to be measured by a suitable detector (Sect.3.5) in case of a spectrometer. In contrast, in a spectrograph the whole spectrum is recorded at once on a photographic plate detector. The sample to be analyzed generally has to be mounted in front of a spectrograph or spectrometer. This already implies that the choice of a specific system depends on the experiment itself (see Sect.3.6 for various examples). It is certainly correct that monochromators have to be considered more as general purpose systems. Therefore we essentially describe monochromator systems in this section.

As pointed out already, there exist certain constraints which have to be considered when designing a monochromator for use with SR. For convenience we summarize the most important constraints in Table 3.6, along with some implications depending on the SR light sources. Table 3.6 clearly shows that, in contrast to conventional spectroscopy, at SR facilities the light source, the beam line with the relevant optical components, the monochromator and the experimental set-up itself have to be considered as a whole in order to maximize the flux of monochromatized photons onto the sample. In Sect. 3.1.3 general ideas on beam lines and beam transport have been presented. In summarizing and extending that discussion a bit, there are a few main factors which govern the attainable flux $N_s(\lambda)$ at the sample:

- The source with a photon flux $N_k(\lambda)$ per horizontal angle and spectral band-with $K = \lambda/\Delta\lambda$ and with characterizing horizontal and vertical emittances ε_{sx} and ε_{sy}.
- The beam line and beam line optics, which transform the source emittances into the beam line emittances ε_{Bx} and ε_{By} at the monochromator, having a transmission coefficient T_B due to reflection efficiencies of the mirrors.
- The beam line acceptance angle, where in general only the horizontal angle Θ_x is of importance due to finite beam line diameters and size of optics.
- The monochromator with horizontal and vertical acceptances α_x, α_y, resp., and a transmission coefficient T_M.
- The sample with its effective area. In these terms the monochromatized flux $N_s(\lambda)$ falling onto the sample is given by [3.170]

$$N_s(\lambda) = N_k(\lambda) \cdot \Theta_x\, T_B\, T_M\, (\alpha_x/\varepsilon_{Bx}) \cdot (\alpha_y/\varepsilon_{By}), \tag{3.28}$$

where $\alpha_i/\varepsilon_{Bi}$ are always smaller than one. Assuming an infinite vertical slit which is appropriate due to the collimation of SR, $N_K(\lambda)$ for an horizontal angel Θ_x can be derived from (1.6). Quantitative estimates on the various other quantities of (3.28) were given by BROWN et al. [3.170].

Traditionally the various VUV monochromators fall into two classes, namely into normal incidence monochromators (NIM) for the spectral range from about 6 eV to about 50 eV photon energies and into grazing incidence monochromators (GIM) for the range of about 30 eV to 600 eV. Nowadays GIM's at fairly large grazing angles are also frequently used in the photon energy range down to 10 eV. Nevertheless, in the following we shall subdivide the presentation into NIM's in Sect. 3.3.2, GIM's in Sect. 3.3.3 and new design concepts in Sect. 3.3.4.

Table 3.6 Constraints on the design of monochromators for use with SR (a similar table is given in [3.155])

Constraints	Implications
1) The light source is fixed in position and in many cases also the experiment	Constant deviation optics is required for the system: beam-line - monochromator-experiment. In the grazing incidence regime in general a complex coupled motion is required for high resolution
2) SR beam is coming horizontally	Horizontal reflections and dispersion planes are more convenient than vertical ones
3) SR beam is highly polarized with the electric vector in the horizontal orbit plane	Vertical reflections and dispersions are preferred due to higher intensity (no Brewster-case) and polarization
4) The SR source width is several times its height [typ. (5-10 x 1-2mm^2]	Vertical dispersion gives higher resolution for monochromators without entrance slit and in general allows for better optical coupling (intensity, slit-efficiency factor)
5) High energy machines present a large flux of hard X-rays (with λ of few Å or less)	First optical components of beam-lines can be severely radiation-damaged (blistering of grating and mirror coating materials, cracking of glass-mirrors) and generally a coating of components with cracked hydrocarbons can occur, importance of easy-to-replace-components
6) Radiation hazards exist at high energy machines	Remotely controlled equipment might be necessary (DESY), access to the experiment is restricted, long beam pipes are necessary (at DESY about 40 m)
7) High collimation of SR and long beam pipes give nearly parallel light	Allows for special monochromators to work without entrance slits and requires condensing mirror at slit-systems
8) At synchrotrons the light intensity can considerably fluctuate with time due to different fillings	Suitable intensity monitors are required for compensation, beam-splitter technique might be necessary for high accuracy experiments
9) The vertical beam position may be unstable depending on accelerator electron optics and energy	Short time fluctuations (due to beam oscillations) determine the attainable resolution in monochromators without entrance slits, the effects can be reduced by either using slits close to the source point or by large demagnifying optics
10) UHV in storage rings and HV in synchrotrons	Often the vacuum requirements on beam-lines and monochromators disallow for instance small motors and lubrications inside vacuum systems and in turn determine the complexity of scanning systems
11) One beam pipe often feeds several instruments	The instrument has either to be removed as a whole or at least part of the incident beam has to continue through the instrument

3.3.2 Normal Incidence Monochromators

Three NIM mountings are commonly used with SR sources, the standard NIM, the Seya-Namioka monochromator and the so-called modified Wadsworth monochromator. These three types are the conventional mounts of concave spherical gratings with the focusing properties determined by the second order terms of the series expansion of (3.11), as discussed in Sect. 3.2.2. But monochromators based on newer optical designs with spherical gratings have also been built; for instance, a mount using higher order focusing derived from a generalized focusing condition [3.171]. For a long time conventionally ruled concave gratings have been used in these instruments. Nowadays, they are more and more replaced by holographically produced gratings due to the various reasons discussed in Sect. 3.2.2. In Fig. 3.26 we give a schematic diagram of some of the important NIM's being used for SR light sources. In Tab. 3.7 further information is provided including references to the original work. In the literature a variety of other NIM's is discussed [3.10-13,16,43] which, to the authors' knowledge, have not been used with SR light sources.

The standard NIM is a monochromator operating in a mode close to the Rowland, the focusing conditions being determined by Eqs. 3.15 and 3.17. They are accurately given by Eqs. 3.14 and 3.15. The monochromator itself has only one optical component, namely the grating. This is the same with the other NIM's to be discussed. In order to vary the wavelength the grating has to be rotated and, for maintaining focusing, it has to be translated at the same time. Thus the basic principle of the scanning mechanism of a standard NIM consists in a movement of the grating along the bisector of the angle which is subtended by the entrance and exit slit at the center of the grating with a simultaneous rotation about a vertical axis tangent to its center [3.10]. The most widely used system is that of the McPherson company [3.172] which uses an angle of .15 degrees. This system has the capability of high resolution with its ultimate value depending on the grating's constant and on the

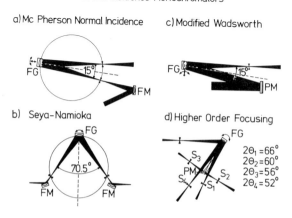

Fig. 3.26 Schematic diagrams of various normal incidence monochromators for use with synchrotron radiation. For explanations and references see text

radius of curvature. With a 3 m focal length a resolution of 0.03 $\overset{\circ}{A}$ independent of wavelength has been achieved with 10 μm wide slits and the 1200 gr/mm grating illuminated over an area of about 82 x 60 mm^2 [3.173,174]. This NIM (McPherson) is installed at the DORIS storage ring SR laboratory (compare Sect.3.1 and Figs.3.2a and b) and has proven as an invaluable instrument for high resolution spectroscopy [3.174]. For a proper optical matching to the source a focusing mirror has to be installed in front of the entrance slit. The magnification of this optical system should be chosen in such a way that the image of the orbiting electrons just fills the width of the entrance slit. If the slit widths is smaller than the image of the source, one looses intensity but the resolution is not affected. Usually one has to make a compromise due to the requirements of different resolution, i.e. slit width, for different experiments and also due to problems in finding an appropriately figured mirror (see Sect.3.2.1) in VUV quality.

The focusing conditions of the Wadsworth mounting are approximately expressed by (3.19, and 20). It requires the light source to be at infinity for an optimum performance in producing a point like stigmatic focus. Thus, this mounting exploits one of the particular properties of SR, namely the small divergence. SKIBOWSKI and STEINMANN [3.175] were the first to build a Wadsworth monochromator at the DESY laboratory in 1967. Since that time a large number of such monochromators has been designed with slight modifications [3.155,176-181]. It is appropriate to say that the Wadsworth monochromator is the work horse of the NIM's. The basic layout of such an instrument is shown in Fig. 3.26; features of particular monochromators are given in Tab. 3.5. The instrument operates without a real entrance slit which actually is substituted by the electron beam. Wavelength scanning is achieved by a simple rotation of the grating. In order to compensate for the severe defocusing when working with a fixed exit slit, one uses an excentric pivot for rotation of the grating similar to the Johnson-Onaka mounting [3.10]. The optimum off-axis position depends on the grating parameters and the sum B of the angle of incidence and refraction. This is shown in Fig. 3.27 for a 2 m concave grating. The straight line is drawn as a best fit to the focusing curve for the desired angle B and the required pivot length can be derived from the plot Fig. 3.27 [3.155].

The attainable resolution of the Wadsworth monochromator is determined by the remaining defocusing and the finite image size of the source point which is horizontally about 5 times as wide as vertically for most SR source. Thus, by vertical mounting of the dispersion plane one increases the resolution [3.176,177]. In addition, the degree of light polarization is improved by the s-reflection. Typical resolutions are between 0.5 and 2 $\overset{\circ}{A}$ for vertical mounting and standard gratings. For grating protection often premirrors are introduced [3.155,176-180], which also offer the possibility of giving the monochromatized light a convenient direction in the laboratory. Of course, the latter goal can also be achieved by using mirrors behind the exit slit in connection with refocusing [3.176,180]. Type I holographic

Table 3.7 Survey on normal incidence monochromators with concave gratings and with fixed exit beams for SR experiments

Type	grating motion	focal length and dispersion plane	resolution	reference	remarks
standard normal incidence monochromator	rotation and translation	0.3 m horiz.	low	3.21, p. 49	
	"	1 m vert.	medium	3.20, 185	
	"	3 m horiz.	high	3.21, p. 50; 3.173	
	"	10 m	very high	3.21, p. 49	under construction
Seya-Namioka	"	0.5 m horiz.	low	3.20	
	"	1 m vert.	medium	3.21, p. 49; 3.81, 119	
	"	1 m horiz.	medium	3.20, 3.91	
		1 m		3.101, 171	with asymmetrical illumination of grating and asymmetric beam length
Wadsworth	rotation and translation	1 m horiz.	low	3.155, 175	
	"	0.5 m vert.	low/medium	3.176, 3.180	premirrors are used and refocussing toroidal mirror [3.180]
	"	1 m vert.	medium	3.174,	
	"	1.5 m horiz.	medium	3.177, 179, 263, 264 3.178	holographic, Au coated grating was used
		corrected holographic grating		3.181	
Pouey higher order focussing	rotation	4 exit beams with different focal lengths	medium/high	3.21, p. 4; 3.101 3.21, p. 34	correction for astigmatism

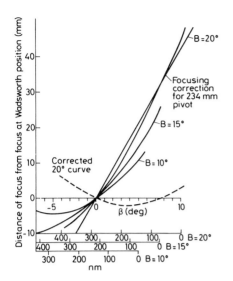

Fig. 3.27 Horizontal focusing correction for a modified Wadsworth monochromator by means of an excentrix pivot for the grating rotation. α and β are the angles of incidence and refraction, resp. with $\alpha + \beta = B$. The defocusing is plotted vs wavelength in nanometer [3.155]

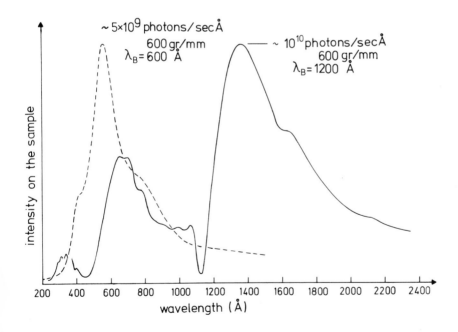

Fig. 3.28 The intensity behind the exit slit of a 1 m focal distance Wadsworth monochromator at the DORIS SR laboratory is plotted vs wavelength for two different gratings, a holographic grating blazed at 1200 Å and a ruled grating blazed at 600 Å (With permission of G. Zimmerer)

diffraction gratings have been used with great success [3.178]. Fig. 3.28 shows two
spectra obtained with a ruled grating and a holographic grating in a vertically
dispersing mounting at the DORIS storage ring [3.179]. Also, by using corrected
holographic gratings, the characteristic of a monochromator has been considerably
improved [3.181].

The Seya-Namioka monochromator (Fig. 3.26) with its real entrance and exit
slits is designed to have the simplest scanning mechanism i.e. a single rotation of
the grating about an axis through the center of the grating. SEYA and NAMIOKA
[3.99,100] have shown that deviations from the Rowland circle can be minimized by
choosing the angle between the incident and diffracted beam to be 70.5°. Due to the
fairly large angle of incidence the most severe aberration of this monochromator is
its astigmatism, if a point light source is used near the entrance slit. This draw-
back can be corrected for by utilizing additional focusing mirrors [3.182] or by
using appropriately corrected holographic gratings. By an asymmetrical illumination
of the grating one can also improve the degree of defocalization and thus improve
the resolution [3.171]. In the standard Seya-Namioka mounting the distances r and r'
of the slits to the center of the gratings (see Fig. 3.18) are equal. A corrected
Seya-Namioka mounting [3.101,183] is obtained by making r and r' slightly unequal.
Using SR the aberrations of the standard mounting can be reduced by long focal
length and astigmatic source optics when illuminating the entrance slit [3.81].
With a vertically dispersing instrument at the Stanford SR source a resolution of
0.18 $\overset{\circ}{A}$ has been obtained with a one meter 1200 1/mm grating and 20 μm slits. Nowadays, in general, vertically dispersing instruments are used and planned with a
layout as that given schematically in Fig. 3.26.

A fourth type of mounting included in Fig. 3.26 is due to POUEY [3.84,101]
and is based on higher-order focussing derived from a generalized focussing condition. Some of the characteristics of this mounting are [3.101]:

- A fixed angle 2 Θ subtended by the fixed slits at the grating which can be
  chosen between 10° and 170° depending on the grating
- The luminosity resolution product reaches a maximum at low Θ values (between
  14° and 17°) for 1200 1/mm grating and increases with the number N of lines
  for Θ values larger than 35°
- Optical correction allows an improved resolution even for the simple rotation
  wavelength scanning mechanism
- To one object (or image distance) correspond two Θ values above and below
  35°

These characteristics as well as others [3.101] lead to the design and construction of a monochromator with four exit slits used with SR at the LURE laboratory.

Also other monochromators operating with a simple rotation of the grating and
working with corrected holographic gratings at near normal incidence have been
designed [3.184].

We are not aware, if such systems have been built for use with SR. A similar system operating at grazing incidence will be described in Sect. 3.3.4.

3.3.3 Grazing Incidence Monochromators

A survey on grazing incidence monochromators (GIM) was given by one of the present authors in 1973 [3.17]. Several general aspects of GIM's have been discussed in that paper as, for instance, the importance and effects of the acceptance window, the size of the source, the distance from the source, higher orders, straylight, etc. In view of this discussion and the various comments and remarks, which we have included in the previous paragraphs, we confine ourselves in this subsection to a description of the various types of grazing incidence monochromators which have been used in conjunction with synchrotron radiation. Some types, which have been abandoned again due to the design of more convenient instruments, are still included in the discussion, since they illustrate the development of the techniques to the present day's highly sophisticated grazing incidence wavelength scanning systems. However, it is neither intended nor possible within limited space to give a complete discussion of all instruments. Similar to the report on NIM's, we outline the principle of operation and give some comments on their performance. Figs. 3.29, 30 and 32 give a survey of plane grating monochromators, Rowland-type monochromators and Non-Rowland monochromators with focusing gratings, respectively. A comparison of the different instruments is given in Table 3.8.

a) Plane Grating Monochromators

Figure 3.29 a shows the simplest possible instrument [3.56,186] with only one optical surface, a plane grating. Wavelength change is achieved by means of rotating a soller slit system around an axis parallel to the grating grooves. Of course, only very moderate resolution is attainable and higher-order suppression is only given in limited wavelength ranges. The latter can be improved by installing a plane mirror in front of the grating, a provision which at high energy SR sources is anyhow necessary for protecting the sensitive gratings from direct exposure to SR [3.56]. It should be realized that this mounting belongs to the "historical ones" which are no longer used.

The instrument shown in Fig. 3.29b uses a plane grating and a spherical mirror [3.187]. The grating is rotated by small angles so that the sum of entrance and exit angles is fixed. The exit beam is spatially fixed which allows the installation of complex experiments. With one spherical mirror two different optical arrangements are possible, if different values for the sum of entrance and exit angles are allowed with a fixed mirror. The two arrangements yield the same position of the exit beam. Only one of the two possibilities is drawn in Fig. 3.29 b. Suppression of higher orders is achieved only in a limited wavelength range by a suitable choice of the

Table 3.8 Survey on different mounts of grazing incidence monochromators for use with SR experiments

Fig. No.	Type	Accelerator (Working group)	Number of Reflections	Grating	Exit Beam	Resolution	References
3.29a	plane grating	DESY (Nowak)	1	plane	moving	low	3.186
3.29b	plane grating	INS-SOR (Miyake et al.)	2	plane	fixed	medium	3.187
3.29b	plane grating	DNPL (West et al.)	2	plane	fixed	medium	3.188
3.29b	distant source	DNPL (Howells et al.)	2	plane	fixed	medium	3.189
3.29c	plane grating	DESY (Kunz et al.)	3	plane	fixed	medium	3.190
3.29c 3.46	distant source	DORIS (Eberhardt et al.)	3	plane	fixed	high	3.191
3.30a	Rowland	all labs	2	spherical	moving	high	3.197
3.30b	Rowland	Glasgow/DNPL (Codling et al.)	4	spherical	fixed	high	3.194
3.30c	Rowland	ACO (Jaegle et al.)	5	spherical	fixed	medium	3.195
3.30d	Rowland-Vodar	Stoughton (Pruett et al.)	3	spherical	fixed	high	3.196
3.30e	Rowland	Bonn (Thimm et al.)	2	spherical (two)	fixed	high	3.197
3.30f	spherical grating	Stanford/Stoughton (Brown et al.)	4	spherical	fixed	medium	3.170
3.31a	parallel illumination	DESY (Haensel et al.)	1	spherical	moving	medium	3.1
3.31b	distant source	NBS (Madden et al.)	1	toroidal	fixed	medium	3.200

Plane Grating Monochromators

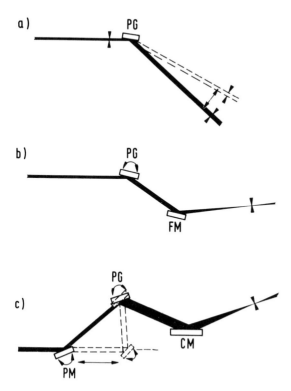

a) PG

b) PG

FM

c) PG

CM

PM

Fig. 3.29 Three types of plane
grating monochromators. The
scanning motion of the elements is
indicated. (PG: plane grating,
PM: plane mirror, FM = CM:
focussing mirror). Synchrotron
light is coming from the left. The
monochromators have been realized
a) at DESY by NOWAK [3.186]. b)
at INS-SOR by MIYAKE et al.
[3.187], c) at DESY by Kunz et al.
[3.190]. (From [3.17].)

angles of incidence at the grating and the concave mirror. For best higher order
reduction it is preferred to work in negative grating orders. It also determines
the choice of the two possible mounting configurations. On the basis of this design
a vertically dispersing monochromator has been built at Daresbury Laboratory [3.188]
with provisions made to test both mounting possibilities. The operation of the instrument proved successful in practice also for the position which has less higher-
order reduction capability. Therefore a new instrument has been designed at
Daresbury implementing the option of using two different focusing mirrors by
means of a precision translation stage [3.189]. Also with this system it is common
to use a grazing incidence premirror.

The instrument in Fig. 3.29 c was built to operate at the DESY synchrotron
[3.190]. It utilizes a plane mirror, a plane grating and a paraboloid mirror for
focussing and produces a fixed exit beam. The motion of the premirror and the
grating is coupled in a fairly complex way. In its simplest mode of operation the
premirror travels along the incident beam and rotates at the same time in such a
way as to illuminate always the center of the grating. Simultaneously the grating

rotates so that it remains parallel to the premirror. The zero-order beam emerges parallel to the incoming beam while light at a fixed angular separation from the zero-order beam is accepted by the exit slit. This type of wavelength change is favorable for the suppression of higher orders, since the longer wavelengths are taken at the steep angles. Indeed with a 1200 lines/mm grating and the other parameters (e.g. coating material of the components) properly chosen higher-order light can be efficiently suppressed over the whole range of operation. Moreover, this instrument is set up to always operate at the blaze maximum independent of wavelength. With additional "non-parallel" modes of operation, where the order suppression is less effective, the DESY instrument covers the energy range 15-280 eV. The maximum intensity is around 100 eV with a resolution of 1:400 to 1:800. Since the system operates without entrance slit, the width of the electron beam in the synchrotron is the limiting factor for the resolution. This complex coupled motion is accomplished, if a complicated mechanical coupling can be used within the vacuum system. This is, however, impossible for UHV monochromators as they are installed at high-energy storage rings such as e.g. DORIS. Therefore the design of the plane grating monochromator in Fig. 3.29c had to be modified in such a way that a series of 6 plane premirrors at different angles of incidence can be used alternatively instead of the single traveling and rotating mirror in Fig. 3.29c [3.191]. Each pre-mirror allows a wavelength scanning by a simple rotation of the grating. Due to a careful selection of the mirror coating materials (see Sect.3.2.1 [3.73]) and appropriate angles of incidence onto the premirror and the grating the monochromator offers mutually overlapping wavelength ranges with extremely small higher-order contributions. Since the source width at DORIS is smaller and much more stable than at the DESY synchrotron and due to a high quality manually ground parabolic mirror [3.192] the resolution ($E/\Delta E$) reaches about 3000 at some wavelengths. In general, such a high resolution can only be obtained with Rowland type mountings.

b) Rowland Mountings

Figure 3.30 schematically shows six different instruments based on the Rowland circle mounting, Sect. 3.2.2, (3.16 and 17), [3.197]. Entrance slit, grating and exit slit are all located on the Rowland circle to minimize aberrations. In principle, this mounting provides the best resolution, for grazing incidence monochromators.

The simplest version of a Rowland mounting is illustrated in Fig. 3.30a. The disadvantage of this conventional instrument is the travelling exit slit. Apart from a few exceptions, mainly absorption measurements are possible with the samples in front of the spectrometer. For absorption type experiments the Rowland spectrograph is extremely useful, since the photographic plate detectors allow the whole spectrum to be taken at a single exposure. Rowland instruments can be operated to fairly short wavelength of about 10 Å which already approaches the domain of crystal spectrometers. However the standard regime is for wavelengths above ∼25 Å. The usual way to illumi-

Rowland Monochromators

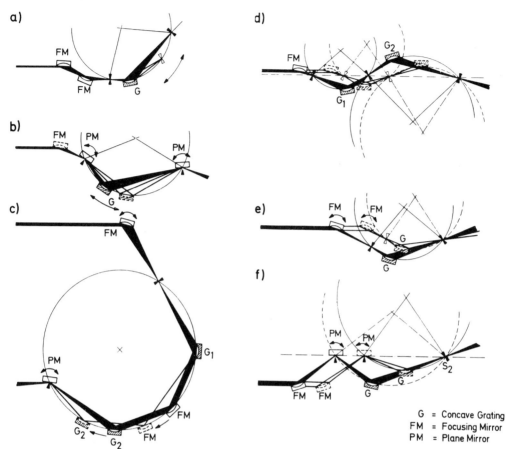

a) b) c) d) e) f)

G = Concave Grating
FM = Focusing Mirror
PM = Plane Mirror

Fig. 3.30 Six types of monochromators are shown which operate according to the Rowland principle. These mountings have been realized a) at several laboratories with slightly varying premirror arrangements, b) at NINA (before at GLASGOW) by Codling et al. [3.194], c) at ACO by Jaeglé et al. [3.156,195], d) at TANTALUS I by Pruett et al. [3.196], e) at BONN II by THIMM et al. [3.197,198], f) at SPEAR by BROWN et al. [3.170]. (From [3.197].)

nate a conventional Rowland instrument with SR is by means of a focusing premirror at grazing incidence. Again, in order to maximize the available intensity behind the exit slit, one has to take care of a proper matching of the image of the source to the slit. We note that for grazing incidence spherical optics the spherical aberration induced increase in focal size is of limiting importance.

The instrument shown in Fig.3.30b was constructed by CODLING and MITCHELL [3.194] at Reading and used at Glasgow and Daresbury. It combines the advantages of a Rowland mounting with the property of having a fixed exit beam. Entrance and exit slits are fixed while the grating slides along the Rowland circle. The directions of the incoming and outgoing beams are changed by a rotating mirror-slit combination. A third focusing mirror is needed to illuminate the entrance slit. A disadvantage of the present arrangement lies in the fact that the longer wavelengths pass the instrument at more grazing angles. This can cause considerable problems with higher orders.

A new type of grazing incidence spectromonochromator [3.156,195] has been built by JAEGLÉ and coworkers for use with SR of the LURE laboratory and for plasma research. The layout of the instrument is based on a two-grating principle. Two gratings, G_1 and G_2, with the same radius of curvature are mounted on the same Rowland circle. A mirror FM, with the same radius, is also mounted on the Rowland circle between the two gratings. A prefocusing concave mirror in front of the entrance slit allows to vary the grazing incidence angle on the first grating G_1. The instrument can be used in a one or two-grating mode and provides a fixed exit beam. In the two grating mode, shown in Fig. 3.30c the prefocusing mirror and G_1 are fixed. The wavelength scanning is achieved by a coupled displacement of the intermediate mirror and the second grating G_2 along the Rowland circle. The exit slit is a special rotating system with a mirror and a single slit jaw. The advantage of the two grating mode lies in its order-sorting capability and its resolution. In the one-grating mode the intensity throughput is about 100 times higher [3.156]. However, in this mode the monochromatized radiation is not always free of higher orders. In this mode wavelength scanning is achieved by a rotation of the prefocusing mirror which is connected with the displacement of G_1 along the Rowland circle [3.195]. Apparently this instrument works quite satisfactory, since three more copies are under construction [3.156]. It is however a question, how the complicated driving mechanism behaves under UHV requirements.

The instrument shown in Fig. 3.30d is a double-Vodar monochromator which has been installed at the Stoughton storage ring [3.196]. The two spherical gratings, G_1 and G_2, together with the entrance slit of the second sub-instrument (exit slit of the first one) are connected rigidly by a bar. This unit is constrained in such a way that the exit slit and the second grating move on straight lines which meet at the position of the exit slit. The focussing mirror and the first entrance slit move together along the incoming beam so that the first grating is always illuminated. The two gratings must have the same number of lines/mm but not necessarily the same curvature. The two Rowland sub-instruments operate in tandem yielding a fixed exit beam. The unpredictable aspect of this instrument lies in the gratings one of which is used in negative order.

The monochromator given in Fig. 3.30e is in the testing stage at the Bonn synchrotron [3.197,198]. The original idea was to restrict the number of reflections to two, which means that it is not possible to generate a fixed exit direction of the beam. However, the position of the exit slit is fixed. The toroid mirror moves along the incoming beam and rotates at the same time. Simultaneously the coupled entrance slit grating unit undergoes a complicated motion. The coupled motion is determined by the following constraints: 1) Rowland condition has to be fulfilled and 2) the toroid mirror has to illuminate the grating with its focus being at the entrance slit. The focusing in the direction perpendicular to the plane of dispersion is only fulfilled approximately. It is intended to use this instrument for photoelectron spectroscopy on gases.

The instrument shown in Fig. 3.30f is one of the first bakable UHV systems built for use at the Stanford storage ring laboratory [3.170]. The optical principle is related to that introduced by SALLE and VODAR [3.199]. A one meter Rowland circle rotates about the exit slit S_2 to give a fixed exit beam. The grating is mounted on an arm connecting the entrance slit PM (similar to that described for the Jaeglé mounting Fig.30c) and the grating and pivots about the entrance slit-mirror combination together with M_1 which translates parallel to the incoming beam [3.170]. The necessary high precision is achieved by means of a linear air bearing which is refenced to a ground marble slab. The monochromator delivers a spectral bandwidth of 0.15 Å at a scanning range 20 eV to zero order. With a 600 1/mm grating blazed at $1^0 31'$ the total photoyield of a Au emitter has been measured up to 700 eV (Fig. 3.31). The focussing all-metal premirror (not shown in Fig.3.30f) had a measured rms roughness of about 165 Å for the lower curve in Fig. 3.31 whereas a new superpolished Cu mirror had a rms roughness of about 30 Å. Note the drastic intensity increase for energies above the carbon K edge. As pointed out earlier this is also due to hydrocarbon build-up on used mirrors. In the last 3 years this instrument has been utilized for a variety of experiments. A second copy of this monochromator (named "Grasshopper") is presently tested at the PSL laboratory [3.119] with important improvements incorporated (e.g. variable slits).

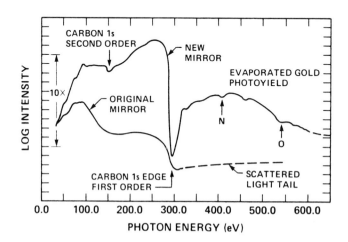

Fig. 3.31 The spectral dependence of the photoyield of Au is shown as measured behind the exit slit of a grazing incidence monochromator. The effects of surface roughness of a used and a new mirror (with 165 Å and 30 Å rms roughness, respectively and of hydrocarbon contamination are seen. [3.170]

c) Non-Rowland Monochromators

The principle of two further instruments which are using the electron beam as the entrance aperture is sketched in Fig.3.32. Part a) shows how it is possible to change the alignment of a conventional Rowland instrument in such a way that it can be used without an entrance slit [3.1,154]. The focal curves for the illumination of a grating with parallel light are lemniscates [3.93]. The grating of a Rowland mono-chromator can be tilted so that the path of the exit slit intersects this focal curve, at a wavelength position where optimum resolution is desired. This mounting yields high intensity because the light undergoes only a single reflection. However there are problems due to the direct illumination of the grating and due to higher order contributions.

At the NBS an instrument has been built with only a single optical element, having a fixed exit beam (Fig.3.32b). With the special geometry at the NBS synchro-tron it could be shown that combinations of entrance angles and exit angles exist (in negative order), where a slight rotation of the grating causes only second order focusing errors. With a grating ruled on a toroidal surface an almost stigmatic focus could be achieved.

Non Rowland Monochromators

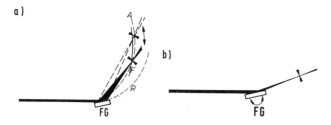

a)   b)

FG   FG

3.3.4 New Concepts

The various mountings discussed in the preceeding sections can be realized with mechanically ruled gratings. In this section we briefly want to describe new mounting concepts utilizing holographic and transmission gratings.

In Fig. 3.33a the principle of a toroidal holographic grating monochromator is shown. A type III Jobin Yvon grating is used (see Sect.3.2.2) to achieve the whole monochromator action by a simple rotation of the grating at a grazing angle of inci-dence of the light. The aberrations inherent with ruled gratings are largely com-pensated for by the special arrangement of lines obtainable only with the holographic production techniques [3.120]. A small 30 cm focal distance instrument is success-fully operating at the ACO storage ring [3.118] in the energy range 15 eV to 100 eV with a high efficiency and a relatively good resolution (0.7 Å for 100 μm slits). New designs have been worked out to extend the energy range to higher energies and to improve on the resolution. Prototype instruments have been built and are being tested at the ACO ring [3.120]. Three monochromators are under construction with toroidal gratings with entrance length of about 1.1 m and exit length of 1.8 m and an angle of 150° between the beams. Two interchangeable gratings are held in special flip over grating holders as indicated in Fig. 3.33a. With a 1800 gr/mm grating a useful energy range 40 eV to 150 eV is obtained at a calculated resolution of 0.2 to 0.3 Å. With 450 gr/mm grating the useful energy range is about 10 eV to 40 eV at a resolution of 0.8 Å [3.119]. These instruments give promise for improved intensities in the monochromatized beams.

The energy range 300 eV to 1000 eV is still the most problematic range to work at due to the various reasons given above. A monochromator design based on a transmission grating has been suggested for use with SR. The layout of Fig. 3.33b is thought [3.42,130] to work also in this problematic energy range. However, since mirrors have to be used for focusing purposes anyhow, contamination, surface roughness and other problems still remain. At present, it appears that more experience with these dispersing elements has to be collected. A first transmission grating

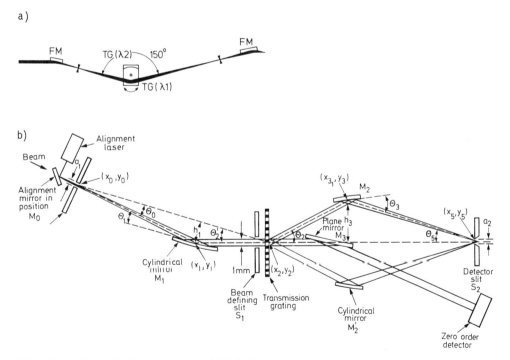

Fig. 3.33 Part a) depicts a toroidal holographic grating monochromator and b) a transmission monochromator [3.130]

monochromator for work with light from resonance lines is in operation at energies
below 40 eV. The whole crude monochromator essentially consists of the transmission
grating which is tilted for varying the wavelength [3.202].

3.4 X-Ray Monochromators

3.4.1 Plane Crystal Instruments

PARRATT [3.203] in 1959 appears to have been the first to note the excellent matching
of a double crystal monochromator (Fig. 3.34) to the SR source. Double crystal monochromators in the parallel mode keep the direction of SR fixed while generating only
a parallel displacement of the beam. They are nowadays most widely used for high
resolution spectrometers with groove cut crystals which are ideally suited for continuous scanning of a spectrum [3.204]. CAUCHOIS and coworkers (3.205) were the
first to use a fairly simple monochromator in connection with SR at Frascati consisting of one single crystal.

BEAUMONT and HART (3.132), and BONSE et al. (3.206) have discussed and tested
several plane crystal arrangements with SR, as shown in Fig. 3.35. While the former
have concentrated their effort to calculate resolution and to test several quite
different multireflection arrangements the latter (3.206) have attacked especially
the problem of higher order rejection. The classical highly dispersive (+,+) setting
of the double crystal monochromator (Fig.3.34), as it was first proposed by
EHRENBERG and MARK (3.207) in 1927, could never be used with sufficient intensity,
since its angular acceptance in the dispersive direction is of the order of the
width of the single crystal reflection curve (see Fig.3.22). For this arrangement
the high brightness of SR is of crucial importance and very high resolution can be
obtained by selecting the appropriate reflections (see Table 3.3).

When scanning wavelength with a (+,+) spectrometer as shown in Fig. 3.34b a fairly
complicated mechanism is necessary to achieve a continuous scan. The second crystal
has to be rotated around the first one and the detector again around the second crystal. In addition, first and second crystals must be rotated around an axis lying in
their surfaces, all with different angular speeds. Before we discuss how to overcome these problems we want to describe the (+,-) setting of the double crystal
monochromator.

The (+,-) arrangement is shown in Fig. 3.34a. In this case the arrangement is not
any more a perfect angle selector. Different energies are transmitted for different
angles of incidence. In practice this spectrometer is equivalent to a spectrometer
consisting of a single crystal with the advantage that the second crystal deflects
the radiation back to its original direction. Thus scanning of a spectrum involves

(a)

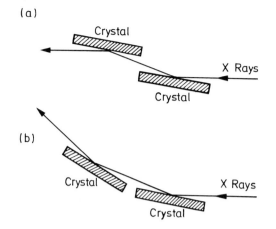

(b)

Fig. 3.34 Double crystal monochromators (a) in the parallel
(+,-) mode and b) in the highly
dispersive antiparallel (+,+)
mode

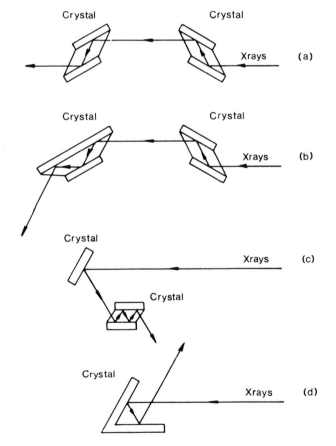

Fig. 3.35 Monochromators
as proposed and tested by
BEAUMONT and HART [3.132]

only small lateral displacements of the beam, but no angular deflection. The resolution of this arrangement is usually determined by the geometrical arrangement. The
size of the aperture in the dispersive direction, the distance from the source and
source characteristic as described in Sect. 3.1.3 determine an acceptance angle
$\Delta\Theta$, which according to

$$\frac{\Delta\lambda}{\lambda} = \frac{\Delta\varepsilon}{\varepsilon} = \Delta\Theta \; \text{cotg} \; \Theta \tag{3.29}$$

defines the resolution. Detailed estimates of the resolution based on (3.29) have
been made e.g. by PIANETTA and LINDAU (3.39) and BONSE et al. [3.206]. If $\Delta\Theta$ is
very small, however, in addition to the geometrical effect the width of the single
crystal reflection curve should also be taken into account. Typical values for this
(+,-) arrangement are $\Delta\varepsilon/\varepsilon = 10^{-4}$ to 10^{-3} (3.206). LINDAU and PIANETTA (3.39), however, have obtained $2.5 \cdot 10^{-5}$ by using very narrow collimators. Such a good collimation, on the other hand costs a lot of intensity.

One particular problem encountered with this type of monochromator is a phenomenon
which KINCAID (3.204) calls "glitches". Always, when the rotation of the monochromator leads to additional oblique lattice planes coming into the reflecting position
for the same wavelength as the main lattice planes, an umweganregung to the intensity
in the main channel takes place, with several complex geometry determined effects on
the measured signal. Often these glitches can be eliminated during data evaluation.
But it should also be possible to control these unwanted structures, perhaps even
eliminate them, by using specially cut groove crystals or spectrometers made from
two individually cut crystals [3.208].

The (+,-) arrangement is widely applied for EXAFS spectrometers [see e.g. 3.204].
The main goal in EXAFS is to measure the Extended X-ray Absorption Fine Structure
above the X-ray edges. These structures are a modulation of the absorption coefficient with a periodicity length of several tens of eV extending up to 1000 eV above
the edge (see Sect.3.6.2). Only moderate resolution is needed in order to resolve
this structure. In this case a vertically deflecting double focusing mirror which
gives a 1:1 image of the SR source is tolerable in front of the monochromator. The
increased angular divergence reduces the resolution to about $5 \cdot 10^{-4}$ which, however,
is sufficient. In return a gain in flux density on the sample by about two orders of
magnitude is obtained [3.209]. Higher-order radiation usually contributes only to
the flat structureless background. This background can cause troubles when quantitative amplitude information is desired.

A very convenient arrangement of a monochromator with four reflections and zero-
beam deviation is shown in Fig. 3.35a. This arrangement is a combination of the
(+,-) and the (+,+) double crystal monochromator. This instrument combines highest
resolution with simple scanning and independence of position fluctuations of the
source. Such an instrument was first built at Daresbury [3.210a] and is presently
installed at DESY. According to BONSE and HART [3.210a] such multiple reflections

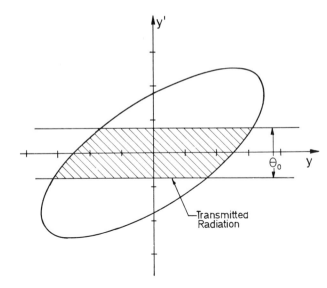

Fig. 3.36 **Fig. 3.36** Phase space ellipse in combination with a well aligned antiparallel (+,+) double crystal monochromator. The monochromator defines an angular interval Θ_0. Note that this interval is invariant with shearing which has to be applied for obtaining the figure at different locations along the beam path Z. (See also Fig.3.7)

serve to suppress the tails in the single crystal reflection curves (Fig.3.22) and thus the resolution function is improved. Some of the other proposed monochromators (Fig.3.35) serve this purpose while others give intentional deflections to the mono-chromatic beam in order to avoid the direct beam of SR in cases of difficult shielding and even others have high mechanical stability. Quite obviously, additional arrangements are possible and will find applications under special circumstances.

As an example of the phase space methods introduced in Sect. 3.1.3 we show in Fig. 3.36 the acceptance for the antiparallel (+,+) spectrometer at the source ellipse for a vertically dispersing instrument. This spectrometer generates a pure angular slit with a width determined by the single crystal reflection curve. If properly aligned the slit picks the central position of the SR emission pattern [3.39, 40].

3.4.2 Higher-Order Rejection

Unwanted effects in all the monochromators of Fig. 3.35 and also the focusing instruments described in Sect. 3.4.4 are the admixture of higher-order wavelengths namely in addition to λ also $\lambda/2$, $\lambda/3$, etc. In special cases like e.g. the Si or Ge (111) reflections the second-order reflection is avoided due to special structure factors, e.g. (???) has zero intensity. This, however, does not provide a general method for higher-order rejection. As in the equivalent cases with ruled gratings in the vacuum ultraviolet the problem of higher orders depends on the type of experiment performed. Higher orders can be suppressed already in the monochromator or in the detector or just by the special experimental arrangement in use. There are, however, cases in which a fairly small contribution of higher-order radiation can be enhanced and thus can have disturbing effects on the experimental results. Therefore we shall describe here one of the methods to suppress higher-order radiation in more detail. BONSE et

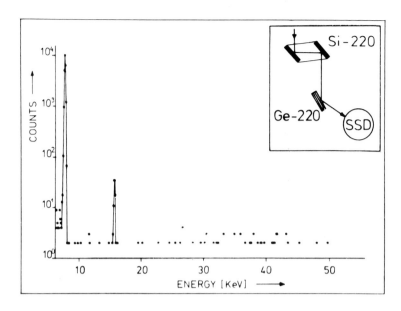

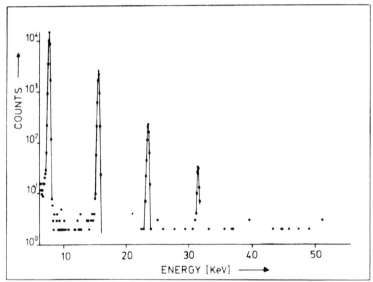

Fig. 3.37 Spectrum of the monochromator as shown in the insert decomposed into higher harmonics by solid state detector (SSD). Top: the harmonics are suppressed by maximizing the fundamental by small angular rotations of the Ge crystal; bottom: for comparison the spectrum obtained when maximizing the first harmonic [3.206]

al. [3.206] suggest to use a spectrometer as it is shown as an insert of Fig. 3.37. Such a monochromator uses reflections from two different types of crystals, Si and Ge in this case. Small wavelength shifts and shifts of the effective angle of diffraction occur, because the index of refraction differs from 1 between the inside and the outside of the crystal. Thus, for a fixed direction crystal reflection does not occur exactly for the harmonics of the fundamental frequency but for slightly detuned frequencies. This detuning depends on the wavelength, the angle of incidence and on the electronic density of the monochromator crystals. The angle of incidence on the Ge crystal in Fig. 3.37 can be slightly detuned in such a way that the first-order radiation coming from the Si crystal is still reflected while the higher orders are not. In Fig. 3.37 it is demonstrated [3.206] that an alignment which maximizes the fundamental frequency suppresses higher harmonics while in an alignment in which the first harmonic is maximized also the higher harmonics and the fundamental frequency are present.

A similar method was applied by BONSE et al. [3.206] in connection with their interferometer. In this case Si (1$\bar{1}$0) reflection are used first in the Bragg (reflection) case, second in the Laue (transmission) case. Since the radiation penetrates into the crystals at different angles with respect to the surface there is again a separation of orders possible.

MATERLIK et al. [3.211] have built a groove cut monochromator of the type shown in Fig. 3.34a which has the first reflector cut at an angle which is oblique compared to the lattice planes (see Fig.3.23). A slight angular detuning again serves to maximize the fundamental and suppresses higher harmonics. In addition, such arrangements serve to concentrate the beam (see Sect.3.2.2).

Another means of suppressing harmonics is reflection from mirrors (see Sect.3.2.1). With an appropriate choice of the angle of incidence harmonics can be effectively suppressed. Problems arise with scanning monochromators for which also the angle of incidence onto the mirror needs to be scanned. This technique up to now is mainly applied with focusing instruments for which the mirror also serves to achieve focusing in one or two planes when it has the appropriate curved shape [3.212-215].

3.4.3 Bent Crystal Monochromators

If a crystal, cut with the reflecting lattice planes parallel to the surface, is bent in one direction, it focuses radiation energing from a source point on the Rowland circle back to an image lying again on the Rowland circle [3.58,212,213]. In contrast to a grating monochromator this condition is fulfilled only for one wavelength (and higher harmonics) at specular reflection. For work involving SR the distance to the crystal is large and consequently the image point would lie at the same distance from the crystal. Scanning of the spectrum is prohibitively complicated. No such monochromators have been built up to now.

ROSENBAUM et al. [3.213,214] were the first to build a focusing monochromator with asymmetric distances between the source and the crystal and the crystal and the focus at the DESY synchrotron. Two more instruments applying practically the same principle have been built in the meantime by WEBB et al. [3.215-217] at SPEAR (Fig. 3.38) and by HASELGROVE et al. [3.218] at Daresbury. All these instruments are used for small angle diffraction from biological samples. Therefore the band pass of the

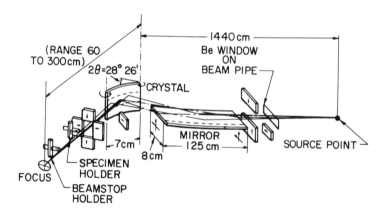

(RANGE 60 TO 300cm)

$2\theta = 28°\ 26'$

CRYSTAL

1440 cm

Be WINDOW ON BEAM PIPE

7cm

8 cm

MIRROR 125 cm

SOURCE POINT

FOCUS

SPECIMEN HOLDER

BEAMSTOP HOLDER

Fig. 3.38 Schematic of a double focusing mirror-crystal monochromator as it is applied for small angle diffraction from biological samples at SPEAR [3.215-217]

monochromator is not of such a critical importance. It is easily verified that with a white source like SR a crystal bent to any radius will reflect radiation like a mirror. If the bending radius deviates from the appropriate Rowland radius different portions of the crystal will reflect different wavelengths. Thus focusing at a shorter distance (and correspondingly a demagnification of the source) can be achieved at the cost of resolution.

With the focal distance of 1.5 m aimed at by ROSENBAUM et al. [3.213] and the source distance of 37 m the wavelength spread with useful apertures would have been prohibitively large. As mentioned above, GUINIER [3.37] has shown that the focal distance can be reduced when cutting the crystal surface at an angle σ with respect to the lattice planes as shown in Fig. 3.23. Then the source and the image lie still on the Rowland circle, but the distance to the source g and the distance to the focus f are given by

$$g = R \sin (\theta + \sigma), \quad f = R \sin (\theta - \sigma), \tag{3.30}$$

where R is the radius of curvature and θ the Bragg angle. For $\lambda = 1.5$ Å, $\theta = 13°$ with quartz $(10\bar{1}1)$ reflecting planes, $\sigma = 7°$ was chosen [3.213]. This would give a focus at 11.3 m with g = 37 m distance and a bending radius R = 108 m [3.213]. Further bending to a radius of about 30 m is necessary to bring down the focal length to 1.5 m. As a consequence the band width of the radiation reflected by the 45 mm long crystal is about $3 \cdot 10^{-3}$ Å. This is not to be confused with the transmitted band width for one direction relative to the lattice planes which depends only on the single crystal reflection curve. This band width is estimated to be $\Delta\lambda = 2.4 \cdot 10^{-4}$ Å [3.213]. An increase of this band width by introducing artificial mosaic spread or gradients in the lattice constant as mentioned in Sect. 3.2.2 would

Table 3.9 Detectors for the vacuum ultraviolet

	Approx. spectral range [Å]	References	Remarks, examples
1. Photographic plates or Films	visible – X-rays	3.10,12,219	see also Table 3.11, Eastman-Kodak SWR 101,104, Kodak-Pathé SC-5,SC-7, Ilford Q
2. Photon Counters 2.1 Geiger-Müller Counter 2.2 Proportional Counter 2.3 Ionization chamber and channel-detector	300 – X-rays	3.10,12,3.220	see also Table 3.11, useful for absolute intensity measurements, primary standards
3. Ionization chambers 3.1 Single Ion Chamber 3.2 Double Ion Chamber	1000 – 250 (50) 1000 – 250	3.10 3.10,222,223	especially for absolute intensity measurements, see also Table 3.11, useful as primary intensity standards
4. Open and closed photodiodes	2500 – X-rays	3.10,12	useful as secondary standards
5. Photoelectron-multipliers 5.1 Phosphor with standard multiplier (S-11)	visible – X-rays ~3500 – (50)	3.10,235	see also Table 3.10 Sodium salicylate, Liumogen, Terphenyl, Coronen etc.
5.2 Szintillation by accelerated photo-electron and multiplier	visible – X-rays	3.235,10	

Table 3.9 Detectors for the vacuum ultraviolet (continued)

5.3 Multiplier with transparent envelope or window	see Table 3.10	facilitates operation since solar blind
5.4 Windowless open multipliers	2500 - X-rays	photocathodes determine spectral characteristic
5.4.1 Divided dynodes	"	see Table 3.10, standard type multiplier: RCA,EMI,EMR,Johnston etc.
5.4.2 Strip dynode	"	
1. Crossed electric and magnetic field	"	Bendix
2. Electric field	"	
5.4.3 Channel dynodes	"	
1. Channeltron	3.227,228,231	nowadays most widely used with capabilities up to µm anode current
2. Microchanneltron plates	"	the most advanced available systems today
6. Miscellaneous		
6.1 Thermopiles	visible - (500)	basically for total absolut intensity measurements, useful as primary standards
6.2 Photovoltaic devices	visible - 250	

Column 2 reference numbers: 3.233; 3.10,233; 3.10,233; 3.10; 3.233a; 3.227,228,231; 3.228,231; 3.10,236; 3.10

Table 3.10   Some Properties of Multipliers

Code No. Table 3.9	Typ	Spectral range [Å]	High-voltage [kV]	typical gain	typical dark current	size of cathode	beakable
5.0	EMR, EMI, RCA etc. side on		< 1.5	$\sim 10^6$	3×10^{-9} A	~ 10 mm Ø	no
5.3	EMR 541 F	$\sim 3000 - 1450$	< 2.5	$\sim 10^6$	4×10^{-11} A	~ 25 mm Ø	no
5.3	EMR 541 G	$\sim 1900 - 1450$	< 2.5	$\sim 10^6$	6×10^{-12} A	~ 25 mm Ø	no
5.3	EMR 542 G	$\sim 1900 - 1050$	< 2.5	$\sim 10^6$	6×10^{-12} A	28 mm	no
5.4.1	Johnston MM 1	$\sim 2500 -$	< 5	$\sim 10^8$ at 3.5 KV	$< 10^{-13}$ A	~ 32 mm Ø	350°C
	MM 2	$\sim 2500 -$ X-rays	< 5	$\sim 10^6$ at 3 KV	$< 10^{-13}$ A	13 mm Ø	350°C
5.4.1	Balzers SEV 117	$\sim 2500 -$ X-rays	< 3.5	$\sim 10^8$ at 3.5 KV		10×6 mm^2	400°C
5.4.2.1	Bendix M 306	$\sim 1500 -$ X-rays	< 2	$> 10^6$	10^{-12} A $1\frac{cts}{s}$	15×18 mm^2	170°C
5.4.2.2	"Schnell" ∎	$\sim 1200 -$ X-rays	< 3.5	5×10^6 at 3 KV		0.5×10 mm^2	
5.4.3.1	Bendix EM 401 channeltron	$\sim 1500 -$ X-rays	< 4	5×10^7 at 3 KV	$< 0.05\frac{cts}{s}$	1 mm Ø	< 300°C
5.4.3.1	Mullard B 419 BL	$1500 -$ X-rays	< 5.5	5×10^7 at 3.5 KV	$< 0.1\frac{cts}{s}$	10 mm Ø 2 mm ↓Ø channel	200°C
5.4.3.1	Galileo 4510 WL	$1500 -$ X-rays	< 4.5	$> 10^7$	$< 0.1\frac{cts}{s}$	$24 - 24$ mm^2	200°C
5.4.3.2	Mullard G 40 - 25	$1500 -$ X-rays	< 4.5	10^6	$< 0.01\frac{cts}{s}$	25 mm Ø	200°C

be highly desirable in order to increase the reflected intensity in such monochromators.

Focusing in the perpendicular direction and higher-order rejection is achieved by
a mirror bent to a radius of 1000 m giving total reflection at a glancing angle
around 4 mrad (see Sect.3.2.1). The focal spot has a size of 200 μm with a gain in
speed at the photographic plate of more than two orders of magnitude over classical
instruments [3.213,214] already at a synchrotron like DESY.

3.5 Photon Detectors

In this section we present a survey on photon detection systems which are currently
utilized in the VUV and in the X-ray region. Several systems to be described operate
in both spectral ranges due to the underlying basic physical processes, e.g. photoionisation, photoemission, etc. Therefore some detectors appear twice in our tables,
namely, in Table 3.9 which summarizes VUV detectors and in Table 3.11 which gives
an overview of X-ray detectors.

3.5.1 Detectors for the Vacuum Ultraviolet

The last few years have seen an enormous progress in detection systems for use in
the VUV. We only mention the very small, but easy and reliably to operate channeltron and microchanneltron plate systems with a time response fast enough to allow
for the investigation of excitation and decay processes (see Sect.3.6) in the subnanosecond regime. Further more, by utilizing monochromatized SR the absolute detection efficiency of vacuum diodes have been calibrated. The wealth of information is
summarized in a first Table 3.9 which provides an overview over various detection
systems. More detailed information on photomultiplier systems is presented in
Table 3.10. It is almost unnecessary to say that the latter table is by no means
complete.

The photographic detection was invented by SCHUMANN in 1892 [3.10] who realized
that for VUV photons in order to be detected with silverhalide crystals the geletin
base of the photoplates had to be diluted. (The strong hydrocarbon absorption in
the VUV is evident from Fig. 3.31.) The resulting "Schumann plates" made by hand
were extremely sensitive to mechanical damage. Eastman Kodak produced the first
Schumann emulsion on a filmbase (Eastman Kodak SWR) about 50 years later, followed
by Kodak Pathé with more sensitive films (SC-5 and SC-7). At present the most advanced VUV emulsions are Eastman Kodak 101 and the slower, but finer grained emulsion 104. A comprehensive report on the VUV performance of these emulsions has
recently been given by VAN HOOSIER and collaborators [3.219].

Photon counters and ionization chambers allow absolute intensity measurements of VUV radiation. For wavelengths shorter than about 100 Å proportional counters or Geiger-Müller counters [3.10,12] (see also Sect.3.5.2) can well be utilized as standard detectors provided the corrections are known which have to be applied to account for absorption in the window materials of the moderately pressurized (several Torr depending on the filing gas [3.10]) counting cells. In principle photon counters can be used to wavelengths well above 1000 Å. However, in the spectral region ~1000 Å to 200 Å one has problems in finding a suitable radiation transmitting window material to withstand the differential pressure. In addition, the energy resolution and sensitivity is reduced. Especially for the range 100 Å to 300 Å SASAKI et al. [3.220] constructed a photon counter with an incorporated channeltron multiplier (to be discussed later-on) to gain sensitivity.

Rare-gas ionization chambers are commonly used at wavelength above 250 Å up to ~1000 Å. This method, developed by SAMSON [3.10] is based upon the fact that one photon absorbed in a rare gas produces just one electron-ion pair which can be measured by means of a collector electrode and a high sensitivity electrometer amplifier. At wavelengths below ~250 Å corresponding to twice the ionization potential of He (2x24.58 eV) one has to consider the effect of secondary ionization by ejected photoelectrons and of multiple photoionization. However, by measuring the pressure dependence of the detector gas on the ion current the first effect can be quantitatively determined. If, in addition the photoionization, yield is known, the absolute spectral intensity can be evaluated. The most appropriate gas for this application is He because its double ionization cross section can effectively be neclected in the range 250 to 20 Å [3.221]. By this method [3.222] the absolute photon flux of a grazing incidence monochromator at DESY was determined in the energy range 20 eV to 170 eV with an experimental set-up shown in Fig.3.39 [3.223].

Below 50 eV the ionisation chamber is operated in the well-known 2-chamber mode [3.10] and above 50 eV the two chambers are connected to facilitate the measurements at low gas pressures. Behind the ionization chamber photocathodes are mounted at normal incidence to be calibrated as secondary detector standards in the range 30 to 170 eV. The total photoelectric current leaving the cathodes is measured. The cathodes were prepared as thin films from standard high purity materials except for Al_2O_3 which was evaporated from 99.5% Al_2O_3 ceramic. The cathodes were then kept at normal atmospheric pressure, but free from dust, for a few days before the measurements were performed. The absolute photoelectric yield of a 1000 Å thick Au film and a 150 Å thick Al_2O_3 film on polished stainless steel is shown in Fig.3.40a. Fig.3.40b depicts the photoyield of a 2000 Å thick CsI film and of a 1800 Å thick LiF film. For comparison other experimental data points have been included [3.224]. In particular we mention the yield data of SALOMON and EDERER [3.225] of an Al_2O_3 cathode prepared by anodic oxidation. The Al_2O_3 calibrated diode which has proven to be stable to within 5-10%, is now regularly available from the NBS. The photo

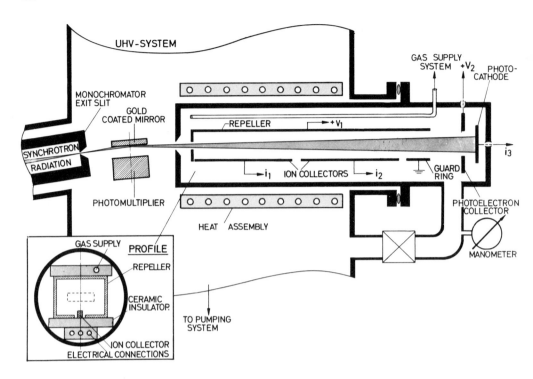

Fig. 3.39 An ionisation chamber behind the exit slit of a grazing incidence mono-chromator is utilized to calibrate a secondary transfer standard diode [3.233]

yield of Au depends on contamination [3.226]. However, after heating in an UHV system it appears to be reproducible to within 20% [3.144b,226]. The photoyield of LiF and in particular CsI is less stable and reproducible with preparation conditions than that of Al_2O_3 and Au. However, it should be noted that the absolute yield of the alkali halides is higher by a factor of 10 to 50. Thus, if a sensitive diode or detector is needed, one should consider these materials. The photoelectric yield of various other meterials has been reported [3.144b,10,12,224-228] in the energy range 10 eV to several hundred eV. Notably the work of LUKIRSKII and coworkers in the high energy range has to be mentioned [3.224b,229].

The photoelectric yield also depends on the angle of incidence of light with respect to the surface of the emitter. This is shown in Fig. 3.41 for an evaporated (in situ in UHV) Au film for photon energies ranging from 22 eV to 190 eV [3.230]. The photo-yield $Y(\theta)$ normalized to the yield $Y(0°)$ at zero degree angle of incidence shows a drastic increase with increasing angle for photon energies above about 90 eV. With increasing energy one also observes a shift of the maximum to more grazing angles. This behavior is well understood and can be explained on the basis of the step model for photoemission [3.144b,223]. At oblique angles of incidence, the linear absorption coefficient μ used in (3.27) has to be generalized in order to include the refraction of the intensity flux in an absorbing medium

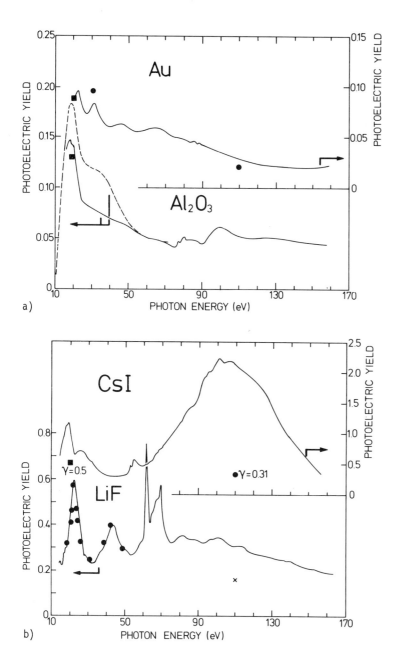

Fig. 3.40 Part a) depicts the absolute photoelectric yield of evaporated Au and
Al₂O₃ at normal incidence. For comparison other experimental data are included.
(Au: ■[3.224a], ●[3.224b]) Al₂O₃: dashed line [3.225], ■[3.224c]). Part b) shows
the yield of CsJ and LiF. (CsJ: ■[3.224d], ●[3.224b], LiF: ●[3.10], p. 227,
x [3.224b]) [3.223]

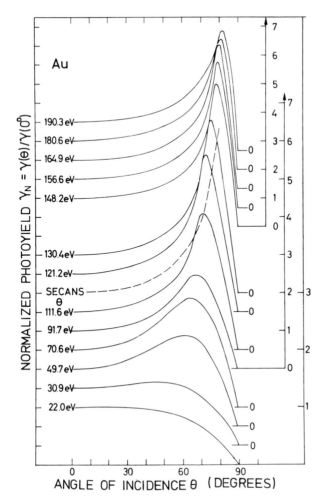

Fig. 3.41 The normalized photoyield of in situ evaporated Au is seen vs. angle of incidence. For high photon energies it can be approximated by a secans θ dependence (dashed line). At low photon-energies refraction effects dominate and the secans θ law is not valid

$$\mu' = \frac{4\pi}{\lambda} \, Jm \, \{(\varepsilon_1 + i\varepsilon_2) - \sin^2 \Theta\}^{1/2} \qquad (3.31)$$

It can be shown [3.144b,223] that the relative spectral behavior of the photoyield and to a certain extent also the absolute spectral response is given by

$$\gamma(E) \approx P\{1-R(E,\Theta)\}M(E) \cdot \frac{\mu'(E,\Theta) \cdot L(E)}{\mu'(E,\Theta)L(E)+1} \qquad , \qquad (3.32)$$

where P is an escape probability factor for the photoelectrons, $R(E,\Theta)$ is the reflectivity at the photon energy E and angle Θ and $M(E)$ is a smoothly varying multiplication factor approximately given by $M(E)\sim E/\varepsilon$ with ε the average energy of the released photoelectrons.

The data of Fig. 3.41 can be quantitatively fitted [3.230] with (3.31) and (3.32) with known optical constants [3.48]. In fact, one can work out the average escape length of the photoelectrons from $\gamma_n = \gamma(\Theta)/\gamma(0^0)$ since P and M(E) cancel. At high energies the absorption is weak and the refractive index approaches one. Therefore one can approximate (3.31) by

$$\mu' = \mu/\cos\Theta \qquad (3.27')$$

with μ given by (3.27). Eq. 3.32 then becomes

$$\gamma(E) \approx P(1-R) \cdot M \cdot \mu \cdot L \cdot \sec\Theta \qquad (3.32')$$

From Fig. 3.41 we see that the high energy data are well described by the $\sec\Theta$ law [3.229] up to angles close to the inflection point of the curves which is determined by the onset of the total external reflection (see Sect.3.2.1). At low energies the enhancement in the photoyield is limited due to refraction effects in the absorbing medium. Clearly, for a detector design the angle of incidence dependence of the yield has to be considered.

A wide variety of photoelectron multiplier systems is in use as seen from Tables 3.9 and 3.10. The first multiplier systems were built with mechanically formed metal dynodes (mostly CuBe) mounted separately (5.4.1 in Table 3.9 and 10). Then strip dynodes have been designed which consist of semiconducting thin films (SnO_2,C) on glass plates with two of the plates in parallel. An electric field is generated across the continuous strip dynodes by an applied potential which gives rise for an electron avelanche to move along the dynodes (see Fig.3.42a). The main advantages of continuous dynodes are the reduced number of necessary different potentials and vacuum feed throughs and the smaller size. Various photomultiplier systems of this kind have been described in detail by SAMSON [3.10].

The latest achievement with this type of detectors, the channeltrons, also have continuous dynodes. The basic channel electron multiplier (CEM) consists of a semi-conducting glass channel having an internal diameter of only a few mm and a ratio of length to diameter in the order to 50:1. The channel has a resistance of typically 10^9 which allows an output pulse of the order of 10^8 electrons to be collected by an anode at an applied potential of 3-4 kV. In general, the channel of the CEM is bent in order to inhibit the acceleration of positive ions towards the photocathode. These could otherwise generate further electrons and thus spurious output signals. The output characteristic of CEM's is determined basically by space charge effects [3.227,228,231]. In order to enlarge the size of the photo sensitive area of the cathode to a size larger than the basic channel dimensions, CEM's are built with large cones up to several cm^2 added to the front end of the CEM. The striking advantages of CEM's include: extremely small size, fairly large count rate capability, the requirement of (for specific types) only two leads, the compatibility with UHV requirements.

Recently, TIMOTHY and LAPSON [3.227,228] gave a comprehensive report on the use of CEM's as secondary standards in the VUV spectral range. They compared and tested various CEM models with different photocathode materials for the CEM to investigate their efficiency and spectral response. The absolute detection efficiency of an MgF_2-coated CEM and an uncoated CEM (Mullard B 419 BL) is shown in Fig. 3.42b in the wavelength range from 44 Å to 1216 Å. As expected, at an angle of incidence of 45^0 [3.228], the efficiency of the coated CEM is slightly higher in the range 70 Å to 900 Å than the uncoated one. For both types, the absolute efficiency is higher than 10% (see Fig.3.42b).

Very many channels with diameters as small as ~10 μm have been manufactured in a way to form an area of microchannels [3.231] covering a plate of up to 10 cm diameter with thickness (i.e. channel length) of about one mm. Such devices allow the detection of photons and electrons at an extremely fast time response (≈50ps) with

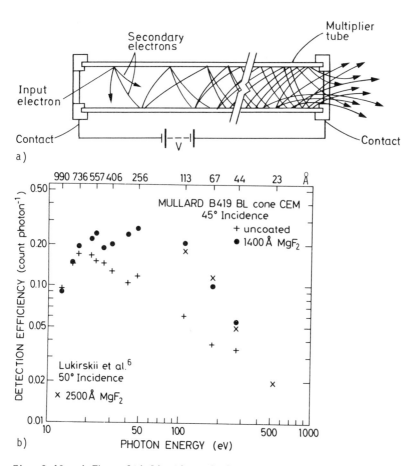

Fig. 3.42 a) The multiplication of electrons in a continuous dynode multiplier is shown. b) The absolute detection efficiency of a MgF_2 coated and an uncoated channeltron is plotted vs. photon energy [3.228]

high uniformity in the sensitive area (see Sect.3.6.1). Even curved microchannel plates (MCP) are available from different suppliers with a spherical or elliptical figure. The present state of the art are MCP's in high-gain chevron configuration located in parallel to each other. By using a resistive anode encoder readout behind the MCP, opposite to the light source or electron source, it can be utilized as a position sensitive detector. A spatial resolution of about 60 µm has been reported [3.232].

3.5.2 X-Ray Detectors

We cannot give here a complete survey on all properties of X-ray detectors but rather would like to discuss those properties which are closely related to the special conditions of a SR source, namely high intensity, higher harmonic rejection and pulsed structure. Various kinds of detectors are listed in Table 3.11. General information on detectors is given [3.239-243] and the special problems involving SR are treated [3.244,245].

For any detector which is used in the counting mode it should be considered that the maximum count rate which is tolerable under certain predetermined count-loss and resolution conditions, N_{max}, has to be multiplied by the duty cycle δ, namely the fraction δ of time during which radiation is emitted. This results in an effective maximum count rate

$$N_{max}^{eff} = N_{max} \cdot \delta \tag{3.33}$$

For 30 keV photons from a synchrotron like DESY at 7 GeV, with a repetition period of acceleration of 20 ms, hard X-rays are emitted only for the last 2 ms, giving $\delta \approx 0.1$.

Another kind of problem arises with storage rings like DORIS or SPEAR when operating in the single bunch mode (see Sect.1.2.3). With DORIS a 150 - 400 ps long light pulse is regularily emitted every 1 µs. There is no detector available which can resolve two or more events arriving within 400 ps. Thus, assuming Poisson statistics for the probability of multiple emission from one bunch to the experiment we obtain for the actual count rate

$$N = n(1-e^{-N_0/n}), \tag{3.34}$$

where N_0 is the true event rate and n is the number of bunches per second. We recognize that $N = N_0$ for low count rates while it approaches n for very high event rates. In this case every light pulse leads to one counted event. We have assumed that the detector is capable of resolving events originating from adjacent bunches. With e.g. $N_0/n = 0.1$ a count loss of 5% results.

The most interesting and frequently used detectors are the lithium drifted silicon and germanium semiconductor counters [Si(Li) and Ge(Li)]. They have the highest energy resolution due to the fact that only 3 - 4 eV are needed for producing one electron-hole pair. The useful count rate is limited to about 50,000 Hz, if no deterioration of the energy resolution is tolerated. If the count rate is too high the energy resolution suffers, because of an incomplete collection of the charges. In Table 3.11 we summarize the available detectors quoting both a "dead time" and "time resolution". The "dead time" is just the minimum time between adjacent pulses under which significant count loss or loss of information (pile-up) occurs, while "time resolution" is the time accuracy with which an event can be located from the steep rising part of the pulse. The numbers quoted are only typical, since the technical development of detectors and the perhaps even more important subsequent electronics is going at a very fast pace.

Scintillation counters usually have a very poor energy resolution. This is due to the fact that about 400 eV are needed for NaI(Tl) and 2000 eV for plastic scintillators in order to produce one photoelectron at the multiplier cathode [3.239-244]. Especially the organic scintillators, however, have very short decay times down to 0.5 ns. This is an important property for experiments with very high count rates.

Proportional counters [3.239,241] are especially useful in the soft X-ray region up to ∼6 keV. With very thin windows they can have a high overall efficiency combined with an energy resolution which is just sufficient to allow for a certain suppression of higher harmonics.

Experiments which take full advantage of the very high intensities in the primary beam, like absorption measurements (EXAFS)(see Sect.3.6.2), cannot be operated with digital electronics. Ne-He (9:1) [3.204a] filled or just plain air filled [3.246] ionization chambers are used. There are flow systems and sealed chambers in usage. The current is collected and amplified by a dc amplifier. Such amplifiers have usually a sensitivity limit of 10^{-14} A. For the common gases an average of 25-35 eV is needed for producing one ion pair. At high intensities, typically above 10^6 photons/s the noise is determined by photon statistics only.

The efficiency of all the systems described above depends on matching the size of the sensitive part of the detector including its shape to the mean free path of the detected photons. With higher X-ray energy the mean free path in general increases and the size or the pressure of the detector have to be increased too. As a consequence, e.g., for Si(Li) and Ge(Li) detectors the energy resolution is becoming worse [3.240] and also the dead time which depends on the collection time for the charges produced is increased.

Table 3.11 lists also the most important position sensitive detectors. Photographic film is used in topography, where nuclear emulsions [3.239] with highest resolution are needed (down to 0.3 μm). For other applications like the registration of

Table 3.11 X-Ray Detectors [3.239-245]. The numbers quoted are typical values.

	References	FWHM energy resolution [eV]	dead time	time resolution [ns]	Remarks
1 Crystal scintillation counters	3.239-241,244	~4000	0.3 µs	10	low efficiency
2 Plastic scintillation counters	3.240,243,244	~8000	2-4 ns	1	
3 Si(Li)	3.240,244,247	200	10 µs*	10	some have even better resolution
4 Ge(Li), and intrinsic Ge					
5 Proportional counters	3.239,240,242,244	1000	0.5 s	25	especially for low energies
6 Geiger-Müller counters	3.239,240,244	--	0.1 ms		
7 Ionization chambers	3.204,244	--	--	--	for more than 10^7 photons/s
Position sensitive detectors:					
8 Photographic film (nuclear emulsions)	3.239	--	--	--	reduced sensitivity with higher energy, shrinkage of film, resolution 0.3 µm
9 Linear semiconductor detector	3.247	--	--	--	resolution > 0.150 mm
10 Linear proportional chamber	3.248,249	~1000	0.5 µs	25	0.15 - 0.7 mm FWHM resolution
11 Multiwire proportional chamber	3.214,250,252		1 µs		
12 Spherical drift chamber	3.521		200 ns		
13 X-ray TV camera	3.253-256	--	--	--	

*This is a typical value. Smaller values are possible if a deterioration of the energy resolution is tolerated

diffraction patterns films with lower resolution (up to 100 μm) but higher sensitivity can be applied. With photographic recording no real time experimentation is
possible and absolute intensity determination is usually a problem. Another difficulty arises with non-uniform film shrinkage when precision measurements are required.

There are several attempts to divide ionization detectors into small subsections
which allow for a localization of individual events. Linear solid state detectors
[3.247] and proportional counters [3.248,249] are used in such a way. The collector
is subdivided into small sections and the signal is fed into a parallel delay line.
The signals travel to both ends and the times of arrival are monitored and translated to a position for the event. Resolutions of 1 mm or less are possible.

Multiwire proportional chambers [3.214,250,251] serve as two dimensional detectors, some are equipped with crossed wires making also use of the positive signal
from the ion current. Separation of the wires can be as low as 1 mm. By a careful
measurement of the charge on neighboring wires interpolation calculations can give
a positional accuracy of about 0.15 mm. Readout of individual wires into the computer allows very high count rates. This is, however, quite expensive. Slower but
cheaper systems are working again with the help of delay lines [3.252]. A special
type of two dimensional detector is the spherical drift chamber developed by CHARPAK
and coworkers [3.51] (see Fig. 3.43). This detector is centered at a sample and col-
lects the electronic charge at one point developed along a radial path corresponding
to one scattering angle. This instrument has a high conversion probability for the
photon and does not suffer from the position uncertainties of inclined paths in
planar chambers. In summary, many of the techniques which originally have been
developed for high energy physics can usefully be applied also in the X-ray region.

Finally we want to mention the use of TV cameras using a fluorescent screen and
an image intensifier [3.253-256]. Such a system can be used with high and low intensity signals. They are of considerable interest, since they make use of the highly
developed TV techniques. On the other hand, cathode homogeneity, geometric distortion

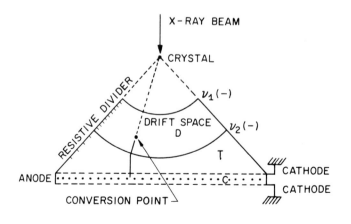

Fig.3.43 Spherical
drift chamber[3.251]in
principle.The negative
charges produced by the
photon at its conversion
point drift to one point
at the multiwire detector
at the bottom irrespective
of where along its path
the photon is converted
[3.7].

and noise constitute problems which have to be overcome. Further the computer interface needs considerable care, since the TV output originally is an analog signal.

3.6 Typical Experimental Arrangements

After having discussed the individual optical components to build spectrographs and
monochromators for spectroscopy and various types of detector systems efficient for
wavelengths from the visible to the X-ray region, we now turn to a brief description
of complete experimental arrangements. The purpose of this section is to make the
reader acquainted with some typical experimental set-ups which can be found in more
or less similar arrangement in several SR laboratories. Simply for reasons of convenience we heavily use material from our laboratory. However, we also include some
unique experimental set-ups to illustrate the possibilities of SR spectroscopies
and to show up possible future trends.

3.6.1 Experiments in the Vacuum Ultraviolet

A general summary on experimental methods in the VUV is given in Table 3.12. For
convenience, some references are included to provide an easy access to an in-depth
information on the various physical and instrumental aspects. For references we
also refer the reader to the bibliography on SR [3.257] and to the proceedings of the
recent VUV conferences [3.2,3,5,9,108,109].

a) Absorption, Reflection, Ellipsometry

Absorption and reflection spectroscopy on solids are now standard techniques to
determine the optical constants by means of a Kramers-Kronig analysis of absorption
or/and reflectivity data or directly from appropriate reflectance measurements.
Moreover, these techniques provide standard means of locating optical transitions.
There is no need to give special references. As mentioned earlier (Sect.3.2.3) ellipsometry is more difficult to use for the determination of optical constants due to
lack of easy to operate VUV polarizers [3.160-169].
   In Fig.3.44 we show an apparatus for absorption measurements of transition metal
and rare earth metal vapors [3.258,259] in the energy range of the outermost core
excitations (50 to 200 eV). By comparing the spectra of free atoms and those in
solids important correlation effects can be studies. The central instrument is a
standard Rowland spectrograph which allows the detection of the full absorption
spectrum of the vapor in front of the spectrograph at a single exposure. A spectrograph is prefered here, since it is difficult to maintain very stable conditions in

Table 3.12 Experiments in the vacuum ultraviolet

	References	Important Properties of SR						
		Tunability	Brightness	Pulses	Polarization	Total Intensity	Calculable	Remarks
Absorption	standard	+				(+)		
Reflection	see Chap.7	+			(+)	(+)		polarization for determination of optical constants
Ellipsometry	3.160-161	(+)				(+)		
Fluorescence Luminescence	3.260 3.261-263	+		+	+	+		pulses for fluorescence decay experiments
Photoionization Photofragmentation	3.264,265	+				+		
Photoemission	3.39,3.159, 3.267	+	+	(+)	+	+		tunability for yield spectro-scopies and pulses for the time of flight spectroscopy
Radiometry	3.268-270	(+)					+	
Microscopy	3.68,90,125, 126,274,275	(+)	+			+		tunability to work below and above absorption edges

an absorption oven operating up to about 2500°C. The heating element is a Ta tube powered with 40 kW. The metal vapor is confined to the oven by means of thin VUV filters (Sect.3.2.3) and by a buffer gas. The SR is collimated by a toroidal mirror and after passing the oven it is focussed onto the entrance slit of the spectrograph.

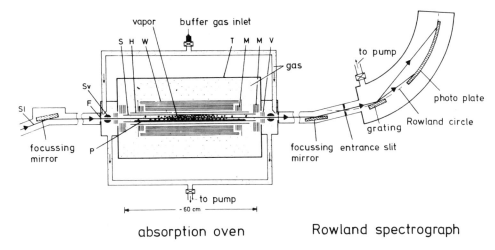

absorption oven   **Rowland spectrograph**

Fig. 3.44 Experimental arrangement to measure the VUV absorption of metal vapors.
[P: liquid sample, SI: Synchrotron radiation, F: window, Sv: rotatable beam stop,
S: Ta boat, H: heating tube (Ta), W: heat-baffle (W), M: heat baffle (Mo), V: heat
baffle (stainless), T: vacuum system] [3.258]

b) Luminescence, Fluorescence

The investigation of secondary processes is developing rapidly especially with
the higher fluxes from storage rings which become more and more available now [3.21,
22]. In soft X-ray fluorescence experiments [3.260] the valence electron structure
of a variety of materials has been studied including organic materials which are
easily decomposed, if high energy electrons are used for the excitation of primary
core vacancies. At the DORIS laboratory SR, pre-filtered by a grazing incidence
mirror, is the exciting source giving rise to fluorescence radiation in the range
up to 600 eV to be detected by a standard Rowland spectrometer.

Fluorescence and luminescence experiments with secondary radiation energies up
to 50 eV have been reported at various laboratories [3.261]. The investigation
capability of decay processes [3.262] in insulators and molecules is greatly enhanced by use of the pulse structure of storage rings (Chap.1). For example, at the
DORIS ring operating in the multi-bunch mode the time structure due the single electron and positron bunches corresponds to light flashes of 250 ps duration separated
by about 8 ns. An experimental arrangement for luminescence studies is shown in
Fig. 3.45 [3.263]. SR from DORIS is monochromatized by a Wadsworth monochromator
(W) and focused onto the exit slit. The refocused light forms a small image on
the sample under investigation. This light spot is then used as the entrance slit
for a second Seya Namioka (SN) monochromator which analyses the luminescence radiation (PM). A simultaneous reflection measurement (RE) is also included as well as a
monitor for the incoming primary radiation (R). Materials studies include solidified

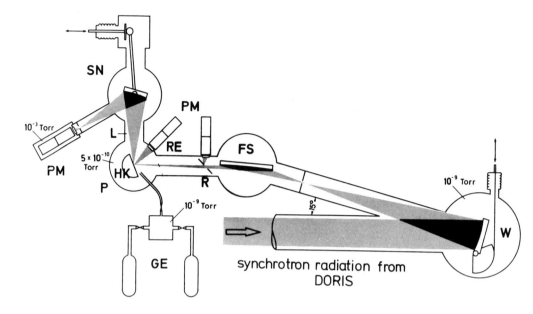

Fig. 3.45 An experiment to study luminescence excitation in the VUV. For abbreviations see text [3.263]

rare gases (by means of a He cryostat (HK)) and organic molecules in rare gas matrices.

c) Photoionization, Photofragmentation

The investigation of the decay of excited molecular states is performed by fluorescence and/or fragmentation. The latter is accomplished by ion mass spectrometry. In current experiments on small organic molecules [3.264] and atmospheric gases [3.265] normal incidence monochromators are utilized to initially excite or ionize the molecules. In commercial quadrupole mass spectrometers the electron beam ionization is replaced by that with monochromatic light.

d) Photoemission

Photoemission experiments have become the major tool for investigating the electronic properties of solids, molecules and atoms [3.5,6,27,39,109,148-150]. The intense, polarized and continuously tunable SR allows to obtain conventional photoelectron energy distributions, both angle-integrated and angle-resolved, as well as various yield-type measurements on solids and solid surfaces [3.159]. The photon energy range provided by normal-and grazing incidence monochromators allows the investigation of valence and core electrons. Also a crystal monochromator has already been utilized for photoelectron spectroscopy [3.39] at 8 KeV. In Fig. 3.46 we show an

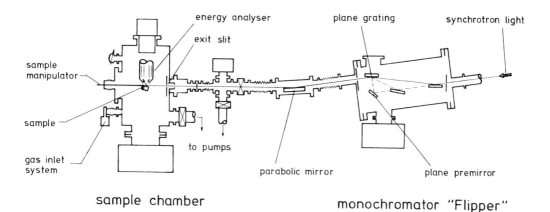

energy analyser   plane grating   synchrotron light

exit slit

sample manipulator

sample

gas inlet system

to pumps

parabolic mirror   plane premirror

sample chamber   monochromator "Flipper"

Fig. 3.46 An experiment to investigate the photoemission of solids with excitation in the VUV [3.191]

experimental arrangement for photoemission studies in the energy range 15 eV to 300 eV [3.191]. The monochromator is a plane grating instrument exploiting the collimation of SR at the DORIS ring. The instrument has been briefly discussed in Sect. 3.3.3. The UHV experimental chamber is built around the exit slit of the monochromator to allow photoemission measurements with highest possible radiation density. Of course, such an arrangement also allows small samples to be used. The energy analysis of photoemitted electrons is made with a commercial double stage cylindrical mirror analyzer [3.266]. For sample preparation and characterization several tools are incorporated which are standard in modern systems, e.g. ion-sputtering, crystal cleaver, evaporating source, sample heating and cooling, Auger electron spectroscopy, low energy electron diffraction etc.

A very different photoelectron spectrometer has been built and utilized by BROWN and collaborators [3.267] to explore the angular dependence of the energy distribution (Fig.3.47). Their time-of-flight spectrometer for photoelectrons is based on the pulsed structure of the SR (at SPEAR pulse width of 1.5 ns and a repetition period of 780 ns). The sample is irradiated with monochromatized light from the "Grasshopper" monochromator described in Sect. 3.3.3 [3.170]. The time of flight spectrometer essentially consists of a drift tube combined with a retardation sector and a fast channelplate as the detector. The transit time along the 17.5 cm long drift distance is approximately 100 ns for a 10 eV electron as measured with a sensitive delayed coincidence timing system [3.267]. The spectrometer can be pivoted around the sample allowing for angularly resolved photoelectron spectroscopy. The energy resolution is determined by the light pulse structure and corresponds to 0.1 eV energy resolution at 10 eV electron energy. First spectra have been obtained with this interesting new instrument. However, the complex electronics needed to take and analyze the data

TIME OF FLIGHT ELECTRON ENERGY SPECTROMETER

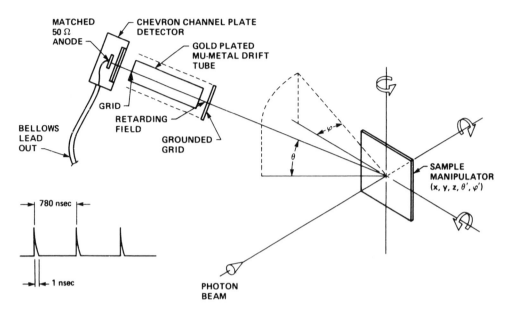

Fig. 3.47 Schematic of the time of flight spectrometer for photoelectron energy analysis [3.191]

outweigh somewhat the simplicity of the spectrometer itself. The system yet has to prove its capabilities to be preferred over others.

e) Radiometry

SR serves as a radiometric standard in the VUV, since it can exactly be calculated, if the accelerator or storage ring parameters are well known. SR is also used for calibration of rocket spectrometers including detectors [3.268,269]. In Fig. 3.48 a radiometer is shown which is designed for calibrating transfer standards of the spectral radiance and irradiance for wavelength between 600 Å and 3500 Å [3.270]. The main optical components are a concave mirror with an aperture stop and two Seya-Namioka monochromators which use a pinhole as a common entrance slit. One of the monochromators is operated at a wavelength in the visible while the other is scanned through the VUV region. By measuring the ratio of the detector current with the concave mirror first in position A and subsequently in position B (Fig.3.48) the spectral radiance of the transfer standard can be calibrated. The ratio recording system is necessary in order to cancel effects due to fluctuations in the SR emitted by the synchrotron [3.270]. The calibration accuracy obtained is ~2%. The calibration of transfer standards to shorter wavelength is of particular importance also for plasma diagnostics.

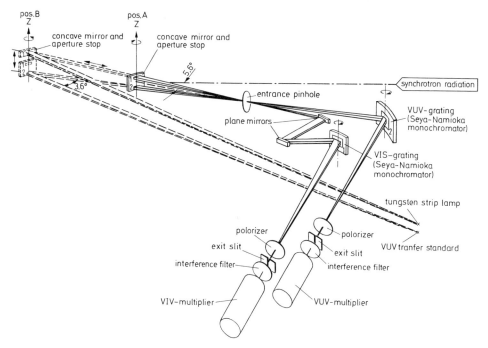

pos.B
Z

concave mirror and
aperture stop

pos.A
Z

concave mirror and
aperture stop

5.6°

3.6°

entrance pinhole

plane mirrors

synchrotron radiation

VUV-grating
(Seya-Namioka
monochromator)

VIS-grating
(Seya-Namioka
monochromator)

tungsten strip lamp

polorizer

exit slit

interference filter

polorizer

exit slit

VUV tranfer standard

interference filter

VIV-multiplier

VUV-multiplier

Fig. 3.48 Layout of an experiment to calibrate VUV transfer standards with synchrotron radiation [3.270]

f) Microscopy

In the last years considerable progress in microscopy techniques with soft X-ray radiation has been achieved in particular through the application of high brightness synchrotron radiation, high-resolution photosensitive resist materials and scanning electron microscopes. Various methods for microscopy have been introduced since 1972. They can be subdivided into two classes: namely, into a class which allows "real-time microscopy" at a moderate resolution (but still better than with light microscopy) and a class, which gives potentially the highest resolution in a non-real-time mode. To the former belong the techniques of scanning microscopy [3.90,126], the application of mirror optics [3.271] and of Fresnel zone plates [3.125]. To the latter belong X-ray holography [3.272] and contact microscopy [3.68,126]. Since scanning microscopy makes more efficient use of the radiation impinging onto the sample the scanning technique has to be preferred for studies on living biological objects because of radiation damage considerations. SCHMAHL and coworkers [3.273] who built the first imaging microscope for soft X-ray synchrotron radiation using zone plates as imaging elements (Sect.3.2.2), were able to record X-ray pictures on photographic film of live 3T3-mouse cells at a moderate resolution. With the technique of contact microscopy various biological specimens have been investigated with a resolution as good as 100 Å [3.68,126,274,275]. For contact microscopy no optical components are required for imaging. In this technique specimens are placed on top of thin films of photosensitive resist material [275a]. The intensity absorbed in the resist depends on the absorption profile of the sample. Therefore a replica in relief of the specimens absorption structure is obtained via a development procedure. This can be magnified and photographed with a scanning electron-microscope. Basically, it is the same process which is used as a step in the production of electronic microcircuits [3.68,126]. Due to the good resolution obtainable with the short wavelength spectrum of SR first provisions are being made to use SR for the fabrication of further miniaturized microcircuits in the future.

3.6.2 Experiments in the X-Ray Range

We give here a survey on different experimental techniques which have been pursued in the X-ray region using SR. We include some references in which instrumental details can be found. The information is summarized in Table 3.13. General references which contain recent information on this topic are the proceedings of the 1977 Orsay meeting on SR instrumentation [3.9], the 1976 Quebec meeting on SR facilities [3.5], the Stanford [3.7], and the DESY [3.8] SR handbooks. In most cases we shall stress only the most important points. In a few cases we give a special example of the experimental arrangements used.

a) Single Crystal Diffraction

Structural research with X-rays is one of its oldest and most important applications. Due to the high brightness of SR sufficient intensity can be collimated onto very small crystals. This should simplify structure determination of difficult to grow crystals from organic materials like e.g. proteins [3.276,277]. In principle tuning of the wavelength to the vicinity of the absorption edge of one of the constituent atoms should be a means to manipulate individual scattering amplitudes due to anomalous dispersion. This would make substitution of atoms unnecessary for solving the problem of phases in determining complicated crystal structures. The instrument consists of a crystal monochromator followed by a diffractometer operating in a vertical plane because of the polarization of SR.

b) Small Angle Diffraction

This technique is typically used for structures with large unit cells like muscles or retina [3.213-218,278]. It takes advantage again of the high brightness of SR. Narrow fans with good collimation are needed to have good angular resolution with small samples. Focusing bent crystal monochromators are used as described above (Sect.3.4.3). The diffraction spots are measured with photographic film or with one- or two-dimensional detectors. The latter type of detectors are preferred since kine-

Table 3.13 Types of experiments in the X-ray region

Methods	References	Important Properties of SR				Remarks
		Tunability	Brightness	Pulses	Total Intensity	
Methods for structural investigation						
1) Single crystal diffraction	3.276,277	(+)	+			tunability only for anomalous dispersion
2) Small angle diffraction	3.213-218,278		+	+		
3) Small angle scattering	3.34		+	+		
4) Mössbauer scattering	3.280-282	+	+	+		
5) Energy dispersive diffraction	3.283-285	+	+			
6) Interferometry	3.206,286	+	+			
7) Absorption (EXAFS)	3.204a,209,210a,287,288	+	(+)		+	brightness allows simple monochromators, pulses could be of interest
8) Topography	3.289-292	+	+			
9) Standing wave excited fluorescence	3.294-296			+		
Methods for investigating electron states						
10) Fluorescence excitation	3.297-299	+			+	
11) Compton scattering	3.300		+		+	
12) Resonant Raman scattering	3.302,303	+			+	with wigglers also non-resonant process of interest
13) Photoelectron spectroscopy (XPS)	3.40,304,305				+	

matic measurements like muscle contraction are of major interest [3.213,214]. Fig.
3.38 shows such an instrumental arrangement [3.218].

c) Small Angle Scattering

This method is applied for the determination of the outer dimensions and possibly
shapes of large organic molecules in solution (and other complexes of corresponding
sizes) [3.34]. Since the method does not need a good monochromatization of the
radiation ($\Delta\lambda/\lambda \sim 1/10$ is already more than sufficient) high intensity can be available if it is possible to filter out such a broad band. As a first attempt of filtering use of the high energy cut-off of total reflection (see Sect.3.2) and the low
energy cut off of an appropriate absorber is suggested [3.34]. This method could
yield $\Delta\lambda/\lambda \approx 0.3$. With the available intensities time dependent phenomena could be
investigated if the detectors can be made fast enough. According to STUHRMANN [3.34]
ZVI KAM [3.279] has proposed experiments to obtain information on molecules which
goes beyond the time and space averaged charge distribution. By measuring scattering
fluctuations on sufficiently small samples in short times correlation functions are
obtained with information on the actual shapes of individual molecules.

d) Mössbauer Scattering

A 1 mrad wide beam of SR from DORIS operating at 4 GeV, 100 mA contains about 1000
photons in a 10^{-8} eV wide energy interval at ~14 keV photon energy (Mössbauer level
of Fe^{57}). Attempts are being made [3.280-282] to filter out these photons from the
background of white SR. If this can be achieved without too much losses, a well
collimated beam of 0.1×1 $mrad^2$ angular spread would become available with an intensity surpassing that attainable in the same solid angle from natural Fe^{57} sources by
about two or more orders of magnitude. This beam would be pulsed as the SR source
and would be nearly 100% polarized. Possible applications are interferometry, investigation of coherent nuclear Bragg scattering, anomalous transmission of Mössbauer radiation through single crystals, structural investigations on biological samples containing Fe^{57} markers etc. The following means of filtering have been suggested
[3.280-282]:

 α) Premonochromatization by ordinary Bragg reflection resulting in a 1 eV wide
   band,
 β) $90°$ horizontal nuclear Bragg reflection from a Fe^{57} or Fe^{57} compound crystal
   making use of the M1 character of the nuclear transition and thus suppressing
   the electronic dipole scattering,
 γ) utilizing the time structure of SR by gating the detectors to be sensitive
   only for time delayed (up to 100 nsec half width) nuclear scattering. Alternatively it is possible to use a detector which consists of a Fe^{57} fluorescent target and a photomultiplier which is gated off during the excitation
   process.

The methods β) and γ) have efficiencies which are in the order of 10. Therefore a considerable overall loss of photons has to be tolerated in the course of the mono-chromatization process. The whole undertaking is therefore tedious and difficult. If the efforts made towards this goal will be combined one day with those towards the production of a periodic undulator producing several orders of magnitude more intensity at 14 keV (or any other appropriate Mössbauer energy) a really intense Mössbauer beam could become available.

e) Energy Dispersive Diffraction

This method [3.283-285] takes full advantage of the continuous spectral distribution of SR. Bragg scattering from powder materials is described by the Bragg equation

$$2d_H \sin\theta_0 = 12400/\varepsilon_H, \tag{3.35}$$

where d_H is the lattice plane spacing for the planes with indices H measured in Å and ε_H is the photon energy [eV] fulfilling the Bragg condition at a fixed Bragg angle θ_0. The experimental arrangement is shown in Fig. 3.49. The white beam of SR hits the sample, while Si(Li) or Ge(Li) detectors (see Sect.5.2) are employed for the energy analysis of the ε_H peaks.

The speed of the method is presently only restricted by the counting limit of ~50000 Hz of the detection system. Even with this speed the method is by a factor of about 100 faster than angular scan methods using monochromatic radiation from commercial tubes and has a better signal to noise ratio. The real advantages lie in the simultaneous appearance of all reflections which makes possible the study of structural changes in a very simple way. Large changes can be observed with a time

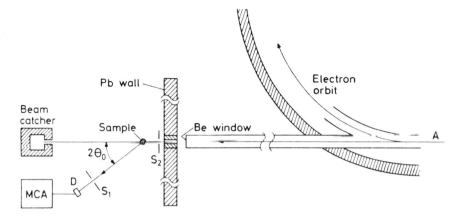

Fig. 3.49 Principle of the energy-dispersive X-ray diffraction method, (S$_1$, S$_2$: collimators, D: semiconductor detector). For one measurement the scattering angle $2\theta_0$ remains fixed [3.283]

constant of 1 s already while fairly accurate spectra allowing for the investigation of small changes are obtainable in 10 s. If faster changes shall be investigated a repetitive technique can be applied. As a demonstration Fig. 3.50 shows a phase transition of manganese from the low temperature α phase into the high temperature β phase. Absolute structure determination is achieved, if the spectral distribution of SR and the detector sensitivity is taken into account. These can be calibrated with known structures.

Applications to high pressure induced phase transitions have also been made [3.285]. In this case the high brightness of SR comes into play, since the small

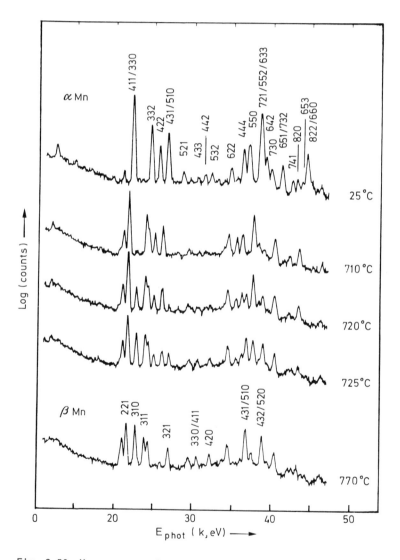

Fig. 3.50 Manganese powder patterns obtained by energy dispersive diffraction showing a phase transformation. The temperatures are indicated at the spectra. Each spectrum was obtained in 300 s at DESY [3.285]

diamond windows of these instruments cut off most of the beam. Further attempts were made to investigate small angle scattering from biological samples. In this case radiation damage plays a dominant role and the method has the disadvantage that a large fraction of the radiation which is impinging onto the sample is not used for obtaining information. Other methods may be more efficient in this respect.

f) Interferometry

This method is described in several publications by BONSE et al. [3.206,286,204a, 285a]. An interferometer (Fig. 3.51) consists of a series of plane single crystal beam splitters which accept only very well collimated beams. Again the high gain in brightness of SR is the important factor. This experiment serves in principle to obtain the complementary information to an absorption experiment, namely the real part of the forward scattering amplitude, which is directly related to the index of refraction. Such a measurement of the scattering amplitude near absorption edges, where it shows its dispersive behavior (anomalous scattering) is a means to obtain the primary data necessary for the application of this method to the solution of the phase problem (see Sect.3.6.2a). This method offers the possibility to measure the dispersion correction directly for the substance under investigation.

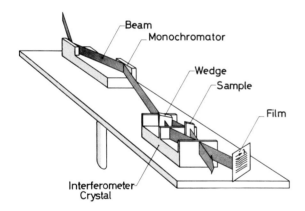

Fig. 3.51 Laue interferom-eter. SR is monochromatized by a double crystal groove-cut crystal. The Laue inter-ferometer is cut from a ·single crystal consecutively splitting the beam, joining it again and finally super-imposing it for interference. A wedge serves to produce interference fringes. The sample material is inserted only halfways into one of the beams

g) Absorption (EXAFS)

The popularity of the absorption measurements in a range up to 1000 eV above absorp-tion edges (mainly K edges) is due to the new interpretation of the Kronig structure [3.204a,287] and the simplicity with which these measurements can be performed with SR in combination with easily tunable double crystal monochromators in the (+,-) mode (see Sect.3.4.1). The Kronig, or extended X-ray absorption fine structure (EXAFS), is due to the scattering of the internally excited photoelectrons in the immediate environment of specific atoms and the interference of the scattered waves with the primary wave (see also Sect.4.2.3). EXAFS spectrometers can serve a large number of different experiments in which cryostats, ovens, high pressure cells, stop-flow devices, ultrahigh vacuum sample chambers etc. are used at the sample position [3.204a,209,210a,288]. A fully equipped EXAFS station must also provide the necessary computer programs to transform the spectra into nearest neighbor distances. The theoretical limitations, however, are still somewhat unclear and there is a danger of overinterpretation of the data by unexperienced users. The methods are still in the stage of development.

A promising future development appears to be the observation of secondary processes which occur after absorption, e.g. fluorescence, Auger electrons and photoelectrons. This makes the technique more selective to certain types of atomic constituents or to special regions of the sample as, for instance, the surface. SR beams condensed by focusing mirrors provide the necessary intensity for such investigations [3.209]

h) Topography

In topography experiments a sample crystal is positioned in the undispersed white beam of SR producing Laue reflections [3.289-292]. Any defects inside the crystals are projected onto the film and are observable as fine structures in the Laue reflections, Fig. 3.52 gives an example. The advantages of SR are

α) The large distance from the source which are possible because of the good collimation of SR. This distance allows either a good resolution if the film is near the sample or a larger distance between the film and the sample which is needed when applying heat, magnetic fields, low temperatures etc. to the sample.

β) The very short exposure time in the order of seconds to minutes, which can be even further reduced, allow for the investigation of non-stationary phenomena like motion of dislocations under stress, motion of domain structure with

Fig. 3.52 X-ray reflection topograph of ferromagnetic domains in terbium below the Curie temperature of 222 K obtained at NINA, 4 GeV, 4 mA, 3 minutes exposure, 3121 reflection. Domain walls run horizontally, i.e. perpendicular to the c axis [3.290]

applied magnetic fields etc. The application of high speed film techniques with image converters appears to be feasible [3.293].

γ) The experimental conditions are simplified, since no tedious prealignment of the samples is needed compared to the use of characteristic X-rays from X-ray tubes.

δ) The white beam leads to a creation of many Laue reflections at the same time. This allows for a stereoscopic observation of the same defect from different angles.

In addition to the Laue transmission arrangement also reflection topographs are of interest. In this case special filters between the sample and the recording film can be helpful in order to reduce the fluorescent radiation from the sample. Although this causes troubles and the conditions differ from case to case the problems appear to be solvable.

i) Standing Wave Excited Fluorescence

A group from Aarhus, Denmark, made first tests with this method [3.294,295] at the DESY synchrotron [3.296]. After producing a parallel, monochromatic, higher order free beam with a spectrometer a standing-wave pattern in the Bragg case is set up in a single crystal of Si (but also any other crystal of high quality could be used) (Fig.3.53). While the rocking curve of this crystal is scanned, the standing waves

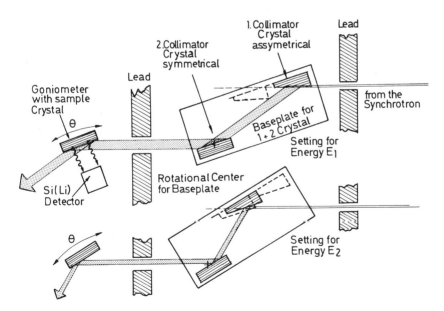

Fig. 3.53 Schematic drawing of the triple-crystal spectrometer for standing wave excited fluorescence. Shown are two settings for different energies [3.296]

move with their maxima from the lattice plane position to the inter-lattice plane
position. The fluorescence signal of impurity atoms like As is observed with a
solid state detector at 90° horizontally in order to suppress scattered light
thus making use of the polarization of SR. A signal ratio of 7 to 1 was observed
when scanning through the rocking curve thereby confirming the substitutional nature
of the As impurities. An ultimate goal of this method would be to apply it to surface physics. Since the standing wave pattern extends beyond the crystal surface
into the vacuum the distance of absorbed atoms relative to the crystal surface could
be measured. Such an experiment would certainly require the very high intensity
from a storage ring.

j) Fluorescence Excitation

The excitation of soft X-ray fluorescence is now applied for many years at DESY for
the investigation of substances which are easily damaged by electron bombardment
[3.297,298]. Although the relative damage of electron and radiation excitation is
different in different cases, there is clear evidence that the excitation with radia
tion is usually less harmful per fluorescence quantum. The soft X-ray experiments
involve the fairly tedious techniques of grating monochromators.

In the X-ray region proper the fluorescence excitation can compete with that of
strong X-ray tubes if a favorable geometry is used. SPARKS and coworkers [3.299]
have used SR of SPEAR because of the very high photon energies available for looking
for the fluorescence of superheavy elements. They used a graphite crystal bent to a
radius of 10 cm collecting SR from an angle of 2 mrad.

k) Compton Scattering

COOPER et al. [3.300] apply radiation filtered by a crystal monochromator at 10 and
20 keV at Daresbury and DESY for a Compton scattering experiment. They find that a
gain by a factor of $\sim10^3$ over the use of a conventional X-ray tube is obtained.
There appears to be some contradiction when comparing this factor with calculations.
COOPER et al. [3.300] used a solid state detector with a resolution of about 250 eV.
There is obviously an apparent mismatch between the monochromaticity of the primary
beam (~1 eV) and the resolution of the detector. A considerable improvement of the
experiment should become possible by increasing the bandwidth of the primary monochromator to 20 eV (see Sect.3.2.2) and by analyzing with a curved secondary monochromator with equivalent bandwidth.

Another direction of future development would be a transfer of the Compton experiment to the 100 keV range where interpretation is easier because of the validity
of the impuls approximation. Such an experiment could compete only with conventional
techniques like γ ray Compton scattering, if the special advantages of SR come into

play like gain of intensity by increasing the usable band width, time structure,
circular polarization for investigating magnetic properties etc.

1) Resonant Raman Scattering

X-ray Raman scattering involves, in principle, the same instrumentation as Compton
scattering. The only difference is that a resolution of the order of the width of
the electronic transitions involved is required. This leads to a resolution of
0.2 - 1 eV which can be achieved only at the expense of enormous intensity loss.
When, however, an electronic transition, like e.g. the K edge of Cu, which can be
excited by the primary photon is approached, the cross section for a Raman scattering
from e.g. the Cu $L_{2,3}$ excitation can be boosted up by several orders of magnitude
(see also Sec.4.8.2). This was shown by SPARKS et al. [3.301] using characteristic
radiation and varying Z, the atomic number of his scatterer. EISENBERGER et al.
[3.302,303] could show that SR due to its tunability is an excellent tool for such
an experiment. The resolution of the $L_{2,3}$ excitation as measured with a Ge(Li)
solid state detector [3.302,303] is yet far from being sufficient enough to obtain
structural details of this transition.

m) Photoelectron Spectroscopy (XPS)

The pilot experiment of LINDAU et al. [3.39,304,305] on photoelectron spectroscopy
(X-ray photoemission spectroscopy, XPS with 8 keV photons) has been mentioned already in connection with their monochromator design in Section 3.4.1. Although the
count rate in this experiment was extremely low, there are some arguments in favor
of such an instrument. They could show that the resolution of their monochromator is
extremely good, in the order of 0.2 eV. The escape depth of photoelectrons in the
8 keV range is large thereby making volume properties accessible for photoelectron
spectroscopy. Volume investigations may even become feasible in the presence of
oxide layers. If the technique should not be limited to very few cases, it will be
necessary to increase the photon flux considerably compared to the plane crystal
monochromators used now [3.30]. It is not yet clear, if bent crystal monochromators
can be used without a serious reduction of the resolution. But such instruments at
smaller distances from the source should allow a large horizontal acceptance. On
the other hand, powerful rotating anode X-ray generators can also be equipped with
bent crystal monochromators outweighing the high brightness of SR by an even larger
acceptance [3.306] (see Sect.1.3). Rough estimates give about equal available intensities for the strongest available storage rings and X-ray sources. An effort to
obtain experience with an improved photoemission equipment at a storage ring appears
to be justified, because of the further possibility to gain several orders of magnitude, when the periodic wigglers will become available (see Sect.1.2.2).

3.7 Acknowledgements

The authors wish to thank numerous colleagues who have provided information for this chapter. Special thanks go to G. Materlik who helped to clarify many details concerning the X-ray instrumentation and who also did a careful reading of this part of the manuscript and made valuable suggestions. We are also indepted to E. Thumann, A. Paetz and H. Dahl for their careful typing of the manuscript and to R. Fisseler, V. Fischer, W. Knaut, J. Schmidt, and M. Sörensen for their help in preparing the figures.

References

3.1   R. Haensel, C. Kunz: Z. Angew. Physik 23, 276 (1976)
3.2   G.V. Marr, I.H. Munro (eds.): *Proc. Intern. Symposium for Synchrotron Radiation Users* (Daresbury Nucl. Phys. Lab. Report DNPL/R26, 1973)
3.3   E.E. Koch, R. Haensel, C. Kunz (eds.): *Vacuum Ultraviolet Radiation Physics* (Pergamon-Vieweg, Braunschweig 1974)
3.4   E.E. Koch: In *Problems of Elementary Particle Physics*, Proc. of the 8th All Union School of High Energy Particle Physics (Yerevan, 1975) p. 502
3.5   J.Wm. McGowan, E.M. Rowe (eds.): *Synchrotron Radiation Facilities*, Proc. Quebec Summer Workshop (University of Western Ontario, London, Ontario 1976)
3.6   A.N. Mancini, I.F. Quercia (eds.): Proc. of the Internat. Summer School on Synchrotron Radiation Res., Alghero 1976 (Intern. College on Appl. Phys. and Istituto Nazionale di Fisica Nucleare, Catania 1976) Vol. 1 and 2
3.7   K.O. Hodgson, H. Winick, G. Chu (eds.): Synchrotron Radiation Research and the Stanford Synchrotron Radiation Project (SSRP Report, Stanford No. 76/100 Aug. 1976)
3.8   E.E. Koch, C. Kunz (eds.): *Synchrotronstrahlung bei DESY* (DESY, Hamburg, Juli 1977)
3.9   F. Wuilleumier and Y. Farge (eds.): *Synchrotron Radiation Instrumentation and Developments*, Proc. Orsay Meeting Sept. 12-14, 1977, Special Issue Nuclear Instrum. and Methods (North-Holland, Amsterdam 1978) Vol. 152
3.10 J.A.R. Samson: *Techniques of Vacuum Ultraviolet Spectroscopy* (Wiley and Sons, New York 1967)
3.11 N. Damany, J. Romand, B. Vodar (eds.): *Some Aspects of Vacuum Ultraviolet Radiation* (Pergamon Press, Oxford 1974)
3.12 A.N. Zaidel, E.Y.A. Schreider: *Vacuum Ultraviolet Spectroscopy* (Ann-Arbor-Humphrey Science Publishers, Ann Arbor, London 1970)
3.13 A.C.S. van Heel (ed.): *Advanced Optical Techniques* (North Holland, Amsterdam, 1967)
3.14 L.V. Azaroff (ed.): *X-ray Spectroscopy* (McGraw Hill, New York 1974)
3.15 S. Flügge (ed.): *Handbuch der Physik*, Vol. 30 (Springer Verlag, Berlin, Göttingen, Heidelberg 1957)
3.16 C.W. Stroke: In *Handbuch der Physik*, Vol. 29, ed. by S. Flügge (Springer Verlag, Berlin, Göttingen, Heidelberg 1967) p. 426
3.17 C. Kunz: [Ref. 3.2, p. 68]
3.18 D.L. Ederer, S.C. Ebner: A Users Guide to SURF (NBS, Washington 1974)
3.19 P. Dagneaux, et al., Ann. Phys. 9, 9 (1975)
3.19a P.M. Guyon, C. Depautex, G. Morel: Rev. Sci. Instrum. 47, 1347 (1976)

3.20 Synchrotron Radiation Center Users Handbook (Physical Sciences Laboratory, University of Wisconsin, Stoughton 1973)
3.21 Synchrotron Radiation a Perspective View for Europe, prepared by ESF (European Science Foundation, Strasbourg, France 1978)
3.22 An Assessment of the National Need for Facilities Dedicated to the Production of Synchrotron Radiation (Report to the National Academy of Sciences, Washington D.C. 1976)
3.23 E.M. Rowe, R.A. Otte, C.M. Pruett, J.D. Steben: IEEE Trans. NS-16, 159 (1969)
3.24 E.M. Rowe, F.E. Mills: Particle Accel. 4, 221 (1973)
3.25 C. Gähwiller, F.C. Brown, H. Fujita: Rev. Sci. Instrum. 41, 1275 (1970)
3.26 G.W. Rubloff, H. Fritzsche, U. Gerhard, J. Freeouf: Rev. Sci. Instrum. 42, 1507 (1971)
3.27 F.C. Brown: Solid State Phys. 29, 1 (1974)
3.28 M.L. Perlman, E.M. Rowe, R.E. Watson: Physics Today 27, 30 (July 1974)
3.29 E.M. Rowe, J.H. Weaver: Scientific American (June 1977) p. 32
3.30 R. Jean, J. Rauss: Le Vide 111, 123 (1964)
3.31 R.C. Wolgast, J.W. Davis: Nucl. Sci. 16, 954 (1969)
3.32 Design Study for a Dedicated Source of Synchrotron Radiation (Science Research Council, Daresbury Laboratory, DL/SRF/R2, 1975) p. 52
3.33 E.E. Koch, C. Kunz, E.W. Weiner: Optik 45, 395 (1976)
3.34 H.B. Stuhrmann: Quarterly Reviews of Biophysics. II, I, 71 (1978)
3.34a W. Ebeling, G.W. Bennett: DESY Internal Rep. S1-7016 (1970)
3.35 W. Gudat, C. Kunz, J. Karlau: Appl. Opt. 13, 1412 (1974)
3.36 H. Dietrich, C. Kunz: Rev. Sci. Instrum. 43, 434 (1972)
3.37 I. Fankuchen: Nature, London 139, 193 (1937; A. Guinier: C.R. Acad. Sci. Paris 223, 31 (1946))
3.38 G.K. Green: Spectra and Optics of Synchrotron Radiation, BNL Report 50522, (Brookhaven, April 1976)
3.39 P. Pianetta, I. Lindau: J. Electr. Spectr. Rel. Phenom. 11, 13 (1977)
3.40 J. Hasting: [Ref. 3.5, pp. 8-19]
3.41 R. Lopez-Delgado, H. Szwarc: Opt. Commun. 19, 286 (1976)
3.42 H.W. Schnopper, L.P. Van Speybroeck, J.P. Delvaille, A. Epstein, E. Källne, R.Z. Bachrach, J. Dijkstra, L. Lantwaard: Appl. Opt. 16, 1088 (1977)
3.43 *Diffraction Gratings-Ruled and Holographic-Handbook*. ed. by Jobin-Yvon Company (Longjumeau, France 1976)
3.44 E. Spiller: Space Optics, ISBN-309, Nat. Acad. Sci. (1974)
3.45 A.V. Vinogradov, B.Ya. Zeldovich: Appl. Opt. 16, 89 (1977)
3.46 V. Rehn: private communication
3.47 A mirror testing program is underway at the DESY Laboratory
3.48 H.J. Hagemann, W. Gudat, C. Kunz: J. Opt. Soc. Am. 65, 742 (1975) and DESY SR-74/7, May (1974)
3.49 J.H. Weaver: Phys. Rev. B 11, 1416 (1975)
3.50 J.H. Weaver, D.W. Lynch, C.G. Olson: Phys. Rev. B 7, 4311 (1973)
3.51 J.H. Weaver, D.W. Lynch, C.G. Olson: Phys. Rev. B 10, 501 (1974)
3.52 J.T. Cox, G. Hass, J.B. Ramsey: J. Opt. Soc. Am. 64, 423 (1974)
3.53 J.F. Osantowski: J. Opt. Soc. Am. 64, 834 (1974)
3.54 See e.g. M. Born, E. Wolf: *Principles of Optics*, 2nd ed., (Pergamon Press, London 1964)
3.55 K.L. Kliewer, R. Fuchs: Theory of Dynamical Properties of Dielectric Surfaces, in *Advances in Chemical Physics*, Vol. 27: Aspects of the Study of Surfaces, Ed. by I. Prigogino and S.A. Rice (Wiley, London 1974)
3.56 J. Römer: Thesis, University of Hamburg (1970)
3.57 C.H. Pruett: [Ref. 3.5, pp. 8-3]
3.58 A.H. Compton, S.K. Allison: *X-Rays in Theory and Experiment* (Van Nostrand, New York 1935)
3.59 R.W. Hendrik: J. Opt. Soc. Am. 47, 165 (1957)
3.60 B.L. Henke: Phys. Rev. A 6, 94 (1972)
3.61 E. Spiller: Appl. Opt. 13, 1209 (1974)
3.62 B.P. Feuerbacher, W. Steinmann: Opt. Commun. 1, 81 (1969)
3.63 J.G. Endriz, W.E. Spicer: Phys. Rev. B 4, 4144 and 4159 (1971)
3.64 L.R. Canfield, R.G. Johnston, R.P. Madden: Appl. Opt. 12, 1611 (1973)

3.65   K. Lindsey: Proc. Symp. X-Ray Optics, Mullard Space Science Lab. (April 1973)
3.66   R.L. Johnston: Ph.D. Thesis, Imperial College of Science and Technology, London (1975)
3.67   M.F. Verhaeghe, Optica Acta, 19, 905 (1972)
3.68   W. Gudat: [3,8] p. 279
3.69   O.S. Heavens: *Thin Film Physics* (Methuen & Co. Ltd, London 1970)
3.70   H. Mayer: *Physik dünner Schichten*, Vols I and II, Wissenschaftliche Verlags-gesellschaft Stuttgart (1950/55) H. Mayer: *Physics of Thin Films*, Complete Bibliography, Part I and II, (Wissenschaftliche Verlagsgesellschaft, Stuttgart 1972)
3.71   W.J. Choyke, R.F. Forich, R.A. Hoffman: Appl. Opt. 15, 2006 (1976)
3.72   V. Rehn, J.L. Stanford, A.D. Baer, V.O. Jones, W.J. Choyke: Proc. Taormina Research Conference on Recent Developments in Optical Spectroscopy of Solids, Taormina (1976), and V. Rehn, V.O. Jones: to be published
3.73   DESY collaboration: to be published
3.74   R.P. Haelbich, C. Kunz: Opt. Commun. 17, 287 (1976)
3.75   E. Spiller: Appl. Opt. 15, 2333 (1976)
3.76   A.P. Lukirskii, E.P. Savinov: Opt. Spectr. 14, 152 (1963)
3.77   A.P. Lukirskii, E.P. Savinov, O.A. Ershov, Yu.F. Shepelev: Opt. Spectr. 16, 168 (1964)
3.78   A.P. Lukirskii, E.P. Savinov, O.A. Ershov, V.A. Fomichev, I.I. Zhukova: Opt. Spectr. 19, 237 (1965)
3.79   O.A. Ershov, I.A. Brytov, A.P. Lukirskii: Opt. Spectr. 22, 66 (1967)
3.80   R.Z. Bachrach, S.A. Flodstrom, R.S. Bauer, V. Rehn, V.O. Jones: [3.9], p. 135
3.81   J.L. Stanford, V. Rehn, D.S. Kyser, V.O. Jones: [Ref. 3.3, p. 783]
3.82   V. Rehn, J.L. Stanford, A.D. Baer, V.O. Jones, W.J. Choyke: Appl. Opt. 16, 1111 (1977)
3.83   see e.g. S.M. Elson, R.H. Ritchie: Phys. Stat. Sol. (b.) 62, 461 (1974) J.O. Porteus: J. Opt. Soc. Am 53, 1394 (1964) and references therein
3.84   M. Pouey: [Ref. 3.11, p. 287]
3.85   J.W. Berthold III, S.F. Jacobs: Appl. Opt. 15, 2344 (1976)
3.86   J.W. Berthold III, S.F. Jacobs, M.A. Norton: Appl. Opt. 15, 1898 (1976)
3.86a   B. Niemann, D. Rudolph, G. Schmahl: private communication
3.87   O.H. Wyatt, D. Dew-Hughes: *Metals, Ceramics and Polymers* (Cambridge University Press, London 1972)
3.88   H. Wolter: Ann. Physik 6c, 94 (1952) and 6, 286 (1952)
3.89   C.E. Winkler, D. Korsch: Appl. Opt. 16, 2464 (1977)
3.90   P. Horowitz, J.A. Howell: Science 178, 608 (1972)
3.91   e.g. D.E. Eastman, W.D. Grobman, J.L. Freeouf, M. Erbudak: Phys. Rev. B 9, 3473 (1974)
3.92   H.A. Rowland: Phil. Mag. 35, 397 (1893)
3.93   H.G. Beutler: J. Opt. Soc. Am. 35, 311 (1945)
3.94   T. Namioka: J. Opt. Soc. Am. 49, 446 (1958)
3.95   T. Namioka: J. Opt. Soc. Am. 51, 412 (1961)
3.96   W. Werner: Appl. Opt. 6, 1691 (1967)
3.97   W. Werner: Appl. Opt. 16, 2078 (1977)
3.98   J.E. Mack, J.R. Stehn, B. Edlen: J. Opt. Soc. Am. 22, 245 (1932) and 23, 184 (1933)
3.99   M. Seya: Sci. Light, 2, 8 (1952)
3.100   T. Namioka: J. Opt. Soc. Am. 49, 951 (1959)
3.101   M. Pouey: [Ref. 3.3, p. 728] M. Pouey, J. Opt. Soc. Am. 64, 1616 (1974)
3.102   T. Namioka: J. Opt. Soc. Am. 51, 4 (1961)
3.103   T. Namioka: J. Opt. Soc. Am. 51, 13 (1961)
3.104   R.C. Chase, J.K. Silk: Appl. Opt. 14, 2096 (1975)
3.105   E.G. Loewen, M. Nevière, D. Maystre: Appl. Opt. 16, 2711 (1977)
3.106   W.R. Hunter: [Ref. 3.3, p. 683] D.J. Michels, T.L. Mikes, W.R. Hunter: Appl. Opt. 13, 1223 (1974)
3.107   R.P. Haelbich, C. Kunz, D. Rudolph, G. Schmahl: [3.9], p. 127 and R.P. Haelbich, Internal Report DESY F41-76/05 (1976)
3.108   Conf. Digest, IIIrd Int. Conf. on VUV Rad. Phys., Tokyo, (1971) ed. by J. Naka

3.109 Proc. Vth Int. Conf. on VUV Rad. Phys., Montpellier (1977) ed. by. M.C. Castex, M. Pouey, and N. Pouey
3.110 R.L. Johnson: [3.9] p. 117
3.111 For further work on "conical diffraction" see e.g. P. Vincent, M. Nevière, D. Maystre: [3.9]; D. Maystre, M. Nevière, W.R. Hunter: [Ref. 3.109, Ext. Abst. Vol. III p. 74]
3.112 A. Cornu: Ann. Phys. Chem. (6. Reihe) 6, 114 (1875)
3.113 A. Labeyrie: Thesis, University of Paris (1966)
3.114 D. Rudolph, G. Schmahl: Umschau in Wissenschaft und Technik 67, 225 (1967)
3.115 T.A. Shankoff: Appl. Optics 7.10 (1968) U.S. Patent 3, 567, 444
3.116 N.K. Sheridon: Appl. Phys. Lett. 12, 316 (1968)
3.117 G. Schmahl, D. Rudolph: Holographic Diffraction Gratings in *Progress in Optics*, Vol. 14, Ed. by E. Wolf, (North Holland, Amsterdam 1976)
3.118 Y. Petroff, P. Thiry, R. Pinchaux, D. Lepere: [Ref. 3.105, Ext. Abst. Vol. III, p. 70]
3.119 C.H. Pruett, E.M. Rowe, T.R. Winchard, F.H. Middleton: [3.9] p. 57
3.120 C. Depautex, P. Thiry, R. Pinchaux, Y. Pétroff, D. Lepère, G. Passereau, J. Flamand: [3.9] p. 101; D. Lepère, G. Passereau, A. Thevenon: [Ref. 3.109, Ext. Abst. Vol. III, p. 67]
3.121 J. Cordelle, J. Flamand, G. Pieuchard, A. Labeyrie: p. 117, In *Optical Instruments and Techniques* ed. by J. Home-Dickson Oriel Press (1970)
3.122 T. Namioka, H. Noda, M. Seya: Sci. Light (Tokyo) 22, 77 (1973); H. Noda, T. Namioka, M. Seya: J. Opt. Soc. Am. 64, 1031 (1974)
3.123 Further references on most recent developments are contained in [3.109]
3.124 G. Schmahl, D. Rudolph: Optik 29, 577 (1969); B. Niemann, D. Rudolph, G. Schmahl: Opt. Commun. 12, 160 (1974)
3.125 B. Niemann, D. Rudolph, G. Schmahl: Appl. Opt. 15, 1882 (1976)
3.126 E. Spiller, R. Feder: X-Ray Lithography in *X-Ray Optics*, ed. by H.J. Queisser, Topics in Appl. Phys., Vol. 22 (Springer, Berlin, Heidelberg, New York, 1977)
3.127 J. Kirz: J. Opt. Soc. Am. 64, 301 (1974)
3.128 W.A. Kleinhaus: Appl. Opt. 16, 1701 (1977)
3.129 J.H. Dijkstra, L.J. Lantwaard: Opt. Commun. 15, 300 (1975)
3.130 H.W. Schnopper, L.P. von Speybroeck, J.P. Delvaille, A. Epstein, E. Källne, R.Z. Bachrach, J.H. Dijkstra, L.J. Lantwaard: [Ref. 3.5, pp. 8-71]
3.131 K.P. Beuermann, R. Lenzen, H. Bräuninger: Appl. Opt. 16, 1425 (1977)
3.132 J.H. Beaumont, M. Hart: J. Phys. E 7, 823 (1974)
3.133 R.W. Gould, S.R. Bates, C.J. Sparks: Appl Spectr. 22, 549 (1968)
3.134 H. Maier-Leibnitz: private communication
3.135 A. Harmsen: private communication
3.136 C. Vogel. F. Wuilleumier, C. Bonnelle: C.R. Acad. Sc. Paris 269, 1255 (1969)
3.137 C. Vogel, M. Dpeyrat: C.R. Acad. Sc. Paris 270, 105 (1970)
3.138 M.W. Charles: J. Appl. Phys. 42, 3329 (1971)
3.139 F. Kohlrausch: *Praktische Physik*, Vol. 2, 21 ed., ed. by H. Ebert, E. Justi (Teubner Verlagsgesellschaft, Stuttgart 1962) p. 522
3.140 A.E. Sandström: [Ref. 3.15 p. 78]
3.140a D.W. Berreman, J. Stamatoff, S.J. Kennedy: Appl. Opt. 16, 2081 (1977)
3.141 E.S. Gluskin, E.M. Trakchtenberg, A.S. Vinogradov: [3.9] p. 133
3.142 R.G. Buckley, D. Beaglehole: Appl. Opt. 16, 2495 (1977)
3.143 A. Malherbe: Appl. Opt. 13, 1275 (1974)
3.144 a) T. Inagaki, E.T. Arakawa, R.N. Hamm, M.W. Williams: Phys. Rev. B 15, 3243 (1977); b) W. Gudat: Thesis, University of Hamburg (1974) DESY Report SR-/4// (1974); c) B.F. Sonntag: "Dielectric and Optical Properties of Rare Gas Solids"; *Rare Gas Solids*, Vol. 2, eds. M.K. Klein and J.A. Venables (Academic Press, New York 1977), Chapt. 18
3.145 R.P. Haelbich, M. Iwan, E.E. Koch: "Optical Properties of Some Insulators in the Vacuum Ultraviolet Region", Physikdaten/Physics Data ZAED, Karlsruhe, Federal Republic of Germany, Vol. 8-1 (1977)
3.146 M. Altarelli, D.L. Dexter, H.M. Nussenzweig: Phys. Rev. B 6, 4502 (1972)
3.147 see e.g. J.H. Hubbel: "Survey of photon-attenuation-coefficient measurements 10 eV to 100 GeV", Atomic Data 3, 241 (1971); B.L. Henke, E.S. Ebisu: "Low energy X-ray and electron absorption within solids", AFORS Report 72-2174,

Univ. Hawaii (1973); B.L. Bracewell, W.J. Veigele; in *Developments in Applied Spectroscopy*, Vol. 9, ed. by E.L. Grove and A.J. Perkins (Plenum, New York 1971) p. 375

3.148 see e.g. C. Kunz; In *Optical Properties of Solids–New Developments*, ed. by B.O. Seraphin, (North-Holland, New York 1976)

3.149 E.E. Koch, C. Kunz, B. Sonntag: Phys. Reports 29c, 153 (1977)

3.150 B. Feuerbacher, R.P. Godwin, T. Sasaki, M. Skibowski, J. Opt. Soc. Am. 58, 1434 (1968)

3.151 J.C. Lemonnier, Y. LeCalvez, G. Stephan, S. Robin; C.R. Acad. Sci. (Paris) 264, 1599 (1967)

3.152 R. Haensel, P. Rabe, B. Sonntag: Sol. State Commun, 8, 1845 (1970)

3.153 W.R. Hunter; In *Physics of Thin Films*, Vol. 7, ed. by G. Hass, M.H. Francombe and R.W. Hoffman, (Academic Press, New York 1973) p. 43

3.154 R. Haensel, C. Kunz, T. Sasaki, B. Sonntag: Appl. Opt. 7, 301 (1968)

3.155 M. Howells, C. Norris, G.P. Williams: J. Phys. E 10, 259 (1977)

3.156 P. Jaeglé, P. Dhez, F. Wuilleumier: Rev. Sci. Instrum. 48, 978 (1977)

3.157 F.C. Brown, R.Z. Bachrach, M. Skibowski: Phys. Rev. B 13, 2633 (1976)

3.158 W. Gudat: [Ref. 3.109, Vol. 2, p. 76]

3.159 see e.g. *Photoemission and the Electronic Properties of Surfaces*, ed. by B. Feuerbacher, B. Fitton, R.F. Willis (John-Wiley & Sons, London 1978)

3.160 D.E. Aspnes: In *Optical Properties of Solids – New Developments*, ed. by B.O. Seraphin (North-Holland, Amsterdam 1976)

3.161 M. Schledermann, M. Skibowski: Appl. Opt. 10, 321 (1971)

3.162 G. Hass, W.R. Hunter: Appl. Opt. 17, 76 (1978)

3.163 H. Damany: J. Opt. Soc. Am. 55, 1558 (1965); H. Damany: Optica Acta, 12, 95 (1965)

3.164 S.P.F. Humphreys-Owens:·Proc. Phys. Soc. London 77, 949 (1961)

3.165 R.N. Hamm, R.A. MacRae, E.T. Arakawa: J. Opt. Soc. Am. 55, 1460 (1965); G. Rosenbaum, B. Feuerbacher, R.P. Godwin, M. Skibowski: Appl. Opt. 7, 1917 (1968)

3.166 M.B. Robin, N.A. Keubler, Yoh-Han Pao: Rev. Sci. Instrum. 37, 922 (1966)

3.167 D.L. Steinmetz, W.G. Phillips, M. Wrick, F.F. Forbes: Appl. Opt. 6, 1001 (1967)

3.168 W.C. Johnson, Jr.: Rev. Sci. Instrum. 35, 1375 (1964)

3.169 U. Heinzmann: J. Phys. E 10, 1001 (1977)

3.170 F.C. Brown, R.Z. Bachrach, N. Lien: [3.9] p. 73

3.171 M. Lavollée: [Ref. 3.3, p. 730]

3.172 GCA-McPherson Company, US Patent 3 090 863

3.173 V. Saile, P. Gürtler, E.E. Koch, A. Kozevnikov, M. Skibowski, W. Steinmann: Appl. Opt. 15, 2559 (1976)

3.174 V. Saile: [3.9] p. 59

3.175 M. Skibowski, W. Steinmann: J. Opt. Soc. Am. 57, 112 (1967)

3.176 W. Gudat: Thesis, University of Hamburg (1970), DESY Report F41-70/8

3.177 E.E. Koch: Thesis, University of Munich (1972)

3.178 G.W. Grime, J.H. Beaumont, J.B. West: Appl. Opt. 14, 2317 (1975)

3.179 R. Broadmann, U. Hahn, G. Zimmerer: private communication

3.180 M. Lavollée, R. Lopez-Delgado: Rev. Sci. Instrum. 48, 816 (1977)

3.181 C. Depautex, M. Lavollée, G. Jezequel, C. Lemonnier, J. Thomas: [3.9] p. 69

3.182 N. Rehfeld, U. Gerhardt, E. Dietz: Appl. Phys. 1, 229 (1973)

3.183 M. Pouey: Opt. Commun. 3, 158 (1971)

3.184 M. Pouey: Appl. Opt. 13, 2739 (1974)

3.185 C.H. Pruett, N.C. Lien, I.D. Steben: [Ref. 3.108, p. 31]

3.186 K. Nowak: Thesis, University of Hamburg (1968)

3.187 K.P. Miyake, R. Kato, H. Yamashita: Sci. Light, Tokyo 18, 39 (1969)

3.188 J.B. West, K. Codling, G.V. Marr: J. Phys. E 7, 137 (1974)

3.189 M.R. Howells, D. Norman, J.B. West, J. Phys. E.11, 199 (1978)

3.190 H. Dietrich, C. Kunz: Rev. Sci. Instrum. 43, 434 (1972); C. Kunz, R. Haensel, B. Sonntag: J. Opt. Soc. Am. 58, 1415 (1968)

3.191 W. Eberhard, G. Kalkoffen, C. Kunz: [3.9] p. 81, see also W. Eberhard: Thesis, University of Hamburg (1978)

3.192 The mirror was delivered from Optical Surfaces Ltd., London

3.193 Standard mountings are in operation at most SR laboratories

3.194 K. Codling, P. Mitchel: J. Phys. E 3, 685 (1970)
3.195 P. Jaeglé, P. Dhez, F. Wuilleumier: [Ref. 3.3, p. 788]
3.196 C.H. Pruett, N.C. Lien, J. Steben: [Ref. 3.108, p. 31]
3.197 K. Thimm: J. Electr. Spectr. Rel. Phenom. 5, 755 (1974), and [Ref. 3.2,
       p. 30]
3.198 G. Puester, K. Thimm: [3.9] p. 95
3.199 M. Salle, B. Vodar: C.R. Acad. Sci. Paris, 230, 380 (1950)
3.200 R.P. Madden, D.L. Ederer, K. Codling: Appl. Opt. 6, 31 (1967); R.P. Madden,
       D. Ederer: J. Opt. Soc. Am. 62, 722 (1972)
3.201 H. Sugawara, T. Sagawa: [Ref. 3.3, p. 790]
3.202 S.A. Flodström, R.Z. Bachrach: Rev. Sci. Instrum. 47, 1464 (1976)
3.203 L.G. Parratt: Rev. Sci. Instrum. 30, 297 (1959)
3.204 a) B.M. Kincaid: SSRP Report No. 75/03 Stanford (1975); b) G. Materlik:
       Thesis, Universität Dortmund (1975)
3.205 Y. Cauchois, C. Bonnelle and G. Missoni: Comptes Rendus 257, 409 (1963);
       R. Barchewitz, M. Montel, C. Bonnelle: Comptes Rendus 264, 363 (1967)
3.206 U. Bonse, G. Materlik, W. Schröder: J. Appl. Cryst. 9, 223 (1976)
3.207 W. Ehrenberg, H. Mark: Z. Physik 42, 807 (1927)
3.208 G. Materlik: private communication
3.209 J.B. Hastings, B. Kincaid, P. Eisenberger: [3.9] p. 167
3.210 a) J. Bordas, I.H. Munro, R. Pettifer, G. Worgan: unpublished; b) U. Bonse,
       M. Hart: Appl. Phys. Lett. 7, 238 (1965)
3.211 G. Materlik, V.O. Kostroun: to be published in Rev. Sci. Instrum. (1979)
3.212 J. Witz: Acta. Cryst. A 25, 30 (1969)
3.213 G. Rosenbaum, K.C. Holmes, J. Witz: Nature 230, 434 (1971); J. Barrington
       Leigh, G. Rosenbaum: J. Appl. Crystallogr. 7, 117 (1974)
3.214 J. Barrington Leigh, C. Rosenbaum: Ann. Rev. Biophysics and Bioengineering
       5, 239 (1976)
3.215 N.G. Webb, S. Samson, R.M. Stroud, R.C. Gamble, J.D. Baldeschwieler: Rev.
       Sci. Instrum. 47, 836 (1976)
3.216 N.G. Webb: Rev. Sci. Instrum. 47, 545 (1976)
3.217 N.G. Webb, S. Samson, R.M. Stroud, R.C. Gamble, J.D. Baldeschwieler: J.
       Appl. Cryst. 10, 104 (1977)
3.218 J.C. Haselgrove, A.R. Faruqui, H.E. Huxley, U.W. Arndt: J. Phys. E 10, 1035
       (1977)
3.219 M.E. Van Hoosier, J.D.F. Bartoe, G.E. Brueckner, N.P. Patterson, R. Tousey:
       Appl. Opt. 16, 887 (1977)
3.220 T. Sasaki, T. Oda, H. Sugawara: Appl. Opt. 16, 3115 (1977)
3.221 T.A. Carlson: Phys. Rev. 156, 142 (1967)
3.222 J.A.R. Samson, G.N. Haddad: J. Opt. Soc. Am. 64, 47 (1974)
3.223 W. Gudat, W. Lenth, C. Kunz: [Ref. 3.3, p. 725] and W. Lenth, Thesis,
       University of Hamburg (1975)
3.224 a) R.P. Madden: Calibration methods in the UV and x-region of the Spectrum,
       Intern Symp. Munich (1968); b) A.P. Lukirskii, E.P. Savinov, I.A. Brytov,
       Yu.F. Shepelev: Bull. Acad. Sci, USSR, Phys. Ser. 28, 774 (1964); c) W. Pong:
       J. Apply. Phys. 40, 1733 (1969); d) P.H. Metzger: J. Phys. Chem. Solids
       26, 1879 (1965)
3.225 B. Saloman, D.L. Ederer: Appl. Opt. 14, 1029 (1975)
3.226 L.R. Canfield, R.G. Johnston, R.P. Madden: Appl. Opt. 12, 1611 (1973)
3.227 J.G. Timothy, L.B. Lapson: Appl. Opt, 13, 1417 (1974)
3.228 L.B. Lapson, J.G. Timothy: Appl. Opt. 15, 1218 (1976)
3.229 M.A. Rumsh, A.P. Lukirskii, V.N. Shchemelev: Dokl. Akad. Nauk, SSSR, 135,
       55 (1959); Sov. Phys. Dokl. 5, 1231 (1960)
3.230 W. Gudat: unpublished results
3.231 see e.g. Acta Electronica Vol. 14, No. 2 (1971); microchannel plates, Valvo
       Technical Note 760928 (1976); B.R. Sandel, A. Lyle Broadfoot, D.E. Shemansky:
       Appl. Opt. 16, 1435 (1977); P.J.K. Langendam, M.J. Van der Wiel: J. Phys.
       E 10, 870 (1977)
3.232 J.L. Wiza, P.R. Henkel, R.L. Roy: Rev. Sci. Instrum. 48, 9 (1977)
3.233 see e.g. the Handbooks and Application Notes of the Companies RCA, EMR, EMI,
       Johnston, Bendix, Galileo Electro Optics
3.233a W. Schnell, G. Wiech: Microchimica Acta (Wien), Suppl. 7, 323 (1977)

166

3.234 J.K. Branch, D.W.O. Heddle, M.J.H. Mogridge: J. Phys. E 4, 9 (1971)
3.235 J.A.R. Samson, G.N. Haddad: J. Opt. Soc. Am. 64, 1346 (1974); U. Backhaus: Thesis, Universität Hamburg (1974)
3.236 R.G. Johnston, R.P. Madden: Appl. Opt. 4, 1574 (1965)
3.237 A.L. Robinson: Science 199, 39 (1978)
3.238 H. Wolf: J. Vac. Sci. Technology 12, 983 (1975)
3.239 S. Flügge (ed.): *Handbuch der Physik*, Vol. XLV (Springer, Berlin, Göttingen, Heidelberg 1958)
3.240 W.J. Price: *Nuclear Radiation Detectors*, 2nd ed. (McGraw Hill, New York 1964)
3.241 J.B. Birks: *The Theory and Practice of Scintillation Counting* (Pergamon Press 1964)
3.242 N.A. Dyson: *X-Rays in Atomic and Nuclear Physics* (Longman Group Ltd., London 1973)
2.243 P.W. Nicholson: *Nuclear Electronics* (John Wiley and Sons, London, Sidney, Toronto 1974)
3.244 G.S. Brown: [Ref. 3.5, pp. 10-17]
3.245 R. Fourme, R. Kahn: [Ref. 3.5, pp. 10-83]
3.246 P. Rabe, G. Tolkiehn, A. Werner: private communication
3.247 F. Cappellani, G. Restelli: In *Semiconductor Detectors*, ed. by G. Bertolini and A. Coche (North Holland, Amsterdam 1968) p. 187
3.248 C. Cork et al: J. Appl. Crystallogr. 7, 319 (1973); J. Appl. Crystallogr. 7, 319 (1973)
3.249 C.J. Borkowski, M.K. Kopp: Rev. Sci. Instrum. 46, 951 (1975); 39, 1515 (1968)
3.250 A. Gabriel, Y. Dupont: Rev. Sci. Instrum. 43, 1600 (1972)
3.251 G. Charpak, Z. Hajduk, A. Jeavous, R. Stubbs, R. Kahn: Nucl. Instr. Methods 122, 307 (1974)
3.252 V. Perez Mendez, S.I. Parker: IEEE Trans. NS-21 45 (1945)
3.253 J.-I. Chikawa, I. Fujimoto: Appl. Phys. Lett. 13, 387 (1968)
3.254 T.C. Minor, J.R. Milch, G.T. Reynolds: J. Appl. Crystallogr. 7, 323 (1974)
3.255 S. Kozaki, M. Hashizume, K. Kohra: Jap. J. Appl. Phys. 11, 1514 (1972)
3.256 U.W. Arndt, D.J. Gilmore, S.M. Boutle: Adv. Electronics Electron Phys. 33B, 1069 (1972)
3.257 Synchrotron Radiation: A Bibliography, ed. by G.V. Marr, I.H. Munro and J.C.C. Sharp, Daresburg (1972) and (1974)
3.258 R. Bruhn, B. Sonntag, H.W. Wolff: [Ref. 3.109, Vol. 1, p. 20]
3.259 B. Sonntag: [Ref. 3.109, Vol. 1, p. 9]
3.260 N. Kosuch, E. Tegeler, G. Wiech, A. Faessler: [3.9]; A. Faessler: [Ref. 3.3, p. 801]
3.261 see e.g. G. Zimmerer: [3.6]
3.262 R. Lopez-Delgado, A. Tramer, J.H. Munro: Chem. Phys. 5, 72 (1974); O. Benoist d'Azy, R. Lopez-Delgado, A. Tramer: Chem. Phys. 9, 327 (1975); N. Schwentner: [3.6]
3.263 R. Haensel, U. Hahn, N. Schwentner: [Ref. 3.8, p. 239] and R. Haensel, U. Hahn, N. Schwentner, B. Jordan, G. Zimmerer, [Ref. 3.8, p. 250]
3.264 W. Lohr, H.W. Jochims, H. Baumgärtel: Ber. Bunsenges Phys. Chem. 79, 901 (1975); D. Reinke, R. Kraessig, H. Baumgärtel: Z. Naturforsch. 28a, 1021 (1973)
3.265 H. Hertz, H.W. Jochims, H. Schenk, W. Sroka: Chem. Phys. Lett. 29, 572 (1974)
3.266 Made by Physical Electronics Industries
3.267 R.Z. Bachrach, F.C. Brown, S.B.M. Hagström: J. Vac. Sci. Technol. 12, 309 (1975); R.Z. Bachrach, S.B.M. Hagström, F.C. Brown: [Ref. 3.3, p. 795]
3.268 P.J. Key: Metrologia 6, 97 (1970)
3.269 E. Pitz: Appl. Opt. 8, 255 (1969)
3.270 D. Einfeld, D. Stuck, B. Wende: p. 114 of [Ref. 3.109, p. 114]; see also DESY SR 77/03
3.271 R.P. Haelbich, C. Kunz: [Ref. 3.8, p. 344]
3.272 S. Aoki, J. Ichihara, S. Kikuta: Jap. J. Appl. Phys. 11, 1857 (1972)
3.273 G. Schmahl, D. Rudolph, B. Niemann: [Ref. 3.109, Vol. 3, p. 40]
3.274 E. Spiller, R. Feder, J. Topalian, D. Eastman, W. Gudat, D. Sayre: Science 191, 1172 (1976)

3.275 R. Feder, E. Spiller, J. Topalian, A.N. Broers, W. Gudat, B.J. Panessa, Z.A. Zadunaisky, J. Sedat: Science 197, 259 (1977)
3.275a H.M. Smith, ed. Holographic Recording Materials, Topics in Applied Physics Vol. 20 (Springer, Berlin, Heidelberg, New York 1977)
3.276 J.C. Philips, A. Wlodawer, M.M. Yevitz, K.O. Hodgson: Proc. Nat. Acad. Sci. USA 73, 128 (1976)
3.277 A. Harmsen, R. Lebermann, G.E. Schulz: J. Mol. Biol. 104, 311 (1976)
3.278 K.C. Holmes: [Ref. 3.3, p. 809]
3.279 Zvi Kam: Macromolecules, in press
3.280 S.L. Ruby: J. Physique C6, Suppl. 12, 35, CG-209 (1974)
3.281 R.L. Cooper. G.L. Miller, K.W. West: Nuclear Resonance Excited by Synchrotron Radiation (Activity Report, SSRP, Stanford, Jan-June 1977) p. VII-43
3.282 E. Gerdau, M. Mueller, R. Rüffer, W. Trautsch, H. Winkler: [Ref. 38, p. 297]
3.283 B. Buras, J. Staun Olsen, L. Gerward: Nucl. Instrum. Methods 135, 193 (1976)
3.284 J. Bordas, M. Glazer, I.H. Munro: Nature 262, 541 (1976)
3.285 B. Buras, J. Staun Olsen, L. Gerward, G. Will, E. Hinze: J. Appl. Crystallogr. 10, 431 (1977)
3.285a U. Bonse, G. Materlik: Z. Physik B24, 189 (1976) and U. Bonse and W. Graeff in: X-Ray Optics, H.-J. Queisser ed., Topics in Applied Physics Vol. 22 (Springer, Berlin, Heidelberg, New York 1977) p. 93
3.286 U. Bonse, G. Materlik: in Anomalous Scattering, ed. by S. Ramaseshan, S.C. Abrahams (Mungsgaard, Kopenhagen 1975) p. 107
3.287 D.E. Sayers, F.W. Lytle, E.A. Stern: In Advances in X-ray Analysis, ed. by B.L. Henke, J.B. Newkirk and G.R. Mallett, Vol. 13, (Plenum Press, New York 1970) p. 248; D.E. Sayers, E.A. Stern, F.W. Lytle: Phys. Rev. Lett. 27, 1204 (1971); E.A. Stern: Phys. Rev. B10, 3027 (1974)
3.288 B.M. Kincaid, P. Eisenberger, K.O. Hodgson, S. Doniach: Proc. Natl. Acad. Sci. (USA) 72, 2340 (1975)
3.289 T. Tuomi, K. Naukkarinen, E. Laurila, P. Rabe: Acta Polytechnica Scand. Ph-100 (1974); T. Tuomi, K. Naukkarinen, P. Rabe: Phys. Stat. Sol (a) 25, 93 (1974)
3.290 B.K. Tanner, M. Safa, D. Midgley: J. Applied Cryst. 10, 91 (1977)
3.291 J. Bordas, A.M. Glazer, H. Hauser: Phil. Mag. 32, 471 (1975)
3.292 M. Hart: J. Appl. Cryst. 8, 436 (1975)
3.293 W. Hartmann, G. Markewitz, M. Rettenmaier, H.J. Queisser: Appl. Phys. Lett. 27, 308 (1975)
3.294 J.A. Golovchenko, B.W. Batterman, W.L. Brown: Phys. Rev. B10, 4239 (1974)
3.295 S. Kjaer Andersen, J.A. Golovchenko, G. Mair: Phys. Rev. Lett. 37, 1141 (1976)
3.296 J.A. Golovchenko, E. Uggerhøj, G. Mair, S.K. Andersen: [Ref. 3.8, p. 291]
3.297 N. Kosuch, E. Tegeler, G. Wiech, A. Faessler: Chem. Phys. Lett. 47, 96 (1977)
3.298 N. Kosuch, E. Tegeler, G. Wiech, A. Faessler: DESY SR-77/19 and [3.9] p. 113
3.299 C.J. Sparks, Jr., S. Raman, H.L. Yakel. R.V. Gentry, M.O. Krause: Phys. Rev. Lett. 38, 205 (1977)
3.300 M. Cooper, R. Holt, P. Pattison, K.R. Lea: Commun. Phys. 1, 159 (1976)
3.301 C.J. Sparks: Phys. Rev. Lett. 33, 262 (1974)
3.302 P. Eisenberger, P.M. Platzman, H. Winick: Phys. Rev. Lett. 36, 623 (1976)
3.303 P. Eisenberger, P.M. Platzman, H. Winick: Phys. Rev. B13, 2377 (1976)
3.304 I. Lindau, P. Pianetta, S. Doniach, W.E. Spicer: Nature 250, 214 (1974)
3.305 I. Lindau. P. Pianetta, K.Y. Yu, W.E. Spicer: Phys. Rev. B13, 492 (1976)
3.306 K. Siegbahn: J. Electron Spectr. 5, 3 (1974)

4. Theoretical Aspects of Inner-Level Spectroscopy

A. Kotani and Y. Toyozawa

With 9 Figures

Theoretical study of inner-level spectroscopy has recently made remarkable progress, stimulated by new and precise information obtained from experimental observations with the use of synchrotron radiation. Inner core electrons, being initial (final) states of photoabsorption (emission), are *well localized* and their properties are already known, as those inherent in each atom, irrespective of the chemical surroundings. On the other hand, the states of the *outer* electrons as final (initial) states of photoabsorption (emission) are of more interest since they are related with the interatomic bond, whose nature is different for different materials.

Inner-level spectroscopy provides us with information on outer electrons projected on the symmetry space which depends on the symmetry of the chosen core state through the optical selection rule *(selection of symmetry)*. For compound systems the spectral contribution from each atomic species can be almost singled out by tuning to its own excitation energy *(selection of position)*. The difference in the outer electron states of an atom in various aggregate systems such as molecules and solids, as regards the spacial extension and the nature of bond, can be known by comparing the spectra of these materials originating from a *common* core state.

Another aspect of particular importance in inner-level spectroscopy is the *final state interaction* due to the core hole left behind, which is coupled not only with the excited electron but also with the surrounding medium. This interaction gives rise to *simultaneous excitations* of various elementary excitations, and electronic and atomic *relaxations*, which are of interest as dynamical many body problems.

In this chapter we describe theoretical aspects of inner-level spectroscopy, with emphasis on the basic understanding of the underlying physics. The similarities and differences in the spectra of atoms, molecules and solids are discussed in connection with their single-electron properties. More space will be devoted, however, to descriptions of the spectroscopy in solids including many body effects, local versus band characters and relaxation processes, due partly to the limitation of space and partly to the interest of the authors. For readers who want to know more details about the spectra in atoms and molecules, we refer to review articles by FANO and COOPER [4.1], ROBIN [4.2] and others [4.3,4].

In Sect. 4.1, we give briefly definitions of various quantities, basic concepts and relations in radiative processes. Sect. 4.2 is devoted to the independent particle description of oscillator strength. A unified treatment of inner core absorption spectra is introduced for atoms, molecules and condensed states based on the orthogonalization and the cancellation theorems. In Sect. 4.3 the electron-hole interaction in insulating solids is discussed as a simplest example of final state interaction. In Sect. 4.4, we deal with configuration interactions giving rise to Auger process and Fano effect. In Sects. 4.5 and 6, many-body final state interactions are discussed in insulators and metals, respectively. The former includes simultaneous excitations of phonons and plasmons accompanying photoabsorption of exciton, and relaxations of the exciton-phonon system, while the latter is devoted to the Fermi edge singularity in simple metals. In Sect. 4.7, final state interactions between the core hole and incomplete d- and f- shells are considered. In Sect. 4.8, X-ray scattering is surveyed. Finally, topics of recent and future interest are summarized in Sect. 4.9.

4.1 Basic Concepts and Relations in Radiative Processes

4.1.1 Polarizability and Dielectric Function

Let us consider a *homogeneous* material system described by Hamiltonian H. When an external electric field

$$\underline{E}^{(0)}(\underline{r},t) = \underline{E}^{(0)}(\underline{k},\omega)e^{i(\underline{k}\cdot\underline{r}-\omega t)} + c.c.$$

is applied to the system, a polarization

$$\underline{P}(\underline{r},t) = \bar{\underline{P}}(\underline{k},\omega)e^{i(\underline{k}\cdot\underline{r}-\omega t)} + c.c.$$

is induced due to the perturbation $H'(t) = - \int_V \underline{P}(\underline{r}) \cdot \underline{E}^{(0)}(\underline{r},t)d\underline{r}$, where $\underline{P}(\underline{r})$ is a polarization operator. In this chapter, we choose the volume of the system V as unity for simplicity of description. *Polarization tensor* $\underline{\underline{\alpha}}(\underline{k},\omega)$ is defined by the relation

$$\bar{\underline{P}}(\underline{k},\omega) = \underline{\underline{\alpha}}(\underline{k},\omega)\underline{E}^{(0)}(\underline{k},\omega) \quad , \tag{4.1}$$

and expressed with the use of the *Kubo formula* [4.5] as

$$\alpha_{mn}(\underline{k},\omega) = \int_{-\infty}^{\infty} dt \; (\tfrac{i}{*\hbar})\theta(t)e^{i\omega t-st}<[P_{km}(t),P_{-kn}]> \quad , \quad (s \to +0) \tag{4.2}$$

where $\underline{P}_k(t)$ is given by $\underline{P}_k(t) \equiv \exp(iHt/\hbar)\underline{P}_k\exp(-iHt/\hbar)$ with $\underline{P}_k \equiv \int \underline{P}(\underline{r})\exp(-i\underline{k}\cdot\underline{r})d\underline{r}$, $\theta(t)$ is the step function, and $<\cdots>$ means statistical average.

Dielectric function $\underline{\underline{\epsilon}}(\underline{k},\omega)$ is defined by

$$\underline{D}(\underline{k},\omega) = \underline{E}(\underline{k},\omega) + 4\pi\underline{P}(\underline{k},\omega) = \underline{\underline{\epsilon}}(\underline{k},\omega)\underline{E}(\underline{k},\omega) \quad , \tag{4.3}$$

where $\underline{D}$ and $\underline{E}$ are, respectively, electric displacement and electric field *within the matter*. In order to relate $\underline{\underline{\epsilon}}$ with $\underline{\underline{\alpha}}$, one must know the relation between $\underline{E}^{(0)}$ and $\underline{E} \equiv \underline{E}^{(0)} + \underline{E}^{(i)}$, $\underline{E}^{(i)}$ being an induced electric field. Two methods of relating them [4.6] are given in the following.

a) *Self-consistent field method*

In the calculation of (4.2), the Hamiltonian H is replaced by an *effective* Hamiltonian H_{eff} which is obtained by removing from H a *long range part* of charged particle interactions in the system. The long range part is considered to have been included as the source of the self-consistent field $\underline{E}^{(i)}$, so that $\underline{E}^{(0)}$ in H' should be replaced by $\underline{E}$, and the polarizability $\underline{\underline{\alpha}}_{eff}$ thus obtained relates $\bar{\underline{P}}$ with $\underline{E}$ by $\bar{\underline{P}} = \underline{\underline{\alpha}}_{eff} \underline{E}$. Hence, we obtain

$$\underline{\underline{\epsilon}}(\underline{k},\omega) - 1 = 4\pi\underline{\underline{\alpha}}_{eff}(\underline{k},\omega) \quad . \tag{4.4}$$

b) *Direct method for longitudinal part*

As far as the *longitudinal* part of response to a *longitudinal* external field $E_{||}^{(0)}$ is concerned, we have

$$1 - \frac{1}{\epsilon_{||}(\underline{k},\omega)} = 4\pi\alpha_{||}(\underline{k},\omega) \quad , \tag{4.5}$$

where $\alpha_{||}$ is given by (4.2) with the *true* Hamiltonian H. This is easily obtained from (4.3) and relations $\bar{P}_{||} = \alpha_{||} E_{||}^{(0)}$ and $E_{||}^{(0)} = D_{||}$.

In general, the method a) is more convenient than the method b), since the long range part which is sometimes intractable has been removed from the Hamiltonian. But the determination of H_{eff} is not always possible and, if possible, it contains inevitably some approximation.

---

*$\hbar$ = h/2π(normalized Planck's constant)

4.1.2 Absorption Coefficient and Oscillator Strength

A propagation of radiation in a polarizable medium is described by Maxwell's equations

$$\underline{\nabla} \times \underline{E} = - \frac{1}{c} \frac{\partial \underline{H}}{\partial t} \quad , \quad \underline{\nabla} \times \underline{H} = \frac{\underline{\underline{\epsilon}}}{c} \frac{\partial \underline{E}}{\partial t} \quad . \tag{4.6}$$

For the moment, the dielectric function $\underline{\underline{\epsilon}}(\underline{k},\omega)$ of the medium will be approximated by

$$\underline{\underline{\epsilon}}(\omega) \equiv \lim_{\underline{k} \to 0} \underline{\underline{\epsilon}}(\underline{k},\omega) \quad ,$$

since the wave length of the radiation is far longer than the interatomic or interelectronic spacing of the medium. From now on we confine ourselves to the isotropic or cubic medium, where $\underline{\underline{\epsilon}}$ and $\underline{\underline{\alpha}}$ are scalar quantities ϵ and α. From (4.6) a plane wave solution of $\underline{E}$ propagating in the x-direction is written as $\underline{E} = \underline{E}_0 \exp[-i\omega\{t - [(\tilde{n}(\omega)/c)x\}]$, where $\tilde{n}(\omega) \equiv n(\omega) + i\kappa(\omega) = [\epsilon(\omega)]^{1/2}$, n and κ being the *index of refraction* and the *extinction coefficient*, respectively. The *absorption coefficient* $K(\omega)$ is defined by $K(\omega) = -(dI/dx)/I$ in terms of the energy flow I of the radiation. Since $I \propto |\underline{E} \times \underline{H}| \propto \exp(-2\omega\kappa x/c)$, we obtain

$$K(\omega) = \frac{2\omega}{c} \kappa(\omega) = \frac{\omega}{cn} \text{Im} \{\epsilon(\omega)\} \quad . \tag{4.7}$$

The quantity $\epsilon(\omega)$ is expressed with the use of (4.2) and (4.4) (i.e. by the method a)) as follows

$$\epsilon(\omega) - 1 = 4\pi A \underset{\mu}{v} \sum_{\nu} \left[\frac{(\underline{P}_0)_{\mu\nu}(\underline{P}_0)_{\nu\mu}}{\hbar(\omega_{\nu\mu}-\omega-is)} + \frac{(\underline{P}_0)_{\mu\nu}(\underline{P}_0)_{\nu\mu}}{\hbar(\omega_{\nu\mu}+\omega+is)} \right]_{s \to +0} \quad , \tag{4.8}$$

where $\omega_{\nu\mu} \equiv (\epsilon_\nu - \epsilon_\mu)/\hbar$, ϵ_μ being the energy eigenvalue of H_{eff}, $A\underset{\mu}{v}$ means the statistical average for the states μ. For the system consisting of *electrons* of density n_e, $\underline{P}_0$ is given by

$$\underline{P}_0 \equiv e \sum_{j=1}^{n_e} \underline{r}_j \quad .$$

The *oscillator strength* is defined by

$$f_{\mu\nu} \equiv \frac{2m\omega_{\nu\mu}}{\hbar e^2} |(\underline{P}_0 \cdot \hat{n})_{\nu\mu}|^2 \quad , \tag{4.9}$$

where $\hat{n}$ is the unit vector in the direction of E. By using (4.7), (4.8) and (4.9), the absorption coefficient is expressed as

$$K(\omega) = \frac{2\pi^2 e^2}{mcn} \, Av \sum_{\mu} \sum_{\nu} f_{\mu\nu} \delta(\omega - \omega_{\nu\mu}) \quad . \tag{4.10}$$

4.1.3 Dispersion Relations and Sum Rules

The polarizability $\alpha(\underline{k},\omega)$ is analytic in the upper half of the complex ω-plane, because of the factor $\theta(t)\exp(i\omega t)$ in the integrand of (4.2). The appearance of $\theta(t)$ means that the response $\bar{P}(t)$ does not precede the external perturbation $\underline{E}^{(0)}(t)$ namely it represents the *causality*. With use of the analyticity of α and a generally acceptable condition $\lim_{|\omega|\to\infty} \alpha = 0$, we obtain the identity

$$\alpha(\underline{k},\omega) = \frac{1}{i\pi} P \int_{-\infty}^{\infty} \frac{d\omega'}{\omega'-\omega} \alpha(\underline{k},\omega') \quad , \tag{4.11}$$

where P stands for the Cauchy principal value. By separating real and imaginary parts, we find

$$\text{Re}\,\{\,\alpha(\underline{k},\omega)\} = \frac{1}{\pi} P \int_{-\infty}^{\infty} \frac{d\omega'}{\omega'-\omega} \, \text{Im}\,\{\,\alpha(\underline{k},\omega')\} \quad ,$$

$$\text{Im}\,\{\,\alpha(\underline{k},\omega)\} = -\frac{1}{\pi} P \int_{-\infty}^{\infty} \frac{d\omega'}{\omega'-\omega} \, \text{Re}\,\{\,\alpha(\underline{k},\omega')\} \tag{4.12}$$

These are called *dispersion relations* or *Kramers-Kronig relations*. Of course, the same relations also hold for $\alpha_{eff}(\underline{k},\omega)$. Then, replacing $\alpha_{eff}(\omega)$ by $[\epsilon(\omega) - 1]/4\pi$ and using the relation $\epsilon(-\omega) = \epsilon^*(\omega)$ representing the reality of $\underline{D}$ and $\underline{E}$, we get the dispersion relations for $\epsilon(\omega)$

$$\text{Re}\,\{\epsilon(\omega)\} - 1 = \frac{2}{\pi} P \int_{0}^{\infty} \frac{\omega' d\omega'}{\omega'^2-\omega^2} \, \text{Im}\,\{\epsilon(\omega')\} \quad ,$$

$$\text{Im}\,\{\epsilon(\omega)\} = \frac{2\omega}{\pi} P \int_{0}^{\infty} \frac{d\omega'}{\omega^2-\omega'^2} \, \text{Re}\,\{\epsilon(\omega')\} \quad . \tag{4.13}$$

The dielectric function satisfies the following sum rule

$$\int_{0}^{\infty} \omega \text{Im}\,\{\epsilon(\omega)\} \, d\omega = \frac{\pi}{2} \omega_p^2 \quad , \tag{4.14}$$

where $\omega_p{}^2 \equiv 4\pi n_e e^2/m$. This is equivalent to the so-called *f-sum rule* (known as the *Thomas-Reiche-Kuhn sum rule* in the case of atoms)

$$\sum_\nu f_{\mu\nu} = n_e \tag{4.15}$$

because of (4.7) and (4.10). The relation (4.15) is nothing but the $\mu\mu$ - element of the operator identity

$$\sum_{j=1}^{n_e} [z_j, p_{zj}]/i\hbar = n_e \quad ,$$

as is easily seen from the equation of motion $p_{zj} = m\dot{z}_j = i[H, mz_j]/\hbar$ with its matrix element

$$(p_{zj})_{\nu\mu} = i(\epsilon_\nu - \epsilon_\mu) m (z_j)_{\nu\mu}/\hbar \tag{4.16}$$

by considering the definition (4.9) with $\hat{n}$ taken parallel to z-axis.

Other important sum rules [4.7] are given below.

$$\int_0^\infty \omega \, \text{Im} \left\{ \frac{1}{\epsilon(\omega)} \right\} d\omega = -\frac{\pi}{2} \omega_p{}^2 \quad , \tag{4.17}$$

$$\int_0^\infty [n(\omega) - 1] \, d\omega = 0 \quad , \tag{4.18}$$

$$\int_0^\infty \omega \, \kappa(\omega) \, d\omega = \frac{\pi}{4} \omega_p{}^2 \quad . \tag{4.19}$$

For large ω such that the distribution of oscillator strength

$$\text{Av} \sum_\mu \sum_\nu f_{\mu\nu} \, \delta (\omega - \omega_{\nu\mu}) \propto \omega \, \text{Im} \{\epsilon(\omega)\}$$

is almost exhausted, the first equation of (4.13) gives the asymptotic formula:

$$\lim_{\omega \to \infty} \epsilon(\omega) = 1 - \omega_p{}^2/\omega^2 \quad ,$$

which means that electrons behave as if they were free for electromagnetic waves
of sufficiently high frequency.

4.2 Distribution of Oscillator Strength

In order to get a basic understanding of core electron absorption spectra, the
distribution of oscillator strength is studied within the scope of the independent
electron approximation. We begin with a brief survey of atomic spectra, and then
present a unified treatment of the *gross contour* of *intra-atomic* origin and the
fine structures of *inter-atomic* origin in the spectra of *condensed* systems.

4.2.1 Absorption Spectra in Atoms

The Schrödinger equation for the single electron state $\phi_{n\ell m}(\underline{r}) = u_{n\ell}(r) Y_{\ell m}(\Omega_{\underline{r}})/r$
in an atom is given by

$$\left[-\frac{\hbar^2}{2m} \nabla^2 + V(r) \right] \phi_{n\ell m} = \varepsilon_{n\ell m} \phi_{n\ell m} \quad , \tag{4.20}$$

where $V(r)$ is a central potential. The principal quantum number n takes integral
values for bound states (i.e. Rydberg states) and continuous values for states above
the ionization threshold. The oscillator strength for optical transitions of an
electron in the $(n\ell)$ shell is expressed from (4.9) as

$$\sum_{n',\ell'=\ell\pm1} f_{n\ell,n'\ell'} = \sum_{n',\ell'=\ell\pm1,m'} \frac{2m(\varepsilon_{n'\ell'}-\varepsilon_{n\ell})}{\hbar^2} \underset{m}{Av} |<\phi_{n'\ell'm'}|(\underline{r}\cdot\hat{n})|\phi_{n\ell m}>|^2 \tag{4.21}$$

$$= \sum_{n',\ell'=\ell\pm1} \frac{2m(\varepsilon_{n'\ell'}-\varepsilon_{n\ell})}{3\hbar^2} \frac{\ell+\ell'+1}{2(2\ell+1)} [R(n\ell,n'\ell')]^2 \quad ,$$

where

$$R(n\ell,n'\ell') = \int_0^\infty r u_{n\ell}(r) u_{n'\ell'}(r) dr \quad . \tag{4.22}$$

The radial equation for $u_{n\ell}(r)$ is obtained from (4.20) in the form

$$\frac{d^2 u_{n\ell}}{dr^2} + \frac{2m}{\hbar^2} \left[\varepsilon_{n\ell} - V(r) - \frac{\ell(\ell+1)\hbar^2}{2mr^2} \right] u_{n\ell} = 0 \tag{4.23}$$

When the photon energy $\hbar\omega$ is higher than the ionization threshold $\hbar\omega_t$, the oscillator strength per unit energy, $f(n\ell \to \varepsilon, \ell \pm 1)$, for the transition $(n\ell) \to (\varepsilon \equiv \hbar(\omega - \omega_t), \ell \pm 1)$ has been calculated by COOPER and his co-workers [4.1,8,9] for various atoms and several sets of $(n\ell)$ by using appropriate model potential $V(r)$. Main features of the calculated $f(n\ell \to \varepsilon, \ell \pm 1)$, which are also in qualitative agreement with experimental data [4.1], are summarized in the following.

(i) *Wide extension of spectrum*: The energy extension of *inner shell* absorption spectrum is generally wider than that of *outer shell* electrons, when measured from their respective thresholds $\hbar\omega_t$.

(ii) *Giant and subgiant bands*: The oscillator strength $f(n\ell \to \varepsilon, \ell + 1)$ is generally larger than $f(n\ell \to \varepsilon, \ell - 1)$ (by one or two orders of magnitude in typical cases). The former absorption band will be called the *giant band* and the latter the *subgiant band*, as will turn out to be convenient in analyzing the spectra of condensed systems in comparison with those of constituent atoms.

(iii) *Delayed edge*: Sometimes, the maximum point of the giant band appears much beyond $\hbar\omega_t$, as is most remarkable for the d-shell excitation. This is attributed to the effect of the *centrifugal potential* $\ell'(\ell' + 1)\hbar^2/(2mr^2)$ appearing in (4.23): When $\ell' = \ell + 1 \geq 2$, the radial function $u_{\varepsilon\ell'}$ with the kinetic energy $\varepsilon \simeq 0$ is kept far out by the strong centrifugal force, so that its small overlap with $u_{n\ell}$ results in small $R(n\ell,\varepsilon\ell')$. As $\hbar\omega$ increases, $u_{\varepsilon\ell'}$ penetrates gradually into the core, resulting in the *gradual* rise in f. With further increase in $\hbar\omega$, f passes a maximum and decreases gradually due to the cancellation of R coming from the ·change of phase relation between $u_{n\ell}$ and $u_{\varepsilon\ell'}$.

(iv) *Resonance near threshold*: As the atomic number Z increases, the potential $V(r)$ becomes deeper, so that the giant band shifts towards the threshold. For sufficiently large Z, a fairly deep potential well appears inside the small but broad centrifugal barrier. As ε is increased across the barrier height, $u_{\varepsilon\ell'}$ penetrates rapidly into the core due to the resonance by the potential well, giving a *sharp* rise in f slightly above the threshold.

(v) *Spectral dip due to sign-reversal of R(n\ell,\varepsilon\ell')*: When $u_{n\ell}$ has nodes, i.e. when $n > \ell + 1$, $R(n\ell,\varepsilon\ell')$ changes its sign above the threshold. Sometimes, this is reflected in the spectrum as a conspicuous dip following a peak near the threshold.

When $\hbar\omega$ is lower than $\hbar\omega_t$, there appear discrete absorption lines of Rydberg series. In the limit of $n' \to \infty$, the oscillator strength per energy is connected continuously with that of the ionization continuum. Except for some atoms with small Z, the total oscillator strength of the discrete lines is far smaller than that above $\hbar\omega_t$.

4.2.2 A Unified Picture for Spectra in Atoms, Molecules and Solids

We give here an alternative approach, the pseudopotential method, which facilitates a unified treatment [4.10] of core electron absorption spectra in atoms, molecules and solids. The clue is the *cancellation of oscillator strength*, a counterpart of the well-known *cancellation of potential*. In place of (4.20), we now write the Schrödinger equation for a system consisting of more than one atom as

$$\left[- \frac{\hbar^2}{2m} \nabla^2 + v(\underline{r}) \right] \phi_\alpha = \varepsilon_\alpha \, \phi_\alpha \quad , \tag{4.24}$$

Here v(r) is approximately given by the superposition of potentials V(r) of all the constituent atoms. Since any *empty* state ϕ_α must be orthogonal to the filled *core* states $\phi_{\nu i}$, it is written in the form

$$\phi_\alpha(\underline{r}) = \chi_\alpha(\underline{r}) - \sum_{\nu,i}' <\phi_{\nu i}|\chi_\alpha>\phi_{\nu i}(\underline{r}) \quad , \tag{4.25}$$

where Σ' extends over all the *core* states $\nu \equiv (n\ell m)$ of all the atoms i. Inserting (4.25) into (4.24) and imposing a condition that $\chi_\alpha(\underline{r})$ be as smooth as possible, we obtain the so-called *pseudo potential* v_p, to be used in the effective Hamiltonian for χ_α, as the following integral operator

$$v_p(\underline{r},\underline{r}') = v(\underline{r}')[\delta(\underline{r} - \underline{r}') - \sum_{\nu,i}' \phi_{\nu i}^*(\underline{r}')\phi_{\nu i}(\underline{r})] \quad , \tag{4.26}$$

where the second term in the square bracket arises from the orthogonalization. As seen from the fact that v_p would vanish if $\phi_{\nu i}$'s formed a complete set, v_p is generally much weaker than the true potential v. This is the cancellation of potential, which is well-known in solid state physics [4.11]. When ϕ_α is an excited state of the photoabsorption, the effective Schrödinger equation for χ_α is expressed as

$$\left(- \frac{\hbar^2}{2m} \nabla^2 + v_p + v_c \right) \chi_\alpha = \varepsilon_\alpha \chi_\alpha \quad , \tag{4.27}$$

where we have added the Coulomb potential v_c due to the core hole left behind. In the absorption spectra the orthogonalization gives another important effect, the cancellation of oscillator strength.

a) Cancellation of Oscillator Strength - Giant and Subgiant Bands

Insofar as we consider the *gross structure* of the absorption spectra in the energy region several tens of eV above the threshold, the effects of v_p and v_c are not of primary importance. Disregarding them, we obtain the solution of (4.27) as plane waves $\chi_{\underline{k}}(\underline{r}) = \exp(i\underline{k} \cdot \underline{r})$ with energies $\varepsilon_{\underline{k}} = \hbar^2 k^2/(2m)$. With the use of (4.21), the oscillator strength per unit energy is given by

$$f(n\ell \rightarrow k) = \frac{\sqrt{2}m^{5/2}}{3\pi^2\hbar^5}\hbar\omega(\hbar\omega - \hbar\omega_t)^{1/2} \sum_{\ell'=\ell\pm 1} \frac{\ell+\ell'+1}{2(2\ell+1)}[R(n\ell,k\ell')]^2 \quad , \tag{4.28}$$

where

$$R(n\ell,k\ell') \equiv \int_0^\infty ru_{n\ell}(r)v_{k\ell'}(r)dr \quad , \tag{4.29}$$

$$v_{k\ell'}(r) \equiv w_{k\ell'}(r) - \sum_{n'}' <u_{n'\ell}|w_{k\ell'}>u_{n'\ell'}(r) \quad , \tag{4.30}$$

$$w_{k\ell'}(r) = \sqrt{4\pi}\, i^{\ell'} j_{\ell'}(kr) \quad , \tag{4.31}$$

$j_{\ell'}(kr)$ being the spherical Bessel function. The overlap between inner shell wave functions of different atoms has been neglected. Within this treatment, $f(n\ell \rightarrow k)$ is common to isolated atom, molecule and solid, and most of the features (i) $\sim$ (v) for atomic spectra mentioned before are reinterpreted qualitatively from the present viewpoint.

The oscillator strengths for the *unorthogonalized* spherical wave states, i.e. $f(n\ell \rightarrow k, \ell \pm 1)$ obtained by putting $v_{k\ell'} = w_{k\ell'}$, are shown schematically by thin and thick solid curves in Fig. 4.1. The maximum appears at higher energy for $\ell' = \ell + 1$ than for $\ell' = \ell - 1$, corresponding to the larger centrifugal force whose effect is incorporated into (4.31). The average extension of the spectra beyond $\hbar\omega_t$ is of the same order of magnitude as binding energy $|\varepsilon_{n\ell}|$ of the initial state, since $R(n\ell,k\ell')$ is appreciable for $v_{k\ell'}(r)$ with wave length $2\pi/k$ of the same order of magnitude as the orbital radius of the initial state. This explains the feature (i). Furthermore, if $n > \ell + 1$ nodes appear in the oscillator strength versus energy curves, feature (v).

The effect of the orthogonalization, represented by the second term of (4.30), gives rise to the cancellation of oscillator strength, a counterpart of the cancellation of potential. In general, the number of filled shells $(n'\ell')$ to which the final state $(k\ell')$ has to be orthogonalized, and hence the degree of cancellation of oscillator strength, is greater for $\ell' = \ell - 1$ than for $\ell' = \ell + 1$. For instance,

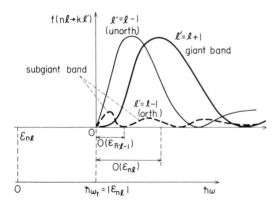

Fig. 4.1 Schematic behavior of oscillator strengths for transitions from the $(n\ell)$-shell to the $\ell' = \ell \mp 1$ components of plane wave states [4.10]

in the case of 4d-shell photoabsorption of I, Xe and Cs, one has four filled p-shells (2p - 5p). The orthogonalization with them reduces the oscillator strength f(4d → kp) by one or two orders of magnitude, while there is no filled f-shell for the (kf) states to be orthogonalized with. This explains the features (ii) and (iii), namely the appearance of the giant band with delayed edge and the subgiant band which is much smaller. Furthermore, the subgiant band is found to oscillate, as shown by the broken curve in Fig. 4.1, originating from that the orthogonalized spherical wave (4.30) has nodes near the atomic core. The separation between the first node of the subgiant band and $\hbar\omega_t$ is of the order of binding energy $|\varepsilon_{\overline{n}',\ell-1}|$ of the outermost occupied electron with ℓ - 1.

The characteristics as depicted in Fig. 4.1 can be seen most clearly in the spectra of (nd) core excitation in closed shell insulators such as rare gas solids and alkali halides [4.12-14].

b) Pseudo Potential and Energy Band Effect

Next we consider the effect of the pseudo potential v_p. For an isolated atom, where $v(\underline{r}) = V(r)$, the pseudo potential for the final state with ℓ' is given by

$$v_p^{(\ell')}(\underline{r},\underline{r}') = V(r')[\delta(\underline{r} - \underline{r}') - \sum_{n',m'}' \phi_{n'\ell'm'}^*(\underline{r}')\phi_{n'\ell'm'}(\underline{r})] \quad . \tag{4.32}$$

As we increase the atomic number 7, $v_p^{(\ell')}$ as well as V become deeper, until a new shell $(n''\ell')$ begins to be occupied. Consequently the peak of the giant band is expected to shift towards the threshold, since the effect of the centrifugal repulsion is reduced due to $v_p^{(\ell')}$. The 4d → εf giant band for the sequence of I, Xe and Cs (with Z = 53, 54 and 55, respectively) shows clearly this tendency [4.12-14], which is a typical example of the *resonance near threshold*, feature (iv). When a new shell is filled up, we have to include it under the summation in (4.32), so that $v_p^{(\ell')}$ becomes shallow again.

In molecules and solids, there appears, in addition to the shift of the giant band mentioned above (intra-atomic effect), another effect (inter-atomic) which comes from the pseudo-potential (4.26) instead of (4.32). The solution χ_α of the effective Schrödinger equation are now expressed as *molecular orbitals* for molecules and as *Bloch functions* with associated energy band structures for solids. This gives rise to fine structures, characteristic of the respective atomic arrangement, superimposed on the subgiant and giant bands. It is in this part of the spectra for which the comparison with band theory has been successful [4.15,16] and [Ref. 4.17, p. 57]. The energy scale of these fine structures are of the order of eV, in contrast to those of the giant (several tens of eV) and subgiant ($\sim$ 10 eV) bands. Considering the case of solids, we write the Bloch function and its energy as $\chi_{\mu k}$ and $\varepsilon_\mu(\underline{k})$, respectively, where μ is a band index and $\underline{k}$ stands for a momentum. The oscillator strength for the excitation of an electron in the $(n\ell)$ shell is written as

$$f(\omega) = \frac{2}{m\hbar\omega} \sum_{\mu\underline{k}}{}'' |p_{\mu\underline{k},n\ell}|^2 \, \delta(\hbar\omega - \varepsilon_\mu(\underline{k}) + \varepsilon_{n\ell}) \quad , \tag{4.33}$$

where

$$|p_{\mu\underline{k},n\ell}|^2 = \underset{m}{Av}|<\phi_{\mu\underline{k}}|i\hbar\underline{\hat{n}} \cdot \underline{\nabla}|\phi_{n\ell m}>|^2 \quad . \tag{4.34}$$

Here $\phi_{\mu\underline{k}}$ is given by (4.25) with χ_α replaced by $\chi_{\mu\underline{k}}$, and $\sum_{\mu\underline{k}}''$ extends over all *unoccupied* Bloch states.

If we disregard μ- and $\underline{k}$-dependences of $|p_{\mu\underline{k},n\ell}|^2$, (4.33) is proportional to the density of states of unoccupied bands, $\rho(\varepsilon) \equiv \sum_{\mu\underline{k}}'' \delta(\varepsilon - \varepsilon_\mu(\underline{k}))$, with $\varepsilon = \hbar\omega + \varepsilon_{n\ell}$. In the three dimensional $\underline{k}$-space, the stationary points of $\varepsilon_\mu(\underline{k})$ are classified into four types: minimum (expressed as M_0), saddle point which is minimum in two directions (M_1), saddle point which is minimum in one direction (M_2) and maximum (M_3). Corresponding to each stationary point, $\rho(\varepsilon)$ has the so-called *van Hove singularity* [4.18] represented by $(\varepsilon - \varepsilon_{M0})^{1/2}$, $-(\varepsilon_{M1} - \varepsilon)^{1/2}$, $-(\varepsilon - \varepsilon_{M2})^{1/2}$ and $(\varepsilon_{M3} - \varepsilon)^{1/2}$, respectively. Furthermore, in metals at zero temperature $\rho(\varepsilon)$ has a singularity called *Fermi edge* which is represented by the step function $\theta(\varepsilon - \varepsilon_F)$, since the states below the Fermi energy ε_F are occupied. It should be emphasized that the inner core absorption spectrum contains direct information on the density of states of unoccupied bands owint to the *negligible dispersion* of the inner core state; if the initial states are valence states $\phi_{v\underline{k}}$ with appreciable dispersion $\varepsilon_v(\underline{k})$, $f(\omega)$ reflects the joint density of states $\tilde{\rho}(\hbar\omega) \equiv \sum_{j,\underline{k}}'' \delta(\hbar\omega - \varepsilon_j(\underline{k}) + \varepsilon_v(\underline{k}))$.

The $(j,\underline{k})$-dependence of $p_{j\underline{k},n\ell}$, i.e. the matrix element effect, causes the deviation of $f(\omega)$ from $\rho(\varepsilon)$. In particular, it gives rise to a fine structure in the energy range of $\hbar\omega - \hbar\omega_t \gtrsim 100$ eV, as will be discussed in the next subsection.

Besides, by taking advantage of the matrix element effect, we can obtain the information on the symmetry of Bloch function $\phi_{\mu k}$, because the integral (4.34) selects that symmetry component of $\phi_{\mu \underline{k}}$ to which the transition is allowed from the core state $\phi_{n\ell}$ under consideration.

c) Effect of Coulomb Attraction

By the Coulomb potential v_c, an infinite number of bound states appear below the ionization threshold; they are Rydberg states in atoms and molecules, and excitons in insulating solids. Usually, v_c is smaller in solids because of dielectric screening. Details about the exciton will be mentioned in Sect. 4.4. In metals, such a bound state is not stable because v_c is screened by conduction electrons resulting in a short range potential. However, the dynamical process of the screening is reflected in absorption spectra as a singularity at the Fermi edge, which will be discussed in Sec. 4.7.

The fact that v_c is small especially in solids warrants our successive approximation described above - a) orthogonalized plane wave, b) pseudo potential, and then c) the Coulomb effect.

4.2.3 Extended X-Ray Absorption Fine Structure (EXAFS)

In the inner core absorption spectra of molecules and solids a series of oscillations with period of order 50 eV is observed up to 1000 eV above the threshold. This oscillation arises from the effect of potential $v(\underline{r})$ on the matrix element (4.34), and we call it the *extended X-ray absorption fine structure* (EXAFS); it is to be distinguished from the fine structure in the close neighborhood of the threshold due to the band density and exciton effects. The photoexcited electron with such a large kinetic energy is perturbed only weakly by the atomic potential, so that it is convenient to describe the final state as a linear combination of an unperturbed spherical wave outgoing from the excited atom and scattered waves from surrounding atoms [4.19]. Then the oscillation of the absorption spectra is interpreted as the interference between the *outgoing* and *scattered* waves at the position of the excited atom. Namely, when these waves are combined constructively there arises an absorption maximum and a minimum when they are combined destructively.

Let us consider the scattering of a spherical wave $h_\ell^{(1)}(kr)Y_{\ell m}(\Omega_r)$, which is outgoing from a central excited atom, by the neighboring atom at $\underline{R}_j$. Approximating the spherical wave by a plane wave with the same amplitude at $\underline{r} = \underline{R}_j$, we obtain the scattered wave in the form

$$h_\ell^{(1)}(kR_j)Y_{\ell m}(\Omega_{\underline{R}_j})S(\theta)\frac{e^{ik|\underline{r}-\underline{R}_j|}}{|\underline{r}-\underline{R}_j|} \quad ,$$

where $S(\theta)$ is the scattering amplitude. Thus, the wave function consisting of both the outgoing and scattered waves is given, in the neighborhood of $\underline{r} = 0$, by

$$h_\ell^{(1)}(kr)Y_{\ell m}(\Omega_{\underline{r}}) - i \frac{e^{i(2kR_j - \frac{\ell\pi}{2})}}{kR_j^2} \times Y_{\ell m}(\Omega_{\underline{R}_j})S(\pi)\exp(- ik \frac{\underline{R}_j \cdot \underline{r}}{R_j}) \quad , \tag{4.35}$$

where we have replaced $h_\ell^{(1)}(kR_j)$ by its asymptotic form and used the plane wave approximation for the scattered wave again. Substituting (4.35) into (4.34) (in place of $\phi_{\mu k}$) and assuming the excitation of the inner shell with $\ell = 0$, we can express $f(\omega)$ of (4.33) as

$$f(\omega) = f_0(\omega)[1 + g(k)] \quad , \tag{4.36}$$

where

$$g(k) = \frac{|S(\pi)|}{kR_j^2} \sin(2kR_j + \varphi) \quad , \tag{4.37}$$

$$S(\pi) = \sum_{\ell=0}^{\infty}(- 1)^\ell \frac{e^{2i\delta_\ell}-1}{2ik} \equiv |S(\pi)|e^{i\varphi} \quad , \tag{4.38}$$

and the average has been taken over the direction $\hat{n}$ by assuming unpolarized X-rays. Here $f_0(\omega)$ is the oscillator strength in the absence of the scattered wave and δ_ℓ is the phase shift due to the scattering. From (4.37) the period of EXAFS is found to be given by $\Delta k = \pi/R_j$, which corresponds to $\Delta(\hbar\omega) = (\pi\hbar/R_j)[2(\hbar\omega - \hbar\omega_t)/m]^{1/2}$ because of $\hbar(\omega - \omega_t) = \hbar^2 k^2/(2m)$. By assuming $R_j \sim 2.5\text{Å}$ (nearest neighbor distance for Cu), for instance, we obtain the typical values of the period $\Delta(\hbar\omega) \sim 50$ eV for $\hbar(\omega - \omega_t) \sim 100$ eV and $\Delta(\hbar\omega) \sim 150$ eV for $\hbar(\omega - \omega_t) \sim 1000$ eV. The expression of $g(k)$ can be improved by the following prescription: We introduce the phase shift δ_1' representing the effect of the potential of the excited atom itself, which is different from that of other atoms due to the existence of the core hole left behind Furthermore, we take account of the effect of inelastic scatterings as a damping factor $\exp(-\gamma R_j)$, and the effect of the atomic vibrations as the Debey-Waller factor $\exp(-2\sigma_j^2 k^2)$. Summing up the contribution from each constituent atom, we finally obtain

$$g(k) = \frac{1}{k} \sum_j \frac{N_j}{R_j^2} |S(\pi)|\sin\left(2kR_j + 2\delta_1' + \varphi\right) e^{-2\sigma_j^2 k^2} e^{-\gamma R_j} \quad , \tag{4.39}$$

where N_j is the number of atoms in the j th shell located at a distance R_j from the central atom.

Since the period of EXAFS reflects the distance R_j, EXAFS provides us with information about the atomic arrangement of materials as low-energy-electron-diffraction (LEED) or conventional X-ray diffraction does. However, in comparison with LEED and X-ray diffraction, EXAFS has many advantages: It is particularly sensitive to the *local environment* of atoms, and the surroundings of *each species* of atom can be separately studied by tuning to the corresponding absorption edge. Furthermore, unlike LEED, complicated multiple scatterings average out and give no serious effect on the line shape except for some special cases [4.20,21]. By taking notice of these facts, with the development of synchrotron source as well, there has recently been much interest in theoretical and experimental studies of EXAFS, and its application to structural investigations of various materials has been proposed. For instance, EXAFS is expected to be especially useful in the study of noncrystalline systems [4.22] and complex systems containing various chemical elements like biological molecules and catalysts [4.23]. An application of EXAFS to the study of surface structure of adsorbed atoms is also suggested [4.24].

4.3 Electron-Hole Interactions

When an electron is photoexcited in insulating or semiconducting solids, it interacts with a hole left behind. Due to the Coulomb attraction between them, the electron-hole pair is well described as a composite particle called *exciton*. We also discuss effects of exchange, dipole-dipole and spin-orbit interactions.

4.3.1 General Treatment of Excitons

To begin with we write the Hamiltonian of an interacting many electron system as

$$H = \sum_i h(\underline{r}_i) + \frac{1}{2} \sum_i \sum_j v_c(\underline{r}_i - \underline{r}_j) \quad , \tag{4.40}$$

where $h(\underline{r}_i) \equiv - (\hbar^2/2m)\nabla_i^2 + v_p(\underline{r}_i)$ is the single electron part and $v_c(r_i - r_j)$ is the Coulomb interaction $e^2/|\underline{r}_i - \underline{r}_j|$. It is assumed that the system is an insulator consisting of only two nondegenerate bands, μ and ν. We write the ground state as $|g> = \prod_{\underline{k}} a_{\mu\underline{k}}^+ |0>$, and a single particle excited state with total momentum K as

$$|k\underline{K}\rangle = a_{\nu\underline{k}}^{+} a_{\mu,\underline{k}-\underline{K}} |g\rangle \quad ,$$

where $|0\rangle$ is the vacuum and $a_{\xi\underline{k}}^{+}$ ($\xi = \mu$ or ν) is an operator creating an electron in the Bloch state $\phi_{\xi\underline{k}}$ with Hartree-Fock energy

$$\varepsilon_{\xi}(\underline{k}) = \langle\xi\underline{k}|h|\xi\underline{k}\rangle + \sum_{\mu'\underline{k}'} (\langle\xi\underline{k},\mu'\underline{k}'|v_c|\mu'\underline{k}',\xi\underline{k}\rangle - \langle\mu'\underline{k}',\xi k|v_c|\mu'\underline{k}',\xi\underline{k}\rangle) \quad .$$

Here we have abbreviated

$$\int d\underline{r}_1 d\underline{r}_2 \phi_{\lambda_1\underline{k}_1}^{*}(\underline{r}_1)\phi_{\lambda_2\underline{k}_2}^{*}(\underline{r}_2) v_c(\underline{r}_1 - \underline{r}_2)\phi_{\lambda_3\underline{k}_3}(\underline{r}_2)\phi_{\lambda_4\underline{k}_4}(\underline{r}_1)$$

as $\langle\lambda_1\underline{k}_1,\lambda_2\underline{k}_2|v_c|\lambda_3\underline{k}_3,\underline{k}_4\rangle$.

The present treatment is applicable both to a valence exciton and an inner core exciton. In the latter case one has only to regard $\varepsilon_{\mu}(\underline{k})$ as dispersionless.

While H is diagonal in total momentum $\underline{K}$, it is not in relative momentum $\underline{k}$ due to the electron-hole Coulomb interaction

$$\langle k'\underline{K}|H|k\underline{K}\rangle = \delta_{\underline{k}'\underline{k}} [\varepsilon_{\nu}(\underline{k}) - \varepsilon_{\mu}(\underline{k} - \underline{K})] - \langle\nu\underline{k}', \mu\underline{k} - \underline{K}|v_c|\mu\underline{k}' - \underline{K}, \nu\underline{k}\rangle$$

$$+ \langle\mu\underline{k} - \underline{K}, \nu\underline{k}'|v_c|\mu\underline{k}' - \underline{K}, \nu\underline{k}\rangle \quad . \tag{4.41}$$

The true excited state $|\lambda\underline{K}\rangle$ must therefore be expressed by the linear combination of $|k\underline{K}\rangle$ as

$$|\lambda\underline{K}\rangle = \sum_{\underline{k}} f_{\lambda\underline{K}}(\underline{k})|k\underline{K}\rangle \quad . \tag{4.42}$$

It is convenient to use, as a basis set of representation, Wannier functions which are related with Bloch function by

$$\phi_{\xi}(\underline{r} - \underline{R}_n) = \frac{1}{\sqrt{N}} \sum_{\underline{k}} e^{-i\underline{k}\cdot\underline{R}_n} \phi_{\xi\underline{k}} \quad , \tag{4.43}$$

or

$$a_{\xi n}^{+} = \frac{1}{\sqrt{N}} \sum_{\underline{k}} e^{-i\underline{k}\cdot\underline{R}_n} a_{\xi\underline{k}}^{+} \tag{4.44}$$

for corresponding Fermion operators, where N is the number of atoms. Then (4.42) is rewritten as

$$|\lambda \underline{K}> = \sum_{\ell} F_{\lambda \underline{K}}(\underline{R}_{\ell})|\ell \underline{K}> \quad , \tag{4.45}$$

where

$$F_{\lambda \underline{K}}(\underline{R}_{\ell}) = \frac{1}{\sqrt{N}} \sum_{\underline{k}} e^{i\underline{k} \cdot \underline{R}_{\ell}} f_{\lambda \underline{K}}(\underline{k}) \quad , \tag{4.46}$$

$$|\ell \underline{K}> \equiv \frac{1}{\sqrt{N}} \sum_{m} e^{i\underline{K} \cdot \underline{R}_{m}} a^{+}_{\nu, m+\ell} \, a_{\mu, m} |g> \quad . \tag{4.47}$$

The secular equation is written in the form [4.25]

$$\sum_{\ell'} <\ell \underline{K}|H|\ell' \underline{K}> F_{\lambda \underline{K}}(\underline{R}_{\ell'}) = E_{\lambda \underline{K}} \, F_{\lambda \underline{K}}(\underline{R}_{\ell}) \quad , \tag{4.48}$$

where

$$<\ell \underline{K}|H|\ell' \underline{K}> = \frac{1}{N} \sum_{\underline{k}\underline{k}'} e^{i\underline{k}' \cdot \underline{R}_{\ell}} <\underline{k}' \underline{K}|H|\underline{k}\underline{K}> e^{-i\underline{k} \cdot \underline{R}_{\ell}'}$$

$$= \frac{1}{N} \sum_{\underline{k}} [\varepsilon_{\nu}(\underline{k}) - \varepsilon_{\mu}(\underline{k} - \underline{K})] \, e^{i\underline{k} \cdot (\underline{R}_{\ell} - \underline{R}_{\ell}')} - \delta_{\underline{R}_{\ell} - \underline{R}_{\ell}'} w_{c}(\underline{R}_{\ell}) \tag{4.49}$$

$$+ \delta_{\underline{R}_{\ell}, 0} \delta_{\underline{R}_{\ell}', 0} \sum_{m} e^{i\underline{K} \cdot \underline{R}_{m}} w_{x}(\underline{R}_{m}) \quad ,$$

$$w_{c}(\underline{R}_{\ell}) \equiv <\nu \underline{R}_{\ell}, \, \mu 0 |v_{c}| \mu 0, \, \nu \underline{R}_{\ell}> \quad , \tag{4.50}$$

$$w_{x}(\underline{R}_{m}) \equiv <\nu \underline{R}_{m}, \, \mu 0 |v_{c}| \nu 0, \, \mu \underline{R}_{m}> \quad . \tag{4.51}$$

In the right-hand side of (4.49), the first term represents the effective single particle Hamiltonian, the second term the electron-hole Coulomb interaction, and the third term the exchange interaction. We study in the next subsection the solution of the secular equation (4.48) in two limiting situations.

4.3.2 Wannier and Frenkel Excitons

When the Coulomb interaction is subject to a strong dielectric screening and, furthermore, the dispersion of $\varepsilon_{\nu}(k)$ is large, the electron-hole pair is bound weakly. In this limiting situation, the exciton is called the *Wannier exciton*

[4.26]. In the opposite situation, it is called the *Frenkel exciton* [4.27], where the pair is bound so tightly that one can put $F_{\lambda \underline{K}}(\underline{R}_\ell) = \delta_{\underline{R}_\ell, 0}$.

a) Wannier Exciton

Since $F_{\lambda \underline{K}}(0)$ is small, we neglect the exchange interaction. We expand $\varepsilon_\nu(\underline{k}) - \varepsilon_\mu(\underline{k} - \underline{K})$ in the neighborhood of its extremum point $\underline{k} = \underline{k}_m(\underline{K})$ in the form

$$\varepsilon_\nu(\underline{k}) - \varepsilon_\mu(\underline{k} - \underline{K}) = \varepsilon^{(m)}(\underline{K}) + (\underline{k} - \underline{k}_m) \frac{\hbar^2}{2\underline{\mu}(\underline{K})} (\underline{k} - \underline{k}_m) + \cdots\cdots \quad , \tag{4.52}$$

and regard $\underline{R}_\ell$ as a continuous variable $\underline{R}$, then the secular equation (4.48) reduces to the following effective mass equation

$$\left[-\nabla \frac{\hbar}{2\underline{\mu}(\underline{K})} \nabla - v_c(\underline{R})/\varepsilon \right] \tilde{F}_{\lambda \underline{K}}(\underline{R}) = \tilde{E}_{\lambda \underline{K}} \tilde{F}_{\lambda \underline{K}}(\underline{R}) \quad , \tag{4.53}$$

where

$$\tilde{E}_{\lambda \underline{K}} \equiv E_{\lambda \underline{K}} - \varepsilon^{(m)}(\underline{K}) \quad , \tag{4.54}$$

$$\tilde{F}_{\lambda \underline{K}}(\underline{R}) = F_{\lambda \underline{K}}(\underline{R}) \, e^{-i\underline{k}_m \cdot \underline{R}} \quad . \tag{4.55}$$

In (4.53), we approximated $w_c(\underline{R})$ as $v_c(\underline{R})$ and took into account that it is screened by the dielectric constant ε of the crystal. When the effective masses m_μ and m_ν of the μ and ν bands are scalar quantities, $1/\underline{\mu}(\underline{K})$ is also a scalar quantity $1/\mu \equiv 1/m_\mu + 1/m_\nu$ independent of $\underline{K}$. In this case the equation of the relative motion (4.53) is quite the same as that of the hydrogen atom, and we obtain the energy $\tilde{E}_{\lambda \underline{K}}$ as

$$\left. \begin{array}{ll} \tilde{E}_n = -\dfrac{R}{n^2} \quad , \quad R \equiv \dfrac{\mu e^4}{2\varepsilon^2 \hbar^2} & \text{for bound states} \\[4mm] E_k = \dfrac{\hbar^2 k^2}{2\mu} & \text{for ionization continuum.} \end{array} \right\} \tag{4.56}$$

b) Frenkel Exciton

Putting $F_{\lambda \underline{K}}(\underline{R}_\ell) = \delta_{\underline{R}_\ell, 0}$ in (4.48), we obtain from (4.49) the energy eigenvalue

$$E_{\underline{K}} = <0\underline{K}|H|0\underline{K}> = E_\nu(0) - E_\mu(0) - w_c(0) + w_x(0) + \sum_{m(\neq 0)} w_x(\underline{R}_m)e^{-i\underline{K}\cdot\underline{R}_m} \quad , \quad (4.57)$$

where $E_\xi(0) = (1/N) \sum_{\underline{k}} \epsilon_\xi(\underline{k})$. Here the energy $E_\nu(0) - E_\mu(0) - w_c(0) + w_x(0)$ corresponds to the excitation energy in an *isolated* atom, and the last term of (4.57) represents the effect of resonant transfer of the excitation due to the interatomic interaction. We now study the nature of the resonant transfer. In the expression of $w_x(\underline{R}_1 - \underline{R}_2) \equiv \int d\underline{r}d\underline{r}' \; \phi_\nu^*(\underline{r} - \underline{R}_1) \; \phi_\mu^*(\underline{r}' - \underline{R}_2)e^2/|\underline{r} - \underline{r}'| \times \phi_\nu(\underline{r}' - \underline{R}_2)\phi_\mu(\underline{r} - \underline{R}_1)$, we expand $1/|\underline{r} - \underline{r}'|$ in the vicinity of $\underline{r} = \underline{R}_1$ and $\underline{r}' = \underline{R}_2$, then the leading term yields the *dipole-dipole interaction*

$$W_x(\underline{R}_{12}) = \frac{|\underline{\mu}|^2}{R_{12}^3} - \frac{3|\underline{\mu}\cdot\underline{R}_{12}|^2}{R_{12}^5} \quad , \quad (4.58)$$

where $\mu \equiv \int \phi_\nu^*(\underline{r})(e\underline{r})\phi_\mu(\underline{r})d\underline{r}$ is the dipole moment arising from the intraatomic excitation. In the limit of $\underline{K} \to 0$, a careful summation over $\underline{R}_m$ gives

$$\lim_{\underline{K}\to 0} \sum_{m(\neq 0)} w_x(\underline{R}_m) \; e^{-i\underline{K}\cdot\underline{R}_m} = -\frac{4\pi}{3\Omega} [\mu^2 - 3(\underline{\mu} \cdot \hat{\underline{K}})^2] \quad (4.59)$$

in cubic crystal, where Ω is the volume of the unit cell, $\hat{\underline{K}}$ a unit vector in the direction of $\underline{K}$, and μ is taken as real. The result (4.59) shows that the energy of exciton, $E_{\underline{K}}$, around $\underline{K} = 0$ depends on the *direction* of $\underline{K}$: (4.59) gives $8\pi\mu^2/3\Omega$ for $\hat{\underline{K}} \parallel \underline{\mu}$, i.e. for *longitudinal* exciton, whereas $-4\pi\mu^2/3\Omega$ for $\hat{\underline{K}} \perp \underline{\mu}$, i.e. for *transverse* exciton. The difference of energies between the longitudinal and transverse excitons with $\underline{K} = 0$, i.e. $E_{\ell 0} - E_{t0} = 4\pi\mu^2/\Omega$ is called *longitudinal-transverse splitting* (LT splitting) [4.28] (see Sect.4.3.4).

The LT splitting originates from the effect of the depolarizing field which is induced only by the longitudinal mode. Another derivation of the LT splitting is given here by applying the methods a) and b) of Sect. 4.1. With the stand-point of method a), the eigenvalue of H_{eff} in the limit of $\underline{K} \to 0$ is E_{t0}, since the long range interaction bringing about the depolarizing field is absent in H_{eff}. Therefore, using (4.2) and (4.4), and taking the limit of $\underline{k} \to 0$, we obtain

$$\epsilon_{||}(\omega) - 1 = \frac{8\pi\mu^2}{\Omega} E_{t0}/[E_{t0}^2 - (\hbar\omega)^2] \quad . \quad (4.60)$$

With the method b), on the other hand, we obtain from (4.2) and (4.5)

$$1 - \frac{1}{\epsilon_{||}(\omega)} = \frac{8\pi\mu^2}{\Omega} E_{\ell 0}/[E_{\ell 0}^2 - (\hbar\omega)^2] \quad , \quad (4.61)$$

where we have used that the eigenvalue with $\underline{K}(\rightarrow 0)\|\underline{\mu}$ of the true Hamiltonian H is $E_{\ell 0}$. The LT splitting is readily given by eliminating $\epsilon_{\|}(\omega)$ from (4.60) and (4.61) in the form

$$E_{\ell 0} - E_{t0} = 4\pi\mu^2/\Omega + 0 \ (\mu^4/\Omega^2 E_0) \quad .$$

Besides, it should be kept in mind that $\hbar\omega = E_{t\underline{K}}$ is the *pole* of $\epsilon_{\|}(\underline{k},\omega)$, whereas $\hbar\omega = E_{\ell\underline{k}}$ is the *zero* of $\epsilon_{\|}(\underline{k},\omega)$. The LT splitting of the Wannier exciton $|\lambda 0\rangle$ is estimated to be $|F_{\lambda 0}(0)|^2 4\pi\mu^2/\Omega$ by the lowest order perturbation.

It is to be noted that LT splitting is smaller for an exciton coming from deeper core since μ^2 is inversely proportional to the transition energy $\hbar\omega$ (for fixed oscillator strength f) according to (4.9). This is a special case of the more general feature of dielectric function valid for continuous spectra as well: For high energy transitions, Im $\{\epsilon(\omega)\}$ and Im $\{-\epsilon(\omega)^{-1}\}$ give essentially the same spectra since the former is small due to (4.7) and (4.10) and the dispersion of Re $\{\epsilon(\omega)\}$ in this region is negligible in view of (4.13).

4.3.3 Optical Absorption Spectra

The exciton state $|\lambda 0\rangle$ is excited directly by photoabsorption. The absorption coefficient is written from (4.9) and (4.10) as

$$K(\omega) = \frac{4\pi^2\omega}{cn} \sum_\lambda |\langle\lambda 0|\underline{P}_0 \cdot \hat{\underline{n}}|g\rangle|^2 \delta(\hbar\omega - E_{\lambda 0}) \quad . \tag{4.62}$$

Here we have used the stand-point method a) of Sect. 4.1, so that $E_{\lambda 0}$ represents the energy of the transverse exciton. For the Frenkel exciton, $K(\omega)$ is easily obtained by

$$K(\omega) = \frac{4\pi^2\omega}{cn} \frac{(\underline{\mu}\cdot\hat{\underline{n}})^2}{\Omega} \delta(\hbar\omega - E_{t0}) \quad , \tag{4.63}$$

and we calculate $K(\omega)$ for the Wannier exciton in the following. By using (4.16) and (4.42), the transition matrix element is expressed as

$$\langle\lambda 0|\underline{P}_0 \cdot \hat{\underline{n}}|g\rangle = -i(\frac{\hbar}{m}) \frac{e}{E_{\lambda 0}} \sum_{\underline{k}} f_{\lambda 0}{}^*(\underline{k})\underline{P}_{-\nu\mu}(\underline{k}) \cdot \hat{\underline{n}} \quad , \tag{4.64}$$

where

$$\underline{p}_{\nu\mu}(\underline{k}) \equiv \int \phi_{\nu\underline{k}}^* (\underline{r}) \ (-i\hbar\underline{\nabla}) \ \phi_{\mu\underline{k}} \ (\underline{r})d\underline{r} \quad .$$

We expand $\underline{p}_{\nu\mu}(\underline{k})$ in the vicinity of $\underline{k} = \underline{k}_m(0)$ as

$$\underline{p}_{\nu\mu}(\underline{k}) \cdot \hat{n} = \underline{p}_{\nu\mu}^{(m)} \cdot \hat{n} + (\underline{k} - \underline{k}_m(0)) \cdot \underline{\nabla}_k(\underline{p}_{\nu\mu}^{(m)} \cdot \hat{n}) + \ldots \ldots \quad . \qquad (4.65)$$

When $p_{\nu\mu}^{(m)} \neq 0$, the transition is denoted as *first class* (or *dipole allowed*) and when $p_{\nu\mu}^{(m)} = 0$, it is denoted as *second class* (or *dipole forbidden*) [4.29].

a) First Class Transition

Approximating $\underline{p}_{\nu\mu}(\underline{k})$ as $\underline{p}_{\nu\mu}^{(m)}$, we obtain with the use of (4.46), (4.55), (4.62) and (4.64)

$$K(\omega) = \frac{4\pi^2 e^2}{m^2 c n \omega \Omega} |\underline{p}_{\nu\mu}^{(m)} \cdot \hat{n}|^2 \sum_\lambda |\tilde{F}_{\lambda 0}(0)|^2 \ \delta(\hbar\omega - E_{\lambda 0}) \quad . \qquad (4.66)$$

Thus, only the s-states, of the relative motion, where $\tilde{F}_{\lambda 0}(0)$ is nonzero, contribute to $K(\omega)$. For bound states, discrete lines appear at $\hbar\omega = E_{n0} = \varepsilon^{(m)}(0) - R/n^2$ with intensities proportional to $|\tilde{F}_{n0}(0)|^2 = \Omega/(\pi a^3 n^3)$, a being $\varepsilon\hbar^2/\mu e^2$. In the limit of large n, they form quasicontinuous spectrum with finite intensity $|\tilde{F}_{n0}(0)|^2 |dE_{n0}/dn|^{-1} = \Omega/(2\pi R a^3)$. For continuum above $\varepsilon^{(m)}(0)$, $|\tilde{F}_{\underline{k}0}(0)|^2$ is given by $\Omega \pi a \ \exp(\pi a)/\sinh(\pi a)$, where $\alpha \equiv 1/ak$, so that we have the intensity

$$\lim_{\hbar\omega \to \varepsilon^{(m)}(0)} \sum_{\underline{k}} |\tilde{F}_{\underline{k}0}(0)|^2 \ \delta(\hbar\omega - E_{\underline{k}0}) = \Omega/(2\pi R a^3) \quad ,$$

which coincides with that of the quasicontinuum calculated above. Namely, the absorption spectra above and below $\varepsilon^{(m)}(0)$ are connected smoothly, as shown in Fig. 4.2a.

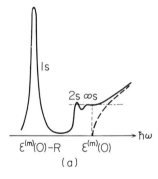

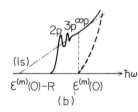

Fig. 4.2   Exciton effect on the line shape near the fundamental absorption edge for the first class (a) and second class (b) transitions. The spectra for vanishing Coulomb interaction are shown with dashed curves

b)   Second Class Transition

In this case we use the second term of (4.65) and obtain

$$\sum_{\underline{k}} f_{\lambda 0}^{*}(\underline{k}) \left[\underline{k} - \underline{k}_m(0)\right] \cdot \nabla_{\underline{k}}(\underline{p}_{\nu\mu}^{(m)} \cdot \hat{n}) = \frac{1}{\sqrt{\Omega}} \nabla_{\underline{R}}\tilde{F}_{\lambda 0}(0)(\nabla_{\underline{k}}\underline{p}_{\nu\mu}^{(m)} \cdot \hat{n}) \quad .$$

Therefore $K(\omega)$ is given by (4.66) only by replacing $p_{\nu\mu}^{(m)}$ and $|\tilde{F}_{\lambda 0}(0)|^2$ with $\nabla_{\underline{k}}p_{\nu\mu}^{(m)}$ and $\nabla_{\underline{R}}\tilde{F}_{\lambda 0}(0)$, respectively. Since $\nabla_{\underline{R}}\tilde{F}_{\lambda 0}(0)$ is nonzero only for p-type relative motion, we have discrete lines for $n \geq 2$ with intensities proportional to

$$|\nabla_x\tilde{F}_{np_x 0}(0)|^2 = \Omega/(\pi a^5)(1/n^3 - 1/n^5) \quad .$$

For large n the spectral intensity of quasicontinuum is given by

$$|\nabla_x\tilde{F}_{np_x 0}(0)|^2 |dE_{n0}/dn|^{-1} = \Omega\{\hbar\omega - [\epsilon^{(m)}(0) - R]\}/(2\pi R^2 a^5) \quad ,$$

the extrapolation of which intersects the abscissa at the position of the forbidden n = 1 line. For $\hbar\omega \geq \epsilon^{(m)}(0)$, we have

$$|\nabla_{\underline{R}}\tilde{F}_{\underline{k}0}(0)|^2 = \Omega\pi\alpha (1 + \alpha^2)\exp(\pi\alpha)/\sinh(\pi\alpha) \quad ,$$

so that

$$\sum_{\underline{k}} |\nabla_{\underline{R}}\tilde{F}_{\underline{k}0}(0)|^2 \delta(\hbar\omega - E_{\underline{k}0})$$

coincides with the intensity of the quasicontinuum in the limit of $\hbar\omega \to \varepsilon^{(m)}(0)$. The line shape is shown schematically in Fig. 2b.

4.3.4 Effects of Spin and Orbital Degeneracies

When the electron spin is taken into account, the formulation in Sect. 4.3.1 holds by reinterpreting the band indices μ and ν to include the spin components σ and τ (= ↑ or ↓). The exciton state $|\lambda \underline{K} \tau\sigma\rangle$ should be described as an eigenstate of the total spin, $\underline{S} = \underline{\tau} - \underline{\sigma}$, of the electron-hole pair, i.e. as either of a triplet state with S = 1 or a singlet state with S = 0. Writing $|\lambda \underline{K} \tau\sigma\rangle$ as

$$|\tau\sigma\rangle \equiv (1/\sqrt{N}) \sum_{m,\ell} F_{\ell \underline{K}}(R_\ell) \exp(i\underline{K} \cdot \underline{R}_m)\, a_{\nu\tau,m+\ell}^{+}\, a_{\mu\sigma m}|g\rangle \quad ,$$

three components of the triplet are $|\uparrow\uparrow\rangle$, $(|\uparrow\uparrow\rangle - |\downarrow\downarrow\rangle)/\sqrt{2}$, $|\downarrow\downarrow\rangle$ and the singlet is $(|\uparrow\uparrow\rangle + |\downarrow\downarrow\rangle)/\sqrt{2}$. In the expression (4.49) of $\langle\ell K|\underline{H}|\ell'K\rangle$, the exchange term depends on the spin component. The integrals with respect to $\underline{r}$ and $\underline{r}'$ in (4.51) should be regarded as including the spin scalar product, so that $w_x(\underline{R}_m)$ vanishes for the triplet, while for the singlet it is twice of $w_x(\underline{R}_m)$ in the spinless case. Since $w_x(0) > 0$, the energy of the singlet is higher than that of the triplet. Only the singlet can be excited by optical absorption, because the ground state $|g\rangle$ is also a singlet.

Next we consider the orbital degeneracy. The state $|\ell \underline{K}\rangle$ should be now written as $|\ell \underline{K}\nu\mu\rangle$ and we have [4.25,30]

$$\langle \ell \underline{K}\nu\mu|H|\ell'\underline{K}\nu'\mu'\rangle = \frac{1}{N} \sum_{\underline{k}} [\delta_{\mu\mu'}\varepsilon_{\nu\nu'}(\underline{k}) - \delta_{\nu\nu'}\varepsilon_{\mu\mu'}(\underline{k})]\, e^{i\underline{k}\cdot(\underline{R}_\ell - \underline{R}_{\ell'})}$$

$$- \delta_{\underline{R}_\ell,\underline{R}_{\ell'}} [w_c(\underline{R}_\ell)]_{\nu\mu\nu'\mu'} + \delta_{\underline{R}_\ell,0}\delta_{\underline{R}_{\ell'},0} \sum_{m} e^{i\underline{K}\cdot\underline{R}_m} [w_x(\underline{R}_m)]_{\nu\mu\nu'\mu'}$$

in place of (4.49). We study the effect of the orbital degeneracy on the dipole-dipole interaction, which is now expressed as

$$- \frac{4\pi}{3\Omega} [\underline{\mu}_{\nu\mu} \cdot \underline{\mu}_{\mu'\nu'} - 3(\underline{\mu}_{\nu\mu} \cdot \hat{\underline{K}})(\underline{\mu}_{\mu'\nu'} \cdot \hat{\underline{K}})] \tag{4.67}$$

instead of (4.59). In the case that the μ's are degenerate p_x, p_y, p_z-states and ν is an s-state in cubic crystal, for example, (4.67) becomes $-4\pi\mu^2/(3\Omega)(\delta_{ij} - 3\hat{K}_i\hat{K}_j)$, so that it is diagonalized, for every direction of $\hat{\underline{K}}$, by one longitudinal wave ($\underline{\mu} \parallel \hat{\underline{K}}$) and two transverse waves ($\underline{\mu} \perp \hat{\underline{K}}$). The eigenvalue is $8\pi\mu^2/(3\Omega)$ for the former and $-4\pi\mu^2/(3\Omega)$ for the latter.

Finally, we discuss the effect of spin-orbit and exchange interactions [4.31] in the cubic system considered above. If the exchange interaction is disregarded, the spin orbit interaction is diagonalized by the *j-j coupling scheme*, i.e. it has the energy $-\lambda/3$ for $|j_c j_v> = |1/2, 3/2>$, and $2\lambda/3$ for $|1/2, 1/2>$, where j_c and j_v are, respectively, the total angular momenta of electron and hole, and $\lambda(>0)$ is the spin-orbit splitting. The intensity ratio of the two absorption lines is 2:1. When the exchange interaction is switched on, they are mixed with each other. Since the total angular momentum $\underline{J} = \underline{j}_c + \underline{j}_v$ is conserved within our subspace, we construct them from the j-j scheme. Then, the states with $J = 2$ and $J = 1$ arise from $|1/2, 3/2>$ and those with $J = 1$ and $J = 0$ arise from $|1/2, 1/2>$. The states with $J = 2$ and $J = 0$ are not affected by the exchange interaction, since they consist of $S = 1$ states alone. Also, they cannot be excited by photons. The remaining two sets of $J = 1$ are mixed states between $S = 0$ and $S = 1$, and in the limit of the strong exchange interaction they tend to pure $S = 0$ and $S = 1$ states (i.e. *L-S coupling scheme*) with the ratio of absorption intensities 1:0. When the exchange interaction is treated as perturbation for transverse Wannier excitons with $\underline{K} = 0$, the energies of the two $J = 1$ states are given by

$$E_\pm = \frac{1}{2} \left[\Delta + \frac{\lambda}{3} \pm \sqrt{(\Delta - \frac{\lambda}{3})^2 + \frac{8}{9}\lambda^2} \right] \tag{4.68}$$

and the relative intensity of each absorption line is

$$I_+ = \cos^2(\gamma - \phi) \quad , \quad I_- = \sin^2(\gamma - \phi) \quad , \tag{4.69}$$

where

$$\gamma \equiv \tan^{-1}\sqrt{2} \quad , \quad \phi \equiv \frac{1}{2}\tan^{-1}[2\sqrt{2}\,\Delta/(3\lambda - \Delta)]$$

and

$$\Delta \equiv 2|F(0)|^2 \left[\int \phi_s(\underline{r})\phi_{p_x}(\underline{r}')v_c(\underline{r} - \underline{r}')\phi_s(\underline{r}')\phi_{p_x}(\underline{r})d\underline{r}d\underline{r}' - \frac{4\pi\mu^2}{3\Omega} \right]$$

The most systematic study of the exciton states associated with s-like conduction band and p-like filled band in cubic and wurtzite crystals has been presented in [4.32,33] on the basis of group-theoretically constructed effective Hamiltonian with exchange and spin-orbit interactions and finite $\underline{K}$ effect, under electric, magnetic and stress fields.

4.4 Configuration Interactions

Let us consider optical excitation of an electron from an inner core state c_1 to a discrete state α which corresponds, for instance, to a Rydberg state in atoms (or molecules) or an exciton state in solids. The excited state, written as $|\alpha\rangle \equiv a_\alpha^+ a_{c_1} |g\rangle$ with the many electron representation, is assumed to be coupled by the Coulomb interaction with the other excited states containing a different core hole. We discuss the effect of such a *configuration interaction* on the absorption line shape.

4.4.1 Auger Process

We assume that the state coupled with $|\alpha\rangle$ is $|k\rangle \equiv a_k^+ a_{c_3}^+ a_{c_2} |\alpha\rangle$, where c_2 and c_3 are core states occupied in the ground state $|g\rangle$, and k is in the ionization continuum, as shown in Fig. 4.3a. All the single electron states are treated to be nondegenerate for simplicity. When energies of the core states satisfy the relation $\varepsilon_{c_2} - \varepsilon_{c_1} \geq \varepsilon_t - \varepsilon_{c_3}$, $\varepsilon_t \equiv \hbar\omega_t + \varepsilon_{c_1}$ being the ionization threshold, then $|\alpha\rangle$ and $|k\rangle$ are mixed resonantly with each other, resulting in a finite life time of $|\alpha\rangle$. This type of configuration interaction is called the *Auger process* [4.34]. The Auger transition with c_1 and c_2 (or c_1 and c_3) having the same principal quantum number is particularly denoted as the *Coster-Kronig transition*, and that with c_1, c_2 and c_3 having the same principal quantum number as the *super Coster-Kronig transition*. The matrix element of the configuration mixing is given by

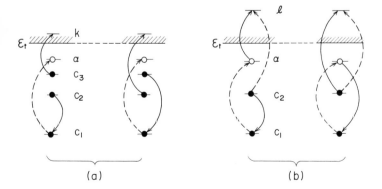

(a)   (b)

Fig. 4.3 Schematic representation of configuration interactions. The transitions giving rise to the Auger process (a) and the Fano effect (b) are shown with solid arrows. Radiative transitions are shown with dashed arrows. The solid circles express the core electrons occupied in the ground state

$$V_{k\alpha} = \int dr dr' \phi_{c_1}^*(r)\phi_k^*(r')v_c(r - r')\phi_{c_3}(r')\phi_{c_2}(r)$$

$$- \int dr dr' \phi_{c_1}^*(r)\phi_k^*(r')v_c(r - r')\phi_{c_2}(r')\phi_{c_3}(r) \qquad (4.70)$$

with the use of single electron wave functions ϕ_{c_1}, ϕ_{c_2} etc.

The optical absorption spectrum due to the dipole transition $H_I = M_{\alpha g}a_\alpha^+ a_{c_1} + $ c.c. is expressed, apart from unimportant factors, as

$$I(\omega) = \frac{1}{\pi} \text{Im} <g|H_I G H_I|g> = \frac{1}{\pi} \text{Im} \{M_{g\alpha}G_{\alpha\alpha}M_{\alpha g}\}$$

$$= \frac{|M_{\alpha g}|^2 \Gamma(\omega)/\pi}{[\hbar\omega - E_\alpha - \Delta(\omega)]^2 + \Gamma(\omega)^2} \qquad , \qquad (4.71)$$

where

$$G = \frac{1}{z - (H_0 + V)} \qquad , \qquad (z = \hbar\omega - is \quad , \quad H_0: \text{ unperturbed Hamiltonian}) \qquad (4.72)$$

$$\Delta(\omega) = P \sum_k \frac{|V_{k\alpha}|^2}{\hbar\omega - E_k} \qquad , \qquad (4.73)$$

$$\Gamma(\omega) = \pi \sum_k |V_{k\alpha}|^2 \delta(\hbar\omega - E_k) \qquad , \qquad (4.74)$$

$$E_\alpha' = \epsilon_\alpha - \epsilon_{c_1} \qquad , \qquad (4.75)$$

$$E_k = \epsilon_k + \epsilon_\alpha - \epsilon_{c_2} - \epsilon_{c_3} \qquad . \qquad (4.76)$$

When the ω-dependence of $\Delta(\omega)$ and $\Gamma(\omega)$ is disregarded, $I(\omega)$ is of Lorentzian line shape.

As is evident from the above calculation, the Auger width 2Γ is due to the life time of the core hole in c_1, and the excited electron α does not play an essential role in it. In actual atoms the value of Γ varies widely, dependent on the species of atoms and core states c_1. As an example, we show, in Fig. 4.4, 2Γ of the K-level, which is obtained by more elaborate calculations [4.35], as a function of the atomic number Z. The Auger width is found to increase monotonically with increase in Z. For other core hole states the level width depends on whether Coster-Kronig and super Coster-Kronig transitions are energetically possible or not.

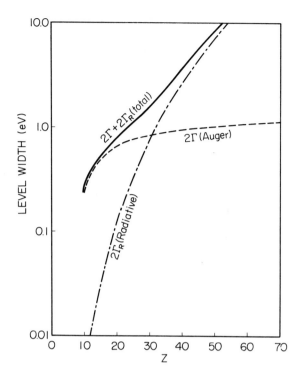

LEVEL WIDTH (eV)

10.0

1.0

0.1

0.01

0   10   20   30   40   50   60   70

Z

$2\Gamma + 2\Gamma_R$ (total)

2Γ (Auger)

$2\Gamma_R$ (Radiative)

Fig. 4.4 K-level width due to the Auger [4.35] and radiative [4.36] processes as a function of atomic number Z

In connection to the core level width, it should be mentioned that the width also originates from a *radiative recombination* between the core hole c_1 and any other occupied electrons c_j. A rough estimation of the radiative width $2\Gamma_R$ can be made by the lowest order perturbation calculation on H_I

$$2\Gamma_R = 2\pi \sum_{j,\underline{k}} |<\alpha|a_{c_1} a^+_{c_j} H_I|\alpha>|^2 \delta(\varepsilon_{c_j} - \varepsilon_{c_1} - \hbar\omega_{\underline{k}})$$

$$= 2\pi \sum_{j,\underline{k}} \frac{2\pi e^2 \hbar}{m^2 \omega_{\underline{k}}} |<\phi_{c_1}|\underline{p} \cdot \hat{\underline{n}}|\phi_{c_j}>|^2 \delta(\varepsilon_{c_j} - \varepsilon_{c_1} - \hbar\omega_{\underline{k}})$$

(4.77)

$$-\left(\frac{e^2}{\hbar c}\right) \sum_j \frac{(\varepsilon_{c_j} - \varepsilon_{c_1})^2}{mc^2} \cdot f_{1j} \quad ,$$

where $\omega_{\underline{k}} \equiv c|\underline{k}|$ is the frequency of the emitted radiation and f_{1j} is the relevant oscillator strength. it should be noted that Γ_R is proportional to square of the energy difference $\varepsilon_{c_j} - \varepsilon_{c_1}$. In Fig. 4.4, the radiative width of the K shell is plotted as a function of Z [4.36]. Here $2\Gamma_R$ is found to be approximately proportional to z^4, which is reasonable since the binding energy of core state is proportional to Z^2 according to the hydrogen model. In contrast, the Auger width changes with Z more gradually, so that the two curves in Fig. 4.4 cross each other in the neighborhood of $Z \simeq 31$. For core states shallower than the K shell the contribution of the Auger process to the total level width becomes more dominant.

Besides the atomic calculations mentioned above, the Auger and radiative processes in some simple metals have been calculated [4.37]. Finally, it is emphasized that the inner core absorption and emission spectra always have a spectral broadening due to the Auger and radiative transitions, which is far larger than that of the valence hole, although we do not take explicit account of it in other sections except Sect. 4.8.

4.4.2 Fano Effect

In the above calculation of the Auger transition, we have disregarded the process that the state $|k> \equiv a_k^+ a_\alpha^+ a_{c_3} a_{c_2} |g>$ is excited directly from $|g>$ by the optical absorption, because such a transition is a two-electron one-photon process and has in general only small oscillator strength. Here we consider another type of configuration interaction $V_{\ell\alpha}$ by which $|\alpha>$ is autoionized into $|\ell> \equiv a_\ell^+ a_\alpha a_{c_1}^+ a_{c_2} |\alpha> = a_\ell^+ a_{c_2} |g>$, where ℓ is in the ionization continuum (the relation $\varepsilon_{c_2} - \varepsilon_{c_1} \geq \varepsilon_t - \varepsilon_\alpha$ being assumed). The situation is depicted in Fig. 4.3b. In this case $|\ell>$ can be excited from $|g>$ by a one-electron optical transition $c_2 \rightarrow \ell$, which we denote by $M_{\ell g}$ following the many electron representation. Then the excitations $M_{\alpha g}$ and $M_{\ell g}$ interfere with each other through $V_{\ell\alpha}$ and gives rise to a modification of absorption line shape, which is called *Fano effect* [4.38].

We next outline the Green's function reformulation [4.39] of the Fano formula. The absorption spectrum is written as

$$I(\omega) = \frac{1}{\pi} \text{ Im} \left\{ M_{g\alpha} G_{\alpha\alpha} M_{\alpha g} + \sum_\ell (M_{g\ell} G_{\ell\alpha} M_{\alpha g} + M_{g\alpha} G_{\alpha\ell} M_{\ell g}) + \sum_{\ell\ell'} M_{g\ell} G_{\ell\ell'} M_{\ell' g} \right\} \quad , \quad (4.78)$$

where each component of G is expressed as

$$G_{\alpha\alpha} = \frac{1}{z-E_\alpha-\Delta-i\Gamma} \quad , \quad G_{\alpha\ell} = G_{\alpha\alpha} V_{\alpha\ell} \frac{1}{z-E_\ell} = G_{\ell\alpha}^* \quad ,$$

$$G_{\ell\ell'} = \frac{\delta_{\ell\ell'}}{z-E_\ell} + \frac{1}{z-E_\ell} V_{\ell\alpha} G_{\alpha\alpha} V_{\alpha\ell'} \frac{1}{z-E_{\ell'}} \quad , \quad (4.79)$$

and $V_{\ell\alpha}$, E_ℓ etc. are given by the corresponding expressions of $V_{k\alpha}$, E_k etc. describe before only by changing suffices c_3 and k to α and ℓ, respectively. Substitution of (4.79) into (4.78) yields

$$I(\omega) = \frac{\pi}{2\Gamma} \sum_{\ell\ell'} |M_{g\ell}V_{\alpha\ell'} - M_{g\ell'}V_{\alpha\ell}|^2 \, \delta(\hbar\omega - E_\ell)\delta(\hbar\omega - E_{\ell'})$$

$$+ \frac{\pi}{\Gamma} | \sum_\ell M_{g\ell}V_{\ell\alpha} \, \delta(\hbar\omega - E_\ell)|^2 \left(1 + \frac{q^2 - 1 + 2q\epsilon \cos\theta}{1+\epsilon^2}\right) ,$$

(4.80)

where

$$qe^{i\theta} \equiv \frac{M_{g\alpha} + P \sum_\ell M_{g\ell}\frac{V_{\ell\alpha}}{\hbar\omega - E_\ell}}{\pi \sum_\ell M_{g\ell}V_{\ell\alpha} \, \delta(\hbar\omega - E_\ell)} ,$$

(4.81)

$$\epsilon \equiv \frac{\hbar\omega - E_\alpha - \Delta}{\Gamma} .$$

(4.82)

Except for some special cases, e.g. the system in magnetic field, θ can be taken to be zero, and $M_{g\ell}$ and $V_{\alpha\ell}$ depend on ℓ only through E_ℓ. Then (4.80) reduces to the result first derived by FANO

$$I(\omega) = I_0(\omega) \frac{(q+\epsilon)^2}{1+\epsilon^2} ,$$

(4.83)

where

$$I_0(\omega) \equiv \sum_\ell |M_{g\ell}|^2 \, \delta(\hbar\omega - E_k)$$

is the background absorption spectrum, namely the spectrum in the absence of the state α.

The line shape $(q + \epsilon)^2/(1 + \epsilon^2)$ is shown in Fig. 4.5 as a function of the normalized photon energy ϵ for various values of the line shape parameter q. It is noted that for $q \sim 0$ there exists only a dip, which is caused by the cancellation of transition matrices between the direct (through $M_{g\alpha}$) and indirect (through $M_{g\ell}$ and $V_{\ell\alpha}$) transitions as seen from the numerator of (4.81). The dip is called *perfect anti-resonance* or *window resonance*. For finite q, the spectrum consists of an asymmetric peak at $\epsilon = 1/q$ and a dip at $\epsilon = -q$. With increase in q, the peak becomes prominent and, in the limit of $M_{g\ell} \to 0$ (so that $q \to \infty$), the line shape is reduced to a symmetric Lorentzian, which is nothing but the result (4.71). Taking account of the coexistence of other absorption processes which does not interfere with the process treated here, $I(\omega)$ is generally expressed in the form

$$I(\omega) = I_0(\omega) \frac{(q+\epsilon)^2}{1+\epsilon^2} + I_0'(\omega) .$$

(4.84)

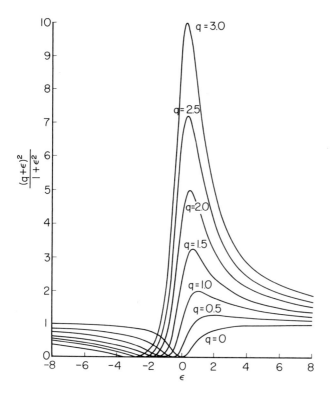

Fig. 4.5 Line shapes given by the Fano formula for various values of q [4.39]. For negative q the abscissa should be reversed

An extension of the Fano formula has also been made by including the interference effect arising from the coexistence of many discrete states and many continua [4.40].

   A typical example of the Fano effect is found in the absorption spectrum of Ar atom [4.41] (as well as solid Ar) in the 26 ~ 29 eV region, where discrete absorption lines of 3s → np (n = 4, 5, 6 ...) excitations are superposed on the continuous spectrum of 3p → εs (and εd) (see Sect.5.5.1). The states c_1, α, c_2 and ℓ in the above calculation correspond to the states 3s, np, 3p and εs (εd), respectively. For the antiresonance associated with 3s → 4p transition, the line shape is reproduced by (4.84) with q = -0.22, Γ = 0.04 eV and $I_0/(I_0 + I_0')$ = 0.86.

   In general the Fano effect is expected to occur when there coexist a quasidiscrete and a continuous absorptions whose excited states interact with each other, irrespective of natures of the excited states and properties af their interaction. Another example of the Fano effect will be given in Sect. 4.7.

4.5 Simultaneous Excitations and Relaxations

Optical transition of an electron generally destroys the balance of forces in the surrounding medium. The final state interaction, which is defined as the potential for the resultant forces of unbalance, gives risè to two important effects, which are intimately correlated with each other. One is the *simultaneous excitations* of various types of elementary excitations in the medium which manifest themselves as sidebands of the optical spectra. Another is the *relaxation*, after optical transition, of the medium towards the new equilibrium state which is sometimes called *relaxed excited state* and is the initial state of the photoluminescence in the ordinary situations. Here we discuss these effects in insulating solids, with discussion of metals left for the next section.

4.5.1 Localized Excitation and Relaxation in Deformable Lattice

We consider the simultaneous excitation of phonons in the photoabsorption of a localized electron [4.42-4.44], which is described as a two-level system with energy separation ε_a. When the electron is in the lower level $\varepsilon_0 \equiv 0$, the Hamiltonian of the system in the harmonic approximation is given by

$$H_g = \sum_j \frac{1}{2} (p_j^2 + \omega_j q_j^2) \quad , \tag{4.85}$$

where q_j is the normal coordinate of the lattice displacement and p_j is the momentum conjugate to q_j. When the electron is excited to the upper level, the system is represented by the Hamiltonian

$$H_e = H_g + \varepsilon_a + V \quad , \quad V = -\sum_j c_j q_j \quad , \tag{4.86}$$

where we have expanded the adiabatic potential difference, $H_e - H_g$, up to the linear term with respect to q_j. The absorption spectrum is expressed as

$$I(\omega) = Av \sum_{n\ n'} |\langle en'|H_I|gn\rangle|^2 \delta(\hbar\omega - E_{en'} + E_{gn}) \quad , \tag{4.87}$$

where $|gn\rangle \equiv |g\rangle|n\rangle$ is an eigenstate (the direct product of electronic ground state $|g\rangle$ and phonon state $|n\rangle$ within the adiabatic approximation) of H_g with energy E_{gn}, $|en\rangle \equiv |e\rangle|n\rangle$ is an eigenstate of H_e with E_{en}, and H_I expresses the operator of transition dipole. The state $H_I|gn\rangle$, where the electron is excited to the upper level with the phonon state $|n\rangle$ unchanged, is not the eigenstate of H_e due to the existence of the final state interaction V, so that it has finite overlaps with various states

$|en'>$. The phonon state $|n'>$ is generally different from $|n>$, namely the electronic excitation is accompanied with the simultaneous emission or absorption of phonons, which is reflected in $I(\omega)$ as phonon sidebands as will be described in more detail in Sect. 4.5.3.

When the electron-phonon interaction is strong enough, the structures of phonon sidebands are smoothed out by the many phonons process involving various combinations of different phonon frequencies. Under this situation, we can calculate $I(\omega)$ by applying the *Franck-Condon principle* with the so-called *configuration coordinate model*. By an appropriate orthogonal transformation from $(\omega_1 q_1, \omega_2 q_2 \ldots.)$ to $(Q_1, Q_2 \ldots..)$, where Q_1 is chosen as $\sum_j c_j q_j/c$, the adiabatic potentials for lower and upper electronic states are written as

$$E_g = \frac{1}{2} \sum_j Q_j^2 \quad , \quad E_e = \varepsilon_a + \frac{1}{2} \sum_j Q_j^2 - cQ_1 \quad , \tag{4.88}$$

where $c^2 \equiv \sum_j (c_j/\omega_j)^2$. Due to the Franck-Condon principle, the normal coordinates of the lattice are approximated to be unchanged in the optical transition of the localized electron. Then, (4.87) is calculated as

$$I(\omega) \propto \int dQ_1 \exp(-\beta Q_1^2/2)\delta(\hbar\omega - \varepsilon_a + cQ_1)$$

$$\propto \exp[-\beta(\hbar\omega - \varepsilon_a)^2/2c^2] \quad , \quad (\beta \equiv 1/k_B T) \quad , \tag{4.89}$$

where the dipole matrix element has been assumed to be independent of Q_1, and the *Boltzman factor* has been introduced for the distribution of initial states. The absorption spectrum turns out to be Gaussian with dispersion $D \equiv c/\beta^{1/2}$. Note that D is just the amplitude of the final state interaction $<V^2>^{1/2}$. In this way we have calculated $I(\omega)$ as a problem of the one-dimensional configuration space Q_1 in place of the multi-dimensional space q_j. We denote Q_1 as the *interaction mode* and show in Fig. 4.6a the adiabatic potentials in the Q_1 direction.

Since the interaction mode Q_1 is not a normal mode, it relaxes by dephasing (even without anharmonicity) after the optical absorption [4.45]. If we assume $Q_1 = 0$ at $t = 0$ (the moment of the photoabsorption), the time evolution $Q_1(t)$ is given by $Q_1(t) = (1/c) \sum_j (c_j/\omega_j)^2(1 - \cos\omega_j t)$, so that it relaxes towards the equilibrium position c, provided ω_j's are distributed continuously, with a relaxation time $\tau_{LR} \sim 1/\omega_{ph}$, ω_{ph} being a characteristic phonon frequency. The relaxation energy of the Q_1 mode, $E_{LR} \equiv c^2/2$, is thereby distributed among the other modes Q_2, Q_3 etc., being dispersed over the whole crystal. In usual situation, the radiative relaxation time $\tau_R = \hbar/2\Gamma_R$ is much longer than τ_{LR}, and hence the optical emission occurs after the distribution of Q_1 has attained the thermal equilibrium around c. The emission spectrum is given by the Gaussian function

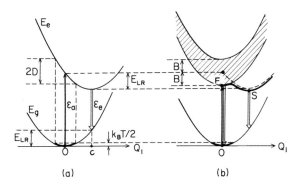

Fig. 4.6 Configuration coordinate models for a localized excitation (a) and the exciton (b)

$$I_e(\omega) \propto \exp[-\beta(\hbar\omega - \varepsilon_e)^2/2c^2] \tag{4.90}$$

with the use of the Franck-Condon principle, again. The shift of $I_e(\omega)$ from the absorption spectrum $I(\omega)$ is given by $\varepsilon_a - \varepsilon_e = c^2$ (= $2E_{LR}$) and it is called the *Stokes shift*. The absorption, relaxation and emission processes are best visualized with the configuration coordinate model shown in Fig. 4.6a.

4.5.2 Host Excitation in Deformable Medium

Let us consider the case of host excitation in insulators. Confining ourselves to one exciton band, for simplicity, we write the Hamiltonian of free exciton as

$$H_x = \sum_{\underline{K}} E_{\underline{K}} \, c_{\underline{K}}^+ c_{\underline{K}} \quad .$$

Denoting the Hamiltonian for the lattice vibrations by H_g, the total energy is given by

$$H_e = H_g + H_x + V \quad , \tag{4.91}$$

where V is the exciton-phonon interaction

$$V = \sum_{\underline{q}\underline{K}} V_{\underline{q}} c_{\underline{K}+\underline{q}}^+ c_{\underline{K}} \quad , \tag{4.92}$$

$$V_q = \sum_{\lambda} \gamma_{\lambda\underline{q}} (b_{\lambda\underline{q}} + b_{\lambda,-\underline{q}}^+) \quad . \tag{4.93}$$

The coupling coefficient $\gamma_{\lambda q}$ depends on the mode λ of lattice vibrations as well as on the internal motion of exciton. The Fröhlich-type interaction with optical mode is a special case of the interaction with a general dielectric medium (with Hamiltonian H_g)

$$V_{\underline{q}} = \left(\frac{4\pi e}{q^2} f_{\underline{q}}\right) \rho_{\underline{q}} \quad , \tag{4.93'}$$

where the first and the second factors are the Fourier component of the Coulomb potential due to the exciton with form factor

$$f_{\underline{q}} \equiv \sum_{\ell} |F(\underline{R}_{\ell})|^2 \left[\exp\left(i \frac{m_h}{m_e + m_h} \underline{q} \cdot \underline{R}_{\ell}\right) - \exp\left(-i \frac{m_e}{m_e + m_h} \underline{q} \cdot \underline{R}_{\ell}\right)\right] \quad ,$$

and that of the polarization charge $\rho(\underline{r}) = -\mathrm{div}\,\underline{P}(\underline{r})$ of the medium, respectively. The absorption spectrum given by (4.87) is rewritten as

$$I(\omega) = \frac{|M|^2}{2\pi\hbar} \int_{-\infty}^{\infty} dt\ \mathop{\mathrm{Av}}_{n} e^{i(\omega + E_{gn}/\hbar)t} \langle gn| c_0 e^{-iH_e t/\hbar} c_0^+ |gn\rangle \quad , \tag{4.94}$$

where we have set $H_I = Mc_0^+ + \text{c.c.}$ The integrand in (4.94) is calculated in the form

$$\mathop{\mathrm{Av}}_{n} e^{i(\omega + E_{gn}/\hbar)t} \langle gn| c_0 \exp\left(-i \frac{H_g + H_x}{\hbar} t\right) T \exp\left[-\frac{i}{\hbar} \int_0^t d\tau V(\tau)\right] c_0^+ |gn\rangle$$

$$= \exp\left[i\left(\omega - \frac{E_0}{\hbar}\right) t - \frac{1}{\hbar^2} \int_0^t dt_1 \int_0^{t_1} dt_2 \mathop{\mathrm{Av}}_{n} \langle gn| c_0 V(t_1) V(t_2) c_0^+ |gn\rangle + O(V^4)\right] \quad , \tag{4.95}$$

so that $I(\omega)$ is written, with neglection of $o(V^4)$ term in (4.95), as

$$I(\omega)^{\cdot} = \frac{|M|^2}{2\pi\hbar} \int_{-\infty}^{\infty} dt\ \exp\left[i\left(\omega - \frac{E_0}{\hbar}\right) t - \frac{1}{\hbar^2} \int_0^t d\tau (t - \tau) g(\tau)\right] \quad , \tag{4.96}$$

where T is the time ordering operator, $V(t) \equiv \exp[iH_g + H_x)t/\hbar] V \exp[-i(H_g + H_x)t/\hbar]$ and $g(\tau) \equiv \mathop{\mathrm{Av}}_{n} \langle gn| c_0 V(\tau) V c_0^+ |gn\rangle$. Substituting (4.92) and (4.93) or (4.93') into the correlation function $g(\tau)$, we can express $g(\tau)$ as

$$g(\tau) = \sum_{\underline{K}} |\gamma_{\underline{K}}|^2 \left[(1 - e^{-\beta\hbar\omega_{\underline{K}}})^{-1} e^{-i\omega_{\underline{K}}\tau} + (e^{\beta\hbar\omega_{\underline{K}}} - 1)^{-1} e^{i\omega_{\underline{K}}\tau}\right] e^{i(E_0 - E_{-\underline{K}})\tau/\hbar} \quad , \tag{4.97}$$

$$g(\tau) = \sum_{\underline{K}} \int_{-\infty}^{\infty} d\omega \left(\frac{4\pi e^2}{K^2}\right) |f_{\underline{K}}|^2 \,\mathrm{Im}\,\{-\epsilon_{||}(\underline{K}, \omega)^{-1}\} (1 - e^{-\beta\hbar\omega})^{-1} e^{-i\omega\tau} e^{i(E_0 - E_{-\underline{k}})\tau/\hbar} \quad , \tag{4.97'}$$

where we have used (4.2) and (4.5) together with the relation $\rho_{\underline{K}} = -i\underline{K} \cdot \underline{P}_{\underline{k}}$. For later use, it is convenient to define the Fourier transform

$$J(\omega) \equiv \frac{1}{2\pi} \int_{-\infty}^{\infty} d\tau \; g(\tau) \; e^{i\omega\tau} \quad . \tag{4.98}$$

In order to see the *gross contour* of exciton absorption spectra given by (4.96), we now define two characteristic times τ_c and τ_a : τ_c is a correlation time with which the correlation function $g(\tau)$ damps and τ_a is the time when the real part of

$$L(t) \equiv (1/\hbar^2) \int_0^t d\tau (t - \tau) g(\tau)$$

becomes unity. $L(t)$ increases as t^2 for small t and as t for large t. The main contribution to $I(\omega)$ comes from the region $|t| \lesssim \tau_a$ in the integral (4.96). We can obtain more explicitly the absorption line shape under the following limiting situations.

a) Slow Modulation Limit

When $\tau_c \gg \tau_a$, $g(\tau)$ scarecely changes in $\tau \lesssim \tau_a$, so that we can put $g(\tau) \simeq g(0)$ in (4.96). Then $I(\omega)$ is given by the *Gaussian* function

$$I(\omega) = \frac{|M|^2}{2\pi\hbar} \int_{-\infty}^{\infty} dt \; \exp\left[i\left(\omega - \frac{E_0}{\hbar}\right) t - \frac{D^2}{2\hbar^2} t^2 \right]$$

$$\tag{4.99}$$

$$= \frac{|M|^2}{\sqrt{2\pi D^2}} \exp\left[-\frac{(\hbar\omega - E_0)^2}{2D^2} \right] \quad ,$$

where

$$D^2 \equiv g(0) = \int_{-\infty}^{\infty} d\omega \; J(\omega) \quad . \tag{4.100}$$

Since $\tau_a \simeq \hbar/D$ from its definition, the condition $\tau_c \gg \tau_a$ can also be written as $\tau_c D/\hbar \gg 1$.

b) Rapid Modulation Limit

When $\tau_c \ll \tau_a$, $g(\tau)$ damps very rapidly, so that in the calculation of $L(t)$ the ω-dependence of $J(\omega)$ is considered to be small enough

$$L(t) = \int_{-\infty}^{\infty} J(\omega) \frac{-i\omega t+(1-e^{-i\omega t})}{(\hbar\omega)^2} d\omega$$

$$\simeq -it \int_{-\infty}^{\infty} \frac{J(\omega)}{\hbar^2\omega} d\omega + J(0) \int_{-\infty}^{\infty} \frac{1-e^{-i\omega t}}{(\hbar\omega)^2} d\omega \qquad (4.101)$$

$$= i \frac{\Delta_0}{\hbar} t - \frac{\Gamma_0}{\hbar} |t| \quad ,$$

where $\Gamma_0 \equiv \pi J(0)/\hbar = D^2 \tau_c/\hbar$ and

$$\Delta_0 \equiv - \int_{-\infty}^{+\infty} \frac{J(\omega)}{\hbar\omega} d\omega \quad .$$

Therefore $I(\omega)$ is given by the *Lorentzian* function

$$I(\omega) = \frac{|M|^2}{\pi} \frac{\Gamma_0}{(\hbar\omega-E_0-\Delta_0)^2+\Gamma_0^2} \quad . \qquad (4.102)$$

The physical meaning [4.46,47] of these results is as follows: The optical excitation energy E_0 of the exciton is modulated by the potential fluctuation $V(t)$ with amplitude D and correlation time τ_c, due to the vibrations of the medium as well as due to the motion of the exciton itself. In order that the exciton can perceive the potential amplitude D with sufficient accuracy, the amplitude must not change within the time $\tau \sim \hbar/D$ from the uncertainty principle. Namely, when the modulation is slow enough to satisfy $\tau_c \gg \tau$ (slow modulation limit), the amplitude D is reflected intact in $I(\omega)$ as its spectral width. In the opposite situation $\tau_c \ll \tau$ (rapid modulation limit), however, the exciton perceive only the fraction τ_c/τ of D, so that the spectral width is reduced to $D\tau_c/\tau = D^2\tau_c/\hbar$. We denote τ_c/τ as the *narrowing factor*.

From (4.97) or (4.97'), $\hbar/\tau_c$ is found to be the larger of the quantities, $\hbar\omega$ and $\overline{E_0 - E_K}$, where $\bar{\omega}$ is a characteristic frequency of lattice vibrations or electric polarization and $\overline{E_0 - E_K}$ is of the order of the half width B of exciton band. Let us assume the situation $D \gg \hbar\bar{\omega}$. Then, the absorption line shape is given by the Gaussian (4.99) or the Lorentzian (4.102) according as $D \gtrless B$. The former is nothing but the absorption band (4.89) for localized excitation (in the *classical limit* $\beta\hbar\omega \ll 1$), as is easily confirmed by noting (4.97), (4.100), and the relations: $q_j = (\hbar/2\omega_j)^{1/2} (b_j + b_j^+)$ and $c_j = -(2\omega_j/\hbar)^{1/2}\gamma_j$. In the other extreme, the *motionally narrowed* width $2\Gamma_0 = (2\pi/\hbar) J(0)$ represents the level broadening of the optically created exciton (with $\underline{K} = 0$) due to phonon scattering as is confirmed by (4.97) and (4.98).

The excitons associated with the valence band in most of inorganic semiconductors belong to the case D << B. The core excitons certainly belong to the case D >> B. It should be noted, however, that the radiative and the Auger broadening becomes more important for inner core excitons.

4.5.3 Sideband Structures

In order to see how the spectral details (4.98) of the energy fluctuation $V(t)$ is reflected in the absorption spectra, it is convenient to expand (4.96), with the use of (4.98), as

$$I(\omega) = \sum_{n=0}^{\infty} I_n(\omega)$$

where

$$I_n(\omega) = \frac{|M|^2}{2\pi \hbar n!} \int_{-\infty}^{+\infty} dt \ \exp\left[-S + i\left(\omega - \frac{E_0 + \Delta_0}{\hbar}\right)t\right]$$

$$\times \int_{-\infty}^{\infty} d\omega_1 \sim d\omega_n \ \frac{J(\omega_1)...J(\omega_n)}{(\hbar\omega_1)^2...(\hbar\omega_n)^2} \ e^{-i(\omega_1+...+\omega_n)t}$$

(4.103)

and

$$S \equiv \int_{-\infty}^{+\infty} \frac{J(\omega)}{(\hbar\omega)^2} \ d\omega \quad .$$

The gross contour of the absorption spectra discussed in the previous subsection turns out to be the superposition of the *multiple sidebands* (4.103) which are the multiple convolutions of the first-order sideband $J(\omega)/(\hbar\omega)^2$ and whose integrated intensities obey the *Poisson distribution*: $|M|^2 \ e^{-S} \ S^n/n!$. The zeroth order line ($n = 0$) is subject to the intensity reduction e^{-S} which is compensated by the accompanying sidebands ($n = 1, 2, ...$).

Let us first consider a localized excitation ($B = 0$) and then the motional effect ($B \neq 0$) of the exciton. In the former case, $J(\omega)$ represents the frequency spectra of the elementary excitations concerned (e.g. phonons), weighted with the coupling coefficient. Phonon sidebands of localized excitation are discussed in detail in [4.44]. As seen from (4.97) and (4.98), the motional effect is to replace ω_K by $\omega_K + (E_{-K} - E_0)/\hbar$, namely, $\hbar\omega$ in $J(\omega)$ now represents the sum of the phonon energy and the recoil kinetic energy of exciton. This results in the spectral broadening of $J(\omega)$, and hence, in the *reduction* of coupling strength S [4.48] as well as in the *narrowing* of the width ($\Gamma_0 \propto J(0)$) of the absorption spectra (note that

$$\int_{-\infty}^{+\infty} J(\omega)\,d\omega = g(0)$$

is independent of the motional effect).

Two examples of sidebands describable by a dielectric formulation will be mentioned. The first is the optical phonon in diatomic ionic crystals. The dielectric function, with neglection of k-dependence, is given by

$$\epsilon(\omega) = \epsilon_\infty + \frac{(\epsilon_0 - \epsilon_\infty)\omega_t^2}{\omega_t^2 - \omega^2} \quad ,$$

so that

$$\text{Im}\left(-\frac{1}{\epsilon(\omega)}\right) = \frac{\pi\omega_\ell}{2}\left(\frac{1}{\epsilon_\infty} - \frac{1}{\epsilon_0}\right)[\delta(\omega - \omega_\ell) - \delta(\omega + \omega_\ell)] \quad . \tag{4.104}$$

Here ω_ℓ and ω_t are, respectively, frequencies of longitudinal and transverse optical phonons. Note that they correspond to the zero and the pole of $\epsilon(\omega)$, similarly to the case of longitudinal and transverse excitons. With the use of (4.97') and (4.98), the first order phonon sideband is given by

$$\frac{J(\omega)}{(\hbar\omega)^2} = 2\pi\left(\frac{1}{\epsilon_\infty} - \frac{1}{\epsilon_0}\right) e^2\hbar\omega_\ell \sum_K |f_K|^2 K^{-2} (\hbar\omega)^{-2}$$

$$\times \left[(n+1)\delta\left(\frac{E_K - E_0}{\hbar} + \omega_\ell - \omega\right) + n\delta\left(\frac{E_K - E_0}{\hbar} - \omega_\ell - \omega\right)\right] \quad , \tag{4.105}$$

where $n = (e^{\beta\hbar\omega_\ell} - 1)^{-1}$ is the number of optical phonons. It consists of phonon emission and absorption terms, which start at $\pm\hbar\omega_\ell$, respectively, and extend towards higher energy due to the recoil kinetic energy of exciton.

The second example is the plasmon sidebands accompanying the photoabsorption of a localized (inner core) exciton. If we assume that valence electrons are responsible for the plasma oscillation, we obtain from (4.8)

$$\underset{\sim}{\epsilon}(\omega) - 1 = 4\pi \sum_\nu \frac{2\omega_{\nu\mu}(\underline{P}_0')_{\mu\nu}(\underline{P}_0')_{\nu\mu}}{\hbar(\omega_{\nu\mu}^2 - \omega^2)} \tag{4.106}$$

where μ stand for the filled valence band and

$$\underline{P}_0' = e \sum_{j=1}^{n_v} \underline{r}_j \quad ,$$

n_v being the density of valence electrons. Expanding the summand of (4.106) with respect to $(\omega_{v\mu}/\omega)^2$ and using the f-sum rule, we can write the leading term of (4.106) as $-4\pi n_v e^2/(m\omega^2)$, which yields

$$\text{Im}\left\{-\frac{1}{\epsilon(\omega)}\right\} = \frac{\pi\omega_{pv}}{2}[\delta(\omega - \omega_{pv}) - \delta(\omega + \omega_{pv})] \quad , \quad \omega_{pv} \equiv \frac{4\pi n_v e^2}{m} \quad . \tag{4.107}$$

The expression (4.107) is valid when the relation $\omega_{v\mu}/\omega_{pv} \ll 1$ is satisfied for $\omega_{v\mu}$ making a dominant contribution to the oscillator strength. From (4.97') and (4.98), $J(\omega)$ is given by

$$J(\omega) = \sum_{\underline{K}} |\gamma_{\underline{K}}|^2 \delta(\omega - \omega_{pv}) \quad ,$$

where $|\gamma_{\underline{K}}|^2 = 2\pi \hbar\omega_{pv} e^2 K^{-2} |f_{\underline{K}}|^2$, and we have made use of $\beta\hbar\omega_{pv} \gg 1$ and $E_{\underline{K}} = E_0$. Substituting $J(\omega)$ into (4.103), we obtain the plasmon sidebands in the form

$$I(\omega) = |M|^2 e^{-S} \sum_{n=0}^{\infty} \frac{S^n}{n!} \delta(\hbar\omega - E_0 - \Delta_0 - n\hbar\omega_{pv})$$

where

$$S = \sum_{\underline{K}} |\gamma_{\underline{K}}|^2/(\hbar\omega_{pv})^2 \quad \text{and} \quad \Delta_0 = -\sum_{\underline{K}} |\gamma_{\underline{K}}|^2/(\hbar\omega_{pv}) \quad .$$

4.5.4 Relaxation in Exciton-Phonon Systems

It has been shown that the exciton motion gives rise to the motional narrowing of the absorption spectrum. The motional effect also plays an important role in the relaxation of the exciton. As shown before, if the exciton is assumed to be localized, the energy of the exciton-phonon system is lowered by E_{LR} due to the distortion of the surrounding lattice. On the other hand, if the exciton is assumed to propagate freely through the undistorted lattice, the energy gain due to the exciton motion is B. Hence, it is expected that the nature of the relaxed excited state depends on whether $B \gtrless E_{LR}$. In fact, for the electron-phonon system with *short range* interaction such as deformation potential, it was shown [4,49,50] that a *discontinuous* change takes place around $B \sim E_{LR}$. The same holds for the exciton-phonon system since the interaction is always short range irrespective of whether or not one deals with an acoustic or optical mode because of charge neutrality [4.51]. If we start from the rigid lattice exciton band, whose bottom is denoted as the *free state* F, and gradually introduce a local lattice distortion Q_1, a bound state splits off from the exciton band only beyond a finite Q_1 value, and hence, with a finite potential barrier as shown in Fig. 4.6b [4.49-53]. When E_{LR}/B is sufficiently small the adiabatic potential for this split-off excitonic state increases monotonically, but otherwise it has a minimum S. The state S is called the *self-trapped state*, where the exciton is trapped by the potential of the lattice distortion induced by the exciton itself. F and S are separated by a potential barrier, and F or S is stable according as $E_{LR} \lessgtr B$. Namely, if we change B or E_{LR} continuously, the lowest relaxed state changes discontinuously at $E_{LR} \simeq B$. Corresponding to this discontinuity, the luminescence spectrum changes in a drastic way, because the emission spectrum due to the annihilation of the F exciton is a sharp resonance line appearing at the absorption edge, while that of the S exciton is a Stokes shifted (by $\sim 2E_{LR} - B$) broad band of Gaussian line shape.

In fact the excitons observed experimentally in inorganic insulators can be distinctly classified into F and S types. For instance, the valence excitons in rare gas solids [4.54], alkali halides [4.55] and AgCl [4.56,57] are S type, while those in AgBr [4.58,56], TlBr and TlCl [4.59] are F type. In the mixed crystal system $AgBr_{1-x} Cl_x$ [4.60], the excitonic luminescence changes from F type to S type rather abruptly around x ~ 0.45. A weak emission line nearly at the position of the first exciton absorption band, coexistent with the strong emission band of S-type, has been observed in rare gas solids [4.61] and alkali halides [4.62] and ascribed to the free exciton. This means that the F state is not unstable but *metastable* because of the finite potential barrier between F and S states. The existence of the metastable F state in alkali halides had been claimed indirectly in connection with the energy transfer to impurities [4.63-65]. The dynamics of excitons in rare gas solids are discussed in detail in [4.54,61]. Inner core excitons should be S type due to the small value of B. The self-trapping of individual particles, electrons and holes, are also describable in a similar way as excitons. In most ionic crystals the hole has a larger effective mass than the electron, so that the self trapping of hole and exciton behaves in the same way, whereas most of electrons are known to be of the F type.

Finally it should be stressed that the criterion between the free and self-trapped states $B \gtrless E_{LR}$ is different from that of the motional narrowing of the absorption spectra $B \gtrless D$. Since $E_{LR}(\sim eV)$ is usually much larger than $1/\beta$, we obtain $D = (2E_{LR}/\beta)^{1/2} < E_{LR}$. Hence, we have a fairly wide range of the coupling constant such that the exciton is nearly free immediately after its radiative creation but becomes self-trapped after lattice relaxation. Furthermore, if we change B continuously, the narrowing of the absorption spectra occurs *gradually* in the vicinity of B ~ D, but the change of the luminescence due to the self-trapping occurs *discontinuously* at B ~ E_{LR}.

4.6 Many Body Effects in Metals

The most important final state interaction in the core electron excitation in a metal is the simultaneous excitations and redistributions of conduction electrons. They cause the so-called *Fermi edge singularity* in absorption and emission spectra.

4.6.1 Friedel Sum Rule and Anderson Orthogonality Theorem

Let us suppose that a core hole is suddenly created at the time $t = 0$. Then conduction electrons are scattered by the potential of the core hole and thereby redistribute so as to screen its charge. If we disregard a finite life-time of the core hole, the system tends to its lowest relaxed state $|g'>$ in the limit of $t \to \infty$. There, the screening is completed and the effect of the scattering is represented by the phase shift $\delta_\ell(\varepsilon)$ of conduction electrons, which appears in the asymptotic form of the radial wave function

$$u_{k\ell}(r) = \sin\left[kr + \delta_\ell(\varepsilon) - \frac{\ell}{2}\pi\right] \quad . \tag{4.108}$$

With the use of $u_{k\ell}(r)$ we can calculate the change of the conduction electron number around the core hole, which should be put equal to unity (i.e. the number of core holes) because of the perfect screening:

$$\frac{2}{\pi} \sum_\ell (2\ell + 1)\delta_\ell(\varepsilon_F) = 1 \quad . \tag{4.109}$$

This is the so-called *Friedel sum rule* [4.66].

According to the *Anderson orthogonality theorem* [4.67], the lowest states of the Fermi sea, $|g'>$ and $|g>$, in the presence and absence of the core hole are orthogonal; more precisely, the overlap integral $<g_N|g_N'>$ in the N conduction electron system is expressed as

$$<g_N|g_N'> = N^{-\sum_\ell (2\ell+1)[\delta_\ell(\varepsilon_F)/\pi]^2} \quad , \tag{4.110}$$

so that

$$\lim_{N\to\infty} <g_N|g_N'> = <g|g'> = 0 \quad .$$

Next we study the relationship between the redistribution of conduction electrons and the photoabsorption spectrum of a core electron (with energy ε_c) to the conduction band. Similarly to (4.87), the absorption spectrum at zero temperature is given by

$$I(\omega) = \sum_e |\langle e|H_I|g\rangle|^2 \, \delta(\hbar\omega - E_e + E_g + \varepsilon_c) \quad , \tag{4.111}$$

where

$$H_I = \sum_k M_k a_k^+ + c.c \quad . \tag{4.112}$$

Here $|g\rangle$ is the ground state (with energy E_g) of unperturbed Hamiltonian $H_g = \sum_k \varepsilon_k a_k^+ a_k$ and $|e\rangle$ is the eigenstate (with energy E_e) of the Hamiltonian $H_e = H_g + V$, V being the core hole potential. If we disregard the effect of the electron created by H_I, the spectrum is just the Fourier transform of the time evolution of the conduction electron redistribution

$$I'(\omega) = \sum_e |\langle e|g\rangle|^2 \, \delta(\hbar\omega - E_e + E_g + \varepsilon_c)$$

$$\tag{4.113}$$

$$= \frac{1}{2\pi\hbar} \int_{-\infty}^{\infty} dt \, e^{i[\omega + (E_g + \varepsilon_c)/\hbar]t} \, \langle g|e^{-i\frac{H_e}{\hbar}t}|g\rangle \quad .$$

With the orthogonality theorem

$$\lim_{t\to\infty} \langle g|\exp(-iH_e t/\hbar)|g\rangle = \langle g|g'\rangle = 0$$

in mind, we now assume

$$\langle g|e^{-i\frac{H_e}{\hbar}t}|g\rangle \sim e^{-i\frac{\tilde{E}_g}{\hbar}t} \, t^{-\gamma} \quad , \tag{4.114}$$

for large t, then (4.113) becomes

$$I'(\omega) \sim \begin{cases} \tilde{\omega}^{-(1-\gamma)} & \text{for} \quad \tilde{\omega} \geq 0 \\ 0 & \text{for} \quad \tilde{\omega} < 0 \end{cases} \tag{4.115}$$

in the neighborhood of the threshold $\tilde{\omega} \equiv \omega - (\tilde{E}_g - E_g - \varepsilon_c)/\hbar$. As expected from (4.110), γ should be a function of the phase shifts. If $1 - \gamma > 0$, $I'(\omega)$ has a one-sided divergence, which is called the *infrared divergence*. It should be mentioned that the spectrum $I'(\omega)$ is directly observable with inner core photoelectron spectroscopy [4.68], where a core electron is excited above the ionization threshold and removed away from the crystal.

A weak coupling version of (4.114) will be derived in the next subsection, and the effect of the excited electron a_k^+ will be discussed in Sect. 4.6.3.

4.6.2 Infrared Divergence

Since the core hole potential scatters conduction electrons below ε_F to above ε_F, the redistribution of conduction electron is equivalent to the excitation of *electron-hole pairs* in the unperturbed Fermi sea. From this viewpoint, $I'(\omega)$ is regarded as the spectrum of the simultaneous excitations of electron-hole pairs accompanying the sudden removal of the core electron. Since the electron-hole pair is describable approximately as a Boson, $I'(\omega)$ can be obtained, as pointed out by HOPFIELD [4.69] (see also [4.70]), in the way similar to the calculation of phonon sidebands. Namely, we obtain in the same way as (4.95)

$$<g|e^{-i\frac{H_e}{\hbar}t}|g> \simeq \exp\left[-i\frac{E_g}{\hbar}t - \frac{1}{\hbar^2}\int_0^t dt_1 \int_0^{t_1} dt_2\ <g|V(t_1)V(t_2)|g>\right] \quad , \qquad (4.116)$$

where

$$V(t) = e^{i\frac{H_g}{\hbar}t}\ V\ e^{-i\frac{H_g}{\hbar}} \quad .$$

Hereafter, we assume, for simplicity, a short range interaction $V = v\sum_{kk'} a_k^+ a_{k'}$, whereby only s-partial waves are scattered, then the second term of the square bracket in (4.116) is calculated as

$$-v^2 \sum_{k>k_F} \sum_{k'<k_F} \frac{-i\frac{(\varepsilon_k-\varepsilon_{k'})}{\hbar}t+1-e^{-i\frac{\varepsilon_k-\varepsilon_{k'}}{\hbar}t}}{(\varepsilon_k-\varepsilon_{k'})^2} = -g^2 \log\left(\frac{iDt}{\hbar}\right) - i\frac{\tilde{E}_g-E_g}{\hbar}t \quad , \qquad (4.117)$$

where $g \equiv -\rho v$, ρ being the density of states of the conduction band at ε_F, and D is a cut off energy of the order ε_F. Hence we obtain

$$<g|e^{-i\frac{H_e}{\hbar}t}|g> = e^{-i\frac{\tilde{E}_g}{\hbar}t}\left(i\frac{D}{\hbar}t\right)^{-g^2} \qquad (4.118)$$

and

$$I'(\omega) = \frac{1}{D\Gamma(g^2)(\hbar\tilde{\omega}/D)^{1-g^2}} \quad , \quad \text{for} \quad \tilde{\omega} \geq 0 \quad , \tag{4.119}$$

where $\Gamma(g^2)$ is the gamma function.

Unlike the usual simultaneous excitation of phonons, the number of excited electron-hole pairs

$$\bar{n} = v^2 \sum_{k>k_F} \sum_{k'<k_F} \frac{1}{(\varepsilon_k - \varepsilon_{k'})^2}$$

is found to diverge due to the *unbounded* increase in low energy pairs, from which stems the divergence of $I'(\omega)$ at $\tilde{\omega} = 0$, i.e. the infrared divergence. As expected from the orthogonality theorem, the exponent of the divergence should be expressed, in a more rigorous calculation [4.71], in terms of the phase shift; the exponent $1 - g^2$ in (4.119) should be replaced by $1 - (\delta/\pi)^2$, which reduces to the former in the case of small v due to the relation $\delta \equiv \delta_0(\varepsilon_F) = -\tan^{-1}(\pi\rho v)$.

Of course, it is also possible to calculate $I'(\omega)$ with the use of the dielectric functional formulation given in the preceding subsection. Then, (4.116) is replaced by

$$<g|e^{-i\frac{H_e}{\hbar}t}|g> \simeq \exp\left[-i\frac{E_g}{\hbar}t + \frac{\hbar}{4\pi^2}\sum_q q^2|\varphi_q|^2 \int_{-\infty}^{\infty} d\omega \frac{-i\omega t+1-e^{-i\omega t}}{(\hbar\omega)^2} \text{Im}\left\{\frac{1}{\epsilon(q,\omega)}\right\}\right] \quad ,$$

where $\epsilon(q,\omega)$ is the dielectric function of electron gas given by

$$\epsilon(\underline{q},\omega) = 1 + \frac{4\pi e^2}{q^2} \sum_{k<k_F} \sum_{|\underline{k}+\underline{q}|>k_F} \frac{1}{\varepsilon_{\underline{k}+\underline{q}} - \varepsilon_{\underline{k}} - \hbar\omega} \quad ,$$

and $\varphi_g = -4\pi e/q^2$. The logarithmic singularity similar to (4.117) arises from the fact that Im $\{1/\epsilon(\underline{q},\omega)\} \propto \omega$ for $\omega \to 0$. Since $\epsilon(0,\omega) = 1 - (\omega_p/\omega)^2$, where $\omega_p \equiv 4\pi ne^2/m$ with the density n of conduction electrons, the plasmon sidebands [4.72,73] can also be treated within the same formalism.

4.6.3 Fermi Edge Singularity

Let us take account of the effect of the electron excited from the core level. With the use of the linked-cluster expansion, the absorption spectrum (4.111) is rewritten as

$$I(\omega) = \frac{1}{2\pi\hbar} \int_{-\infty}^{\infty} dt \; e^{i\left(\omega + \frac{E_g + \varepsilon_c}{\hbar}\right)t} \; e^{-i\frac{H_e}{\hbar}t} <g|H_I e^{-i\frac{H_e}{\hbar}t} H_I|g>$$

(4.120)

$$= \frac{1}{2\pi\hbar} \int_{-\infty}^{\infty} dt \; e^{i\left(\omega + \frac{E_g + \varepsilon_c}{\hbar}\right)t} \left[\sum_{kk'} M_k^+ M_k <g|a_{k'} \; e^{-i\frac{H_e}{\hbar}t} \; a_k^+|g> \right]_c <g|e^{-i\frac{H_e}{\hbar}t}|g> \; .$$

Here $[\;]_c$ means the contribution from connected diagrams, which was first calculated by MAHAN [4.74] with the most divergent term approximation in the form

$$\left[\sum_{kk'} M_k^+ M_k <g|a_{k'} \; e^{-i\frac{H_e}{\hbar}t} \; a_k^+|g> \right]_c = \frac{|M_{k_F}|^2}{it} (iDt)^{2g} \; e^{-i\varepsilon_F t} \; .$$

(4.121)

Disregarding the effect of the electron redistribution, i.e. putting $<g|\exp(-H_e t/\hbar)|g> = \exp(-iE_g t/\hbar)$, MAHAN obtained the Fermi edge singularity $I(\omega) \sim 1/(\hbar\omega - \varepsilon_F + \varepsilon_c)^{2g}$, which is caused by the scattering of the excited electron by the core hole (*called exciton effect in metals*). Taking account of both the redistribution and exciton effects, we obtain from (4.118,120 and 121)

$$I(\omega) = \frac{|M_{k_F}|^2 \; \rho}{\Gamma(1-2g+g^2)\left(\frac{\hbar\omega - E_t}{D}\right)^{2g-g^2}} \quad , \quad (\hbar\omega \geq E_t) \quad ,$$

(4.122)

where the Fermi edge is represented by $E_t \equiv \varepsilon_F - \varepsilon_c - (\tilde{E}_g - E_g)$. By replacing g with δ_0/π, and by including spin and orbital angular momenta, we finally obtain the absorption spectrum from a core state with the orbital angular momentum ℓ

$$I(\omega) = W_{\ell-1}(\omega)\left(\frac{D}{\hbar\omega - E_t}\right)^{\alpha_{\ell-1}} + W_{\ell+1}(\omega)\left(\frac{D}{\hbar\omega - E_t}\right)^{\alpha_{\ell+1}} \quad ,$$

(4.123)

where $W_{\ell\pm1}(\omega)$ is the spectrum for vanishing V, and

$$\alpha_\ell = \frac{2\delta_\ell}{\pi} - 2 \sum_{\ell'=0}^{\infty} (2\ell' + 1)\left(\frac{\delta_{\ell'}}{\pi}\right)^2 \quad .$$

(4.124)

This is the result derived originally by NOZIÈRES and DE DOMINICIS [4.71]. The spectrum (4.123) is asymptotically exact when $\hbar\omega$ approaches E_t. If $\alpha_{\ell-1}$ or $\alpha_{\ell+1}$ is positive, $I(\omega)$ is enhanced and diverges at the Fermi edge, but if both of them are negative, $I(\omega)$ is suppressed and becomes a rounded edge. In the same way, the luminescence spectrum also has the Fermi edge singularity with the same exponent as that of the absorption spectrum. Some extensions of the result (4.123) have also been made by taking account of the electron-hole exchange interaction [4.75,76] and the coexistence of exchange and spin-orbit interaction [4.77], and by extending the validity range of $\hbar\omega$ with the use of spectral convolution [4.78].

According to experimental observations [4.79-81], the spectra of the L_{23} edges of Na, Mg and Al are sharp and peaked, while those of the K edges of Li and Al are broad and rounded. If we assume that the screened potential of the core hole is sufficiently of short range, the s-wave scattering is predominant, i.e. $\delta_0 \gg \delta_1, \delta_2$ etc., so that we obtain $\alpha_0 > 0$ and $\alpha_1 < 0$ from (4.124), together with the Friedel sum rule (4.109). This means that the L edge should be peaked and the K edge should be rounded, in *qualitative* agreement with the experimental observations. However, it has been a subject of controversy whether those line shapes are *quantitatively* reproducible by the MAHAN-NOZIERES-DE DOMINICIS theory alone. It was pointed out by DOW (see [4.82] and references therein) and others [4.83] that the electron-phonon interaction is important in the K edge spectrum of Li, and that one-electron characters, both effects of the band density of states and transition matrix element, are essential for the L_{23} edge of Mg and the K edge of Al. Furthermore, there remains a problem on the interpretation for the momentum transfer dependence of electron energy loss spectra in Li and Mg [4.84,85], which will be mentioned in Sect. 4.8 again.

4.7 Final State Interactions Associated with Incomplete Shells

Rare earth atoms (for $58 \leq Z \leq 69$) have incompletely filled 4f-states bound *inside* the centrifugal potential barrier. Hence a core hole created by photon is subject to a strong final state interaction with the 4f-electrons. Incompletely filled d-shells in transition elements are also in a similar situation. In metals consisting of these atoms, it is interesting to study how the *local versus band characters* are reflected in inner-level spectroscopy.

4.7.1 Multiplet Splitting

In the optical absorption due to $4d^{10}\,4f^N \rightarrow 4d^9\,4f^{N+1}$ transition in rare earths [4.86-91] a large *multiplet splitting* extending over $10 \sim 30$ eV is observed because of the strong exchange interaction between 4f-electrons and a 4d-core hold. The multiplet structure of the $4d^9\,4f^{N+1}$ configuration in trivalent rare earth ions was calculated by DEHMER and his co-workers [4.92-96]. Among the direct integrals (F^2,

F^4), exchange integrals (G^1, G^3, G^5) and spin-orbit coupling parameters (ζ_d, ζ_f) included in the calculation, they adjusted F^2 and G^1 so as to get a best fit to the absorption spectrum observed experimentally, fixing $F^2 : F^4$ and $G^1 : G^3 : G^5$ to those of the Hartree-Fock calculation and setting ζ_d and ζ_f to those given by other experiments. The scaling (reduction) of the interaction integrals from their Hartree-Fock values is justified by taking account of virtual 4d-shell and 4f-shell excitation [4.96]. As an example, the absorption spectrum for Ce^{3+} thus calculated is shown in Fig. 4.7 with discrete lines, the continuous spectrum is an experimental result for Ce metal [4.88]. Rather good correspondence between them shows that the gross feature of the spectrum originates from the multiplet structure and, furthermore, suggests that the multiplet splitting is essentially a local event which is more or less common to atoms, molecules and solids. The latter point was confirmed directly from comparison of the spectra observed for atomic and metallic samples [4.91]. In metals, however, it is also an interesting problem to study theoretically how the conduction electrons behave in the photoabsorption, even though their role is minor.

Similar multiplet splitting is also expected for transition elements. However, the centrifugal potential for d-electrons is smaller than that for f-electrons, so that in transition metals the d-electron can move resonantly from atom to atom by an s-d mixing interaction with an overlapping s-band as well as by a direct d-d transfer, and consequently forms the *d-band*. At the same time, the d-electrons have a tendency of localization, since they are kept away from each other by their repulsive Coulomb interaction. Due to the competition of the two opposite tendencies, the d-electrons show a complicated behavior of partly local and partly itinerant nature in their electrical, optical and magnetic properties.

According to experimental observations [4.97], the M_{23} absorption spectrum of 3d-transition metals consist of a prominent peak, whose tail extends on the high energy

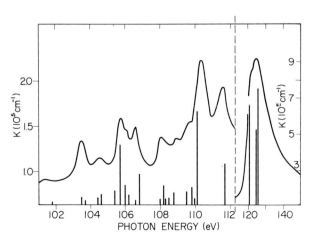

Fig. 4.7 Comparison of the experimentally observed N_{45} absorption spectrum for Ce [4.88] (continuous spectrum), with calculated relative positions and strengths of the $4d^{10}4f(^2F_{5/2}) \rightarrow 4d^94f^2$ transitions (discrete lines) [4.94]

side as much as a few tens of electron volts. Since this width is much larger than the calculated width of the empty part of the 3d-band, it should be ascribed to a superposition of multiplets of atomic $3p^5 3d^{N+1}$ configuration [4.92]. In this connection, electron energy loss spectra corresponding to the $3p^6 3d^N \rightarrow 3p^5 3d^{N+1}$ transition are analyzed successfully on the basis of an atomic picture [4.98,99]. In the analysis, the multiplet splitting of the $3p^5 3d^{N+1}$ configuration is shown to be essentially important, as well as the Fano-type interference effect between a $3p^6 3d^N \rightarrow 3p^6 3d^{N-1}$ εf transition and a $3p^6 3d^N \rightarrow 3p^5 3d^{N+1}$ transition which autoionize into $3p^6 3d^{N-1}$ εf via the super Coster-Kronig transition mentioned in Sect. 4.4 (see also Sect.5.5.1.2). Such an apparent manifestation of atomic characters in metals may be partly due to the strong perturbation coming from the core hole and partly due to the rapidity of the super Coster-Kronig transition. However one should always keep in mind the coexistence of the itinerant nature of d-electrons. It is also pointed out that the M_{23} absorption spectrum in Ni has structures in fair agreement with the d-band density of states [Ref.4.17, p. 66]. In any case, a unified interpretation for the local versus band characters is an important problem in metals with incomplete shell.

4.7.2 Local versus Band Pictures

We introduce here a simple model [4.100] in order to get an insight into some important aspects of the interplay between local and band characters in metals with an imcomplete shell. Let us consider a system consisting of incomplete shells (denoted by d-state, irrespective of the actual symmetry), a conduction band (denoted by s-state), and core states. The d-state is assumed to be a nondegenerate localized state interacting with the s-state through the s-d mixing interaction V. For simplicity, electrons are treated as spinless. A core electron can be excited optically to the s- and d-states with transition matrices M_s and M_d, respectively. It is assumed that in the initial state of the optical transition the d-level of each atom is high enough above the Fermi level ε_F, while in the final state the d-level of the excited atom is lowered down to ε_d due to the interaction with a core hole left behind. It is to be understood that ε_d in our simplified model represents one of the multiplet terms near the absorption edge.

Disregarding the effect of the d-state in the initial state, we express the Hamiltonian for the initial state as $H_g = \sum_k \varepsilon_k a_k^+ a_k$ and that for the final state as $H_e = H_g + \varepsilon_d a_d^+ a_d + V \sum_k (a_k^+ a_d + a_d^+ a_k)$. Although we have neglected the direct interaction of s-electrons with the core hole, the scattering of s-electrons by the hole occurs due to the second order process via the intermediate d-state with matrix element

$$v_{kk'} = V^2/(\varepsilon_f - \varepsilon_d) \quad , \tag{4.125}$$

which will be shown to cause the infrared divergence.

For vanishing V, the absorption spectrum consists of a line spectrum at $\hbar\omega = \varepsilon_d$ ($\hbar\omega$ being measured with the core level as origin) corresponding to a final state

$$|d\rangle \equiv a_d^+ \prod_{k\leq k_F} a_k^+ |0\rangle$$

with the intensity $|M_d|^2$ and a continuous spectrum $|M_s|^2\rho$ above ε_F, corresponding to the final states

$$|k'\rangle \equiv a_{k'}^+ \prod_{k\leq k_F} a_k^+ |0\rangle \quad,$$

as shown with dashed lines in Fig. 4.8a and b. We next study how this spectrum is affected by switching on a weak s-d mixing.

Case A: $\varepsilon_d < \varepsilon_F$

When V is switched on, the lowest excited state $|d\rangle$ for the unperturbed system is coupled, by the second order process ($d \to k$ and then $k' \to d$) with matrix element $V_{kk'}$, to the state $a_{k'}^+ a_k |d\rangle$ ($\varepsilon_k < \varepsilon_F < \varepsilon_{k'}$), which means the excitation of an *electron-hole pair* on the Fermi sea. The repetition of such a process results in the *infrared divergence* at the absorption edge, as well as the shift of the edge, $\tilde{\varepsilon}_d - \varepsilon_d$, due to the simultaneous excitations of infinite number of electron-hole pairs with infinitesimal excitation energies. Namely, the absorption edge is expressed as

$$I(\omega) \sim (\hbar\omega - \tilde{\varepsilon}_d)^{-(1-g^2)} \quad , \quad g \equiv -\rho V^2/(\varepsilon_F - \varepsilon_d) \quad . \tag{4.126}$$

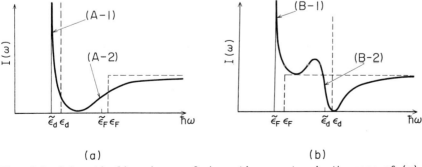

(a)   (b)

Fig. 4.8 Schematic line shapes of absorption spectra in the case of (a) $\varepsilon_d < \varepsilon_F$ and (b) $\varepsilon_d > \varepsilon_F$. The spectra for vanishing s-d mixing are shown with dashed lines [4.100]

It should be noted that the negative power of the divergence is not $2g - g^2$ but $1 - g^2$, because the core electron is excited to the localized d-state, instead of the conduction band, so that the singularity arises only from the redistribution of conduction electrons, see (4.119).

On the other hand, the final states $|k'>$ responsible for the discontinuous Fermi edge at $\hbar\omega = \varepsilon_F$ are mixed resonantly with $a_d^+ a_k |k'>$ by the perturbation V, which causes not only the energy shift $\tilde{\varepsilon}_F - \varepsilon_F$ but also the life time broadening $\hbar/\tau \sim 2\pi\rho V^2$. Thus the Fermi edge is blurred out in the form $|M_s|^2 \rho\{1/2 + (1/\pi) \tan^{-1}[(\hbar\omega - \tilde{\varepsilon}_F)/(\pi\rho V^2)]\}$, as shown in Fig. 4.8a.

According to experiments of N_{45} absorption spectra in La and Ce [4.89], no sharp Fermi edge is observed at the Fermi level determined by photoelectron measurements, and furthermore one or several sharp peaks corresponding to the transition between 4d- and 4f-levels are recognized below the presumed Fermi level. This situation is qualitatively in agreement with the spectrum of Fig. 4.8a.

Case B: $\varepsilon_d > \varepsilon_F$

In this case the spectrum at the Fermi edge $\tilde{\varepsilon}_F$ is given by quite the same singular form as that of simple metals [see (4.122)], only with the replacement of the core hole potential v by $v_{kk'}$ given by (4.125)

$$I(\omega) \sim (\hbar\omega - \varepsilon_F)^{-(2g-g^2)} \quad . \tag{4.127}$$

In the neighborhood of $\hbar\omega \sim \varepsilon_d$, on the other hand, excitations of the core electron to the s- and d-states interfere with each other through the mixing V, resulting in the *Fano effect*. The line shape is given by (4.83) with

$$\epsilon = (\hbar\omega - \varepsilon_d - \Delta)/\Gamma \quad , \quad q = [VM_d/M_s + \Delta]/\Gamma \quad , \tag{4.128}$$

where Δ and Γ are, respectively, real and imaginary parts of the self-energy of the state $|d>$. The spectrum of the case B is shown schematically in Fig. 4.8b with a solid curve.

The calculations described so far show that the *local* [(A - 1) and (B - 2) of Fig. 4.8] and *band* [(A - 2) and (B - 1)] characters of the one electron picture manifest themselves as separate structures in the absorption spectra with interference between them. On the other hand, the absorption threshold always shows *singularity* characteristic of metallic electrons irrespective of its one electron origin — local (A - 1) or band (B - 1). However, this result is limited to the situation where the coupling constant $|g|$ is smaller than unity. When $|g| \gtrsim 1$, the local and band characters become dissolved into a single entity characteristic of many electron system, i.e. only one prominent peak of absorption spectrum tailing off on the high energy side. The absorption edge accompanying the infrared divergence is always well-defined as the lowest state of the many electron systems, but there is no longer a clear-cut difference between the cases A and B, and it is not meaningful whether one denotes the absorption edge as $\tilde{\varepsilon}_F$ or $\tilde{\varepsilon}_d$ on the one electron picture. The negative power of the infrared divergence is now given by the unified form $\alpha = 2(\delta/\rho) - (\delta/\rho)^2$ irrespective of the cases A and B, where δ is the phase shift at ε_F given by $\delta = -\tan^{-1}[\pi\rho V^2/(\varepsilon_F - \varepsilon_d)]$, $(0 \leq \delta \leq \pi)$. When $|g| \ll 1$, the power α is easily checked to reduce to $1 - g^2$ and $2g - g^2$, respectively, corresponding to cases A and B.

For the inner core photoelectron spectrum, we expect, from the present model, to have a threshold singularity of the negative power $1 - (\delta/\pi)^2$, as well as a characteristic spectral structure originating from different final states corresponding to the *occupied* and *unoccupied* d-state [4.101]. Some experimental evidence [4.102-104] has been found for such effects of the incomplete shell. An extention of the theory so as to include the electron spin has also been made [4.105,106].

4.7.3 Correlation Effects in Narrow d-Band

We give two more examples of the phenomena suggesting the importance of the correlation effect of d-electrons in transition metals. The firts is a splitting of 3s core photoelectron spectrum in Fe, Co and Ni, which is caused by the exchange interaction between the core hole and 3d-electrons. FADLEY and SHIRLEY [4.107] observed that the splitting persists above the Curie temperature, so that the origin of this splitting is not the long-range-ordered magnetic moment of 3d electrons but their local moment which is fluctuating from time to time. In view of the itinerant theory of magnetism, the appearance of such a local moment is ascribed to the strong correlation of 3d-electrons, as well as the perturbation due to the core hole. Let the characteristic amplitude of the fluctuating local moment be denoted by S_d and its correlation time by τ_c, then the splitting can be observed only when $J(2S_d + 1) > \hbar/\tau_c$, J being the exchange coupling constant between the 3d- and 3s-spins, and otherwise the splitting is washed away by the rapidity of the fluctuation (*motional narrowing*). These situations correspond to the slow and rapid modulations discussed in Sect. 4.5.

The next example is M_{23} absorption spectrum in Cu-Ni alloys. GUDAT and KUNZ [4.108] observed that the spectrum of the alloy is almost the same as those of pure Cu and Ni metals superimposed proportionally to their composition. This means that d-electrons in the alloy are well localized on each atomic site, contrary to the expectation from the simple band theory of alloys, and also suggests the strong correlation of 3d-electrons and the perturbation caused by the core hole.

4.8 Inelastic X-Ray Scatterings

In addition to the absorption and emission processes treated so far, the scattering of radiation also provides us with important information on the microscopic behavior of materials. X-ray scattering is divided into two classes: elastic scattering, which gives rise to the well-known Bragg reflection in solids, and inelastic scattering, which will be discussed in the following.

When incident radiation with (ω_1, k_1) is scattered inelastically to (ω_2, k_2) by a material system, the energy $\hbar\omega \equiv \hbar\omega_1 - \hbar\omega_2$ and momentum $\hbar k \equiv \hbar k_1 - \hbar k_2$ are transferred to some excitations left in the system, so that the scattering cross section contains both *energy* and *momentum* information about the excitations. This is in contrast to the absorption or emission process, which contains only energy informations. In the case of $\hbar\omega \gg E_B$, E_B being the binding energy of electrons participating in the scattering, the inelastic X-ray scattering is called *Compton scattering*, and otherwise it is called *Raman scattering*. The former provides us with the information on the momentum distribution of electrons in the ground state, while the latter gives the information on the behavior of collective and individual excitations.

Characteristic features of the X-ray Raman scattering are understood from comparison with light (visible) scattering and neutron scattering. In the scattering of light, whose wave length is of order 10^4 Å, $k \equiv |k_1 - k_2|$ is orders of magnitude smaller than the reciprocal of a characteristic length $\lambda_c (\sim 1 \text{ Å})$ of condensed system. Hence, the cross section of the light scattering contains only the energy information about excitations almost uniform in space. On the other hand, the wave length of neutron and X-ray photon is possibly of order λ_c, where the energy of such a neutron is of the order of the thermal energy but that of the X-ray is $\sim$ 10 KeV. Thus the inelastic neutron scattering gives both the energy and momentum information on low energy excitations like phonons and magnons, while the X-ray Raman scattering can provide us with information on relatively high energy excitations like plasmons and the individual excitation of inner core electrons.

4.8.1 Compton and Raman Scatterings

We take account of the electron-radiation interaction of the form $e^2/(2mc^2) \sum_i A(r_i)^2$, as well as the form $e/(mc) \sum_i \underline{p}_i \cdot \underline{A}(r_i)$ responsible for absorption and emission processes, where $\underline{A}(\underline{r})$ is the vector potential of radiation. With these interactions as perturbation, the scattering cross section is expressed as

$$\frac{d^2\sigma}{d\Omega d\omega_2} = \frac{\omega_2^2}{c^4} \left(\frac{1}{2\pi}\right)^3 W_{12} \quad , \tag{4.129}$$

where the transition rate W_{12} is given by

$$W_{12} = \frac{2\pi}{\hbar} \, Av \sum_{i \, f} \frac{(2\pi)^2}{\omega_1 \omega_2} \left(\frac{e^2 \hbar}{m} \right)^2 \delta(E_f - E_i + \hbar\omega_2 - \hbar\omega_1)$$

$$\times \left| \left[\frac{1}{e} <f|\rho_{-k}|i> (\hat{n}_1 \cdot \hat{n}_2) + \frac{1}{m} \sum_n \left(\frac{<f|\underline{p}\cdot\hat{n}_2|n> <n|\underline{p}\cdot\hat{n}_1|i>}{E_n - E_i - \hbar\omega_1} \right. \right. \right.$$

$$\left. \left. \left. + \frac{<f|\underline{p}\cdot\hat{n}_1|n> <n|\underline{p}\cdot\hat{n}_2|i>}{E_n - E_i + \hbar\omega_2} \right) \right] \right|^2 \quad . \tag{4.130}$$

Here, $|i>$, $|n>$ and $|f>$ are initial, intermediate and final states of the material system, respectively, $\hat{n}_1$ and $\hat{n}_2$ are polarizations of incident and outgoing radiations, and $\underline{p} \equiv \sum_i \underline{p}_i$, $\rho_k \equiv e \sum_i \exp(-i\underline{k} \cdot \underline{r}_i)$. The first term in the square bracket of (4.130) comes from the first-order perturbation of the A^2 type interaction, while the second term arises from the second order perturbation of the $\underline{p} \cdot \underline{A}$ type. Disregarding here the second term, which is important only when $\hbar\omega_1 \sim E_n - E_i$ (i.e. resonant Raman scattering), we obtain

$$\frac{d^2\sigma}{d\Omega d\omega_2} = \left(\frac{\omega_2}{\omega_1} \right) \left(\frac{e^2}{mc^2} \right)^2 (\hat{n}_1 \cdot \hat{n}_2)^2 \, S(\underline{k}, \omega) \quad , \tag{4.131}$$

where

$$S(\underline{k}, \omega) = Av \sum_{i \, f} |<f|(\rho_{-k}/e)|i>|^2 \, \delta\left(\omega - \frac{E_f - E_i}{\hbar} \right)$$

$$\tag{4.132}$$

$$= \frac{1}{2\pi e^2} \int_{-\infty}^{\infty} dt \, e^{i\omega t} <\rho_k(t)\rho_{-k}> \quad .$$

$S(\underline{k}, \omega)$ is called the *dynamical structure factor* and it is related with the dielectric function by $S(\underline{k}, \omega) = -\hbar k^2/(4\pi^2) \, Im\{1/\epsilon(\underline{k}, \omega)\}$.

In Compton scattering, where $k \gg 1/\lambda_c$, the scattering occurs almost independently by each electron contained in the system and, furthermore, $\hbar\omega$ is almost equal to the change of the kinetic energy of the excited electron. Hence we get from (4.132)

$$S(\underline{k}, \omega) \sim \int d\underline{p} \, n_{\underline{p}} \, \delta\left(\hbar\omega - \frac{\hbar^2 k^2}{2m} - \frac{\hbar\underline{k}\cdot\underline{p}}{m} \right) \quad , \tag{4.133}$$

where $\hbar^2 k^2/2m$ is the recoil energy, $\hbar\underline{k} \cdot \underline{p}/m$ the Doppler shift, and $n_{\underline{p}}$ is the probability for an electron to have momentum $\underline{p}$ in the initial state $|i>$. In this way the

Compton profile contains the information on the *electron momentum distribution*. Theoretical and experimental study of the Compton profile of conduction electrons in simple metals has been reported, and many-body effects, as well as band effects, on their momentum distribution have been discussed [4.109,110].

Next we mention two examples of the Raman scattering in metals. The collective excitation of the Fermi sea is the plasmon, whose energy spectrum is given by the pole of $S(\underline{k}, \omega)$. Experimental observations of $S(\underline{k}, \omega)$ over a wide range of $\underline{k}$ were made, for simple metals, by inelastic X-ray scattering measurements [4.109,111,112]. The plasmon dispersion relation determined in this way was found to deviate considerably, especially for large $\underline{k}$, from that of the theoretical prediction with the random phase approximation. The reason for this discrepancy has been discussed, based upon the many-body theory of the electron fluid [4.112,113], as well as upon the energy band effect [4.111].

The individual excitation of the Fermi sea is the electron-hole pair, the low energy part of which causes the Fermi edge singularity in the absorption spectra, as mentioned in Sect. 4.6. It is easily seen from the first line of (4.132) and the last paragraph of Sect. 4.3.2 that $S(\underline{k}, \omega)$ reduces, in the limit of $k \to 0$, to the absorption spectrum due to the dipole transition (apart from a factor k^2). Hence, the cross section of the X-ray Raman scattering, originating from the core electron excitation accompanied by excitations of electron-hole pairs on the Fermi sea, is expected to show the Fermi edge singularity for $\underline{k} \to 0$ and for $\hbar\omega \sim \tilde{\varepsilon}_F$ (measured from the core level). For $k \neq 0$, on the other hand, monopole and quadrupole transitions become possible, so that some different types of the Fermi edge singularity are expected to appear. For instance, the K threshold of Li should change from a *rounded* type (for $k \to 0$) to an *enhanced* type (for $\underline{k} \neq 0$) with increase in $\underline{k}$ [4.114]. In order to detect such an effect, experimental observations were attempted, instead of the X-ray Raman scattering, by electron energy loss measurements, where quite the same phenomenon is expected to occur, but the results were in contradiction to the theoretical prediction [4.84,85]. The reason for the discrepancy has not been clarified yet.

4.8.2 Resonant Raman Scattering

A particularly interesting phenomenon in the Raman scattering is a *resonance effect*. When $\hbar\omega_1$ approaches the absorption threshold, the second term of W_{12} in (4.130) is remarkably enhanced due to the smallness of the denominator $E_n - E_i - \hbar\omega_1$. This effect was observed experimentally in the scattering of Cu K_α X-rays incident on various target atoms, Ni, Cu, Zn and so on [4.115]. Here the intermediate states $|n\rangle$ correspond to those where a K-electron is excited virtually above the Fermi level, and in the final state $|f\rangle$ an L-electron makes transition to the K-state [4.116]. Thus the outgoing photon was observed at $\hbar\omega_2 \sim \hbar\omega_1 - \varepsilon_L$, ε_L being the binding energy of the L-electron, and its intensity was found to be consistent with resonant enhancement from one material to another.

In this example $E_n - E_i - \hbar\omega_1$ is always positive. When $\hbar\omega_1$ is larger than the absorption edge, however, the second order perturbation formula (4.130) would break down because of *vanishing* energy denominator. In fact, however, the formula (4.130) still holds even in such a situation provided we replace $E_n - E_i - \hbar\omega_1$ with $E_n - E_i - \hbar\omega_1 + i\Gamma_n$, by taking account that the electron population in the intermediate state decays with a finite life time $\hbar/(2\Gamma_n)$ due to the radiative and/or Auger processes. Experimental observation corresponding to this situation was recently performed by EISENBERGER et al. [4.117] for Cu metal. As shown in Fig. 4.9, they controlled ω_1 continuously around $(\Omega_K + \varepsilon_F)/\hbar$ and observed ω_2 in the neighborhood of $(\Omega_K - \Omega_L)/\hbar$. In order to roughly understand what is expected as ω_1 is changed, we calculate W_{12}, apart from unimportant factors, in the form

$$W_{12} \sim \sum_{k>k_F} \left| \frac{1}{\Omega_K + \varepsilon_k - \hbar\omega_1 + i\Gamma_K} \right|^2 \delta(\Omega_L + \varepsilon_k + \hbar\omega_2 - \hbar\omega_1)$$

$$= \begin{cases} \rho/[\Delta E_0)^2 + \Gamma_K^2] & \text{for} \quad \Delta E_0 \leq \Delta E_R \\ 0 & \text{for} \quad \Delta E_0 > \Delta E_R \end{cases} \tag{4.134}$$

where

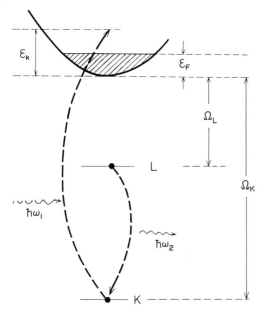

Fig. 4.9 Schematic representation of energy levels in the resonant X-ray Raman scattering for Cu [4.117]

$$\Delta E_R \equiv \hbar\omega_1 - (\Omega_K + \varepsilon_F)$$

$$\Delta E_0 \equiv \hbar\omega_2 - (\Omega_K - \Omega_L) \quad .$$

Here Γ_K and ρ are, respectively, the level width of the K-hole and the density of states of the conduction band. We have neglected the level width of the L-hole, Γ_L, and the effect of the interaction between the conduction electrons and core hole. When the effect of Γ_L is taken into account, the discontinuity of W_{12} at $\Delta E_0 = \Delta E_R$ will be blurred out by the width Γ_L. In any case, the above calculation shows that when $\Delta E_R < 0$ the peak of W_{12} appears at $\Delta E_0 = \Delta E_R$ and when $\Delta E_R > 0$ it is at $\Delta E_0 = 0$. The half width at the half maximum of the peak should be minimized when $\Delta E_R \simeq 0$. The experimental observation is consistent with this trend.

When we take account of various interactions in the intermediate state, such as interactions of conduction electrons with the core hole, phonons and conduction electrons temselves, what is expected on the line shape of the resonant Raman scattering? NOZIÈRES and ABRAHAMS [4.118] studied the effect of conduction electron-core hole interaction with the same model system as that used in the study of the Fermi edge singularity of absorption or emission spectra in Sect. 4.6. In the limit of the small core level width Γ, they obtained the essential part of W_{12} in the following forms

$$W_{12} \sim (\omega_1 - \omega_2)(\hbar\omega_1 - \tilde{\varepsilon}_F)^{-(1+\alpha)} (\hbar\omega_2 - \tilde{\varepsilon}_F)^{-(1+\alpha)} \quad \text{for} \quad \tilde{\varepsilon}_F < \hbar\omega_2 < \hbar\omega_1 \quad (4.135)$$

and

$$W_{12} \sim \theta(\hbar\omega_1 - \tilde{\varepsilon}_F)(\hbar\omega_1 - \tilde{\varepsilon}_F)^{-\alpha} \theta(\tilde{\varepsilon}_F - \hbar\omega_2)(\tilde{\varepsilon}_F - \hbar\omega_2)^{-\alpha}/\Gamma$$

$$(4.136)$$

$$\sim I(\omega_1) I_e(\omega_2)/\Gamma \quad \text{for} \quad \hbar\omega_2 < \tilde{\varepsilon}_F < \hbar\omega_1 \quad .$$

This result is interpreted as follows: Since $\hbar\omega_1 > \tilde{\varepsilon}_F$, a core electron is excited by absorbing $\hbar\omega_1$ above the Fermi level as an energy conserving process. Then the intermediate state thus produced begins to relax due to the redistribution of conduction electrons by the interaction with the core hole, as well as by the life time effect of the core hole. If $\hbar/\Gamma$ is well larger than the relaxation time of the conduction electron redistribution, most part of the outgoing radiation is emitted after the intermediate state reaches its equilibrium distribution. In this process, the scattering is nothing but the absorption followed by the ordinary luminescence, and W_{12} is proportional to the product of the absorption $I(\omega_1)$ and emission $I_e(\omega_2)$ spectra as given by (4.136) for $\hbar\omega_2 < \tilde{\varepsilon}_F$. On the other hand, the spectrum (4.135) appearing for $\tilde{\varepsilon}_F < \hbar\omega_2 < \hbar\omega_1$ corresponds to the radiation emitted before the intermediate state arrives at the equilibrium. In this way the resonant Raman scattering reflects the dynamics of the relaxation in the intermediate states [4.119-122]. The so-called *hot luminescence effect* [4.123] becomes significant when the radiative life time of the core hole is comparable to or smaller than the (dephasing) relaxation time of the intermediate state. As mentioned in Sect. 4.4, the radiative life time is shorter when the core level is deeper. In this respect, the X-ray resonant Raman scattering is expected to be favorable for the observation of hot luminescence effect and hence for the study of relaxation process in various photoexcited states.

4.9 Topics of Recent and Future Interest

Partly as a summary and partly as a supplement of the preceding sections, we give here topics of recent and future interest, some of which have already been mentioned but others have been omitted because of the limited space.

As mentioned in Sect. 4.2, EXAFS is one of the recent topics and is expected to develop in future as a powerful tool for structural determination of various complex systems. The study of van Hove singularity [4.124] of conduction bands is also developing with the experimental information obtained by modulation spectroscopy (see Sect. 7.3.1).

In Sect. 4.3 we have dealt with excitons from band theoretical viewpoint. Recently another approach, a *localized excitation model* [4.125] with use of the ligand field theory, has been proposed for inner shell excitons arising essentially from the intraatomic excitation such as $np^6 \rightarrow np^5 n's$ and $np^6 \rightarrow np^5 nd$ transitions of alkali ions in alkali halides. Calculations have been made for $4p \rightarrow 5s$, $4d$ excitations of Rb^+ ion in Rb halides by using appropriate values of Slater integrals, the spin-orbit interaction constant and the ligand field strength, resulting in good agreement with experimental observations.

Although we have omitted in Sect. 4.4, many-body effects in atoms have also been extensively studied with the use of RPA method, multi-configuration Hartree-Fock method and so on [4.4].

Simultaneous excitations and relaxations in insulators (treated in Sect. 4.5) and metals (in Sect. 4.6) are important objects of the present and future study in the inner level spectroscopy. We supplement here several topics concerning them. Essentially the same idea as the dielectric functional method in Sect. 4.5 has recently been applied to the simultaneous excitation of the valence exciton accompanying the core electron excitation in ionic crystals [4.126-128]. In this case, however, there seems to be a problem that the effect of the intraatomic Coulomb interaction is not incorporated properly in the framework of the dielectric functional method.

Self-trapped states similar to that discussed in Sect. 4.5 also exist in liquid He. Since the bottom of the conduction band in liquid He is by about 1.0 eV higher than the vacuum level, an injected excess electron is stabilized by removing the surrounding atoms, forming the *electronic bubble* [4.129]. A Frenkel exciton with (1s)(2p) configuration in liquid He also relaxes to a self-trapped state as is evidenced by a broad band emission spectrum with large Stokes shift. The relaxed state is presumed to be an excited state of the He_2 molecule accompanying a bubble around it. Further study of these relaxed states is a problem of future interest.

Another interesting phenomenon related with electron-lattice relaxation is the formation of color centers in alkali halides by ultraviolat radiation or electron bombardment. The energy conversion from intrinsic electronic excitation to lattice defect formation is the most basic process in radiation damage. There is some evidence [4.130] that self-trapping of holes or excitons opens a very efficient channel for this electron-lattice energy conversion. The spectral dependence of this efficiency on excitation energy would shed more light on microscopic details of formation mechanism.

The electron correlation (band versus local characters) in metals with incomplete shell is a subject of incessant interest in solid state physics. As mentioned in Sect. 4.7, the inner level spectroscopy is a probe to reveal some aspects of the correlation effect and will contribute to future development in this field. In addition to the topics given in Sect. 4.7, we mention here another recent subject concerning the incomplete shell. In some compounds containing rare earths like Sm, Eu and Tm, the valence of the rare earth ion is fluctuating by a resonance mixing between divalent and trivalent state. In fact, inner shell photoabsorption [4.131] and photoelectron [4.132] spectra show structures originating in the multiplet splitting corresponding to both the divalent and trivalent ions with intensities proportional to their mixing amplitudes. This phenomenon is called the *valence mixing* or *spontaneous interconfiguration fluctuation* [4.133].

Finally we expect that X-ray scattering, in particular the X-ray resonant Raman scattering, will open a new development in inner level spectroscopy with the progress of the synchrotron light source.

References

4.1   U. Fano, J.W. Cooper: Rev. Mod. Phys. <u>40</u>, 441 (1968).
4.2   M.B. Robin: *Higher Excited States of Polyatomic Molecules*, Vol. 1 (Academic Press, New York 1974).
4.3   *Chemical Spectroscopy and Photochemistry in the Vacuum-Ultraviolet*, ed. by C. Sandorfy, P.J. Ausloss, M.B. Robin (D. Reidel Publ. Comp., Dordrecht, Boston 1974).

4.4   *Photoionization and Other Probes of Many-Electron Interactions,* ed. by
      F.J. Wuilleumier (Prenum Press, New York, London 1976).
4.5   R. Kubo: J. Phys. Soc. Japan $\underline{12}$, 570 (1957).
4.6   Y. Toyozawa: In *Polarons in Ionic Crystals and Polar Semiconductors,* ed. by
      J.T. Devreese (North-Holland, Amsterdam, London 1972) pp. 1-27.
4.7   M. Altarelli, D.L. Dexter, H.M. Nussenzveig, D.Y. Smith: Phys. Rev. B$\underline{6}$,
      4502 (1972).
4.8   J.W. Cooper: Phys. Rev. $\underline{128}$, 681 (1962).
4.9   S.T. Manson, J.W. Cooper: Phys. Rev. $\underline{165}$, 126 (1968).
4.10 Y. Toyozawa: In *Conf. Digest of 3rd. Interm. Conf. Vacuum Ultraviolet
      Radiation Physics* (Phys. Soc. Japan, 1971) 30a A3-2.
4.11 M.H. Cohen, V. Heine: Phys. Rev. $\underline{122}$, 1821 (1961).
4.12 R. Haensel, G. Keitel, P. Schreiber, C. Kunz: Phys. Rev. $\underline{188}$, 1375 (1969).
4.13 M. Cardona, R. Haensel, D.W. Lynch, B. Sonntag: Phys. Rev. B$\underline{2}$, 1117 (1970).
4.14 F.C. Brown, C. Gähwiller, H. Fujita, A.B. Kunz, W. Scheifley, N.J. Carrera:
      Phys. Rev. B$\underline{2}$, 2126 (1970).
4.15 A.B. Kunz, N.O. Lipari: J. Phys. Chem. Solids $\underline{32}$, 1141 (1971).
4.16 U. Rössler: Phys. Stat. Sol. (b) $\underline{45}$, 483 (1971).
4.17 F.C. Brown: In *Solid State Physics*, Vol. 29, ed. by F. Seitz, D. Turnbull,
      H. Ehrenreich (Academic Press, New York 1974) pp. 1-73.
4.18 L. Van Hove: Phys. Rev. $\underline{89}$, 1189 (1953).
4.19 D.E. Sayers, F.W. Lytle, E.A. Stern: In *Advances in X-Ray Analysis*, ed. by
      B.L. Henke, J.B. Newkirk, G.R. Mallett (Plenum Press, New York 1970) pp.
      248-271.
4.20 C.A. Ashley, S. Doniach: Phys. Rev. B$\underline{11}$, 1279 (1975).
4.21 P.A. Lee, J.B. Pendry: Phys. Rev. B$\underline{11}$, 2795 (1975).
4.22 D.E. Sayers, E.A. Stern, F.W. Lytle: Phys. Rev. Lett. $\underline{27}$, 1204 (1971).
4.23 F.W. Lytle, D.E. Sayers. E.B. Moore, Jr. : Appl. Phys. Lett. $\underline{24}$, 45 (1974).
4.24 P.A. Lee: Phys. Rev. $\underline{13}$, 5261 (1976).
4.25 R.S. Knox: In *Solid State Physics*, ed. by F. Seitz, D. Turnbull (Academic
      Press, New York 1963) Suppl. 5.
4.26 G.H. Wannier: Phys. Rev. $\underline{52}$, 191 (1937).
4.27 J. Frenkel: Phys. Rev. $\underline{37}$, 17 (1931).
4.28 W.R. Heller, A. Marcus: Phys. Rev. $\underline{84}$, 809 (1951).
4.29 R.J. Elliott: Phys. Rev. $\underline{108}$, 1384 (1957); also in *Polarons and Excitons*,
      ed. by C.G. Kuper, G.D. Whitfield (Plenum Press, New York 1963) pp. 269-
      293.
4.30 G. Dresselhaus: J. Phys. Chem. Solids $\underline{1}$, 14 (1956).
4.31 Y. Onodera, Y. Toyozawa: J. Phys. Soc. Japan $\underline{22}$, 833 (1967).
4.32 K. Cho, S. Suga, W. Dreybrodt, F. Willmann: Phys. Rev. B$\underline{11}$, 1512 (1975).
4.33 K. Cho: Phys. Rev. B$\underline{14}$, 4463 (1976).
4.34 For a review, see W. Bambynek, B. Crasemann, R.W. Fink, H.-U. Freund, H.
      Mark, C.D. Swift, R.E. Price, P.V. Rao: Rev. Mod. Phys. $\underline{44}$, 716 (1972).
4.35 V.O. Kostroun, M.H. Chen, B. Crasemann: Phys. Rev. A$\underline{3}$, 533 (1971).
4.36 J.H. Schofield: Phys. Rev. $\underline{179}$, 9 (1969).
4.37 T. Kobayasi, A. Morita: J. Phys. Soc. Japan $\underline{28}$, 457 (1970).
4.38 U. Fano: Phys. Rev. $\underline{124}$, 1866 (1961).
4.39 A. Shibatani (Kotani), Y. Toyozawa: J. Phys. Soc. Japan $\underline{25}$, 335 (1968).
4.40 F.H. Mies: Phys. Rev. $\underline{175}$, 164 (1968).
4.41 R.P. Madden, D.L. Ederer, K. Codling: Phys. Rev. $\underline{177}$, 136 (1969).
4.42 M. Lax: J. Chem. Phys. $\underline{20}$, 1753 (1952).
4.43 R. Kubo, Y. Toyozawa· Progr. Theor. Phys. $\underline{13}$, 160 (1955).
4.44 A.A. Maradudin: In *Solid State Physics*, Vol. 18, ed. by F. Seitz, D. Turnbul,
      H. Ehrenreich (Academic Press, New York 1966) pp. 273-420.
4.45 Y. Toyozawa: J. Luminescence $\underline{12/13}$, 13 (1976).
4.46 Y. Toyozawa: Progr. Theor. Phys. $\underline{20}$, 53 (1958).
4.47 Y. Toyozawa: In *Proc. 4th Interm. Conf. Vacuum Unltraviolet Radiation Physics,*
      ed. by E.-E. Koch, R. Haensel, C. Kunz (Pergamon Vieweg, Braunschweig 1974)
      pp. 317-330.
4.48 Y. Toyozawa: J. Luminescence $\underline{1, 2}$, 632 (1970).
4.49 Y. Toyozawa: Progr. Theor. Phys. $\underline{26}$, 29 (1961).
4.50 D. Emin: Adv. in Phys. $\underline{22}$, 57 (1973).

4.51 E.I. Rashba: Izv. AN SSSR, ser. fiz. 40, 1 (1976).
4.52 K. Cho, Y. Toyozawa: J. Phys. Soc. Japan 30, 1555 (1971).
4.53 H. Sumi, Y. Toyozawa: J. Phys. Soc. Japan 31, 342 (1971).
4.54 J. Jortner: In *Proc. 4th Intern. Conf. Vacuum Ultraviolet Radiation Physics*, ed. by E.-E. Koch, R. Haensel, C. Kunz (Pergamon Vieweg, Braunschweig 1974) pp. 263-316.
4.55 M.N. Kabler: Phys. Rev. 136, A 1296 (1964).
4.56 H. Kanzaki, S. Sakuragi: Photographic Sci. Eng. 17, 69 (1973).
4.57 K. Murayama, K. Morigaki, S. Sakuragi, H. Kanzaki: J. Phys. Soc. Japan 41, 1617 (1976).
4.58 H. Kanzaki, S. Sakuragi: J. Phys. Soc. Japan 29, 924 (1970).
4.59 J. Nakahara, K. Kobayashi: J.Phys. Soc. Japan 40, 180 (1976).
4.60 H. Kanzaki, S. Sakuragi, K. Sakamoto: Solid State Commun. 9, 999 (1971).
4.61 As a review article, see G. Zimmerer: In *Proc. Intern. Summer School on Synchrotron Radiation Research* (Alghero 1976) of the I.C.A.P. Series (Intern. College on Applied Physics, Catania, Italy), ed. by I.F. Quercia.
4.62 Ch. Lushchik, I. Kuusmann, P. Liblik, G. Liida. N. Lushchik, V. Plekhanov, A. Ratas, T. Soovik: J. Luminescence 11, 285 (1975/76).
4.63 M. Tomura, Y. Kaifu: J. Phys. Soc. Japan 15, 1295 (1960).
4.64 H. Nishimura, M. Tomura: J. Phys. Soc. Japan 39, 390 (1975).
4.65 M. Itoh, Y. Nakai: J. Phys. Soc. Japan 39, 418 (1975).
4.66 J. Friedel: Nuovo Cimento, Suppl. 7, 287 (1958).
4.67 P.W. Anderson: Phys. Rev. Lett. 18, 1049 (1967); Phys. Rev. 164, 352 (1967).
4.68 S. Doniach, M. Sunjić: J. Phys. C 3, 283 (1970).
4.69 J.J. Hopfield: Comments Solid State Phys. 2, 40 (1969).
4.70 K.D. Schotte, U. Schotte: Phys. Rev. 182, 479 (1969).
4.71 P. Nozieres, C.T. De Dominicis: Phys. Rev. 178, 1097 (1969).
4.72 L. Hedin, S. Lundqvist: In *Solid State Physics*, Vol. 23, ed. by F. Seitz, D. Turnbul, H. Ehrenreich (Academic Press, New York 1969) pp. 1-181.
4.73 D.C. Langreth: Phys. Rev. B1, 471 (1970).
4.74 G.D. Mahan: Phys. Rev. 163, 612 (1967); also in *Solid State Physics*, Vol. 29, ed. by F. Seiz, D. Turnbul, H. Ehrenreich (Academic Press, New York 1974) pp. 75-138.
4.75 A. Kato, A. Okiji, Y. Osaka: Progr. Theor. Phys. 44, 287 (1970).
4.76 A. Yoshimori, A. Okiji: Phys. Rev. B16, 3838 (1977).
4.77 Y. Onodera: J. Phys. Soc. Japan 39, 1482 (1975).
4.78 G.D. Mahan: Phys. Rev. B11, 4814 (1975).
4.79 T. Sagawa: In *Soft X-Ray Band Spectra and the Electronic Structure of Metals and Materials*, ed. by D.J. Fabian (Academic Press, New York 1968) pp. 29-43.
4.80 C. Kunz, R. Haensel, G. Keitel, P. Schreiber, B. Sonntag: In *Electronic Density of States*, ed. by L.H. Bennett (US Government Printing Office, Washington, D.C. 1971) p. 275.
4.81 H. Neddermeyer: Phys. Lett. 44A, 181 (1973); Phys. Rev. B13, 2411 (1976).
4.82 J.D. Dow: In *Proc. 4th Intern. Conf. Vacuum Ultraviolet Radiation Physics*, ed. by E.-E. Koch, R. Haensel, C. Kunz (Pergamon Vieweg, Braunschweig 1974) pp. 649-661.
4.83 R.P. Gupta, A.J. Freeman: Phys. Rev. Lett. 36, 1194 (1976).
4.84 J.J. Ritsko, S.E. Schnatterly, P.C. Gibbons: Phys. Rev. 10, 5017 (1974).
4.85 S.G. Slusky, P.C. Gibbons, S.E. Schnatterly, J.R. Fields: Phys. Rev. Lett. 36, 326 (1976).
4.86 T.M. Zimkina, V.A. Fomichev, S.A. Gribovskii, I.I. Zhukova: Sov. Phys. Solid State 9, 1128 (1967).
4.87 V.A. Fomichev, T.M. Zimkina, S.A. Gribovskii, I.I. Zhukova: Sov. Phys. Solid State 9, 1163 (1967).
4.88 R. Haensel, P. Rabe, B. Sonntag: Solid State Commun. 8, 1845 (1970).
4.89 S. Suzuki, I. Nagakura, T. Ishii, T. Satoh, T. Sagawa: Phys. Lett. 41A, 95 (1972).
4.90 S. Suzuki, T. Ishii, T. Sagawa: J. Phys. Soc. Japan 38, 156 (1975).
4.91 H.W. Wolff, R. Bruhn, K. Radler, B. Sonntag: DESY SR-76/14 (1976).
4.92 J.L. Dehmer, A.F. Starace, U. Fano, J. Sugar, J.W. Cooper: Phys. Rev. Lett. 26, 1521 (1971).
4.93 A.F. Starace: Phys. Rev. B5, 1773 (1972).

4.94 J. Sugar: Phys. Rev. B5, 1785 (1972).
4.95 J.L. Dehmer, A.F. Starace: Phys. Rev. B5, 1792 (1972).
4.96 A.F. Starace: J. Phys. B 7, 14 (1974).
4.97 B. Sonntag, R. Haensel, C. Kunz: Solid State Commun. 7, 597 (1969).
4.98 R.E. Dietz, E.G. Mcrae, Y. Yafet, C.W. Caldwell: Phys. Rev. Lett. 33, 1372 (1974).
4.99 L.C. Davis, L.A. Feldkamp: Solid State Commun. 19, 413 (1976).
4.100 A. Kotani, Y. Toyozawa: J. Phys. Soc. Japan 35, 1073 (1973); 35, 1082 (1973).
4.101 A. Kotani, Y. Toyozawa: J. Phys. Soc. Japan 37, 912 (1974).
4.102 S. Hufner, G.K. Wertheim: Phys. Rev. B11, 5197 (1975).
4.103 S. Hüfner, G.K. Wertheim, J.H. Wernick: Solid State Commun. 17, 417 (1975).
4.104 P.C. Kemeny, N.J. Shevchik: Solid State Commun. 17, 255 (1975).
4.105 H. Kaga, A. Kotani, Y. Toyozawa: J. Phys. Soc. Japan 41, 1851 (1976).
4.106 H. Kaga: J. Phys. Soc. Japan 41, 1861 (1976).
4.107 C.S. Fadley, D.A. Shirley: Phys. Rev. A2, 1109 (1970).
4.108 W. Gudat, C. Kunz: Phys. Stat. Sol. (b) 52, 433 (1972).
4.109 P.M. Platzman: In *Elementary Excitations in Solids, Molecules and Atoms*, Part A, ed. by J.T. Devreese, A.B. Kunz, T.C. Collins (Plenum Press, New York, London 1974) pp. 31-64.
4.110 P. Eisenberger, L. Lam, P.M. Platzman, P. Schmidt: Phys. Rev. B6, 3671 (1972).
4.111 P. Eisenberger, P.M. Platzman, K.C. Pandy: Phys. Rev. Lett. 31, 311 (1973).
4.112 P. Eisenberger, P.M. Platzman, P. Schmidt: Phys. Rev. Lett. 34, 18 (1975).
4.113 P.M. Platzman, P. Eisenberger: Solid State Commun. 14, 1 (1974).
4.114 S. Doniach, P.M. Platzman, J.T. Yue: Phys. Rev. 4, 3345 (1971).
4.115 C.J. Sparks, Jr.: Phys. Rev. Lett. 33, 262 (1974).
4.116 Y.B. Bannett, I. Freund: Phys. Rev. Lett. 34, 372 (1975).
4.117 P. Eisenberger, P.M. Platzman, H. Winick: Phys. Rev. Lett. 36, 623 (1976); Phys. Rev. B13, 2377 (1976).
4.118 P. Nozieres, E. Abrahams: Phys. Rev. 10, 3099 (1974).
4.119 V. Hizhnyakov, I. Tehver: Phys. Stat. Sol. 21, 755 (1967); 39, 67 (1970).
4.120 R. Kubo, T. Takagahara, E. Hanamura: In *Physics of Highly Excited States in Solids*, ed. by M. Ueta, Y. Nishina, (Lecture Notes in Physics, Vol. 57), (Springer, Berlin, Heidelberg, New York 1976) pp. 304-319.
4.121 Y. Toyozawa: J. Phys. Soc. Japan 41, 400 (1976).
4.122 Y. Toyozawa, A. Kotani, A. Sumi: J. Phys. Soc. Japan 42, 1495 (1977).
4.123 K. Rebane, P. Saari: J. Luminescence 12/13, 23 (1976).
4.124 D.E. Aspnes, C.G. Olson, D.W. Lynch: Phys. Rev. 14, 2534 (1976).
4.125 C. Satoko, S. Sugano: J. Phys. Soc. Japan 34, 701 (1973).
4.126 T. Miyakawa: J. Phys. Soc. Japan 17, 1898 (1962).
4.127 J.C. Hermanson: Phys. Rev. 177, 1234 (1969).
4.128 A.B. Kunz, J.T. Devreese, T.C. Collins: J. Phys. C 5, 3259 (1972).
4.129 W.B. Fowler, D.L. Dexter: Phys. Rev. 176, 337 (1968).
4.130 Y. Kondo, M. Hirai, M. Ueta: J. Phys. Soc. Japan 33, 151 (1972).
4.131 E.E. Vainshtein, S.M. Blokhin, Yu.B. Paderno: Sov. Phys. - Solid State 6, 2318 (1965).
4.132 M. Compagna, E. Bucher, G.K. Wertheim, D.N.E. Buchanan, L.D. Longinotti: Phys. Rev. Lett. 32, 885 (1974).
4.133 C.M. Varma: Rev. Mod. Phys. 48, 219 (1976).

5. Atomic Spectroscopy

K. Codling

With 16 Figures

The aim of this chapter is to discuss that area of atomic physics which has bene-
fited from the application of synchrotron radiation, namely photoabsorption spectroscopy. More specifically, we will deal exclusively with the extreme ultraviolet
(XUV) spectral region (also sometimes called "VUV" and "soft X-rays"), covering an
energy range of approximately 10-1000 eV (1000-10 Å). In Sect. 5.1 we briefly re-
state the properties of the radiation that make it useful and examine in Sect. 5.2
the types of experiment performed with this unique source. In the same section we
also describe some of the experimental problems associated with using synchrotron
radiation and then discuss in Sect. 5.3 the limitations of photoabsorption spectroscopy as a probe of atomic structure when compared with alternative techniques
such as electron energy loss and beam-foil spectroscopy.

From a theoretical viewpoint, the scientific objective of such experiments may
be expressed in general terms as the illucidation of the extra-nuclear quantum
structure of atoms using as a probe a beam of photons of well-defined but variable
energy. This is summarized in Sect. 5.4. The central Sect. 5.5 is concerned with
the experimental results obtained with synchrotron radiation measuring absorption,
photoelectric emission and mass spectra. These results are shown in the context of
the previously available information and their theoretical implications are discussed. Such studies are not only of fundamental interest, however. The basic atomic
data acquired can be vital in a wide range of scientific activities, such as the
understanding of radiation processes in laboratory and astrophysical plasmas, the
interpretation of astronomical spectra, the production of XUV or soft X-ray lasers,
isotope separation and so on.

Moreover, it has become increasingly clear in recent years that an understanding
of the XUV absorption spectrum of an atom in the tree state is an essential prerequisite to the interpretation of the complex structure observed in the solid state.
The understanding of centrifugal barrier phenomena, which has been mentioned in
Chap. 4 and will be discussed in Sect. 5.5.1, is a case in point.

In a short Sect. 5.6 at the end of the chapter a few ideas are presented on how
research in this field probably will, and maybe should, proceed.

5.1 Atomic Photoabsorption Spectroscopy in the Extreme Ultraviolet

Clearly, the importance of the application of synchrotron radiation to the field of XUV photoabsorption spectroscopy cannot be appreciated unless it is put into some kind of historical context. The growth of the XUV field was dependent upon the discovery of appropriate background sources of continuum, the development of gratings and detectors, and not least improvements in vacuum technology. The early contribution of SCHUMANN and his successors must be acknowledged [5.1]. The region of the spectrum from 10-25 eV (100-500 Å) belongs to Lyman, in recognition of his pioneering work involving the exploitation of concave gratings and development of light sources, see for example the early publications of BOMKE [5.2] and BOYCE [5.3].

Prior to 1950, the workers in the fields of both XUV emission and absorption spectroscopy could almost be counted on the fingers of one hand. Edlén contributed greatly to the field of emission spectroscopy while Beutler did likewise in absorption spectroscopy, see, for example, the more recent reviews of GARTON [5.4], SAMSON [5.5] and MARR [5.6]. Interest in this region of the electromagnetic spectrum was greatly enhanced due to the new era of observational astrophysics, brought about by the use of space satellites or rockets to overcome the problems of atmospheric absorption and to the enormous interest in plasma physics and the hopes, so far unfilled, of achieving controlled nuclear fusion.

Until relatively recently the sources of continuum available were limited to basically three types; the HOPFIELD [5.7] continuum in helium, see also [5.8]; the LYMAN [5.9] continuum emitted by an impulsive heavy current, low inductance "flash tube", see also [5.10]; the BRV source of BALLOFFET et al. [5.11]. All of these sources suffered from one defect or another. The inert gas continua, although relatively pure, are weak and do not extend to the higher photon energies. The Lyman continuum has many emission lines of highly ionized atomic species superimposed upon the continuum, particularly above 25 eV. The BRV source seems the best conventional source but the lifetime of the anode, when run under conditions of high repetition rate, is unacceptably short. Moreover the debris which evaporates from the anode, (typically uranium) is certain to coat any optical component that is in the line of sight.

In 1955 the picture was radically changed when TOMBOULIAN [5.12] used the Cornell 300 MeV synchrotron briefly as a background continuum in the XUV and showed the great potential of such unusual light sources. Within a decade synchrotron sources were in use throughout the world.

The advantages of such sources have been discussed in Chap. 1; they emit a pure continuum extending from the infrared through to the X-ray region. They are very stable in both position and time. The radiation, which is produced in a clean, high-vacuum environment, is highly polarized and pulsed at well-defined frequencies.

As we shall see throughout this chapter, the property that has been extensively and almost exclusively utilized in atomic spectroscopy is the continuum property. More recently the fact that the radiation is highly polarized in the orbital plane has been taken advantage of in electron angular distribution studies (Sect.5.5.2). The fact that the radiation becomes circularly polarized some distance from the orbital plane has not yet been exploited.

5.2   The Basic Experiments in Photoabsorption Spectroscopy

When photons with energy in excess of the first ionization potential are incident upon an atom A, ionization can occur

$$A + h\nu \rightarrow A^+ + e \quad . \tag{5.1}$$

At sufficiently high photon energies, the ion may alternatively be left in an excited state

$$A + h\nu \rightarrow A^{+*} + e \quad . \tag{5.2}$$

If, however, the photon energy is coincident with an excited state of the neutral atom, associated with subshell or inner shell electron excitation (or with double electron excitation), an alternative process is possible

$$A + h\nu \rightarrow A^{**} \quad . \tag{5.3}$$

The subsequent decay of this quasistable state through autoionization will result finally in ionization, as indicated in Fig. 5.1

$$A^{**} \rightarrow A^+ + e \quad . \tag{5.4}$$

An alternative decay channel for the photoexcited state would be fluorescence

$$A^{**} \rightarrow A^{(*)} + h\nu \quad . \tag{5.5}$$

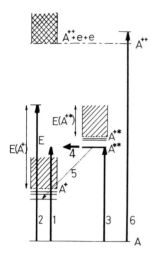

Fig. 5.1 A simplified atomic energy level scheme, showing the possible excitations and decay mechanisms (see text for the meaning of symbols)

Fluorescence decay of photoexcited atoms is much less likely than electron ejection. Hence we are in essence determining photoionization cross sections when we determine the photoabsorption cross sections for atoms in the XUV.

If the energy of the photon is sufficient, double (or multiple) ionization can occur

$$A + h\nu \rightarrow A^{++} + e + e \quad . \tag{5.6}$$

This double electron ejection will quite likely occur as a two-step process. Firstly, the incident photon will eject an inner electron

$$A + h\nu \rightarrow A^{+*} + e \quad . \tag{5.7}$$

Secondly an Auger decay will occur, when an outer electron will fill the vacancy and a second (outer) electron will be ejected (shake-off).

$$A^{+*} \rightarrow A^{++} + e \tag{5.8}$$

Alternatively, electron correlation effects must be responsible for double ionization in those cases where the first process (5.7) is energetically forbidden [5.13].

If a source of pure continuum such as synchrotron radiation is available, the various atomic processes outlined above can be studied in great detail. Since the experimental results in Sect. 5.5 will be discussed without reference to experimental difficulties, it is felt that, whilst describing a particular technique, some of the problems associated with obtaining the final published numbers should be mentione

5.2.1 Photoabsorption Spectroscopy

The aim of atomic photoabsorption spectroscopy is to determine the distribution
of oscillator strength as a function of photon energy. To achieve this end one measures the absorption cross section as a function of wavelength, $\sigma(\lambda)$. The simpler
experiment would be to use a spectrograph to photograph the absorption spectrum
over an extended energy range. In this way one obtains clear evidence of autoionization phenomena and other correlation effects. Although the photographic technique
is occasionally used to determine absolute cross sections, there is no doubt that
accurate cross sections can be obtained only by the more time-consuming methods involving photoelectric detection. In either experiment, a low-to-medium current
synchrotron or storage ring would provide an adequate photon flux except where
one wished to determine the absolute oscillator strength associated with a particularly narrow resonance.

   The intensity of radiation $I(\lambda)$ transmitted through a gas of column length ℓ
is given in the low pressure limit by the expression

$$I(\lambda) = I_o(\lambda) \exp[- \sigma(\lambda) \, n\ell] \tag{5.9}$$

where $I_o(\lambda)$ is the incident wavelength and n the number of atoms per cc. This
expression can be rewritten in the form

$$\sigma(\lambda) = \frac{T}{273} \cdot \frac{760}{p\ell} \cdot \frac{1}{N} \cdot \log_e \left[\frac{I_o(\lambda)}{I(\lambda)} \right] \tag{5.10}$$

where N is Loschmidt's number (2.69×10^{19} atoms/cc), T the absolute temperature
and p the pressure in Torr. Cross sections are usually given in megabahns
(1 Mb = 10^{-18} cm^2). Alternatively the cross section is quoted as a linear absorption coefficient, k, in cm^{-1} at S.T.P. and relates to σ by the expression k = Nσ.

   One may find continuum cross sections quoted in terms of oscillator strength
distribution df/dϵ where

$$\sigma = \frac{\pi e^2}{mc} \cdot \frac{df}{d\epsilon} \quad . \tag{5.11}$$

To determine a cross section, therefore, one requires the temperature, pressure
and path length in addition to $I_o(\lambda)$ and $I(\lambda)$.

   The problems of determining accurate cross sections fall into various categories,
depending upon the element concerned, and the wavelength range under consideration.
The path length of an inert gas can be determined with great precision and errors in
cross section stem from inaccuracies in pressure measurement and from gas flow

problems associated with differential pumping. WEST and MARR [5.14] have recently
measured the cross sections of the inert gases He through Kr up to 300 eV to an
accuracy of ± 5% using a long absorption cell to minimize gas flow problems, and a
monochromator almost free from order overlap and stray light [5.15].

The problem of order overlap has been mentioned in Sect. 3.2. Clearly, if a monochromator output contains a substantial fraction of higher orders or stray light,
these various contributions will be attenuated in relation to the absorption cross
section at those wavelengths and unambiguous determination of the first-order cross
section becomes extremely difficult, if not impossible.

To obtain cross sections for metal vapors the problems are much more severe.
To avoid coating mirrors or slits with the vapor contained in the absorption cell,
it is usual to employ a buffer inert gas to hinder its diffusion. The heat pipe has
been mentioned in Sect. 3.6.1; measurement of path length and vapor pressure seems
reasonably straightforward for energies up to 10 eV (the LiF cut-off). Between 10
eV and 25 eV, thin films must be used to contain the metal vapor and buffer gas,
and differential pumping is required. Above 25 eV, the buffer gas itself will absorb
strongly and this makes the determination of metal vapor cross sections even more
difficult [5.16] Alternatively, if the metal vapor cross section is too large, the
vapor pressure may have to be held at such a low value that the absorption cell will
cease to operate as a heat pipe.

The reliability of earlier cross section data on metal vapors is far from satisfactory. McILRATH and SANDEMAN [5.17] in the case of Ca, and more recently PARKINSON
et al. [5.18] in the cases of Mg and Sr, have determined cross sections close to the
first ionization threshold. They find that their values differ from previous data by
factors of 1.9 to 2.2, due presumably to inaccuracies in vapor pressure versus temperature curves used in the earlier experiments.

The experimental problems mentioned so far relate to measurement of absorption
cross sections of neutral atoms in their ground state. If one wishes to use sources
of synchrotron radiation to determine cross sections for absorption out of excited
or ionized atoms, the problems multiply. One must produce sufficient numbers of
excited or ionized atoms, the excited states must exist for a time compatible with
the synchrotron light pulse and one must somehow determine the number density of
excited or ionized atoms. We will return briefly to such problems in Sect. 5.3.

5.2.2 Photoelectron spectroscopy

The technique of photoelectron spectroscopy (PES) has developed rapidly in recent
years [5.19-21], and has proved particularly useful in molecular spectroscopy (Sect.
6.5) and solid-state spectroscopy. An enormous amount of data has accumulated using
conventional light sources. At low energies the 584 Å (21 eV) He I resonance line is
used to determine the binding energies of the loosely bound (molecular) orbitals

while the characteristic X-rays of Aℓ and Mg are used for the more tightly bound core electrons.

The relevance of PES to the measurement of partial photoionization cross sections in atomic spectroscopy can be seen by referring to Fig. 5.1. Imagine atom A is Ar $(3s^2 3p^6 {}^1S_0$ ground state). A^+ represents an atom with the outer 3p electron missing. A^{+*} represents an Ar atom with a subshell 3s electron missing. If photons of energy E are incident on an assembly of Ar atoms, the Ar^+ ions can be left in either state Ar^+ or Ar^{+*}. To determine the branching ratio or the *partial* cross section for removal of the 3p or 3s electron at photon energy, E, one simply needs to determine the relative numbers of electrons ejected with kinetic energies $E(A^+)$ and $E(A^{+*})$.

The particular contribution of synchrotron radiation to the field of PES is again the total control one has of the incident photon energy, E. One can map out the variation of the partial cross sections of the 3s and 3p electrons as a function of E and show in a much more direct way than is possible with total cross section measurements, the unusual correlation effects that occur. A number of examples will be given in Sect. 5.5.2.

There are considerable experimental problems associated with determining partial cross sections, one of which is the low level of electron counts achieved when using a low current synchrotrons. To obtain signal-to-noise ratios that are comparable with those achieved with a line source, high current storage rings will be required. In order to correct the raw count rates for electrons of energies $E(A^+)$ and $E(A^{+*})$ into meaningful branching ratios, the efficiency of the electron analyzer must be determined as a function of electron energy. The required relative efficiency can be found by using a gas such as helium, whose absolute cross section is accurately known. A comparison of the number of electrons collected with that which would be expected on the basis of the known cross section gives the information required. Furthermore, although one can in principle determine the second and higher order contributions from the optical monochromator by observing those electrons at higher energies that have been produced by the higher order radiation, the problem is by no means straightforward.

If one were using a static electron analyzer placed at the 'magic' angle of $54°44'$ [5.22], then the asymmetry parameter (β) which characterizes the angular distribution of the ejected electrons (3s and 3p in the case of Ar), would not be required. Collection of electrons at any other angle would require a knowledge of β for both the 3s and 3p electrons and the state of polarization of the incident monochromatic radiation. The study of the variation of the asymmetry parameter as a function of photon energy is another area of PES where the inclusion of electron correlation effects may be required before any reasonable agreement is achieved between experiment and theory.

In some respects the experiments on β determination are more difficult than those where partial cross sections are obtained, because one must severely limit the angular acceptance of the analyzer. On the other hand, one does not need to know the electron analyzer efficiency as a function of energy or the contribution of the higher spectral orders from the optical monochromator to the first order signal.

If the incident photon is energetic enough to produce double (or multiple) ionization, then generally speaking, the two (or more) electrons can share the energy in varying amounts. PES will be unable in a simple way to determine the ratio of single-to-double (multiple) ionization. In such circumstances one turns to mass spectrometry to obtain these 'partial' cross sections for ionization.

5.2.3 Mass Spectrometry

Mass spectrometry has a particularly important role to play in molecular spectroscopy, where ionization, autoionization (pre-ionization), dissociation, dissociative ionization, etc., are some of the competing processes that must be differentiated. Although in atoms mass spectrometry appears to be a less important than PES, there are nevertheless two particular areas where it can make a significant contribution.

As mentioned earlier, once the photon energy is sufficiently high, mass spectrometry is required so that one can apportion the (known) total absorption cross section into partial cross sections for single, double, triple ionization. Alternatively, in determining the atomic photoionization cross section for metal vapors, the contribution to ionization by possible dimers can cause problems, particularly close to the first ionization threshold [5.23]. Mass spectrometry is clearly able to unambiguously separate the two contributions.

We mention only one instrumental problem associated with the use of mass spectrometry and a continuum source, such as synchrotron radiation. When a photon ionizes an atom, the electron(s) will take almost all of the available kinetic energy. If the resulting single and double ions are then pulsed out of the interaction region into a time-of-flight drift tube, the energy imparted to the ions by the incident photon will be negligible compared to the pulse voltage (typically 100 volts). If the photon beam emerging from the monochromator exit slit contains a substantial contribution of second order or stray light, then this unwanted contribution may cause the single-to-double ion ratio, for example, to bear no relationship to that which would be observed if first order alone were present. In contrast to PES, there is no way in the experiment envisaged here of using the fact that the ions will have a different kinetic energy, depending on whether they were produced by first or second order radiation, to determine the real first order single-to-double ion ratio.

Although a considerable body of data now exist on ionization cross sections using a "pseudo-photon" source [5.24] very little work has yet been done where mass spectrometry has been coupled to a synchrotron radiation source. With the availability of high current storage rings the situation could alter dramatically in the near future.

5.2.4 Fluorescence

When a hole is created in an inner shell of an atom (the mechanism of electron ejection is not important), there are two ways that the excited state can decay, either by an Auger process or by fluorescence. The radiative transition probability calculated from simple screened hydrogenic wavefunctions is proportional to the fourth power of the atomic number Z, whereas the radiationless transition probability is approximately constant [5.25]. Fluorescence will therefore be important for elements of high atomic number, see also Sect. 4.4.1. Extensive literature exists on the measurement and calculation of K- and L-shell fluorescence yields [5.26].

We will not pursue the subject further except to comment that in the XUV region, fluorescence involving either outer shells or light elements (or both) is an unlikely process when compared to Auger decay (or autoionization) and as far as the author is aware, no atomic fluorescence spectroscopy has been attempted using synchrotron radiation. Some work has been performed on molecules [5.27] and solids [5.28].

5.3 Limitations of Photon Absorption Experiments

When XUV photons are incident on an assembly of ground-state atoms, our capacity to observe resonances (excited states "lying" in the photoionization continuum), or indeed the continuum states themselves, is severely limited by the dipole selection rule. For example, the inert gases have a $^1S_0$ ground state and therefore only those states with $^1P_1^0$ character can be probed. This results in a fundamental limitation to photon absorption experiments that is only recently being overcome.

There are no such limitations with electron scattering experiments, where both optically allowed and forbidden states can be observed. One should perhaps classify these electron collision experiments into two kinds. The criterion for this classification is whether the incident electron is "fast" or "slow" compared to the internal motion of the electrons pertaining to the phenomenon under examination [5.29]. More specifically, the distinction is whether an incident electron is simply an external agent to an excitation, or becomes incorporated into the target for a period of time.

Experiments with forward scattering of fast electrons (in which momentum transfer is small) are sometimes labelled "pseudo-photon" since the fast electrons exert a force which is uniform over the spatial extent of the atom in the same way as implied by the dipole approximation for photoabsorption. Many recent results exist using high energy electrons [5.30], but since one can never quite realize the equivalent of the photon absorption experiment (zero momentum transfer), photon experiments are still extremely important. Moreover, the resolution attained in these electron collision experiments has not generally been of the same order as that achievable with photon absorption, although this is perhaps no longer the case [5.31].

If one views the electron scattering of fast electrons in directions other than the forward direction or if one uses slower electrons, one can observe forbidden as well as allowed transitions. A wealth of resonances have been observed associated with neutral atomic states as well as negative ion states [5.32]. We will not, however, discuss this large and ever-expanding field at the present time.

Highly excited states of atoms have been probed not only by photon absorption and electron scattering but by light and heavy ion bombardment [5.33] and by beam-foil spectroscopy [5.34]. The doubly excited states probed by beam-foil emission spectroscopy are those that, almost by definition, do not autoionize into the adjacent continuum. For example, the $2p^2$ ^3P state of He cannot autoionize into the adjacent continuum. Hence fluorescence to a singly excited state can be observed in an accessible spectral region [5.35]. The power of the beam-foil technique, however, lies in its capacity to produce highly ionized species with relative ease and allow determination of energies and lifetimes which may have important astrophysical implications.

One can overcome the limitation of photon absorption experiments if one can populate excited states of atoms in sufficient numbers and observe absorption out of them. One then probes those states which are forbidden from the ground. One can use shock tubes or electrical discharges to create excited states [5.36,37] but more selective excitation is achieved with tunable lasers [5.38]. Recently LUCATORTO and McILRATH [5.39] have used a high-power laser tuned to the 5896 Å line in Na. The laser proved to be surprisingly efficient in producing Na$^+$ ions and when used with a BRV pulsed continuum source, the absorption spectrum of Na$^+$ was observed in the 40 eV region. The question in all such experiments when coupled to synchrotron radiation sources will be whether the appropriate temporal matching can be created between the excited or ionic state production and the background continuum emission.

5.4 The General Theoretical Framework

Since the theory associated with atomic photoabsorption spectroscopy has been dealt with in an earlier chapter, we shall not dwell on it in any detail. It is sufficient to say that atomic theory developed largely through independent particle models, which have to a large extent proved adequate for describing the gross features of the spectra observed, particularly at high photon energies [5.40].

Over the last 10-15 years, however, experimental progress has presented us with a rapidly expanding body of detailed evidence of phenomena in which electron-electron interactions play a more elaborate role, exceeding the scope of ordinary perturbation treatments. These more complex phenomena, often characterized by the term "correlation effects" have now been probed by many of the techniques mentioned earlier. Synchrotron radiation, as a source of pure polarized XUV continuum, has made a significant contribution to these efforts.

Theoretically, the weakness of the interaction of light with matter enables one to treat the initial step of photoabsorption by first-order perturbation theory. Therefore the complex aspects of particle interactions (correlations) do not affect the entrance channel of photoreactions, but only their central complex and the exit channels. The ensuring conceptual simplifications may be summarized by the statement that a photoprocess amounts to "half a scattering" process [5.41]. For this reason the results of straightforward photoabsorption experiments must be understood in detail before the more complex particle collision experiments can be interpreted. The limitations of the photoabsorption experiments are obvious; one only gains information on how the atomic electrons *as a whole* respond to incoming photons. Photoelectron spectroscopy allows one to single out a particular exit channel and provides detailed information on the complex correlation phenomena occuring during deexcitation of the atomic system. Measurements on the angular distribution of photoejected electrons provide even more information.

Calculations of photoionization cross-sections, Auger rates etc., have been performed using a variety of theoretical procedures incorporating many-electron effects. One can distinguish the correlations of two (or more) electrons that are excited to the extent of moving mostly outside the radius of the ground state atom from those that are confined within the valence of inner shells. FANO [5.42] and associates have studied the former in some considerable detail. A variety of interaction effects which are confined to filled shells have been calculated by many-body theory. These methods study directly the net electron density, instead of attempting to describe the motion of individual electrons. Two somewhat different lines have developed. The AMUSIA group [5.43] and WENDIN [5.44] have utilized the random phase approximation with exchange (RPAE) method. KELLY [5.45] and POE [5.13] have used the many-body perturbation theory (MBPT) approach.

In the 1960's the close coupling method was employed by BURKE [5.46] and SMITH [5.47] to treat one electron accurately. This method has its limitation in that the effects of interaction of this electron with the rest of the atom can only be dealt with in a limited way. Many-electron interactions have also been studied by Multi-Configuration Hartree-Fock (MCHF) [5.48]. Adaptation of the method allowing direct calculation of R-matrices has resulted in extensive applications by BURKE [5.49] and LEE [5.50].

5.5 Experimental Results

The format in this section follows that of Sect. 5.2. We first discuss a number of examples of correlation phenomena observed in photoabsorption experiments using synchrotron radiation. The more sophisticated experiments of photoelectron spectroscopy and mass spectrometry are discussed in later subsections.

5.5.1 Photoabsorption Spectroscopy

This section will quite arbitrarily be separated into two parts. The first will treat discrete resonances in the continuum. Such resonances result from the interaction of excited states due to subshell, inner electron or two-electron excitation, and the adjacent continuum states associated with ionization of the less bound outer electrons. These resonances have widths of 0.1 eV and less. The second will deal with gross structure stretching over many eV (sometimes 20-30 eV). This is a quite arbitrary separation, since both are manifestations of interchannel or intrachannel interaction and indeed the existence or otherwise of discrete resonances (Rydberg states) in one channel can give valuable clues as to the interpretation of broader resonance structure that may be observed at higher photon energies.

a) Discrete Resonances

Synchrotron radiation made its first and perhaps greatest impact in atomic absorption spectroscopy when used in the early 1960's to discover a wealth of discrete resonances in the photoionization continua of the rare gases [5.51]. The broader resonances discussed later could, in principle, have been seen with existing line sources but it is an interesting fact that until synchrotron radiation was employed, much structure was ascribed to autoionizing resonances which was subsequently found not to exist.

The absorption spectra of neutral atoms exhibit numerous Rydberg series of lines, each converging to the ground or to an excited state of the positive ion. Most lie above the first ionization limit and thus correspond to transitions from the ground state to discrete states which are stationary only in the independent-electron model Interchannel interaction, in this case between the discrete states of one channel and the continuum of another channel of equivalent energy, causes the discrete state to autoionize in times of the order of 10^{-13} seconds. This short life time compared to a typical radiative lifetime of 10^{-8} seconds is the reason why such states are probed by absorption rather than emission spectroscopy.

In 1964, SILVERMAN and LASSETTRE [5.52] observed discrete energy losses in the forward scattering of electrons by helium at about 60 eV. The first atomic absorption spectrum obtained with synchrotron radiation, that of He in the 200 Å region [5.53], is shown in Fig. 5.2. The resolution attained was approximately 20 times that of the earlier electron loss data. The considerable structure observed was classified into four series of resonances. The two series approaching the n = 2 state of He^+ are shown, one strong and broad (+), the other weak and narrow (-).

The importance of this helium spectrum in terms of the theoretical understanding of electron correlation effects cannot be overstated because here is the simplest conceivable example of the highly correlated motion of two excited electrons. The (+) and (-) labels relate, in the original assignment of the two series by COOPER et al. [5.54], to the particular mixture of independent-electron states 2snp and 2pns given by

$$\psi \ (sp2n \ \pm) \ = \frac{1}{\sqrt{2}} \left[U(2snp) \ \pm \ U(2pns) \right] \ . \tag{5.12}$$

These correspond in physical terms to the in-step (+) and out-of-step (-) radial oscillations of the two excited electrons. The spectral structure clearly forms two regular series distinguished by widely different rates of formation and decay and extreme departures from the usual picture of quasi-independent electrons is apparent.

Because the Schrödinger equation for two electrons could be solved, the spectrum in Fig. 5.2 stimulated extensive calculations [5.55], but the existence of series emerged *only* from an analysis of the results obtained, much as it did from the experimental observations. A full theory would have to provide an algorithm that yields each series separately and characterizes it by appropriate quantum numbers corresponding to new quasi-constants of the motion. The challenge raised by this situation has stimulated efforts that are still in progress.

We cannot pursue this problem further, except to state that MACEK [5.56] made a major contribution when he formulated the problem in hyperspherical coordinates. The method used, although yielding a grouping of levels into channels by reducing the Schrödinger equation to a single variable form with alternative optical potentials, was not developed sufficiently to reveal the physical nature of each channel or to test the approximations used adequately. LIN [5.57] carried the ideas further, to allow one to distinguish the effects of angular correlations, radial correlations and exchange in determining the properties of a given potential curve for each channel. Apparently the (+) and (-) series can be distinguished by the radial correlations, but angular correlations cannot be completely disregarded.

244

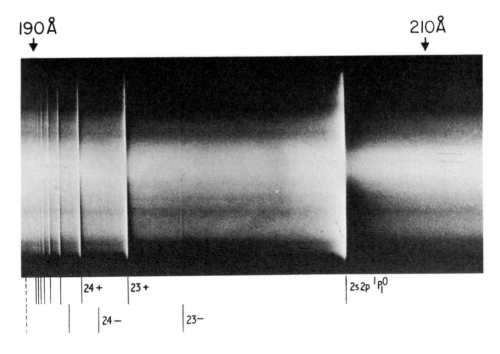

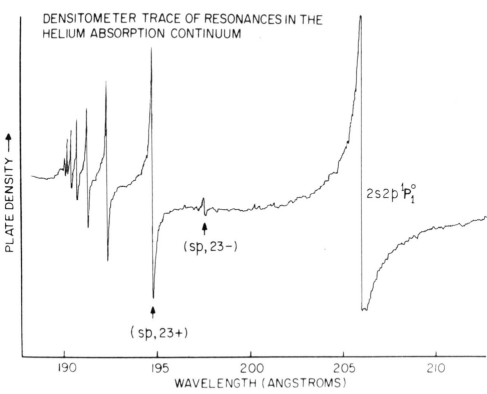

Fig. 5.2 The absorption spectrum of He in the 60 eV (200 Å) region [5.53]

The He spectrum exhibits many effects which can be understood in terms of the FANO [5.58] theory of configuration interaction. The 2s2p $^1P_1^o$ resonance has an asymetric profile and the cross section in the resonance region can be described in terms of the formula

$$\sigma = \sigma_A \cdot (q + \varepsilon)^2/(1 + \varepsilon^2) \qquad (5.13)$$

where q is a line profile index involving three matrix elements, ε is an energy variable and σ_A is the cross section in the resonance wings; see also Sect. 4.4.2. The 2s2p resonance is the simplest example of a single resonance interacting with a single set of continuum states (1sEp) and the cross section should have a zero minimum. Figure 5.3a shows the line profile with its expected zero minimum, a q = -2·8 and Γ = 0·038 eV.

At energies somewhat higher than shown in Fig. 5.2 one observes a further series of resonances approaching the n = 3 state of He$^+$, with q-values that are positive and of very poor contrast. To describe the first of these resonances, 3s3p $^1P_1^o$, a further parameter ρ^2 was introduced [5.59]. It is located at an energy such that it can interact with more than one continuum, 1sEp, 2sEp, 2pEs and 2pEd. Hence the cross section can be described by

$$\sigma = \sigma_A(q + \varepsilon)^2/(1 + \varepsilon^2) + \sigma_B \qquad (5.14)$$

where σ_A is the part of the continuum that does, σ_B the part that does not interact with the discrete state and

$$\rho^2 = \sigma_A/(\sigma_A + \sigma_B) \qquad (5.15)$$

where ρ is a correlation coefficient or overlap integral. SHORE [5.60] gave an alternative parameterization. DHEZ and EDERER [5.61] have studied this resonance, obtaining values q = 1.36, ρ^2 = 0.012 and Γ = 0.13 eV. The low value of ρ^2 derives from the preferential autoionization into the weaker n = 2 continua.

Within this restricted framework one is able to parameterize the line profiles of many of the resonances observed in the photoionization continua of the rare gases. (Whether this parameterization leads to a more fundamental appreciation of the physics involved is a point which will not be dealt with here.) In Figs. 5.3b, c and d are shown three further examples. These are associated with a two-electron, a subshell and an inner shell electron excitation. In the cases of Ne and Ar, the resonances due to excited states nsnp^6 mp $^1P_1^o$ are in the same energy range as those due to two-electron excitation ns^2 np^4 mℓ m'ℓ'. One is therefore able, in a simple absorption experiment, to gauge the relative strengths of one- and two-electron excitation states and determine, at least qualitatively, the importance of correlation effects.

In Fig. 5.3b is shown the Ne resonance $2p^4$ $3s3p$ $^1P_1^o$ for which CODLING et al. [5.62] obtained a rather low value of ρ^2. The argon resonance at 27 eV is due to a single 3s excitation and is the first member of a series of unusual "window" resonances or antiresonances. Although quite understandable within this simple theoretical framework as examples of resonances where $q \sim 0$, their occurrence was nevertheless somewhat surprising when first observed using synchrotron radiation. FANO [5.63] suggested a more instructive formulation of (5.14), i.e.,

$$\sigma = (\sigma_A + \sigma_B) \ [1 + \rho^2 q^2/(1 + \varepsilon^2) - \rho^2/(1 + \varepsilon^2) + 2q \ \varepsilon \ \rho^2/(1 + \varepsilon^2)] \tag{5.16}$$

The second term in the square brackets represents the conventional Lorentz profile, the third term the effect of spectral repulsion and the last term the interference. When $q \approx 0$, a window resonance is observed because absorption through the discrete state fails to compensate the effect of spectral repulsion.

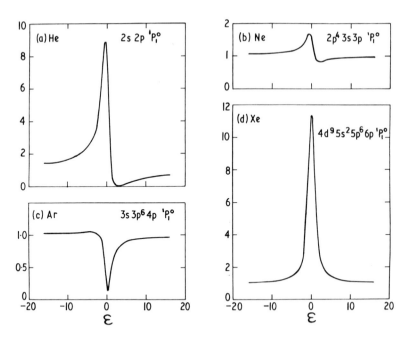

Fig. 5.3 Resonance profiles in the inert gases in the XUV. (a) He, with q = -2.8, Γ = 0.038 eV; (b) Ne, with q = -2.0, Γ = 0.01 eV, ρ^2 = 0.17; (c) Ar, with q = -0.2, Γ = 0.008 eV, $\rho^2 \sim 0.86$; (d) Xe, with q $\sim$ 200, Γ = 0.11 eV, $\rho^2 \sim 0.003$

The parameters of the Ar resonance were obtained by MADDEN et al. [5.64] and are shown in Fig. 5.3 (see also Sect.7.4.1 for the same lines with solid Ar), At that time the q, Γ and ρ^2 were also determined for the second and third members of the Rydberg series $3s\ 3p^6\ np\ ^1P_1^o$. The fact that the values of q and ρ^2 were found to be the same within experimental error and the width, Γ, decreased as the inverse cube of the effective quantum number (n*), was expected at that time. Moreover, once these parameters had been obtained, one could calculate a "branching ratio" for photoionization into this new $3s\ 3p^6\ Ep$ channel relative to the old $3s^2\ 3p^5\ Es$ (Ed) channels.

In now appears that this expectation is in error and that one cannot use the resonance parameters as a useful guide to the branching ratio, except in the limiting case, and that new parameters must be introduced to evaluate the new threshold cross section. The theoretical evidence seems so far to stem from an R-matrix calculation of BURKE and TAYLOR [5.65]. According to them, q and Γ vary systematically through a Rydberg series. FANO [5.66] has reformulated and extended his treatment of autoionization profiles by configuration mixing and by the complementary method of channel interaction (quantum defect theory) and compared the two approaches. Unfortunately neither of the theories appears to be clearly preferable or physically more transparent.

In Fig. 5.3d is shown the resonance profile of the $4d^9\ 5s^2\ 5p^6\ 6p\ ^1P_1^o$ state of Xe. This is an example of the typical resonance shape observed at higher photon energies and is clearly of Lorentz shape, with no sign of asymmetry. EDERER and MANALIS [5.67] determined the parameters to be q $\sim$ 200 and $\rho^2 \sim$ 0.0003. The high value of q reflects the large oscillator strength of the discrete state compared to that of the continuum states of equivalent energy. The low value of ρ^2 shows that autoionization leads to different final states of the Xe ion than direct ionization.

Another aspect of Rydberg states at higher photon energies is obvious but nevertheless worth mentioning. Such highly excited states will tend to decay rapidly by the Auger process, producing large resonance widths. Because high series members will continue to converge rapidly though their width remains constant, very few series members will be observed.

The simple situations described above rarely occur in practice due to configuration mixing of series approaching the same (or only slightly displaced) ion limits. This mixing may be so extensive that it becomes impossible to label excited states as involving a particular electron configuration. Perturbations may be such that one can no longer pick out Rydberg series and determine accurate ionization limits (or binding energies). The above resonance parameterization may still be possible, but meaningless. Many such situations occur in the rare gases and in such cases *ab initio* calculations are really required before one can usefully compare experimental fact and theoretical calculation.

In recent years many metal vapor spectra have been photographed, particularly by CONNERADE and MANSFIELD [3.68,69] using the Bonn synchrotron. An enormous amount of structure has been observed and partially classified. This wealth of new data will be discussed only where it is relevant to the overall picture. Apart from showing the complexites of structure expected in these spectra, their work has confirmed the expectation that the XUV absorption spectra of atoms in the vapor and solid phases can be similar in both their gross features and in their finer details.

b) Gross Features

In this section we attempt to give an overview of the present experimental situation with regard to the gross features in atomic photoionization continua. It is an unfortunate fact that *absolute* XUV cross section data of all elements other than the rare gases are incomplete or suspect (or both). It is inevitable, therefore, that much theoretical effort has been expended on the rare gases where cross sections are thought to be known to an accuracy of ± 5%. However, rather than concentrate on the rare gases as a group, the recent data obtained with synchrotron radiation sources will be discussed in terms of the excitation or ionization of a particular electron, and how the associated spectral features vary as a function of increasing atomic number. One also wishes to study such features along an isoelectronic sequence but production of ions in sufficient quantities is not easy and the determination of accurate absolute cross sections would be extremely difficult.

A considerable literature now exists on the photoionization cross sections of many elements throughout the periodic table. Most data were obtained photographically and in some cases absolute cross sections were determined. Because important correlation effects are now included, theoretical calculations are becoming increasingly reliable and it is important to have experimental accuracies at the 5% level. Such accuracies can only be achieved using photoelectric detection.

A number of experiments were designed to measure the transmittance of thin solid films in the XUV in a rather straightforward manner, with few experimental difficulties. Such experiments probe what are basically *atomic phenomena* and the similarities between the atomic and solid spectra are quite striking. The explanation for this lies in the centrifugal barrier for excited or ionized inner shell electrons of high angular momentum which localizes these electrons. The (initial) core level states and (final) excited states are therefore both isolated from the environment of the atom and the photoabsorption spectra will be similar. Examples of such solid spectra and their explanation in terms of atomic effects will be given.

L-Shell Spectra. The absorption spectra associated with the 2s and 2p excitation and ionization in Ne ($2s^2 2p^6 \, ^1S_0$) and Na ($2s^2 2p^6 3s \, ^2S_{1/2}$) have both been observed with synchrotron radiation [5.62,70,71]. We will not discuss the details of the shar resonance structure associated with the excitation of the 2s and 2p electrons except to state that the spectra exhibit many features that can only be interpreted in terms of configuration mixing.

In describing the continuum absorption due to transitions of the 2p electron to continuum states of d-symmetry (the 2p → Es contribution to the total cross section is small), a single-electron Hartree-Slater calculation reproduces the gross features quite adequately. Figure 5.4 shows the absorption cross section of Na in the 40-500 eV region. The open circles and the solid line are absolute measurements [5.16] using a heat pipe, the closed circles are relative measurements [5.71] and the dotted line a single-electron calculation [5.72].

Spectra have been obtained recently of the structure in the 300-400 eV region involving excitation of the 2p electron in K and Ca. Interesting effects occur which can be explained in terms of shell collapse, but for details the reader is referred to the original work [5.73,74].

M-Shell Spectra. There exist a number of spectra involving the excitation of the 3s, 3p and 3d electrons; we will concentrate on a few of them. The 3p photoionization cross section of Ar ($3s^2\,3p^6\,^1S_0$) has been studied exhaustively both experimentally and theoretically. The various sets of data have been critically reviewed over an

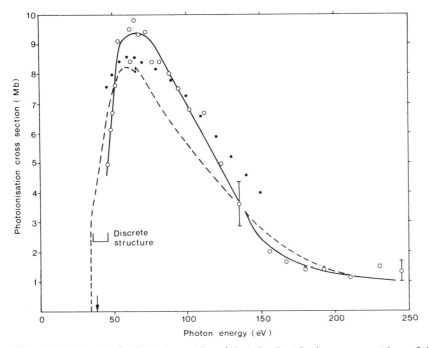

Fig. 5.4 The total photoabsorption (photoionization) cross section of Na in the 40-250 eV photon energy range. Experiments: Solid curve and open circles [5.16]; closed circles [5.71]. Theory: dashed curve, single-electron Hartree-Slater [5.72]

250

extended energy range by WEST and MARR [5.14], but the region of interest for 3p
ionization, from 15-45 eV, is shown in Fig. 5.5.

COOPER [5.75] used a single-electron model to predict the general shape of the
3p → Ed cross section and explained the existence of a minimum, the Seaton-Cooper
minimum, at approximately 50 eV and why the threshold behavior of the 2p electron
in Ne would be expected to be quite different. Cooper predicted a threshold cross
section that was too high and a fall-off too abrupt when compared to experiment.
He pointed out that correlation effects in the 3p shell would be expected to smear
out the rapid drop to give better agreement with experiment.

Subsequently the more sophisticated calculations [5.65,76-79] have born out this
prediction. Shown in Fig. 5.5 are two of these theoretical calculations, the RPAE
calculation of AMUSIA et al. [5.77] and the R-matrix calculation of BURKE and TAYLOR
[5.65]. The latter curve is not plotted in the region of the window resonances
caused by excitation of the 3s electron. We shall return to the interesting behavior
of the 3s photoionization cross section in Sect. 5.5.2.

In the late 1960's, the DESY group studied the transmission spectra of many thin
films in the XUV. In particular, SONNTAG et al. [5.80] obtained the spectra of the

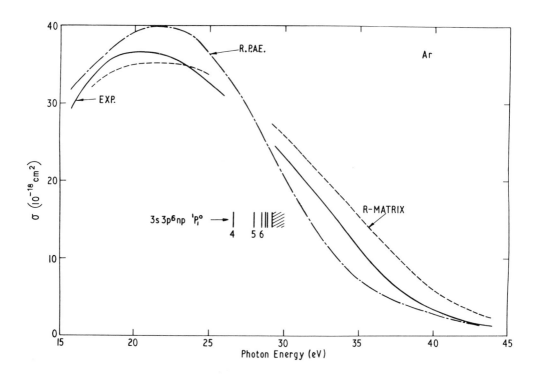

Fig. 5.5 The total photoionization cross section of Ar. The experimental curve is
is a weighted average of all published data [5.14]. The broken curves show RPAE
[5.77] and R-matrix [5.65] theoretical calculations

the transition elements, Ti through Ni in the energy range 30-200 eV. The surprising
feature of these spectra at the time was the slow rise in the cross section at the
$M_{II,III}$ edges and the large widths of the structure above the edges, far larger than
could be explained in terms of the unoccupied d-band width. FANO and COOPER [5.81],
in an addendum to their 1968 review paper [5.40], noted that the absorption near
threshold was preceded by a dip, so that the full absorption profile very much resembled a broad autoionizing line; this implied that the inner shell absorption peak
interfered with the background absorption continuum.

The particular excitations involved in the transition elements are of the type
$3p^6 3d^N \rightarrow 3p^5 3d^{N+1}$. In these elements, as in the rare earths, the key to understanding the inner shell spectra and how they vary with atomic number lies in the appreciation of the importance of the centrifugal barrier term, $\ell(\ell + 1)/r^2$, which can local-
ize those excited (or ionized) electrons with high angular momentum. COMBET FARNOUX
[5.82] performed *atomic* calculations on the transition elements and in the case of
V and Co found that the terms of the $3p^5 3d^{N+1}$ configuration that were responsible
for the major part of the oscillator strength, lay a few eV above the $M_{II,III}$ threshold. These excited states were therefore broadened by autoionization, and the agreement between experiment and theory appeared to be good.

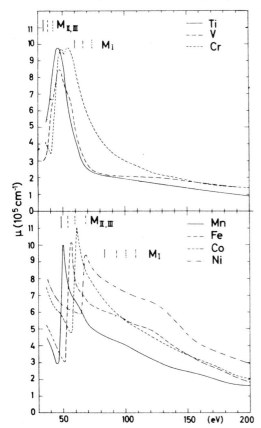

Fig. 5.6 The absorption coefficient
of the transition metals [5.80]

It might be argued that the centrifugal barrier term is known explicitly for a particular electron and hence the spectra one observes should be easily interpretable. However, the height of a barrier depends on the relative contributions of the screening and the barrier terms and the former depends on rather poorly predictable effects such as exchange and correlations. Therefore a comparison between experiment and theory should test definitively whether the appropriate correlation effects have been included. The barrier can clearly depend critically upon the environment of the atom (whether free or in the molecular or solid state) and from this point of view it is important to obtain and understand the atomic, vapor phase, spectrum before one is entirely happy with the interpretation of the molecular and solid state spectra.

CONNERADE et al. [5.83] have recorded the absorption spectrum of Mn vapor due to excitation of the 3p electron and compared the "giant" resonance seen at around 50 eV with the Mn metal and halide spectra [5.84]. The spectrum of Mn metal had also been determined by DIETZ et al. [5.85] using electron energy loss techniques. MANSFIELD and NEWSOM [5.86] have recently obtained the 3p subshell spectrum of Ca between 10-40 eV and discuss the 3p → 3d excitation spectra of elements with Z = 18 to 28.

N-Shell Spectra. We will concentrate here on the spectra observed when the 4d electron is excited. Once again the rare gas Xe has been studied extensively and cross sections are known with fair accuracy. When CODLING and MADDEN [5.87] first observed the discrete spectrum associated with excitation of the 4d electron, it seemed surprising that no series could be assigned to transitions 4d → nf. Moreover a broad absorption feature was observed, centered some 30 eV above the 4d threshold ($N_{IV,V}$ edge) and about 35 eV in half-width [5.88,89]. These two observations are both explained qualitatively in terms of the centrifugal barrier to the nf electrons which excludes them from the core region. This barrier is even more effective for nf electrons than for the d electrons in the transition elements. The resulting small overlap with the initial state 4d electron wavefunction results in a total loss of oscillator strength for the 4d → nf series. Only when the f electron attains sufficient energy to overcome this barrier is there substantial overlap with the 4d wavefunction. This causes a "delayed onset" of oscillator strength.

STARACE [5.76] pointed out the reason for the failure of the Hartree-Slater model calculation of COOPER [5.90] to reproduce the height, position and shape of the experimental curve - the strong exchange interaction between the excited electron and the remaining 4d electrons (or the 4d hole). This intrashell interaction causes the "shape" resonance in the cross section [5.91]. The more sophisticated RPAE calculations of AMUSIA et al. [5.77] and WENDIN [5.92] have since given good agreement with experiment. Figure 5.7 shows the experimental data obtained with DESY synchrotron [5.93] compared with the K-matrix calculation of Starace and the SRPAE calculation of LIN [5.78].

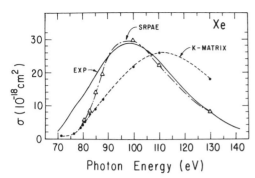

Fig. 5.7 The total photoionization cross section of Xe. Experiment [5.93]; K-matrix [5.76]; SRPAE [5.78]

ZIMKINA et al. [5.94] systematically studied the transmission of solid films of the rare earths in the 50-500 eV range. They observed large absorption features around 150 eV, due to the excitation of the 4d electrons. The later work at DESY [5.95] led DEHMER et al. [5.96] to reconsider the original suggestion of FOMICHEV et al. [5.97], that the sharp structure below the 4d threshold could be explained in terms of transitions $4d^{10} 4f^N \rightarrow 4d^9 4f^{N+1}$ in the triple ionized atom.

The lowering of the centrifugal barrier in the lanthanides (as compared to Xe), due to the increase in Z, means that the 4f orbit now has a small radius which fits *inside* the centrifugal barrier; f orbits of somewhat higher energy remain well outside. The large overlap of the 4f wavefunction with the 4d hole results in a large exchange interaction which spreads the levels of the excited state configurations over 20 eV or more. The higher levels can now autoionize into the $4d^9 4f^N$ Ef continuum and contain such oscillator strength that broad absorption features are observed. This interpretation is supported by the evidence that the strength of the main absorption features decrease in relation to the member of 4f vacancies in the atom until Lu $(4d^{10} 4f^{14})$, no structure is observed. A similar phenomenon explains the weakening of the resonance in the transition elements Ti through to Ni (see Fig.5.6); in this case the 3d shell is filling.

In the calculations of SUGAR [5.98] and DEHMER and STARACE [5.99], scaled Slater integrals were used in order to achieve good agreement with energies of the discrete structure in general and the broad $4d^9 4f$ $^1P_1^o$ resonance in particular. Subsequently STARACE [5.100] performed an *ab initio* RPAE calculation for La $(4d^{10} 4f^0)$ and achieved equally good agreement with experiment, showing that the scaling procedure adopted has been appropriate.

Since that time experiments have been performed on metal vapors Ce $(4d^{10} 4f)$ and Eu $(4d^{10} 4f^6)$ at DESY [5.101] and Bonn [5.102] using photographic techniques. The earlier assumption that the vapor phase and solid state spectra would be very similar has been fully justified. Fig. 5.8 shows the absorption spectra of Ce in the solid and vapor phases, where there is seen to be considerable agreement in both fine structure and 'giant' resonance. This demonstrates that the particular arrangement of the outer 5d and 6s electrons hardly influences the spectra.

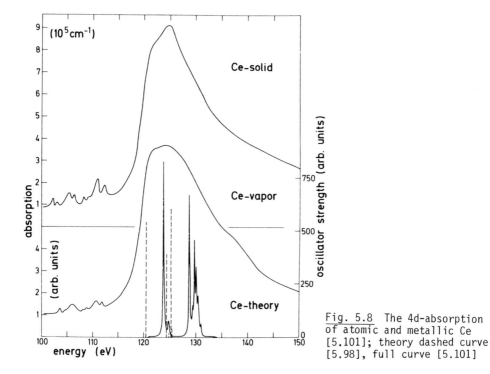

Fig. 5.8 The 4d-absorption of atomic and metallic Ce [5.101]; theory dashed curve [5.98], full curve [5.101]

A considerable effort both experimental and theoretical has been expended on the absorption spectrum of Ba. Ba lies in the periodic table between Xe and La. The apparent similarity between the spectra of atomic Ba and solid La led EDERER et al. [5.103] to conclude that the giant resonance in Ba at about 110 eV was to be associated with the $4d^9$ $4f$ $^1P_1^o$ excited state, pushed into the continuum by the large exchange interaction. HANSEN et al. [5.104] disagreed, stating that the $^1P_1^o$ resonance was in fact below the 4d ionization threshold and that the large resonance was to be associated with the $4d^9$ Ef continuum, much in the same way is in Xe, and in agreement with the earlier WENDIN [5.105] interpretation and the work of FLIFLET et al. [5.106]. Fig. 5.9a and b show the absorption spectra obtained at the DESY [5.107] and Bonn [5.108] synchrotrons.

WENDIN [5.110] appears to have had the final word when he suggests that the residual electron-electron interaction is so strong that the concept of electron configuration breaks down. He concluded that although the $4d^9$ $4f$ $^1P_1^o$ resonance lies below the 4d threshold, the oscillator strength associated with the $4d^9$ $4f$ HF_{av} configuration is pushed above threshold!

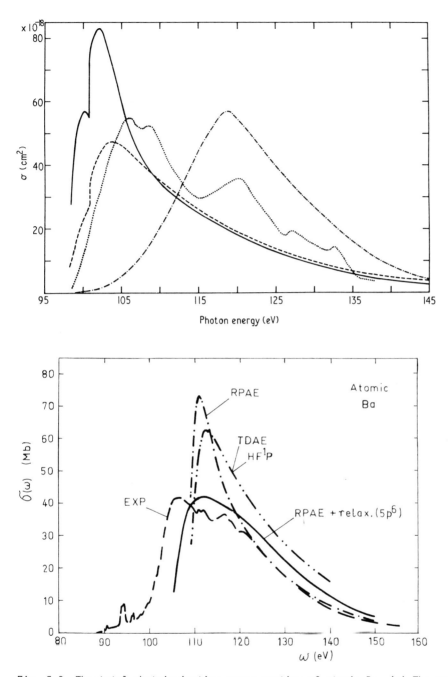

Fig. 5.9 The total photoionization cross section of atomic Ba. (a) The experimental (dashed) curve is taken from [5.107]. The various labelled theoretical curves are taken from [5.105]. (b) The experimental (dotted) curve is taken from [5.108]. Theoretical curves are as follows: full curve, average of length and velocity form RPAE results including effect of the splitting between the $4d^9\ ^2D_{3/2}$ and $^2D_{5/2}$ cores; dashed curve, average of length and velocity ground state correlation results including spin-orbit splitting; dash-dot curve, Wendin [5.105] RPAE result

5.5.2 Photoelectron Spectroscopy

In the previous section we have discussed a number of results from photoabsorption
spectroscopy and found, in general, a fair agreement between experimentally determined cross sections and those calculations which specifically included correlation
effects [5.111]. It is clear, however, that such experiments are somewhat limited in
scope since the results obtained reflect the response of the atomic electrons as
a whole. If we wish to investigate correlation effects in a more definitive way, and
determine the final states of the atomic system after photoionization, then we must
perform experiments in photoelectron spectroscopy (PES), mass spectrometry, or both.

Such experiments require a greater experimental finesse, higher photon fluxes
and a great deal of time. In the past PES has been performed at low energies with
emission line sources, such as the He I and He II resonance lines (UPS) and at
higher energies with characteristic X-ray lines such as the $A\ell\ K_\alpha$ line (XPS).
Since much of the light source energy goes into these emission lines and monochromators are not usually employed, the photon flux levels are much higher than could
possibly be achieved with low current synchrotrons at equivalent resolution.

However, as we shall see, there are a number of experiments in atomic PES that
can benefit from a source of continuously variable energy, even if we accept the
generally poorer resolution of the final data. For example, any threshold phenomenon, such as post collision interaction [5.112], can best be studied with a 'clean'
source of variable energy. When existing experiments are coupled to high current
storage rings, more sophisticated experiments can be envisaged.

The electron correlation phenomena that can be probed by PES may be strong and
direct, such as the excitation (ionization) of two or more electrons (shake-up and
shake-off) or weak and indirect, such as energy level shifts. Such topics as initial
or final state configuration interaction, core rearrangement and virtual Auger processes were admirably discussed by KRAUSE [5.19] and others [5.113]. Although some
preliminary work along these lines is being performed with synchrotron radiation,
these aspects will not be pursued here. We concentrate on partial cross section
measurements and how they give important information on correlation effects. Brief
reference will also be made to angular distribution measurements.

a) Partial Photoionization Cross Sections

Argon L-Shell. Since the discovery of window resonances in the $3p \rightarrow Ed$ continuum
of Ar at approximately 30 eV, due to transitions $3s^2\ 3p^6\ {}^1S_0 \rightarrow 3s\ 3p^6\ np\ {}^1P_1^o$, the
details of the correlation effects giving rise to them and to the associated
$3s \rightarrow Ep$ continuum have held a fascination for experimentalist and theoretician
alike.

In the late 1960's an attempt was made [5.114] to measure the 3s partial photoionization cross section with the Glasgow synchrotron (0.1 mA average current).

Signs of unusual behavior in the 3s cross section as a function of energy were apparent and subsequently the Daresbury NINA synchrotron was used to remeasure this cross section [5.115] with the results shown in Fig. 5.10. The Hartree-Fock prediction [5.116] was that the cross section would rise steadily from a threshold value of 0.07 Mb. The experimental curve exhibits a totally different behavior, with a zero minimum some 10-15 eV above threshold. The calculations including correlation effects between the 3s and 3p subshells [5.65,117,118] show a much better agreement with experiment. AMUSIA explained the zero minimum in physical terms in the following way. The projectile photon induces virtual oscillations in the outer 3p subshell which are in antiphase with the incident electromagnetic wave. When the amplitudes of these oscillations are comparable, the 3s subshell is totally screened and the 3s electron cannot be ionized. The behavior of the 2s electron in Ne is totally different [5.120] since a similar intrashell correlation phenomenon cannot occur.

Xenon N, O Shells. One further example, this time of intershell interaction, is seen in the photoelectron spectrum of Xe in the region of the 4d shape resonance around 100 eV, discussed earlier. Here is an example of an inner shell influencing the behavior of an outer shell [5.121]. In a single-electron picture, it would be expected that the 5s and 5p partial cross sections should be monotonically decreasing at 100 eV. In fact, due to intershell interaction between the 4d and 5s or 5p electrons,

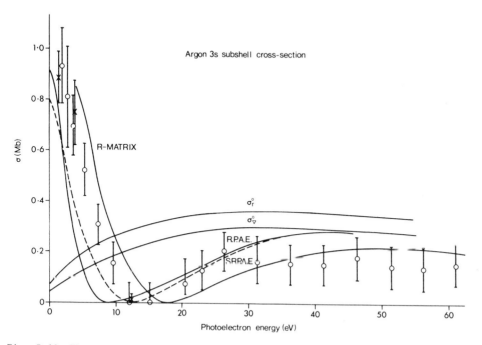

Fig. 5.10 The partial cross section of the Ar 3s electrons [5.115,119]. Theoretical curves: HF-L [5.116]; R-matrix [5.65]; RPAE [5.117]; SRPAE [5.118]

258

the partial cross sections are enhanced. The effect is shown in Fig. 5.11, where calculations of AMUSIA et al. [5.122,123] are shown for comparison.

We now discuss two further aspects of partial cross section measurements as they apply to the outer 5p electrons in Xe. Photoionization of the outer subshell leaves the Xe ion in either the $5p^5(^2P_{3/2})$ or $5p^5(^2P_{1/2})$ state. The branching ratio $\sigma_{3/2}/\sigma_{1/2}$ has been found using line sources to be remarkably constant at 1.55 ($\pm$ 0.08), and therefore well below the statistical ratio of 2.0, over a range of 25 eV above threshold [5.124]. These earlier measurements showed large changes of ratio within autoionizing resonances, due presumably to strong correlation effects.

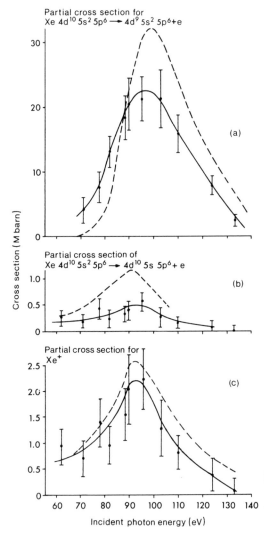

Fig. 5.11 Partial cross sections for the 4d (upper), 5s (middle) and (5s + 5p) electrons (lower) [5.121]. The dashed curves are RPAE calculations [5.122,123]

Recently KEMENY et al. [5.125] have used synchrotron radiation from the Wisconsin storage ring to determine this ratio as a function of energy through the 5s 5p^6 6p ^{1}P$^o_1$ window resonance at 20.95 eV. Using a fitting procedure they estimated that the ratio should reach a value of approximately 9 near the center of the resonance. Fig. 5.12 shows the variation of branching ratio as a function of photon energy together with a theoretical fit [5.126] to the data, and demonstrates the extent of correlation effects within a small energy region. Similar correlation effects are responsible for the variation of asymmetry parameter (see Sect.5.5.2) within a resonance [5.127]. Furthermore, the values of branching ratio provide information which helps to further characterize an autoionizing state such as 5s 5p^6 6p ^{1}P$^o_1$.

In a second experiment WUILLEUMIER et al. [5.128] have measured the same branching ratio over a much extended photon energy range, from 21 to 107 eV. Their results, which were obtained basically with the ACO storage ring, are shown in Fig. 5.13. A minimum occurs at about 30 eV, the ratio rises slightly above statistical at above 60 eV reaching the statistical value at about 90 eV. The open circle point is from DEHMER [see ref. 5.128] and is in excellent agreement with the most recent data. The data disagrees with the dashed horizontal line up to 40 eV, which is taken from SAMSON et al. [5.124]. The point at threshold is calculated by LU [5.129] using the quantum defect theory to extrapolate from the Rydberg series.

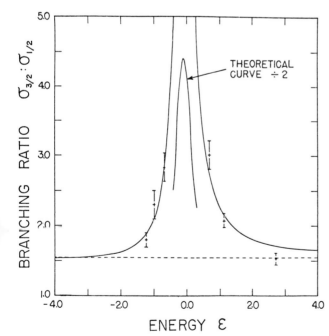

Fig. 5.12 ^{2}P$_{3/2}$: ^{2}P$_{1/2}$ branching ratio for the Xe 5p electrons [5.125] in the neighborhood of the 5s5p^6 6p ^{1}P$^o_1$ resonance. The solid theoretical curve is taken from reference [5.126]

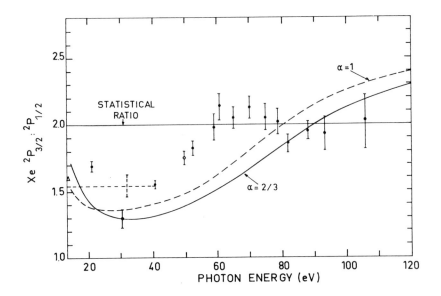

Fig. 5.13 $^2P_{3/2}$: $^2P_{1/2}$ branching ratio for the Xe 5p electrons. Closed circles and open circle [5.128]; dashed horizontal line [5.124]; threshold point [5.129]. Theoretical curves [5.130] are shown with two values of exchange parameter, α

The full and dashed curves are taken from a Dirac-Slater model calculation of DESCLAUX [5.130] and it seems clear that a minimum in the branching ratio is to be expected although the agreement with experiment is by no means perfect.

b) Angular Distributions of Photoelectrons

We discuss briefly the application of synchrotron radiation to the field of electron angular distribution measurements. Apart from its continuum property, we use here the fact that synchrotron radiation is very highly polarized in the orbital plane, although this is not a prerequisite in such experiments. Indeed, one must not only know the state of polarization of the XUV radiation (the radiation is elliptically polarized), but be aware also that the plane of polarization might be rotated on passing through the XUV monochromator [5.131].

The angular distribution of electrons ionized by linearly polarized light is characterized by an expression of the form [5.132]

$$\frac{d\sigma_j}{d\Omega} = \frac{\sigma_j(\varepsilon)}{4\pi} \left[1 + \beta(\varepsilon) P_2(\cos\theta) \right] \qquad (5.17)$$

where σ_j is the photoionization cross sections for channel j, θ is the angle between the electric vector of the incident radiation and the photo-ejected electron and $\beta(\varepsilon)$ is the asymmetry parameter. In general, $\beta(\varepsilon)$ is given by a fairly complex expression involving dipole matrix elements and phase shifts associated with the allowed continuum channels.

Two examples will be given involving those subshell electrons whose partial cross sections have previously been discussed. In the case of Ar, we are interested in the angular distribution of the outer 3p electrons from threshold to well beyond the "Cooper" minimum. (We will not differentiate between the two spin-orbit components in this experiment.)

At threshold and well beyond, any contribution to the total photoionization cross section from the weaker 3p → Es channel is masked by the much stronger. However, the interference between the two channels manifests itself in the form of considerable variation in β with photon energy. Moreover, the form of the theoretical curve is sensitive to the correct inclusion of correlation effects, particularly in the "Cooper" minimum.

The angular distribution of the 3p electrons has been studied extensively in the past with conventional light sources at low energies [5.133-135], but no photon experiment had been able to map out the important and interesting region around the minimum until synchrotron radiation was used for that purpose at NINA [5.115]. Fig. 5.14 shows the results obtained. The solid theoretical curves are taken from KENNEDY and MANSON [5.116]. The dashed curve is an RPAE calculation of AMUSIA et al. [5.136] which includes correlation effects in the ground and excited states. Clearly, the RPAE calculation is in much closer agreement with experiment, although the agreement is still not perfect.

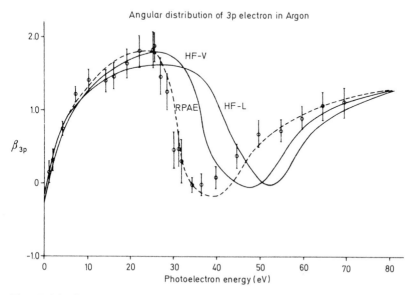

Fig. 5.14 Asymmetry parameter, β, for the Ar 3p electrons. Open circles [5.115]; closed circles [5.135]; HF-L and -V [5.116]; RPAE [5.136]

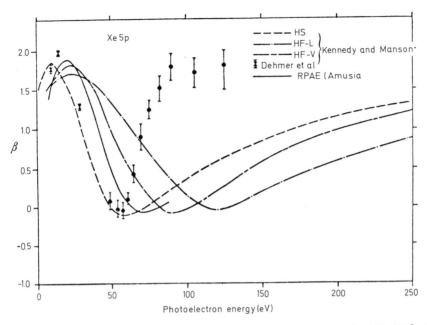

Fig. 5.15 Asymmetry parameter, β, for the Xe 5p electrons [5.131,135]. The broken curves are Hartree-Slater (HS) or Hartree-Fock (HF) calculations [5.116], the full curve an RPAE calculation [5.136]

Finally, in the case of the 5p electrons in Xe, it is apparent once again that electron correlations (intershell interaction) must be included in theoretical models if the experimental data is to be correctly interpreted. The results obtained for β [5.131] are shown in Fig. 5.15 over a resonably extended energy range. Once again the Hartree-Slater and Hartree-Fock calculations [5.116] are considerably in error. When correlations between the outer 5p electrons and the inner 4d electrons are taken into account in an RPAE calculation [5.136], the agreement seems little better. However, a more recent calculation by AMUSIA and IVANOV [5.137] shows almost perfect agreement between experiment and theory. The reason for the earlier disagreement is not known.

5.5.3 Mass Spectrometry

It is perhaps stating the obvious that the details of the photoionization process can be studied either by observing the out-going photoelectrons or the resulting positive ions. The measurement of the variation of the branching ratio for the production of ions in their various states of ionization as a function of incident photon energy can therefore be expected to yield important information on electron correlation effects in atoms which is complementary to that obtained by PES techniques. In situations of double (or multiple) ionization, where the two (or more) electrons share the available energy, mass spectrometry may provide accurate branching ratios where PES cannot.

When the primary ionization process occurs in an inner shell, the sudden creation of a hole leads to electron shake-off and the process has been quite widely studied and is reasonably well understood. When the absorption takes place in an outer shell, the large observed probability of double ionization relative to single ionization cannot be explained by such a process alone. Other correlation effects must be invoked.

The theory of the multiple ionization process has been studied in only a few cases. In He, BYRON and JOACHAIN [5.138], and BROWN [5.139] have used highly correlated Hylleras wavefunctions which yielded agreement with the measurements of CARLSON [5.140] using a conventional X-ray source. However, correlations in the final states were neglected and the dipole length and velocity formulations gave appreciably different results. More recently AMUSIA et al. [5.141] have performed MBPT calculations for He, and CHANG and POE [5.13], and CARTER and KELLY [5.142] have done likewise for Ne and Ar.

The earlier measurements on multiple ionization were performed over an extended energy range using a pseudo-photon source [5.143], but when SCHMIDT et al. [5.144] repeated the measurement of the branching ratio σ^{++}/σ^+ for the inert gases He, Ne and Ar using the ACO storage ring, they found serious disagreement with the earlier work. WIGHT and VAN DER WIEL [5.24] subsequently repeated their measurements and although they are now substantially in agreement with SCHMIDT et al., the reasons for the earlier errors are not known. SCHMIDT et al. used a magnetic field mass analysis and took great care to check on the reliability of the apparatus with respect to its handling of different charged ions. SAMSON and HADDAD [5.146] used discrete XUV lines up to 107 eV and an ionization chamber. Recently HOLLAND et al. [5.147] have used the NINA synchrotron and a time-of-flight mass spectrometer to determine branching ratios for all the inert gases up to about 250 eV.

In Fig. 5.16 are shown the results of various workers for the Ne σ^{++}/σ^+ branching ratio from threshold to 280 eV. There is considerable scatter in the data. The ratios obtained at NINA are low, which would suggest a systematic error. However, the ratios for He and Ar agree well with other workers.

The importance of having a variable energy photon source is illustrated by the work of BREHM and BUCHER [5.148] on atomic Ba. They used the resonance lines of Ne and He at 16.67 and 21.21 eV to determine the σ^{++}/σ^+ ratio and found an amazingly high value at 21.21 eV [2.4] compared with 16.67 [0.25]. Subsequent PES work [5.149] and the theoretical work of HANSEN [5.150] seem to have sorted out the reason for this anomaly in terms of a two-step autoionization process involving first an accidental degeneracy between an excited state of neutral Ba at about 21.2 eV. It would clearly be worthwhile to study the region around 21 eV in detail with a photon source of continuously variable energy.

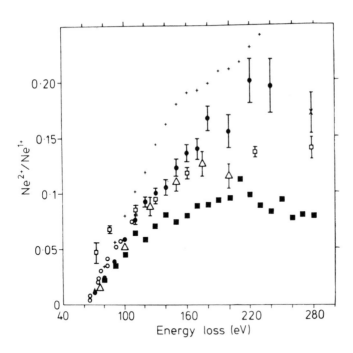

Fig. 5.16 Ratio of double-to-single ionization for Ne. Closed circles [5.24]; crosses [5.143]; open triangles [5.144]; open squares [5.140]; open circles [5.146]; single cross at 280 eV [5.145]; closed squares [5.147]

5.6 Future Work

At this point in time, the aim of the experimenter should not be to obtain total cross sections, partial cross sections etc., for every atom known to exist. Rather it should be to concentrate on a representative number of atoms or ions, some with closed shells, others with open shells, some with low, others with high Z, and to obtain data which are good to an accuracy of a few percent. This reliable data could then be used as a basis to develop a theoretical approach which could predict, with a high degree of confidence, resonance energies and profiles, multiple excitation or ionization branching ratios, angular distribution parameters and states of polari zation of ejected electrons, post-collision interaction phenomena and so on. Althoug great strides have been made recently, both experimentally and theoretically, this situation cannot yet be said to exist.

The role of synchrotron radiation in all such experiments has been made clear in the foregoing chapters. These photoionization experiments, although important, are only one part of the total picture. Our understanding of the ionization process (and thereby the complex correlation phenomena which occur) has advanced due to the application of many techniques, such as electron-atom and ion-atom collisions and nuclear spectroscopy. Such an understanding is surely necessary before the more complex phenomena occurring in solids can be adequately appreciated.

References

5.1   R. Tousey: Appl. Opt. $\underline{1}$, 679 (1962)
5.2   H. Bomke: *Vakuumspektroskopie* (Barth, Leipzig 1937)
5.3   J.C. Boyce: Rev. Mod. Phys. $\underline{13}$, 1 (1941)
5.4   W.R.S. Garton: In *Advances in Atomic and Molecular Physics*, Vol. 2, ed. by D.R. Bates and I. Eastermann (Academic Press, New York 1966) pp. 93-176
5.5   J.A.R. Samson: In *Advances in Atomic and Molecular Physics*, Vol. 2, ed. by D.R. Bates and I. Eastermann (Academic Press, New York 1966) pp. 177-261
5.6   G.V. Marr: *Photoionization Processes in Gases* (Academic Press, New York 1967)
5.7   J.J. Hopfield: Phys. Rev. $\underline{35}$, 1133 (1930)
5.8   Y. Tanaka, R.E. Huffman, J.C. Larrabee: J. Quant. Spectr. Rad. Trans. $\underline{2}$, 451 (1962)
5.9   G. Collins, W.C. Price: Rev. Sci. Instr. $\underline{5}$, 423 (1934)
5.10 W.R.S. Garton: J. Sci. Instr. $\underline{36}$, 11 (1959)
5.11 G. Balloffet, J. Romand, B. Vodar: C.R. Acad. Sci. (Paris) $\underline{252}$, 4139 (1961)
5.12 D.H. Tomboulian, P.L. Hartman: Phys. Rev. $\underline{102}$, 1423 (1956)
5.13 R.N. Chang, R.T. Poe: Phys. Rev. $\underline{A12}$, 1432 (1975)
5.14 J.B. West, G.V. Marr: Proc. Roy. Soc. Lond. $\underline{A349}$, 397 (1976)
5.15 J.B. West, K. Codling, G.V. Marr: J. Phys. $\underline{E7}$, 137 (1974)
5.16 K. Codling, J.R. Hamley, J.B. West: J. Phys. $\underline{B10}$, 2797 (1977)
5.17 T.J. McIlrath, R.J. Sandeman: J. Phy. $\underline{B5}$, L217 (1972)
5.18 W.H. Parkinson, E.M. Reeves, F.S. Tomkins: J. Phys. $\underline{B9}$, 157 (1976)
5.19 F.J. Wuilleumier (ed.): *Photoionization and Other Probes of Many-Electron Interactions*, (Plenum Press, New York 1976) pp. 133-163
5.20 D.A. Shirley (ed.): *Electron Spectroscopy* (North Holland, Amsterdam 1972)
5.21 D.W. Turner, C. Baker, A.D. Baker, C.R. Brundle: *Molecular Photoelectron Spectroscopy* (Wiley, New York 1970)
5.22 J.A.R. Samson, J.L. Gardner: J. Opt. Soc. Am. $\underline{62}$, 856 (1972)
5.23 G.V. Marr, D.M. Creek: Proc. Roy. Soc. Lond. $\underline{A304}$, 233 (1968)
5.24 G.R. Wight, M.J. Van Der Wiel: J. Phys. $\underline{B9}$, 1319 (1976)
5.25 E.V. Condon, G.H. Shortley: *The Theory of Atomic Spectra* (University Press, Cambridge 1964)
5.26 W. Bambynek, B. Crasemann, R.W. Fink, H.U. Freund, H. Mark, C.D. Swift, R.E. Price, P. Venugopala Rav: Rev. Mod. Phys. $\underline{44}$, 716 (1972)
5.27 L.C. Lee, R.W. Carlson, D.L. Judge, M. Ogawa: J. Phys. $\underline{B8}$, 977 (1975)
5.28 K. Feser: Phys. Rev. Lett. $\underline{29}$, 901 (1972)
5.29 M. Inokuti: In [Ref 5.19, pp. 165-185]
5.30 M.J. Van Der Wiel: In [Ref. 5.19, pp. 187-208]
5.31 M. Tronc, G.C. King, R.C. Bradford, F.H. Read: J. Phys. $\underline{B9}$, L555 (1976)
5.32 G.J. Schultz: Rev. Mod. Phys. 45, 378 (1973)
5.33 M.E. Rudd, J.H. Macek: Case Studies Atomic Physics $\underline{3}$, 47 (1972)
5.34 S. Bashkin (ed.): *Beam Foil Spectroscopy, Topics in Current Physics*, Vol. 1 (Springer, Berlin, Heidelberg, New York 1975)
5.35 H.G. Berry, J. Desesquelles, M. Dufay: Phys. Rev. $\underline{A6}$, 600 (1972)
5.36 W.R.S. Garton, W.H. Parkinson, E.M. Reeves: Proc. Phys. Soc. $\underline{80}$, 860 (1962)
5.37 N.P. Penkin, L.N. Shabanova: Opt. Spectr. $\underline{18}$, 535 (1965)
5.38 P. Ewart, A.F. Purdie: J. Phys. $\underline{B9}$, L437 (1976)

5.39 T.B. Lucatorto, T.J. McIlrath: Phys. Rev. Lett. 37, 428 (1976)
5.40 U. Fano, J.W. Cooper: Rev. Mod. Phys. 40, 441 (1968)
5.41 U. Fano: In [Ref 5.19, pp. 1-9]
5.42 U. Fano, C.D. Lin: In *Atomic Physics 4*, ed. by G. zu Putlitz, E.W. Weber and A. Winnacker (Plenum, New York 1975) pp. 47-70
5.43 M.Ya Amusia, N.A. Cherepkov: Case Studies Atom. Phys. 5, 47 (1975)
5.44 G. Wendin: In [Ref. 5.19, pp. 61-81]
5.45 H.P. Kelly: In [Ref. 4.19, pp. 83-109]
5.46 P.G. Burke: In *Advances in Atomic and Molecular Physics*, Vol. 4, ed. by D.R. Bates and I. Estermann (Academic Press, New York 1968) pp. 173-219
5.47 K. Smith: *The Calculation of Atomic Collision Processes*, (J. Wiley, New York 1971)
5.48 A. Weiss: In *Advances in Atomic and Molecular Physics*, Vol. 9, ed. by D.R. Bates and I. Estermann (Academic Press, New York 1973) pp. 1-46
5.49 P.G. Burke, W.D. Robb: In *Advances in Atomic and Molecular Physics*, Vol. 11, ed. by D.R. Bates and B. Bederson (Academic Press, New York 1976) pp. 143-214
5.50 C.M. Lee: Phys. Rev. A10, 584 (1974)
5.51 R.P. Madden, K. Codling: In *Autoionization*, ed. by A. Temkin (Mono Book Corp., Baltimore 1966) pp. 129-151
5.52 S.M. Silverman, E.N. Lassettre: J. Chem. Phys. 40, 1265 (1964)
5.53 R.P. Madden, K. Codling: Astrophys. J. 141, 364 (1965)
5.54 J.W. Cooper, U. Fano, F. Prats: Phys. Rev. Lett. 10, 518 (1963)
5.55 P.G. Burke, D.D. McVicar, K. Smith: Phys. Lett. 12, 215 (1964)
5.56 J.H. Macek: J. Phys. B1, 831 (1968)
5.57 C.D. Lin: Phys. Rev. A10, 1990 (1974)
5.58 U. Fano: Phys. Rev. 124, 1866 (1961)
5.59 U. Fano, J.W. Cooper: Phys. Rev. 137, A1364 (1965)
5.60 B.W. Shore: Rev. Mod. Phys. 39, 439 (1967)
5.61 P. Dhez, D.L. Ederer: J. Phys. B6, L59 (1973)
5.62 K. Codling, R.P. Madden, D.L. Ederer: Phys. Rev. 155, 26 (1967)
5.63 U. Fano: Comments Atom. Molec. Phys. 4, 119 (1973)
5.64 R.P. Madden, D.L. Ederer, K. Codling: Phys. Rev. 177, 136 (1969)
5.65 P.G. Burke, K.T. Taylor: J. Phys. B8, 2620 (1975)
5.66 U. Fano: Private communication
5.67 D.L. Ederer, M. Manalis: J. Opt. Soc. Am. 65, 634 (1975)
5.68 J.P. Connerade, M.W.D. Mansfield: Proc. Roy. Soc. Lond. A335, 87 (1973)
5.69 J.P. Connerade: Proc. Roy. Soc. Lond. A352, 561 (1977)
5.70 J.P. Connerade, W.R.S. Garton, M.W.D. Mansfield: Astrophys. J. 164, 203 (1971)
5.71 H.W. Wolff, K. Radler, B. Sonntag, R. Haensel: Z. Physik 257, 353 (1972)
5.72 E.J. McGuire: Res. Rep. SC-RR-721, Sandia Laboratories (1970)
5.73 M.W.D. Mansfield: Proc. Roy. Soc. Lond. A346, 555 (1975)
5.74 M.W.D. Mansfield: Proc. Roy. Soc. Lond. A348, 143 (1976)
5.75 J.W. Cooper: Phys. Rev. 128, 681 (1962)
5.76 A.F. Starace: Phys. Rev. A2, 118 (1970)
5.77 M.Ya Amusia, N.A. Cheropkov, L.V. Chernysheva: Sov. Phys. JETP 33, 90 (1971)
5.78 H.P. Kelly, R.L. Simmons: Phys. Rev. Lett. 30, 529 (1973)
5.79 C.D. Lin: Phys. Rev. A9, 181 (1974)
5.80 B. Sonntag, R. Haensel, C. Kunz: Solid St. Commun. 7, 597 (1969)
5.81 U. Fano, J.W. Cooper: Rev. Mod. Phys. 41, 724 (1969)
5.82 F. Combet Farnoux: Physica Fennica 9S, 80 (1974)
5.83 J.P. Connerade, M.W.D. Mansfield, M.A.P. Martin: Proc. Roy. Soc. Lond. A350, 405 (1976)
5.84 S. Nakai, H. Nakamori, A. Tomita, K. Tsutsumi, H. Nakamura, C. Sugiura: Phys. Rev. 9, 1870 (1974)
5.85 R.E. Dietz, E.G. McRea, Y. Yafet, C.W. Caldwell: In *Proc. Intern Conf. on VUV Radiation Physics*, ed. by E.E. Koch (Pergamon Vieweg, Braunschweig 1974) pp. 472-474
5.86 N.W.D. Mansfield, G.H. Newsom: Proc. Roy. Soc. Lond. A357 77 (1977)
5.87 K. Codling, R.P. Madden: Phys. Rev. Lett. 12, 106 (1964)
5.88 D.L. Ederer: Phys. Rev. Lett. 13, 760 (1964)
5.89 A.P. Lukirskii, L.A. Britov, T.M. Zimkina: Opt. Spectr. 17, 438 (1964)

5.90   J.W. Cooper: Phys. Rev. Lett. 13, 762 (1974)
5.91   J.L. Dehmer: Physica Fennica 9S, 60 (1974)
5.92   G. Wendin: J. Phys. B6, 42 (1973)
5.93   R. Haensel, G.Keitel, P. Schreiber, C. Kunz: Phys. Rev. 188, 1375, (1969)
5.94   T.M. Zimkina, V.A. Fomichev, S.A. Gribowskii, I.I. Zhukova: Sov. Phys. Solid
       St. 9, 1128 (1967)
5.95   R. Haensel, P. Rabe, B. Sonntag, C. Kunz: Solid St. Commun. 8, 1845 (1970)
5.96   J.L. Dehmer, A.F. Starace, U. Fano, J. Sugar, J.W. Cooper: Phys. Rev. Lett.
       26, 1521 (1971)
5.97   V.A. Fomichev, T.M. Zimkina, S.A. Gribowskii, I.I. Zhukova: Sov. Phys.
       Solid St. 9, 1163 (1967)
5.98   J. Sugar: Phys. Rev. B5, 1785 (1972)
5.99   J.L. Dehmer, A.F. Starace: Phys. Rev. B5, 1792 (1972)
5.100 A.F. Starace: Phys. Rev. B5, 1773 (1972)
5.101 H.W. Wolff, R. Bruhn, K. Radler, B. Sonntag: Phys. Lett. 59A, 67 (1976)
5.102 M.W.D. Mansfield, J.P. Connerade: Proc. Roy. Soc. Lond. A352, 125 (1976)
5.103 D.L. Ederer, T.B. Lucatorto, E.B. Saloman, R.P. Madden, J. Sugar: J. Phys.
       B8, L21 (1975)
5.104 J.E. Hansen, A.W. Fliflet, H.P. Kelly: J. Phys. B8, L127 (1975)
5.105 G. Wendin: Phys. Lett. 46A, 119 (1973)
5.106 A.W. Fliflet, R.L. Chase, H.P. Kelly: J. Phys. B7, L443 (1974)
5.107 P. Rabe, K. Radler, H.W. Wolff: In Proc. Int. Conf. on VUV Radiation Physics,
       ed. by E.E. Koch (Pergaman Vieweg/Braunschweig 1974) pp. 247-249
5.108 J.P. Connerade, M.W.D. Mansfield: Proc. Roy. Soc. Lond. A341, 267 (1974)
5.109 G. Wendin: Phys. Lett. 51A, 291 (1975)
5.110 G. Wendin: J. Phys. B9, L297 (1976)
5.111 K. Codling: Electron Correlation Effects in XUV Photoabsorption Spectroscopy
       of Atoms; Daresbury Synchrotron Radiation Lecture Note Series No. 4 (Daresbury
       Lab. 1976)
5.112 R. Morgenstern, A. Niehaus, U. Thielmann: J. Phys. B10, 1039 (1977)
5.113 D.A. Shirley, R.L. Martin, B.E. Mills, S. Suzer, S.T. Lee, E. Matthias,
       R.A. Rosenberg: In Proc. II Intern Conf. on Inner Shell Ionization Phenomena,
       ed. W. Mehlhorn (Freiburg 1976) pp. 238-257
5.114 M.J. Lynch, A.B. Gardner, K. Codling, G.V. Marr: Phys. Lett. 43A, 237 (1973)
5.115 R.G. Houlgate, J.B. West, K. Codling, G.V. Marr: J. Elect. Spect. and Related
       Phenom. 9, 205 (1976)
5.116 D.J. Kennedy, S.T. Manson: Phys. Rev. A5, 227 (1972)
5.117 M.Ya Amusia, V.K. Ivanov, N.A. Cherepkov, L.V. Chernysheva: Phys. Lett.
       40A, 361 (1972)
5.118 C.D. Lin: Phys. Rev. A9, 171 (1974)
5.119 J.A.R. Samson, J.L. Gardner: Phys. Rev. Lett. 33, 671 (1974)
5.120 K. Codling, R.G. Houlgate, J.B. West, P.R. Woodruff: J. Phys. B9 L83 (1976)
5.121 J.B. West, P.R. Woodruff, K. Codling, R.G. Houlgate: J. Phys. B9 407 (1976)
5.122 M.Ya Amusia: In Proc. Int. Conf. on VUV Radiation Physics, ed. by E.E. Koch
       (Pergaman Vieweg/Braunschweig 1974) pp. 204-24
5.123 M.Ya Amusia, V.K. Ivanov, N.A. Cherepkov, L.V. Chernysheva: Sov. Phys-JETP
       39, 752 (1974)
5.124 J.A.R. Samsom, J.L. Gardner, A.F. Starace: Phys. Rev. A12, 1459 (1975)
5.125 R.C. Kemeny, J.A.R. Samson, A.R. Starace: J. Phys. B10, L201, (1977)
5.126 A.F. Starace: Phys. Rev. A16, 231 (1977)
5.127 J.A.R. Samson, J.L. Gardner: Phys. Rev. Lett. 31, 1327 (1973)
5.128 F. Wuilleumier, M.J. Adam, P. Dhcz, N. Sandner, V. Schmidt, W. Mehlhorm:
       Phys. Rev. A16, 646 (1977)
5.129 K.T. Lu: Phys. Rev. A4, 579 (1971)
5.130 J.P. Desclaux: Private communication
5.131 L. Torop, J. Morton, J.B. West: J. Phys. B9, 2035 (1976)
5.132 J. Cooper, R.N. Zare: Lectures in Theoretical Physics, Vol. 11c, ed. by S.
       Geltman (Gordon and Breach, New York 1969)
5.133 T.A. Carlson, A.E. Jonas: J. Chem. Phys. 55, 4913, (1971)
5.134 A. Niehaus, M.W. Ruf: Z. Physik 252, 84 (1972)
5.135 J.L. Dehmer, W.A. Chupka, J. Berkowitz, W.T. Jivery: Phys. Rev. A12 1966 (1975)

268

5.136 M.Ya Amusia, N.A. Cherepkov, L.V. Chernysheva: Phys. Lett. $\underline{40A}$, 15 (1972)
5.137 M.Ya Amusia, V.K. Ivanov: Phys. Lett. $\underline{59A}$, 194 (1976)
5.138 F.W. Byron, C.J. Joachain: Phys. Rev. $\underline{164}$, 1 (1967)
5.139 R.L. Brown: Phys. Rev. $\underline{A1}$, 586 (1970)
5.140 T.A. Carlson: Phys. Rev. $\underline{156}$, 142 (1967)
5.141 M.Ya Amusia, E.G. Drukarev, V.G. Gorshkov, M.P. Kazachkov: J. Phys. $\underline{B8}$,
       1248 (1975)
5.142 S.L. Carter, H.P. Kelly: J. Phys. $\underline{B9}$, L565 (1976)
5.143 M.J. Van Der Wiel, G. Wiebes: Physica $\underline{54}$, 411 (1971)
5.144 V. Schmidt, N. Sandner, H. Kuntzemuller, P. Dhez, F. Wuilleumier, E. Källne:
       Phys, Rev. $\underline{A13}$, 1748 (1976)
5.145 G.S. Lightner, R.J. Van Brunt, D. Whitehead: Phys. Rev. $\underline{A4}$, 602 (1971)
5.146 J.A.R. Samson, G.N. Haddad: Phys. Rev. Lett. $\underline{33}$, 875 (1974)
5.147 D.M.P. Holland, K. Codling, J. West, G.V. Marr: J. Phys. B (submitted)
5.148 B. Brehm, A. Bucher: Intern J. Mass Spect. Ion Phys. $\underline{15}$, 463 (1974)
5.149 H. Hotop, D. Mahr: J. Phys. $\underline{B8}$, L301 (1975)
5.150 J.E. Hansen: J. Phys. $\underline{B8}$, L403 (1975)

6. Molecular Spectroscopy

E. E. Koch and B. F. Sonntag

With 44 Figures

This chapter is intended to summarize the recent progress of molecular spectroscopy in the vacuum ultraviolet (VUV) brought about by the availability of synchrotron radiation (SR). Though limited in scope by focusing rather narrowly on the results obtained during the last 10 years or so in the various SR laboratories we are aware of the more general, often noted revival of interest in molecular physics in the VUV range in general [6.1-7] impelled by the ever growing sophistication of theoretical techniques to calculate and predict molecular properties of excited states, (e.g. [6.8]) and the various new experimental methods such as photoelectron spectroscopy [6.9], e^- scattering experiments [6.10], fast-beam spectroscopy, multiphoton spectroscopy, etc.

The elucidation of the electronic structure of molecules is an important step towards the understanding of the chemical and physical properties of matter at a microscopic level, providing a link between atomic and solid-state physics. Bombarding a molecule with photons of suitable energy is conceptually one of the most simple and direct methods for obtaining information about its electronic structure. The manifold of excited, superexcited or ionized states resulting from such a treatment (Fig.6.1), can be investigated by absorption, photoelectron, and fluorescence spectroscopy, by mass-spectroscopic techniques as well as by scattering experiments. The various possible processes are sketched in Fig. 6.2. In Table 6.1 we have summarized these experimental methods and the desired information which one hopes to obtain from the experiments.

The general concepts for these processes are briefly discussed in Sect. 6.1. More thorough discussions of the theoretical background each with a different emphasis may be found in the references [6.1-10]. The rapid development of the experimental methods in connection with SR over the last couple of years (see Chap. 3) and the successful analysis of many types of absorption spectra from a large variety of molecules (Sect.6.2) has led to a far more detailed understanding of the behavior of electrons in molecules than previously possible.

It now is generally realized, that dynamical processes in molecular systems, once highly excited states have been prepared by photoabsorption, have rarely been investigated thoroughly enough to shed light on all the various pathways of decay and relaxation mechanisms. Although we are presently just obtaining the first detailed results from SR-experiments in this field (Sect.6.3,6.4 and 6.5) it is our understanding that each of the various spectroscopies will soon form an interesting area of research in its own rights. The major advantages of SR for this kind of experiments are the high intensity and the tunability for the selective excitation of specific states. Further the time structure of SR from storage rings with pulse length down to the picosecond range (see Chap.3) is a very powerful tool for these investigations.

Molecular data are important for progress in other areas of science and technology [6.11] (see also [6.24]). Thus we have seen recently, for example, a tremendous interest in the VUV spectra of halocarbons and their photodecomposition [6.12] since they play a major role in stratospheric reactions. In a wider perspective there is also the possiblity of a new chemistry of molecules in their excited states opening up new areas of research. Finally in order to give another example, we mention the extension of molecular spectroscopy to highly excited states of larger biologically important systems [6.13] which may well be of importance for an explanation of biological activity in terms of electronic structure.

It is the aim of this chapter to summarize these interesting developments and to serve as a guide to the rapidly growing number of original papers. We have covered the literature up to the middle of 1978 quite exhaustively at least in tabular form in the appendix. However, we will use the limited space to discuss in detail only a few examples which we find more interesting. Thus we have tried to emphasize the general underlying concepts rather than to present a complete discussion of all spectra. Although connections to work concerned with solids, in particular the electronic structure of molecular crystals [6.14] and matrix isolation spectroscopy [6.15] will briefly be made, these interesting experiments are not covered in the present review and the reader is referred to the respective literature.

6.1 Concepts

The quality of atoms to form bonds between each other is responsible for the world surrounding us. The detailed understanding of these bonds and the properties of the resulting molecules is one of the most interesting problems. The immense variety of existing molecules spanning the range from simple diatomic molecules to the macromolecules encountered in biology adds to the fascination of this problem. From a very fundamental point of view the basic questions have been answered with the advent of quantum mechanics. The whole problem boils down to solving the Schrödinger equatic

$$H(R_N, r_e)\psi(R_N, r_e) = E\psi(R_N, r_e) \qquad (6.1)$$

where R_N stands for the coordinates of the nuclei and r_e for the coordinates of the electrons. The knowledge of the ground- and excited-state wave functions $\psi(R_N, r_e)$ and of the energies E would enable us to describe all properties and reactions of molecules. The main problem is, that for most molecules the Schrödinger equation cannot be solved exactly because of the large number of particles involved. Models based on reasonable assumptions allowing the many-body problem to be reduced to a solvable task are the only way out. Proceeding along this road substantial progress has been made towards the understanding of the mechanisms of bond formation, molecular properties and chemical reactions. Most calculations are based on the Born-Oppenheimer approximation, in which the nuclear ψ_N and electronic ψ_e wave functions are considered separable, i.e.

$$H = T_N + H_e \quad , \qquad (6.2)$$

where T_N is the kinetic energy of the nuclei. Setting $T_N = 0$ leads to the electronic Schrödinger equation

$$H_e(R_N, r_e)\psi_e(R_N, r_e) = E_e(R_N)\psi_e(R_N, r_e) \qquad (6.3)$$

in which R_N plays the role of a parameter. Solving the electronic Schrödinger equation for all possible positions of the nuclei yields the multi-dimensional energy surfaces $E_e(R_N)$, schematically presented in Fig. 6.1. The total wave function is given by

$$\psi(R_N, r_e) = \psi_N(R_N)\psi_e(E_N, r_e) \quad . \qquad (6.4)$$

ψ_N is an eigenfunction of the Schrödinger equation

$$H_N\psi_N = E\psi_N \qquad (6.5)$$

describing the motion of the nuclei, where

$$H_N = T_N + E_e \quad . \qquad (6.6)$$

The total energy E consists of electronic energy and of vibrational and rotational energy. The eigenstates of the molecule are schematically depicted in Fig. 6.1.

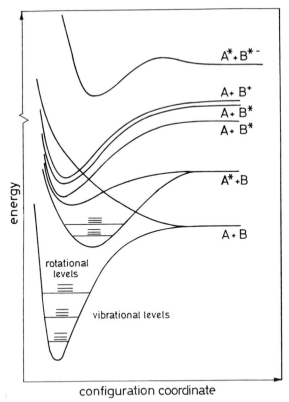

energy

A*₊B*⁻

A₊ B⁺

A₊ B*

A₊ B*

A*₊B

A₊B

rotational
levels

vibrational levels

configuration coordinate

Fig. 6.1 Schematic repre-
sentation of the molecular
potential energy curves

The molecular orbital approximation mathematically expressed in the Hartree-Fock approximation, has provided a framework for quite accurate calculations of a wide variety of molecular phenomena. In this model each electron is considered to move independently in an effective potential created by the nuclei and the other electrons. The electronic wave function ψ_e of the molecule is described by an anti-symmetrized and properly space symmetrized product of molecular orbitals (MO), which are approximated by a linear combination of basis orbitals localized at the atoms. (LCAO-MO approximation.) The effective potential is determined by self-consistent-field (SCF) methods. Within the validity of Koopman's theorem the energies of the molecular orbitals of closed shell systems may be associated with the ionization potentials of the molecule. The molecular orbitals are classified according to the molecular symmetry (symmetry-restricted Hartree-Fock approximation).

Chemical binding energies are small compared to the total energy of the molecules. For accurate predictions of molecular properties and molecular reactions, therefore the calculations and the underlying models must meet very high standards. For small molecules and for special classes of molecules, models and techniques based on the SCF approximation have been developed which allow very precise calculations

of molecular properties. Relativistic - and correlation effects have been incorporated in many cases, see e.g. [6.8a,e]. With the continuous improvement of the
theoretical models and the computing facilities the restrictions imposed on this
approach will be pushed back in the future but still for the majority of molecules
the results of such calculations can only be considered as guidelines. There is
no substitute for reliable experimental data. Furthermore experimental data are
necessary for critical tests of the theoretical models. A series of excellent experimental methods have been developed to determine molecular properties (see e.g.
[6.16] from which in turn the energy surfaces $E_e(R_N)$ and the wave functions
$\psi(R_N, r_e)$ can be constructed to a good approximation. Many of these methods are confined to the investigation of ground-state properties, though in the future the
applications may be extended to the investigation of excited states prepared in sufficient concentration, e.g. by high-power lasers. The importance of excited states
[6.17] becomes clear if one recalls that any change of the molecular state induced
by outer fields or the interaction with other molecules can be described as an admixture of excited states of the unperturbed molecule to the ground state of the
molecule. The interaction with photons is one of the most powerful experimental
techniques to obtain information on both the ground state and the excited states
of molecules. In Fig. 6.2 a schematic picture of such an experiment is shown.

In an ideal experiment all parameters to uniquely characterize the initial state
of the molecule, the incoming photon and the outgoing electrons, ions, neutral fragments and photons would be determined. This, for example, would allow a detailed
analysis of energy dissipation. In reality the limited intensity of the source, and
principle or technical problems render such an experiment not feasible. In most
experiments like absorption, photoemission, mass spectrometry, Raman scattering
and fluorescence only one type of the outgoing particles is detected and even then

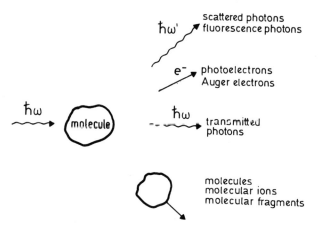

Fig. 6.2 Schematic representation of the interaction of photons with molecules

not all parameters necessary for a unique characterization of its state are determined. By piecing together the results of the different experiments a great deal of our present knowledge on molecules has been obtained. The experimental methods, the parameters determined and the information obtained are compiled in Table 6.1.

The molecular orbitals can be divided in five classes schematically shown in Fig. 6.3, see e.g. [6.18]:

I)   *core orbitals* are concentrated at the nuclei of the constituent atoms and are therefore well represented by atomic orbitals. The weak influence by the neighboring atoms can e.g. be accounted for within the ligand field model [6.19]

II) *valence orbitals* reflect the structure of the total molecule and may extend over the full volume of the molecule

III) *Rydberg orbitals* are bound orbitals mainly located at the periphery of the molecule [6.3]

IV) *shape resonances,* formed by continuum wave functions, are mainly concentrated within the molecule by the influence of the molecular field [6.20]

V)   *continuum wave functions,* with energies large compared to the average molecular potential are only weakly influenced by the molecular field. They are well represented by atomic wave functions close to the atomic nuclei and plane waves outside (orthogonalized plane waves).

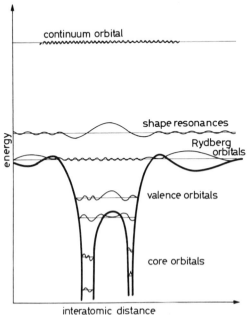

Fig. 6.3 Schematic representation of the effective molecular potential and the different types of molecular orbitals

Table 6.1

photon } parameters:
intensity, energy, polarization,
time-structure, coherence

molecule parameters:
initial state, electronic-, vibra-
tional-, rotational state, kinetic
energy, orientation, external fields

primary processes

scattering experiments

(Bragg-, energy dispersive-, small
angle-, Compton-, and Raman scattering)

parameters determined: number energy,
angular distribution and polarization
of scattered photons

information obtained: scattering cross
sections, excitation energies; electro-
nic-, vibrational and rotational energy
levels; momentum distribution of the
electrons, molecular structure, correlation effects

absorption

parameters determined: number of
absorbed photons

information obtained: total cross sections; excitation energies = E_{final} -
$E_{initial}$; ionization potentials; elec-
tronic-, vibrational- and rotational
energy levels; potential energy curves;
transition probabilities; Franck-Condon
Factors; molecular structure (EXAFS);
correlation effects, e.g. autoionization

secondary processes

fluorescence spectroscopy

parameters determined: number, energy
time dependence, angular distribution
and polarization of fluorescence photons

information obtained: fluorescence
yield; radiative and radiationless decay processes; lifetimes; branching
ratios; electronic, vibrational and
rotational energy levels; dissociation,
predissociation, (fluorescence from
fragments); excitation thresholds;
correlation effects; potential energy
curves

photoelectron spectroscopy

parameters determined: number, energy,
angular distribution and polarization
of outgoing electrons

information obtained: total and partial
cross sections; ionization potentials;
electronic vibrational and rotational
energy levels; orbital symmetry, potential energy curves, transition probabilities, Franck-Condon Factors, correlation
effects, e.g. relaxation, shake-up,
shake-off, Auger transitions

chemical analysis

parameters determined: chemical reactions of photoexcited molecules, molecular ions and molecular fragments (up
to now not used in the VUV)

mass spectrometry

parameters determined: number, composition, kinetic energy, charge and angular
distribution of outgoing molecules and
molecular fragments

information obtained: ionization cross
sections (single, multiple); fragmentation cross sections; ionization thresholds; electronic, vibrational rotational
energy levels; relaxation phenomena; potential energy curves; thermodynamical
data; photochemical reactions; dissociation

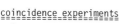

coincidence experiments

primary and secondary processes are investigated
simultaneously

parameters determined: as for the experimental
methods listed above

information obtained: detailed information on
the final state prepared by photoabsorption or
photon scattering; potential energy surfaces,
wave functions, as listed above

This classification of the molecular orbitals finds its counterpart in the partitioning of space into contiguous regions of potential energy schematically sketched in Fig. 6.4:

I)   *atomic potential* within the region of nonoverlapping spheres centered at the constituent atoms

II) *intramolecular potential* in the region between the "inner" atomic spheres

III) *extramolecular potential* at the periphery of the molecule.

The multiple scattering method [6.22], a promising approach to molecular calculations, is based on this division of the molecule in atomic, intramolecular and extramolecular regions.

Photon-induced transitions between all regions and all orbitals listed above are accessible with the help of continuous synchrotron radiation [6.7,24]. Among others such experiments allow the:

I)   Investigation of transitions from core to empty valence orbitals yielding detailed information on the energy, symmetry and the spatial distribution of the valence orbitals.

II) Investigation of transitions from core to Rydberg orbitals allows, e.g., a very accurate determination of ionization potentials. The same holds for transitions from valence to Rydberg orbitals.

III) Investigation of transitions from core orbitals to shape resonances showcasing the effect of the molecular field on the outgoing electron.

IV) Mapping of the occupied valence states by excitation to high-lying continuum states.

V)   Determination of absolute total and partial cross sections over a large energy range. Besides their practical importance this allows the application of various sum rules to check the consistency of the data.

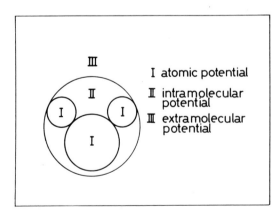

I atomic potential

II intramolecular potential

III extramolecular potential

Fig. 6.4 Division of the molecule into I, atomic; II intramolecular and III, extramolecular regions

VI) Determination of potential energy curves of high excited states by the analysis of the vibrational and rotational progressions.

VII) Investigation of radiative and nonradiative decay processes of well characterized excited states [6.17].

VIII) Investigation of correlation effects, e.g. autoionization, predissociation and multiple excitations. For the latter it is very interesting that by increasing the excitation energy above threshold one can go from the adiabatic to the sudden approximation [6.22].

For the simple molecules the development of sophisticated theoretical models has kept pace with the experimental progress. In many cases, however, one has to have recourse to simpler semi-empirical models. Thus the energy shifts of core levels in different chemical environment are well described by the point charge model [6.9a]. For the assignment of structures in the experimental spectra the Z + 1 analogy model [6.18] and the "constancy of the Rydberg term values model" [6.3] have been proven to be a good starting point. Atomic approximations have been successfully used as guidelines in many cases where high excited continuum states are involved. The concept of doorway states [6.23] opens an access to the understanding of relaxation phenomena. At this point we do not intend to present a complete list encompassing all the models proposed so far and we will not go in any more details here. The relevant models their merits and failures will be discussed in context with pertaining experimental results.

6.2 Absorption Spectroscopy

6.2.1 Valence Spectra of Simple Di- and Tri-Atomic Molecules

The absorption spectra of simple molecules containing two or three light atoms, such as N_2, O_2, CO, NO, CO_2, N_2O and H_2O have been the subject of numerous investigations and much has been achieved in the analysis of vibrational and rotational structures of their electronically excited states [6.1]. Nevertheless, experiments exploiting the advantages of synchrotron radiation have contributed during the last few years to an even better understanding of many features of both the valence- and inner shell spectra of these molecules.

As a starting point of our discussion we reproduce in Fig. 6.5 the absorption cross section of several simple molecules, as reported by LEE et al. [6.25]. Whereas the onset of the valence shell absorption is generally characterized by a number of sharp bands and progressions one observes only weak structure superimposed on broad

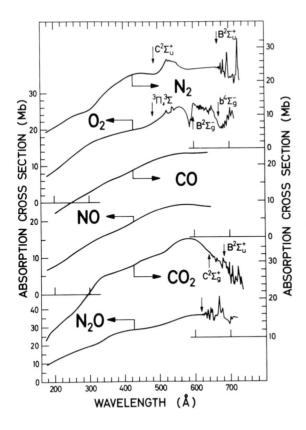

Fig. 6.5 Absolute cross sections (in Mb) of molecular N_2, O_2, CO, NO, CO_2 and N_2O in the 200 to 740 Å region [6.25]

continua towards shorter wavelengths. LEE et al. (6.25) have discussed their consistent set of photoelectrically measured absorption cross sections in comparison to the large number of investigations with line emission sources and either photographic or photoelectrical recording. In this study as well as in a couple of other investigations (see the appendix) the main emphasis was on obtaining reliable absorption cross section values over a large spectral region. Little effort was made in understanding the observed spectral features.

A glance at Fig. 6.5 shows that a number of features, e.g. the structure associated with the C state in N_2 go undetected if only a limited number of lines is available. In fact, using the continuous spectrum of SR and photographic recording CODLING [6.26], and CODLING and MADDEN [6.27] have been able to detect and analyze weak fine structure in the absorption continua of the N_2 and O_2 spectra. This structure is associated with Rydberg series leading to the C-state of N_2^+ at 23.6 eV (see also Fig.6.8) and with four Rydberg series converging to a level in O_2^+ at 24.5 eV and to a level at 24.75 eV, respectively. In N_2 the transitions involve a two electron process where one electron is removed from the $3\sigma_g$ orbital and another electron is excited from the $1\pi_u$ orbital into the unoccupied $1\pi_g$ orbital (shake-up process). For O_2 the transitions involve the excitation of an inner σ_g 2s electron to outer nsσ or ndπ orbitals. The detection and analysis of these weak autoionizing states is a nice example to show the obvious advantages of doing molecular absorption spectroscopy with a continuous and strong SR-light source. In the following we shall discuss the absorption spectra of N_2 and H_2O in the range of valence shell absorption as obtained with SR in some more detail illustrating perturbation effects of Rydberg and valence excitations, line shape analysis and Rydberg assignments in a fairly complex spectrum (H_2O).

6.2.2 Valence and Rydberg excitations in N_2

The absorption spectrum of molecular nitrogen has been studied many times since the initial work by LYMAN [6.28] and HOPFIELD [6.29]. Thus it provides a good example for the constant improvement of the experimental techniques and refinement of the analysis. A concise account of this development and references may be found in the paper by CARROL and COLLINS [6.30] and the detailed analysis given by DRESSLER [6.31] and in the recent critical review on this subject by LOFTHUS and KRUPENIE [6.32]. The frequencies and intensities of transitions to Rydberg levels as well as autoionization processes have been investigated theoretically in considerable detail by various methods, e.g. recently by DUZY and BERRY [6.33], by SHIH et al. [6.34] and by RESCIGNO et al. [6.35], thus leaving the experimentalists with a new challenge to test the validity of these detailed calculations by improved data.

GÜRTLER et al. [6.36] have recently published photoelectrically recorded spectra of the photoabsorption cross section in the range 10 to 35 eV including the 500 Å region which had been studied previously [6.26]. They have used a 3m normal incidence monochromator [6.37] in connection with a storage ring source at a resolution of up to $\Delta\lambda = 0.03$ Å. Their results and analysis are displayed in Figs. 6.6-8.

Roughly speaking the absorption spectrum can be divided into four parts: 1) For wavelengths $\lambda > 795.9$ Å many strong discrete bands with detailed resolved rotational fine structure forming several progressions are observed. For shorter wavelengths the discrete band absorption is superimposed on a background of continuous absorption; 2) In the range ~ 795 to 730 Å sharp Rydberg bands show up with several vibrational progressions, however, without resolvable rotational lines; 3) Below ~ 730 Å the spectrum is somewhat smoother with a fairly simple absorption and "apparent emission" structure; 4) Finally as mentioned above at around 500 Å the quite regularily spaced progression associated with the C state of N_2^+ is observed. Features 1) to 3) are associated with transitions involving electrons from the three valence orbitals $3\sigma_g$ 2p with a binding energy of 15.5 eV, $1\pi_u$ 2p with $E_B = 16.8$ eV and $2\sigma_u$ 2s with $E_B = 18.6$ eV [6.9b].

In the range below the first ionization potential (range 1) the absorption is caused by the two Rydberg series $3\sigma_g \rightarrow np\sigma_u$ ($C'^1\Sigma_u^+$) and $\rightarrow np\pi_u(C^1\Pi_u)$ and the valence excitations $b'^1\Sigma_u^+$ and $b^1\Pi_u$. The nature of these states and their mutual perturbations

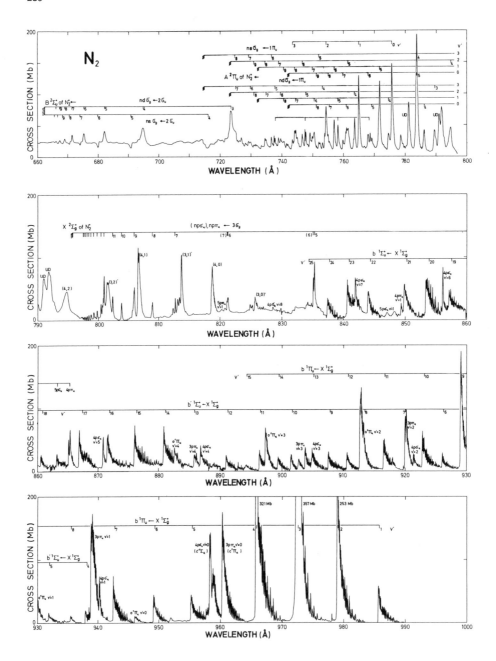

Fig. 6.6 Absolute absorption cross section of molecular nitrogen in the range 990 to 660 Å [6.36]

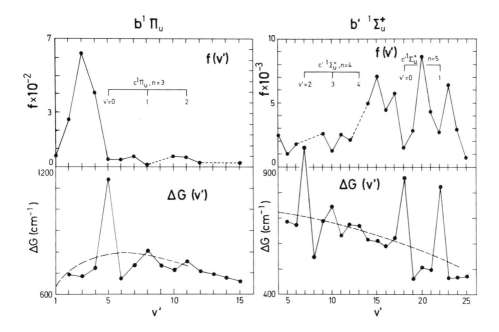

Fig. 6.7 Numerical evaluation for the perturbations of the oscillator strengths $f(v')$ and spacings $\Delta G(v')$ for the progressions of the $b^1\Pi_u$ and $b'\ ^1\Sigma_u^+$ valence states caused by the nearby Rydberg states $c\ ^1\Pi_u$, $n = 3$ ($v' = 0,1,2$) and $c'\ ^1\Sigma_u^+$, $n = 4,5$ ($v' = 2,3,4$), ($v' = 0,1$) respectively. The dashed curves for $\Delta G(v')$ give the results for the unperturbed cases as calculated by LEONI [6.38][6.36]

have been discussed in detail by DRESSLER [6.31]. A quantitative evaluation of the experimentally observed perturbation effects based on the high resolution photoelectrically recorded spectrum is displayed in Fig. 6.7 [6.36], where the oscillator strengths $f(v)$ for the vibrational bands and the change in vibrational spacings ΔG have been plotted as a function of the vibrational quantum number v'. Severe changes of the intensity within the progression and shifts in energy are observed for those energies, where valence and Rydberg excitations are almost degenerate. The deperturbed case as calculated by LEONI [6.38] is shown for comparison by the dashed curves. Experimentally a deperturbation can be achieved by investigating the same valence excitations for solid N_2. In this case the Rydberg states are heavily attenuated and the vibrational progression of the valence states is far more regularly spaced than in the gas phase [6.39].

In the high-resolution absorption experiment [6.36] it was possible to follow the (unresolved π and σ) Rydberg series up to $n = 23$ (see Fig.6.6). According to the calculations by DUZY and BERRY [6.33] for these Rydberg series there should occur a minimum for the oscillator strength of the $np\sigma_u\ c'\ ^1\Sigma_u^+$ series for $n = 8$, mainly caused by the variation in configuration mixing with final state energy. Since from $n = 6$ onwards the $np\sigma$ and $np\pi$ series coalesce in the spectrum this prediction can

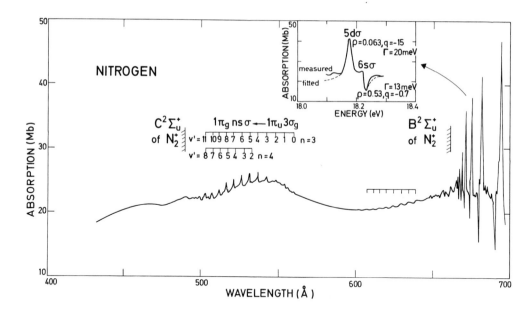

Fig. 6.8 Absolute absorption cross section of molecular nitrogen in the range
700 to 450 Å. The insert shows the 5dσ and 6sσ Rydberg excitation from the 2σ$_u$
molecular orbital in an expanded scale together with a fitted line shape [6.36]

not be verified. Further one has to note that perturbations caused by the close by
valence states have not been incorporated in the calculations.

In the range above the first ionization potential (range 2) two series have been
assigned [6.33], (see also the earlier results cited [6.33]), namely a $1\pi_u \to nd\sigma_g(^1\Pi_u)$
series with a quantum defect $\delta = 1.05$ and a $1\pi_u \to nd\sigma_g(^1\Pi_u)$ series with $\delta = 0.17$.
For the nsσ series the ab initio calculation by DUZY and BERRY [6.33] is in very good
agreement with experiment, whereas the energies for the nd series deviate from the
experimentally determined values. This deviation is due to the quantum defect $\delta = 0.8$
used for the calculations, which is considerably larger than the one determined experimentally ($\delta = 0.17$).

Autoionization manifests itself most clearly in the N_2 absorption spectrum for
wavelength below 730 Å (range 3; Figs.6.6,8). In this range a $nd\sigma_g$ and a $ns\sigma_g$ - type
Rydberg series originating from the 2σ$_u$ orbital converge to the $B^2\Sigma_u^+$ state of the
N_2^+ ion. They interfere with the ionization continua of the $X^2\Sigma_g^+$ and the $A^2\Pi_u$ states
causing characteristic line shapes and dips in the absorption cross section (Hop-
fields "apparent emission" lines [6.29]). PLUMMER et al. [6.40] could show in their
recent determination of partial photoionization cross sections with SR that the interference effects causing the absorption dips are far more pronounced in the partial
cross section for the $A^2\Pi_u$ state than for the $X^2\Sigma_g^+$ state. Based on the analysis of
the line shapes according to the FANO-COOPER theory [6.41] GÜRTLER et al. [6.36] located the energetic positions of the bands unequivocally and obtained parameters for

the 5d and 6sσ peak (Fig.6.8). They found that the 5d peak with a small overlap parameter of 6% is strongly autoionizing with an autoionization rate of 3×10^{13} s^{-1} whereas for the 6sσ peak having an overlap parameter of 53%, a smaller autoionization rate of 2×10^{13} s^{-1} was determined. We shall discuss some further aspects of the N_2 spectrum in connection with the N 1s core levels (Sect.6.2.4) and the photoemission experiments (Sect.6.3).

6.2.3 Rydberg series in the valence absorption spectrum of H_2O and D_2O

In its ground state the molecular orbital configuration of the water molecule is

$$(1a_1)^2 \; (2a_1)^2 \; (1b_2)^2 \; (3a_1)^2 \; (1b_1)^2$$

where the uppermost three orbitals have vertical ionization energies of 12.62 eV, 14.75 eV and 18.54 eV, respectively [6.9]. In the VUV range for photon energies up to 20 eV transitions originating from these uppermost three orbitals are expected. However, in spite of considerable recent theoretical (e.g. [6.42-45] and experimental efforts [6.46]) a number of puzzles concerning the excited states of the water molecule remained. For a discussion of these problems and the relevant literature we refer to [6.1,3].

Recently GÜRTLER et al. [6.46] reported on new high-resolution absorption cross section measurements and a detailed assignment of Rydberg series. An overview of the H_2O absorption spectrum and part of their assignments appear in Fig. 6.9. As originally suggested by MULLIKEN [6.47] and corroborated quantitatively by the recent calculations of the orbital sizes by GODDARD and HUNT [6.42], the simplest view of the excited orbitals of H_2O is to consider each as arising from an $n \geq 3$ Rydberg atomic orbital modified by the molecular field [6.48]. Making use of (i) the close correspondence of the vibrational progressions in photoelectron and Rydberg absorption spectra (ii) a rotational line shape analysis for several bands and (iii) by comparing the assignments to recent ab initio calculations GÜRTLER et al. [6.46] arrived at a number of new and fairly probable assignments. They are summarized in Table 6.2. In particular it was possible to assign the four optically allowed $1b_1 \rightarrow 3d$ transitions unequivocally, and excellent agreement with the results of GODDARD et al. [6.42] can be stated. On the basis of term values a p-type and a s-type Rydberg series leading to the $^2A_1$ and the $^2B_2$ state, respectively, with quantum defects $\delta \approx 0.15$ and $\delta \approx 1.36$ have been assigned. The detailed fine structure observed for several vibrational bands belonging to these Rydberg transitions (see the insert in Fig.6.9) is still a problem. Aside from rotational structure the deviation of the molecular potential from spherical symmetry results in Liehr splitting effects [6.48] of the p-like final states. At the same time a Renner-Teller splitting [6.49] in the nearly linear geometry of the excited states is expected. Without

Table 6.2 Assignment of Rydberg transitions in the absorption spectrum of H_2O and D_2O (All energies are in eV. Subscript v: vertical energies)

Orbital, ionization energy binding properties [6.9b] H_2O	assignment	transitions experiment H_2O	δ	D_2O	δ	[6.42]	[6.45]	theory H_2O [6.43]	[6.44]
$1b_1$	$1b_1 \to 3s$	7.44_v [6.2]				7.30	7.61	7.30	7.22
	$4s$	10.64				10.64			
	$5s$					11.66			
12.62 12.62	$3pa_1$	9.998	0.72	10.011	0.72	10.04	10.06	9.90	9.02
(1st band)	$3pb_1$	10.171	0.67	10.171	0.65	10.16	10.16	10.32	9.48
non bonding	$3da_1$	10.990	0.11	10.966	0.12	11.07			9.61
perpendicular to	$3da_2$	11.041	0.06	11.055	0.07	-			
molecular plane	$3db_2$	11.057	0.04	11.070	0.05	11.17			
	$3da_1$	11.122	-0.02	11.130	0.00	11.17			
	$4pa_1$	11.374	0.69	11.385	0.71	11.42			
	$4pb_1$	11.432	0.60	11.427	0.65	11.48			
	$4da_1$	11.729	0.08	11.752	0.08				
	$4db_1$	11.770	-0.01	11.793	-0.01				
	$5p$	11.890	0.66	11.901	0.70				
	$5d$	12.061	0.05	12.075	0.08				
	$6p$	12.120	0.74	12.155	0.69				
	$6d$	12.235	0.02	12.254	0.04				
	$7p$	12.288	0.52	12.300	0.65				
	$7d$	12.337	0.01	12.365	-0.07				
	$8p$	12.361	0.63	12.386	0.65				
	$8d$	12.399	0.07	12.423	0.03				
	$9p$	12.411	0.77	12.438	0.75				
	$9d$	12.448	-0.03	12.468	0.03				
	∞p	12.612	0.67	12.637	0.70				
	∞d	12.612	0.03	12.638	0.01				
$3a_1$	$3a_1 \to 3s$	9.85				9.82		9.80	9.54
	$4s$	~12.9 [6.2]							
13.987 13.988						11.47		11.21	
(1st band)	$4p$	13.5_v	0.83	13.9_v	0.75	11.92		11.72	
						12.08			
14.87_v 14.96_v	$5p$	14.1_v	0.74	14.2_v	0.70				
weakly O-H bonding	∞	~14.9_v		~15.0_v					
parallel to molecular plane									
$1b_2$	$1b_2 \to 3s$	~13.8_v		~13.8_v					
17.22 17.26	$4s$	16.9_v	1.38	17.0_v	1.35				
strongly O-H bonding	∞	18.9_v		19.0_v					
parallel to molecular plane									

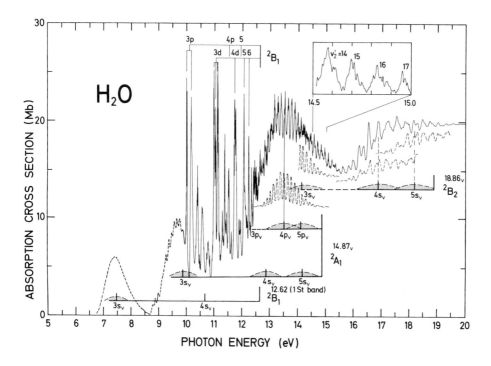

<u>Fig. 6.9</u> Rydberg assignments for the absorption bands in the spectra of H₂O. The dashed curves are the vibrational progressions observed for the $^2A_1$ and $^2B_2$ state in photoelectron spectra [6.9a]. The insert shows the observed fine structure for some of the bands in more detail [6.46]

detailed calculations it seems at present premature to give a more elaborate discussion of this point. We note in passing that ISHIGURO et al. [6.50] obtained a different assignment for a number of transitions in H₂O and D₂O in their photo-graphically recorded spectra taken with a resolution somewhat better than 0.3 Å.

6.2.4 Core-Spectra of Simple Di- and Tri-Atomic Molecules

In contrast to the rich structure observed for the valence shell absorption spectra core spectra are expected to be far more simple. While the former can be thought of as being composed of many transitions from a number of occupied valence orbitals to a large number of virtual MO's, which can be used for the description of excited electrons, core spectra should essentially reflect the manifold of terminating or-bitals. Depending on the energies of the terminating orbitals excitations from core levels give rise to fine structure preceeding and following the ionization thres-holds of the various core levels.

Since the first measurements of the absorption spectrum of the nitrogen molecule in the 30 Å region in 1969 by NAKAMURA et al. [6.51] core spectra of several simple di-atomic and tri-atomic molecules such as O_2 [6.52,53], NO [6.54,55], CO [6.52], NO_2 [6.56] and N_2O [6.57] have been investigated with synchrotron radiation. In most cases, however, the experimental conditions have been far from ideal. The available resolution as well as photographic recording of the spectra frequently limited the usefulness of the data. Furthermore, in some instances the pressure has not been optimized for all portions of the spectra. For instance, the pressure dependence of some features of the N_2 spectrum has been discussed by CONNERADE et al. [6.58]. We mention at this point, that at present the extensive information about core spectra from molecules obtained by high resolution electron energy loss measurements is in general still superior to the optical data, e.g. concerning the determination of absolute oscillator strengths [6.59-61,147].

As expected, most of the far XUV spectra are simpler than the corresponding valence shell spectra and the majority of the absorption structures has been attributed to single electron excitations from an atomic inner core shell to an empty molecular orbital. Transitions due to double excitations have been invoked in several cases in order to assign weak features, e.g. in the N 1s absorption spectrum of NO [6.54,55] and probably also in the O 1s spectrum of O_2 [6.52], in the N 1s and O 1s energy region of the NO_2 absorption spectrum [6.56], and in Li_2 [6.239].

In spite of the relative simplicity of the soft X-ray absorption spectra mentioned above our understanding and the assignment of the details is in many cases far from satisfactory, indicating that the above mentioned simple one electron MO-assignments are not sufficient. We have outlined the general concepts for an interpretation of the core spectra going beyond the simple one electron MO-approach in Sect. 6.1. The equivalent core or Z + 1 analogy model [6.18] has been frequently applied with good success for an assignment of the finer details of the core spectra. For core spectra of di-atomic or tri-atomic molecules shape resonances in the continuum within a few Rydberg above the K-shell edges, as discussed by DEHMER and DILL [6.20], have only recently been investigated in some detail. They used a multiple-scattering model for their calculations of the K-shell photoionization cross section of N in N_2. With this method the photoabsorption cross section can be calculated in a spectral region near the absorption edge, where the usual theory of EXAFS is not applicable. Strong indications for these resonances in the K-shell absorption spectra have recently been reported and discussed by SCHWARZ et al. [6.56] for the core-electron excitation spectrum of NO_2 and by BIANCONI et al. [6.57] for N_2 and N_2O.

a) *N_2*

The K-shell absorption spectrum of nitrogen, which has been studied by several groups with SR [6.51,52,57,58] will be discussed here in some more detail. The photoelectrically determined absorption cross section over the range from ~ 400 eV to 450 eV as recently reported by BIANCONI et al. [6.57] appears in Fig. 6.10. This spectrum had

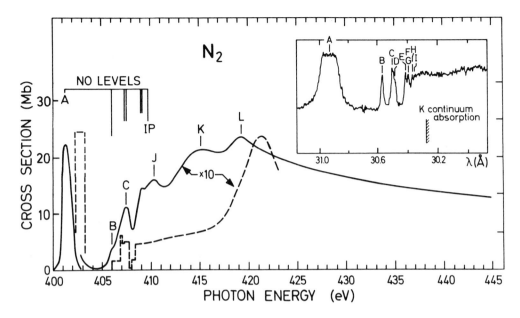

Fig. 6.10 Nitrogen K-edge photoabsorption spectrum [6.57]. The dashed curve gives the oscillator strengths according to the theory of DEHMER and DILL [6.20b]. In the insert the densitometer trace of the N_2 K-absorption as reported by NAKAMURA et al. [6.51] is shown

been previously determined with higher resolution (somewhat better than 0.03 Å) at the fourth order of a 2m grazing incidence spectrograph by NAKAMURA and coworkers [6.51]. Their result is shown in the insert of Fig. 6.10. The latter authors have also presented a discussion of the spectrum with the help of the Z + 1-core analogy model [6.18]. In many cases this model has been very useful for the interpretation of core excitation phenoma in atoms and molecules: If a core electron is excited from the vicinity of a nucleus Z into an unoccupied valence orbital, the shielding of the nuclear charge is thereby reduced by about 1 unit and the number of valence electrons is increased by one. Therefore the molecular states obtained by excitations of an electron from the core around nucleus Z, will be similar to normal states of a molecule with atom (Z + 1) instead of atom Z. Thus, by comparing the K-shell absorption features to the energy levels of the valence shell spectrum of the NO molecule, one obtains the assignment of the various maxima as given in Table 6.3, With the exception of the strong band A, the sharp bands B, C, D, E, F and G are due to Rydberg excitations.

The maximum labelled A in Fig. 6.10 is remarkably intense and broad with respect to the width of the originating N 1s orbital as determined by X-ray photoemission spectroscopy, where a half width of 0.22 eV has been reported [6.9a]. In order to shed some light on this anomaly, CONNERADE et al. [6.58] have studied the line width of this maximum as a function of pressure using the Bonn synchrotron as a

Table 6.3 Assignment of the absorption structure near the K-edge of the nitrogen atom in the N_2 and NO absorption spectrum. (All energies are in eV, and transitions to Rydberg states are marked by the letter R)

N_2

	Experiment[a]		Term value + assignment		
	(b)	(c)	(e)	(f)	(g)
A	400.84	401.3	9.06 $1\pi_g$	7.20	$\pi_g\ 2p$ R
B	405.59	406.1	4.31 $3s$	3.50	$3s\sigma$ R
C	406.50	407.3	3.40 $3s$	2.58	$3p\pi$ R
D	406.72		3.18 $3s+\nu_1$		$3p\sigma$ R
E	407.66		2.24 $3p$		$3d\sigma$ R
F	407.99		1.91 $3p+\nu_1$		$4s\sigma$ R
G	407.99				
H	408.25		1.65 $3d$		$4p\pi_u$ R
I	408.50	408.9	1.40 $3d+\nu_1$		$4p\sigma_u$ R
IP(d)	409.9				∞
J		410.0	-0.1		
K		415.0	-5.1		
L		419.3	-9.1	-10.90	σ_u(c)

NO

	Experiment (h)	Term value (h)	Assignment (h)
A	399.80	10.10	$(\pi 2p)^2$
B	402.25	7.65	double
C	403.87	6.03	excitation
D	406.60	3.9	$3s\sigma$ R
E	407.00	2.9	$3p\pi$ $3p\sigma$ R
F	408.15	1.75	$3d\sigma$, $4s\sigma$ R
	409.9	0	∞

(a) The letters refer to the labeling of the features in Figs. 6.10 and 11; (b)[6.51]; (c) [6.57]; (d) [6.98]; (e) according to ROBIN [6.3]. [6.51] 409.9 was used as the ionization potential; (f) calculated term values [6.20b]; (g) according to NAKAMURA et al. [6.51]; (h) [6.55]

light source. As already suggested by the appearance of this band from the densitometer trace as obtained by NAKAMURA et al. [6.51] they could show that the linewidth is partly due to saturation broadening. However, extrapolation of the halfwidth to zero pressure still yields a width somewhat larger than 0.22 eV. This discrepancy can be resolved by considering the fact that excitation of a 1s electron into an antibonding valence orbital (π_g) yields a much stronger vibrational excitation than

excitation far above threshold as is the case in X-ray photoemission. Meanwhile
KING et al. [6.60] have been able to resolve the vibrational structure of this peak
in a high-resolution electron impact experiment, supporting a π_g 2p assignment.

As far as the relative intensities of the various features in the N_2 spectrum
are concerned the spectra recorded photoelectrically by BIANCONI et al. [6.57]
using SR are probably the most reliable optical data. These authors have discussed
the N_2 core spectrum with respect to the recent photoionization cross section calculated by DEHMER and DILL [6.20] using the multiple scattering method (dashed
curve in Fig.6.10). Note, that in both the calculation and the experiment the core
to valence transition (peak A) is about 10 times as intense as the other absorption
features due to Rydberg transitions. According to [6.20], the large oscillator
strength of this leading peak is caused by a centrifugal barrier effect, resulting
in a shape resonance in the high ℓ-components of the final state wavefunctions.
The existence of the barrier is also known from the work on shape resonances in
electron scattering on N_2 [6.62] where electrons of 2.3 eV could occupy the unbound
π_g 2p orbital to form the temporary negative ion N_2^{-*}. Following the original assignment by NAKAMURA et al. [6.51] all later studies including the electron energy loss
experiments [6.59-61] agree that peak A corresponds to this π_g final state.

b) NO

As another example we discuss the absorption spectrum of the NO molecule in the
30 Å region, as reported by MORIOKA et al. [6.55]. The absorption structure in this
region (Fig.6.11) is similar to that of the N_2 molecule. The features observed are
obviously due to transitions caused by the excitations of the N 1s electron from the
ground state of NO to high lying molecular states including Rydberg states converging
to the ionization threshold at $\sim$ 410 eV. In addition to the densitometer trace of the
NO absorption spectrum, potential energy diagrams for the various known states of
the O_2 molecule are shown, which according to the Z + 1 analogy should correspond
to those of core excited states of NO.

In its ground state the NO molecule has the electron configuration

$$(\sigma \ 1s)_O^2 \ (\sigma \ 1s)_N^2 \ (\sigma \ 2s)^2 \ (\sigma \ 2s)^2 \ (\sigma \ 2p)^2 \ (\pi \ 2p)^4 \ (\pi \ 2p) \ X^2\Pi$$

According to the Z + 1 core anology model [6.18] excitation of electrons from inner
shells to an outer shell MO leads to an electron configuration of the highly excited
NO molecule similar to that of the O_2 molecule. This anology has been exploited by
MORIOKA et al. [6.55] for the analysis of the spectrum. Thus the strong broad band
labeled A in Fig. 6.11 is interpreted as being due to the excitation from the ground
state to an upper state with the electron configuration $(\sigma 1s)_N^{-1} \ (\pi \ 2p)^2$ leading to
four different states $^4\Sigma^-$, $^2\Sigma^-$, $^2\Delta$ and $^2\Sigma^+$. The latter three states can be reached
by allowed transitions from the ground state $X^2\Pi$ of the molecule. They correspond

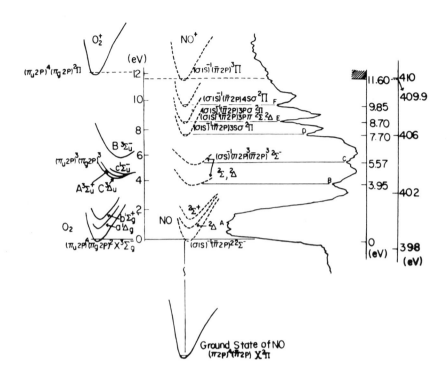

to the X $^3\Sigma_g^-$, a $^1\Delta_g$ and b $^1\Sigma_g^+$ states of the O_2 molecule as indicated in Fig. 6.11.

The absorption bands labeled B and C at 3.95 eV and 5.57 eV above the onset of the first band A can probably be attributed to two-electron excitations such as $(\sigma~1s)_N^{-1}$ $(\pi 2p)^{-1}$ $(\pi 2p)^3$. The resulting states of the excited NO molecule and their O_2 counterparts are given in Fig. 6.11.

The absorption bands, D, E and F have been assigned to Rydberg states. They are similar to the Rydberg series B, C, D, E, F and G observed in the K-absorption spectrum of nitrogen (see Fig.6.10). Based on the frequently noted independence of term values for the Rydberg states of similar molecules [6.3] one arrives at the assignment of the Rydberg states as given in Table 6.3 where comparison is made to the Rydberg assignment for the nitrogen 1s excitations as discussed above. The application of the Z + 1 analogy to these Rydberg states is hampered by the fact that only a few Rydberg lines have been observed for O_2 below the first ionization threshold [6.1]. The valence spectrum of O_2 in the range of interest is rather dominated by the strong Schumann-Runge continuum absorption followed by other dissociation continua, which probably obscure the Rydberg series in that region.

6.2.5 d-Spectra

So far we have only dealt with molecules existing under normal conditions. We will
now turn to molecules, which can only be produced by vaporization at elevated
temperatures. The experimental investigation of the VUV spectra of these molecules
is hampered by (i) the high temperatures required to achieve a sufficient vapor
pressure, (ii) the difficulties to confine and stabilize the vapor column and to
separate the hot vapor region from the high vacuum of the beam pipe and the spectrograph, (iii) the aggressiveness of the liquid and vaporized material, (iiii) the
presence of molecules with different numbers of atoms (dimers, trimers,...).

a) Se_2

Figure 6.12 shows the absorption spectrum of Se vapor at the onset of the Se 3d-
transitions, reported by CONNERADE et al. [6.63]. The temperature of the vapor was
$\sim$ 800 K. At this temperature Se_2 molecules form the dominant constituent though an
appreciable fraction of Se_6 and Se_8 molecules is also present in the vapor.
CONNERADE et al. tentatively ascribed the structures below 58 eV to transitions of
Se 3d-electrons to an antibonding π_g valence state of Se_2. The splitting is attributed to the atomic 3d-4p interaction and the Se 3d spin-orbit interaction. Since
the lowest empty valence states, responsible for the maximum below 56 eV, mainly
stem from Se 4p there is no doubt about the importance of these intraatomic interactions. In spite of this, the interpretation proposed by CONNERADE et al. seems
to oversimplify the situation. The weak structures at 60 eV are probably due to
3d $\rightarrow$ np Rydberg transitions. Towards higher energies the gross features of the
absorption are mainly determined by the delayed onset of the atomic 3d $\rightarrow$ εf
transitions [6.41].

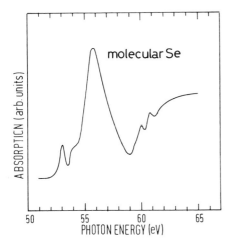

Fig. 6.12 3d-absorption of molecular
selenium CONNERADE [6.63]

b) Te_2

For Te_2 vapor the experimental situation is more favorable than in the case of Se because the concentration of Te_2 is much higher. The Te 4d-spectrum of Te vapor as determined by RADLER et al. [6.64] is presented in Fig. 6.13. The four peaks below 44 eV can be grouped in spin-orbit pairs. The maxima at 39.2 eV and 40.8 eV are due to transitions from the $4d_{5/2}$ level, the maxima at 41.4 eV and 42.9 to transitions from the $4d_{3/2}$ level. The splitting is in good agreement with the spin-orbit splitting of the 4d level of atomic and solid Te [6.65]. The separation of the two spin orbit pairs corresponds to the separation of the lowest 1_g and 0_u^+ excited states of Te_2 [6.66]. The atomic 4d-5p interaction may explain part of the width of these maxima. 4d-np Rydberg transitions are probably the origin of the weak maxima detected between 45 eV and 49 eV. From the IP of 8.3 eV the lowest series limits are estimated at 47.5 eV and 49 eV.

c) I_2

COMES et al. [6.67] have given a detailed analysis of the I 4d-spectrum of I_2 shown in Fig. 6.14. The strong continuum absorption peaking near 93 eV is due to atomic $4d \rightarrow \epsilon f$ transitions [6.67]. For f-symmetric final states the interplay between the attractive electrostatic interaction and the repulsive centrifugal term gives rise to a centrifugal barrier separating the effective atomic potential in an inner and an outer well. 40 eV above threshold the f-symmetric final states penetrate into the inner well and thus give rise to the resonant enhancement of the absorption.

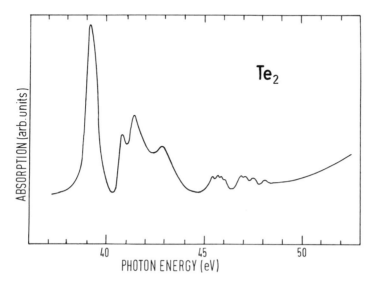

Fig. 6.13   4d-absorption of molecular tellurium [6.64]

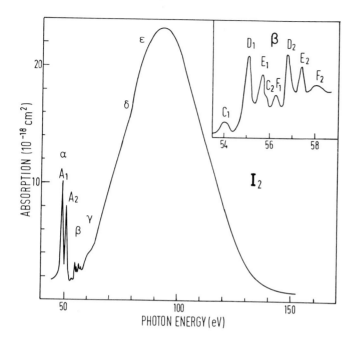

Fig. 6.14   4d-absorption of molecular iodine [6.67]

Table 6.4   Maxima in the spectrum of iodine and their assignment [6.67]

	peak	gas energy [eV]	assignment
α	A_1	49.27	$4d_{5/2} \rightarrow (5p)\sigma_u$
	A_2	50.93	$4d_{3/2} \rightarrow (5p)\sigma_u$
β	C_1	54.05	$4d_{5/2} \rightarrow \sigma_g 6s$
	D_1	55.08	$4d_{5/2} \rightarrow \sigma_u,\pi_u 6p$ and $\sigma_g,\pi_g 5d$
	E_1	55.70	$4d_{5/2} \rightarrow \sigma_u,\pi_u 4f$
	F_1	56.29	$4d_{5/2} \rightarrow 7p,\ldots$
	C_2	55.80	$4d_{3/2} \rightarrow \sigma_g 6s$
	D_2	56.76	$4d_{3/2} \rightarrow \sigma_u,\pi_u 6p$ and $\sigma_g,\pi_g 5d$
	E_2	57.40	$4d_{3/2} \rightarrow \sigma_u,\pi_u 4f$
	F_2	58.01	$4d_{3/2} \rightarrow 7p,\ldots$
γ		~61	$(4d5p) \rightarrow (LEMO)$
δ		~74	$(4d,5s) \rightarrow (LEMO)$
ε		~82	$(4d,5s) \rightarrow$ Rydberg orbitals

The importance of correlation effects has been well established for the corresponding maximum in the Xe 4d-spectrum [6.22,68-71]. The transitions are essentially
atomic because the initial core molecular orbitals are nearly pure 4d-orbitals which
overlap the continuum MO's mainly in the region of the atomic core, where these
can be described by εf atomic orbitals. This is corroborated by the fact that solidification has hardly any influence on the absorption in the region of the 4d εf
maximum [6.67]. The situation is obviously different from the K-spectrum of N_2
where the anisotropic molecular potential is responsible for the prominent shape
resonance [6.20]. We will come back to this significant difference in Sect. 6.2.8.

The assignment of the structures detected in the spectrum is listed in Table 6.4.
Transitions from the spin orbit split $4d_{5/2,3/2}$ levels to the lowest empty molecular
orbitals, mainly originating from I 5p, are responsible for the two strong lines
at threshold. The antibonding character of the σ_u (I5p) orbital explains the considerable vibrational broadening of the two lines. Spin-orbit pairs of Rydberg lines
show up in region β. The broad structures $\gamma,\delta,\varepsilon$ probably arise from the simultaneous excitation of a 4d electron and a valence electron. There is good agreement
between the one-electron energies of the Rydberg orbitals obtained from core and
valence excitations. This demonstrates the usefulness of the one-electron model and
at the same time shows the capacity of core absorption spectroscopy to contribute
to the clarification of the level structure of the empty molecular orbitals.

6.2.6 Alkalihalides

a) *Li 1s-absorption in LiF*

The ionic character of the bond and the simple structure of the Li-halides render
the core excitation spectra interpretable within the framework of simple model concepts. Because of the small number of electrons LiF is particularly accessible to
detailed theoretical calculations. Thus the investigation of the Li 1s-spectrum of
LiF can serve as a critical test for simple and more advanced theoretical models.
The Li 1s-absorption of LiF, as determined by RADLER et al. [6.72,73], is shown
in Fig. 6.15.

Due to the higher concentration of dimers ($\sim$ 50%) there is a considerable overlap
of the spectra of monomers and dimers. The analysis of the vibrational structure,
superimposed on the broad bands, A, B, D and E, however, offers a simple way to dis-
criminate between both contributions. Based on this analysis the bands, A, D, E and
the structured high energy part of band B are ascribed to the monomers whereas dimer
are responsible for the unstructured low energy part of band B, band C and probably
the shoulder on the low energy side of band D. Within the framework of an *ionic*
model RADLER et al. [6.72] assigned band A and the higher energy part of band B to
the transition $Li^+1s^2 \rightarrow Li^+1s2p$ which is split into a σ and a π component by the
molecular field. The same transitions in Li_2F_2 are assumed to be responsible for

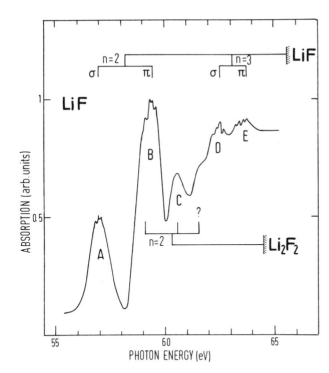

Fig. 6.15 Li-1s excitations of molecular lithiumfluoride [6.72]

the low energy part of band B, band C and the shoulder of band D. The bands D and E are ascribed to the next strongly allowed transitions $Li^+1s^2 \rightarrow Li^+1s3p$. Inserting the energies of bands A and D, and B and E, respectively, into the Rydberg formula a crude estimate of 65.5 eV has been obtained for the Li1s ionization limit in LiF. Ab initio calculations give 65.6 eV (64.5 eV) for the vertical Li1s ionization potential of LiF (Li_2F_2). Both values are 0.8 eV smaller than the values predicted by the simple ionic point charge model. The ionic model fails to predict that there is only one σ level from $Li^+2s,2p$. Ab initio calculations show that this σ level is mainly 2s with a strong admixture of 2p. The possibility to apply the (Z + 1) analogy model [6.18] was also considered. Experimental and calculated term values of BeF, the Z + 1 analogue of core excited LiF (denoted by Li*F) are given in columns 1 and 2 of Table 6.5. Column 8 gives the experimental term values for Li*F. There is reasonable agreement for the n = 3 Rydberg term values, whereas the term values for the n = 2 valence shell levels disagree. The failure of the Z + 1 core analogy model can be explained by (i) different geometries of BeF and Li*F, (ii) different exchange interactions between the electrons in the valence shell and the core shell, which is closed in BeF, but has a hole in Li*F. Correcting for bond length (see column 3 of Table 6.5) and for exchange (column 5 of Table 6.5) results in reasonable agreement for the 2σ* state whereas the term value of the 2π* state and the σ*-π* splitting still deviate considerably from the experimental values. The

Table 6.5 Vertical term values of BeF and Li*F and orbital energies ε of BeF⁺. The experimental values for BeF are based on the data of [6.72, 251-253]

Optical orbital	BeF, R = 1.36 Å		BeF R = 1.56 Å		R = 1.56 Å exchange corr.		Li*F (R = 1.56 Å)	
	exp.	theor.	BeF exp.	BeF⁺ ε theor.	BeF "exp."	BeF⁺ ε theor.	theor.	"exp."
Column	1	2	3	4	5	6	7	8
n = 2σ*	9.1	9.27	9.4	9.40	8.5	8.53	7.77 + δ	8.6
n = 2π*	5.0	4.80	5.5	5.24	5.1	4.86	5.06 + δ	6.1
Δ(σ* - π*)	4.1	4.47	3.9	4.16	3.4	3.67	2.71	2.5
3s	2.95		3.0	2.94	2.9	2.79		
3pσ	2.85		2.8	2.62	2.7	2.53		3.0
3pπ				1.99		1.94		1.9

theoretical term values of Li*F given in column 7 are two low by δ but reproduce the experimental σ*-π* splitting fairly well. The failure of the Z + 1 core analogy model for LiF can be attributed to the very different spatial extension of the cores of Li*F and BeF.

It has been shown that the term values of core excited states are approximately given by the virtual orbital energies of the positive ion of the Z + 1 analogous molecule i.e. BeF⁺ for Li*F [6.18,74]. The applicability of this *equivalent ionic core virtual orbital model* is borne out by the term values of BeF⁺ given in column 4 of Table 6.5 and especially by the exchange corrected values in column 6. The

marked deviation of the π* value is due to the same atomic effect causing the failure of the Z + 1 core analogy model.

b) *Cs-halides*

Because of the large number of electrons there are no calculations for excited
states of Cs-halides. Nevertheless a great deal of information can be extracted from
the spectra by using the ionic model as a guideline. Figure 6.16 shows the Cs^+5p
spectrum of molecular CsCl [6.75]. For comparison the energies of the Cs^+ 5p → 6s,5d
transitions are included [6.76]. The energy separation of bands A and C agrees with
the spin-orbit splitting of the Cs^+5p^5 level. Based on this and on the analysis of
the vibrational structure A and C have been assigned to the transitions
$Cs^+5p^6 → Cs^+5p^56s$. The analysis of the vibrational structure also shows that a different transition contributes to the low energy part of C. Band B, the low energy
part of C and the structured band D have been ascribed to the transitions
$Cs^+5p^6 → Cs^+5p^55d$. In comparison to the transitions in the free ion these transitions are shifted in the molecule by less than 0.5 eV, whereas the transitions to
6s are shifted by approximately 1 eV towards lower energies. This indicates that
the "6s" function extends over a larger region than the "5d" function. Transitions
to higher lying Rydberg orbitals are responsible for the maxima above 16 eV. In

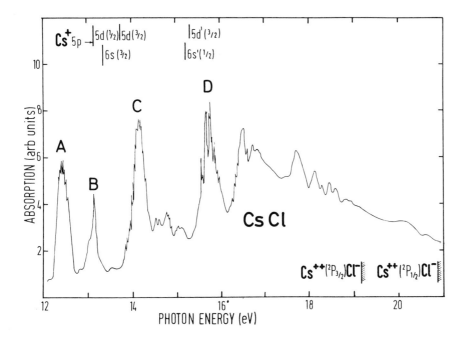

Fig. 6.16   Cs-5p excitations of molecular CsCl [6.75]

this range also transitions from Cl⁻3s Jevel may contribute. Taking polarization into account the ionic model places the Cs^+5p ionization in CsCl at 19.2 eV and 20.9 eV. The strong influence of intraatomic effects is obvious.

Above 76 eV transitions from the Cs^+4d level contribute to the absorption. The spectra of molecular and crystalline CsF, CsCl and CsBr at the Cs^+4d-threshold are presented in Fig. 6.17 [6.73,77-79]. In all molecular spectra there is a weak peak A at the onset followed by four prominent bands B, B'; C,C'D. The separation of B',B and C,C' is determined by the spin-orbit splitting of the 4d-hole. These bands are due to transitions of the Cs^+4d electrons to final states with dominant $Cs^+6s,6p$ parentage. Going from the molecular Cs-halides to the crystalline Cs-halides the small peak A persists whereas the maxima B,C and B',C' coalesce. The strong band E shows up at almost the same energy in all spectra. In contrast to this the other peaks shift considerably. The insensitivity of the peak position to the environment supports the assignment of E to transitions to highly localized f-symmetric final states. This is in agreement with Hartree-Fock calculations for the Cs^+4d^94f states.

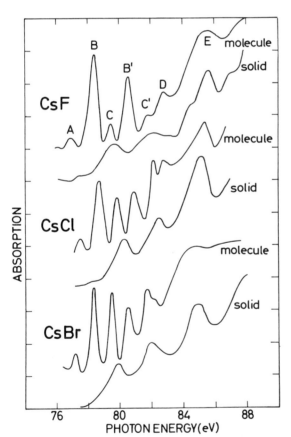

Fig. 6.17 Cs-4d absorption of molecular and solid CsF, CsCl and CsBr [6.78]

6.2.7 Xenon Fluorides

Since their first preparation the xenon fluorides have attracted much interest. NIELSEN and coworkers [6.80-82] have performed a detailed investigation of the valence and core excitation spectra which has contributed considerably to our present knowledge of the energy levels of these molecules. Figure 6.18 shows the absorption of XeF_2 below 35 eV. The vertical ionization potentials of the occupied valence orbitals are included. Forbidden transitions $\pi_u(Xe5p) \rightarrow \sigma_u(Xe5p)$ give rise to a small absorption below 7 eV. The first allowed transitions $\sigma_g(Xe5p) \rightarrow \sigma_u(Xe5p)$, $\pi_g(F2p) \rightarrow \sigma_u(Xe5p)$ are responsible for the broad maximum at 7.85 eV. At higher energies excitations from the highest occupied MO's into nonbonding Rydberg orbitals result in series of sharp lines converging to the lowest IPs at 12.4 eV, 12.9 eV and 13.6 eV. On an enlarged scale (Fig.6.19) the ample vibrational structure of the lower members of the $\pi_{u3/2,1/2}$ ns,nd series is clearly to be seen. Due to ionization and dissociation processes the vibrational structure gets blurred above $\sim$ 12 eV rendering the assignment less certain. Autoionizing higher Rydberg states are probably responsible for the asymmetric lines above the lowest IP. Considerations based on the Xe5p and F2p cross sections led NIELSEN and SCHWARZ [6.80,82] to assign the

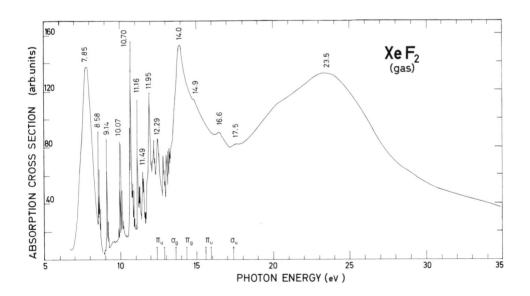

Fig. 6.18 Absorption cross sections of gaseous XeF_2. The vertical ionization potentials are also given [6.80]

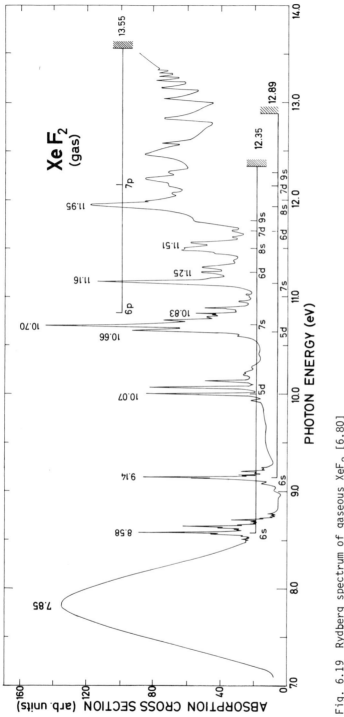

Fig. 6.19 Rydberg spectrum of gaseous XeF$_2$ [6.80]

14.0 eV maximum to the first ionization continuum from π_u(Xe5p) and the broad hump at 23 eV to transitions from the π_g,π_u,σ_g(F2p) levels. This broad hump increases on going to XeF$_4$ and XeF$_6$ whereas the peak at 14 eV decreases in accordance with the decreasing occupation of the Xe5p levels. The potential barrier formed by the F ligands expells the Rydberg states from the inner well and thus causes the disappearance of sharp Rydberg lines on going from linear XeF$_2$, to planar XeF$_4$ and octohedral XeF$_6$.

The low energy spectra are very complicated due to the overlapping transitions from the numerous upper filled molecular orbitals. In contrast to this the spectrum at the Xe4d-threshold presented in Fig. 6.20 shows a much simpler structure. With the help of these data NIELSEN and SCHWARZ [6.80,82] succeeded in establishing a level diagram of the empty molecular orbitals, which is consistent with both the valence and core excitation spectra. The binding energies of the Rydberg orbitals are almost independent on the shell in which the hole resides. This finding underlines the power of the *"Constancy of Rydberg term values model"*, which has been discussed in great detail by ROBIN [6.3] and SCHWARZ [6.18]. The assignment proposed by NIELSEN and SCHWARZ is given in Table 6.6. For the first time these spectra have given clear evidence for a ligand field splitting of a core level. The influence of the ligand field manifests itself in the asymmetry of the 0_{45} maximum and in the splitting of the $4d_{5/2,3/2} \rightarrow 6p\pi$ Rydberg transitions. The splitting of the 4d level obtained by diagonalizing the Hamiltonian

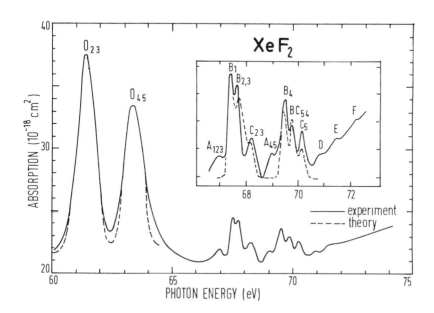

Fig. 6.20 Absorption cross section of molecular XeF$_2$ at the Xe-4d threshold [6.81]

Table 6.6 Transition energies of XeF_2 in electron volts [6.81,82]

Peak	Energy gas	Energy solid	Assignment	
O_{23}	61.38	60.96	$4d_{5/2} \rightarrow 5p$	
O_{45}	63.29	62.94	$4d_{3/2} \rightarrow 5p$	
A_{123}	66.94	69.2	$4d_{5/2} \rightarrow 6s$	
B_1	67.49		$4d_{5/2}(\delta_{5/2}) \rightarrow 6p\pi$	
B_{23}	67.73		$4d_{5/2}(\pi_{3/2},\ \sigma_{1/2}) \rightarrow 6p\pi$	
C_{23}	68.27		$4d_{5/2} \rightarrow 6p\sigma$	
A_{45}	69.00	71.1	$4d_{3/2} \rightarrow 6s$	$4d_{5/2} \rightarrow 7p$
B_4	69.53		$4d_{3/2}(\delta_{3/2}) \rightarrow 6p\pi$	
BC_{54}	69.84		$4d_{3/2}(\pi_{1/2}) \rightarrow 6p\pi$	
C_5	70.20		$4d_{3/2}(\delta_{3/2}) \rightarrow 6p\sigma$	
			$4d_{3/2}(\pi_{1/2}) \rightarrow 6p\sigma$	
D	70.89	72.8	$4d_{3/2}(\delta_{3/2}) \rightarrow 7p\pi$	
E	71.47		$4d_{3/2}(\pi_{1/2}) \rightarrow 7p\sigma;\ 7p\pi$	
F	72.30		$4d_{3/2} \rightarrow ?$	
α	144.91	145.0	$4p_{3/2} \rightarrow 6s$	

$$H = H_o - (-2\lambda/5)\underline{LS} + V_{lig}$$

is presented in Fig. 6.21 (H_o totally symmetric part of H, λ spin-orbit parameter, V_{lig} ligand field). The lower indices of the d_i symbols are also used to identify the initial level of the transitions. Based on this model SCHWARZ has calculated the theoretical spectrum included in Fig. 6.20. For XeF_4 and XeF_6, in addition to the ligand field splitting of the core level, the spin-orbit splitting of the $e_u(Xe5p)$ and $t_{1u}(Xe5p)$ levels has to be taken into account. The intensity ratio of the $4d_{5/2} \rightarrow 5p$ to $4d_{3/2} \rightarrow 5p$ transitions is 1.2 for XeF_2, 1.0 for XeF_4 and 0.5 for XeF_6. By calculations in intermediate coupling SCHWARZ has shown that this steady change is due to the 4d-5p exchange interaction. There are only weak Rydberg transitions for XeF_4 whereas no Rydberg transitions could be detected for XeF_6 [6.83]. As in the case of the valence spectra the Rydberg transitions are suppressed by the potential barrier formed by the F ligands.

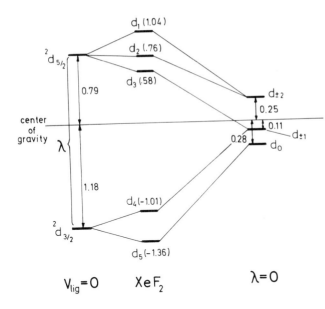

Fig. 6.21 4d level splitting in XeF₂ by ligand field (Vlig ≠ 0, λ = 0), by spin-orbit interaction (Vlig = 0, λ ≠ 0), and by both effects [6.81]

6.2.8 Inner-well Resonances

In the discussion of the xenon fluorides we already encountered the influence of the potential barrier formed by the ligands. The concept of a potential barrier in molecules was introduced by FOMICHEV and BARINSKII [6.20a] and BARINSKII and KULIKOVA [6.20b]. The barrier has been attributed to (i) the electrostatic repulsion of the excited electron by the electronegative ligands, (ii) the requirements of orthogonality to the occupied orbitals localized at the ligands. Recently DEHMER [6.20c] and DEHMER and DILL [6.20d] have pointed out that centrifugal forces acting on high-ℓ components of the final-state wave function may be an essential ingredient. The barrier results in a two-well potential. Consequently the states can be partitioned in "inner well" states, highly localized within the barrier, and outer well states, distributed outside the barrier. Inner-shell photoionization is an excellent tool for studying these molecular effects.

Comparing the S K-absorption of H_2S and SF_6 LAVILLA and DESLATTES [6.84] demonstrated the importance of barrier effects for the first time. The interest was further stimulated by the excellent work of ZIMKINA and coworkers [6.85]. The SF_6 sulfur 2p-spectrum, which is dominated by inner-well resonances, is one of the best examples [6.86-88] (Fig.6.22). BLECHSCHMIDT et al. demonstrated that the S 2p spectrum of solid SF_6 is identical to the gas-phase data [6.87]. This clearly proves that the final states are highly localized inside the molecule. Based mainly on symmetry arguments DEHMER [6.89a] proposed the following assignment: $(A,B,C,D) \rightarrow (a_{1g}, 5_{1u}, t_{2g}, e_g)$. GIANTURCO [6.89b] generally confirmed this assignment by considering the virtual orbitals calculated by a Hartree-Fock LCAO method. SACHENKO et al. [6.89c] employed the

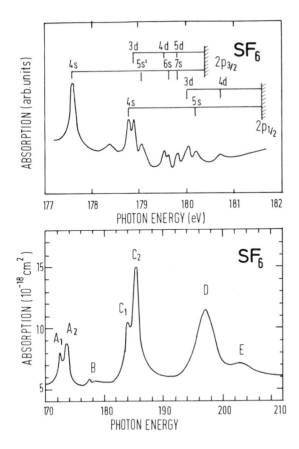

Fig. 6.22 S-2p absorption in molecular SF$_6$ [6.88]

multiple scattering method to compute the photoionization cross section of SF$_6$ above the S2p-threshold. In agreement with the above results they found strong resonances of t_{2g} and e_g symmetry at the energy positions of peak C and D.

The 2p-spectra of both gaseous and solid SiF$_4$ show strong resonances whereas there are no resonances in the spectra of molecular and solid SiH$_4$ [6.90,91]. This under- lines the influence of the ligands and corroborates the concept of inner-well reso- nances. Rydberg states are expected to be primarily distributed outside the potential barrier and therefore transitions from core states to Rydberg states should be sup- pressed. NAKAMURA et al. [6.86] and GLUSKIN et al. [6.88] succeeded to discover the four extremely weak Rydberg series at the S2p threshold of SF$_6$ shown in Fig. 6.22. This observation verifies that Rydberg states are indeed populated, but their ex- treme weakness indicates that they are effectively excluded from the region of the sulfur core levels. In contrast to this HAYES and BROWN [6.90,92] detected sharp Rydberg lines at the onset of the Si2p transitions in SiH$_4$. The H ligands do not give rise to a potential barrier thus neither the existence of inner-well resonances nor the suppression of Rydberg transitions is to be expected. Figure 6.23b shows the spectrum of molecular and solid SiH$_4$ at the Si2p-threshold. The comparison of the

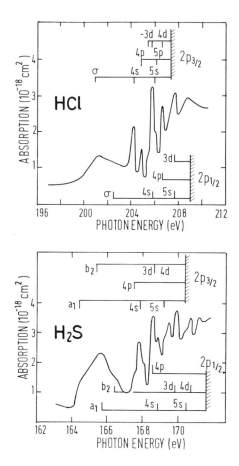

Fig. 6.23a Cl-2p and S-2p absorption spectra of molecular HCl and H₂S reported by HAYES and BROWN [6.90] The assignment given by SCHWARZ [6.93b] is included

spectra of both phases forms a good yard stick for the discrimination between valence and Rydberg excitations. Rydberg states extend far beyond the ligands and thus are drastically modified upon solidification whereas the localized valence excitations are much less affected. Thus transitions of prevailent valence character are responsible for the broad band peaking at 103 eV whereas there is no doubt about the Rydberg character of the lines at higher energies. The core spectra of HCl, H_2S and PH_3 reported by HAYES and BROWN [6.90] show a similar pattern (see Fig.6.23). For a discussion of the core excitation spectra of the simple molecular hydrides the reader is referred to [6.3,93] which are addressed to the question "Rydberg versus valence excitation". Coming back to the inner-well resonances we want to draw the reader's attention to the fact that the F ligands cause pronounced resonances in the 2p-continuum of SiF_4 and SF_6 whereas no indication of corresponding resonances could be detected in the 4d-continuum of XeF_4 and XeF_6. The 4d-spectra of TeF_4 and TeF_6 [6.94] show a maximum which may be due to the influence of the F ligands. The gross features of the molecular spectra above the 4d threshold correspond to those

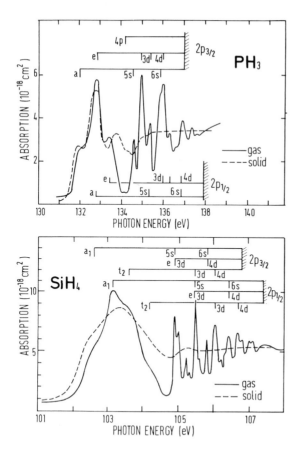

Fig. 6.23b P-2p and Si-2p absorption of molecular (6.90) and solid [6.91] PH_3 and SiH_4. The assignment given by SCHWARZ [6.93b] is included

of the spectra of the free atoms. When comparing the 1s-spectrum of N_2 and the 4d spectrum of I_2 one encounters a similar situation. The main oscillator strength is concentrated in the $\ell \to \ell + 1$, i.e. the 4d → εf channel. For atomic Te, I and Xe there exists a pronounced potential barrier for f-symmetric final states which causes the resonant enhancement of the absorption above the 4d-threshold. From the absence of molecular inner-well resonances we conclude, (i) the atomic potential barrier for f-symmetric final states is conserved in the molecule, (ii) the high centrifugal repulsion prevents the formation of resonant states with $\ell \geq 4$, (iii) d-type resonances are suppressed by the presence of occupied d-orbitals, (iiii) the molecular field due to the ligands fails to effectively couple higher ℓ-components to the f-symmetric wave produced by the excitation of a 4d electron in the essentially atomic field close to the nucleus.

6.2.9 EXAFS

In addition to inner-well resonances core spectra of molecules often show a weak fine structure extending hundreds of volts into the continuum. As an example the

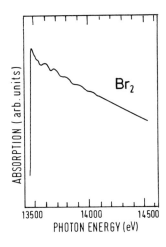

Fig. 6.24 Extended X-ray absorption fine structure above the K-edge of Br_2 [6.95]

Br_2 K-spectrum reported by KINCAID and EISENBERGER [6.95] is given in Fig. 6.24. The investigation of this extended X-ray absorption fine structure (EXAFS) was initiated by KRONIG [6.96] more than 40 years ago. The activities in this field until the beginning 60th have been reviewed by AZAROFF [6.97]. The work was carried on mainly by STERN and coworkers [6.98,99] (see also [6.100,101]) who made substantial contributions to the understanding of the phenomena. They realized that an appropriate Fourier transform of the EXAFS data yields exact information on the geometric structure around the ionized atom. This step turned EXAFS into a very promising method for the determination of interatomic distances in a large variety of substances, e.g. crystalline and amorphous solids, liquids and macromolecules.

The severest limitation has been the low intensity of conventional X-ray sources. The replacement of the X-ray sources by the powerful synchrotron radiation, which made high resolution, high precision EXAFS feasible, stimulated a rapid development of the field. In the framework of the current short-range order theories [6.98-101] the modulation of the absorption can be explained as follows: The outgoing photo-electron, ejected from a core level is partially reflected from the neighboring atoms. The interference of the reflected waves and the initial wave close to the nucleus of the excited atom leads to an oscillatory behavior of the absorption matrix element, as the interference is alternatively constructive or descructive. The assumption underlying the theories currently used for the interpretation of FXAFS render these theories inapplicable to energies within ~ 50 eV of the threshold. There is no principal limitation which inhibits the application of the basic ideas underlying the EXAFS models close to threshold. Several approaches which encompass the whole spectrum above threshold have been outlined, but none has been worked out in detail [6.102]. It is clear that the inner-well resonances discussed above are attributable to the modification of the amplitude of the final-state wave function close to the core, caused by the superposition of the outgoing wave and the waves backscattered by the neighboring atoms. The low energy of the electrons requires the consideration of multiple scattering and a more accurate treatment of the potential. We consider a unified treatment of the absorption spectrum above threshold based on a multiple scattering [6.21] method extremely rewarding.

6.2.10 Valence Shell Spectra of Organic Compounds

Interest in the higher excited states of organic molecules has grown spectacularly in the last few years. High conductivity organic crystals, the growing interest in these compounds for solid state devices [6.103] and the role organic molecules play as building units for biologically important larger molecules (see e.g. [6.13]) have been some of the reasons to focus the attention of spectroscopists on these compounds. The involvement of organic molecules in astrophysical processes and the photoreactions in the upper atmosphere [6.11,12] adds to this interest. Furthermore, the development of high speed MO-programs makes many of the electronic properties even of larger molecules amenable to detailed theoretical studies (e.g. [6.104]). The measurement of their photoabsorption cross sections is of fundamental interest.

However, due to the large number of electronic levels present in larger molecules the spectra are heavily crowded and complex and one cannot expect to achieve an analysis as detailed as is possible for smaller molecules. Nevertheless, electron energy loss spectroscopy [6.3,10,105,106] and, in particular, optical experiments with SR have provided a fair amount of excellent data in the VUV for several groups of similar molecules. Thus the observation of trends in a family of spectra as well as comparison to spectra from the same compound in the solid phase gave considerable insight for the assignment and understanding of the spectral features. Recently, KOCH and OTTO [6.106] have reviewed various aspects of these studies in the VUV range among them (i) the comparison of cross section data in optical and electron energy loss experiments, (ii) general considerations for the assignment of absorption spectra from organic compounds, (iii) the occurence or nonoccurence of collective effects in isolated large organic molecules, and (iiii) the relevance of gas phase work for optical studies on organic molecular crystals. The few examples selected for the present review serve as an illustration of the more recent advances.

a) *Saturated Hydrocarbons: Alkanes, Neopentane*

The spectrum of methane, the first member in the series of alkanes which have bonding σ-orbitals only, was treated by MULLIKEN [6.107] as long ago as 1935. In the meantime much has been added to our knowledge of the spectra of completely saturated hydrocarbons [6.3,108] notably by using synchrotron radiation as a continuum background [6.25,109,110]. For a review on the discussion of these spectra, in particular the question whether the observed structures are due to normal-to-valence (N-V) transitions or Rydberg excitations, we refer to the summary by ROBIN [6.3

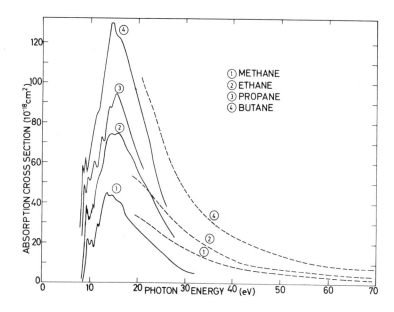

Fig. 6.25 Absorption spectra of methane, ethane, propane and butane in the range 7 to 70 eV after KOCH and SKIBOWSKI [6.109] and LEE et al. [6.25], taken from [6.106]

In Fig. 6.25 a synopsis of the absorption spectra of methane, ethane, propane and butane is presented, as determined by KOCH and SKIBOWSKI [6.109] and LEE et al. [6.25]. It now seems widely accepted that almost all the observed discrete structures superimposed on the continuum absorption at higher energies are due to Rydberg type excitations, although only in very few cases can more than one member of a "series" be assigned with any certainty. In Table 6.7 we reproduce [6.106] the tentative assignments for the observed features mainly based on the interpretation of the spectra, as given in [6.3]. For the comparison with the spectra from the solid phase see [6.106,109]. For a recent detailed theoretical analysis of the propane spectrum we refer to [6.111]

Comparison of the two sets of data reveals considerable discrepancies for the values of the absolute cross sections in the range of overlap of the individual investigations. This has been discussed by BERKOWITZ [6.112] who pointed out correctly that in the range of continuous absorption extreme caution has to be exercised in dispersing synchrotron radiation and analyzing the data, because of the scattered light and order-sorting problems (see also [6.113,114]). It appears, that in several instances new data profiting from the constant improvement of spectroscopic techniques are needed for absolute cross sections.

There seemed to be only little chance of observing extensive Rydberg series to higher ionization limits since - as discussed by HERZBERG [6.1] - in almost all cases the transition in the Franck-Condon range would go to the repulsive part of

Table 6.7 Excitation energies in eV for gaseous and solid alkanes and some tentative assignments [6.106] (sh: shoulder). For the assignment of the propane spectrum see also [6.111]

Methane Symmetry: T_d		Ethane D_{3h}, D_{3d}		Propane C_{2v}		Butane C_{2h}
vapor	solid(a)	vapor	solid	vapor	solid	vapor
9.7 $1t_2 \to 3s$	9.6	9.4 $\begin{cases} 1e_g \to 3p \\ 3a_{1g} \to 3p \end{cases}$	9.6	7.93 $2b_2 \to 3s$		8.65
				8.85 $2b_2 \to 3p$		
				$4a_1 \to 3s$	9.4	
10.4 $1t_2 \to 3s$	11.0sh	10.7 $\to 4p$		9.65 $4a_1 \to 3p$	9.8	
					10.3	
11.7 $1t_2 \to 4s$		11.8		11.0 $1a_2 \to 3p$		11.2sh
				$2b_1 \to 3s$		
13.7	13.8			12.70 $3a_1 \to 3p$		13.2
14.7		14.5	14.5	14.20		14.7
16.7	16.4	16.4		15.60 $2a_1 \to 3s$	15.6	16.5
		18.9	19.0sh			

the potential curve of the upper state and hence would be structureless and diffuse. Indeed, for the alkanes, in the previous work with conventional techniques with the exception of [6.115] fine structure due to Rydberg excitations was not detected. With the help of synchrotron radiation as a continuum background LEE et al. [6.110] have recently been able to detect considerable fine structure in the absorption spectrum of methane in the range 19.4 eV ≤ hν ≤ 22.5 eV. They assigned these structures to two Rydberg transitions originating from the $2a_1$ orbital to 3p and 4p orbitals with term values of 2.59 eV and 1.35 eV, respectively.

This finding is similar to previous observations by KOCH et al. [6.116] for neopentane, $(CH_3)_4C$, another completely saturated hydrocarbon. These authors found extensive fine structure in the VUV absorption spectrum at around 16 eV. In Fig. 6.26 a survey of the absorption spectrum from the onset of strong absorption at about 7.2 eV up to 35 eV is shown. For photon energies below 10 eV the spectrum, as reported by SANDORFY [6.117], is reproduced. At around 16 eV a number of sharp and narrow absorption bands are observed, which are displayed on an expanded scale in the insert. Some of the band shapes are asymmetric indicating Fano-Beutler type absorption profiles [6.41] or antiresonances. Assuming T_d symmetry for neopentane and using information from photoelectron spectroscopy [6.118] the bands at 15.46, 16.59 and 17.00 eV have been identified as due to a Rydberg series with p-character originating from the bonding $2a_1$ orbital [6.116], with term values of T_{3p} = 2.222 eV, T_{4p} = 1.09 eV and T_{5p} = 0.68 eV, respectively. Note that these

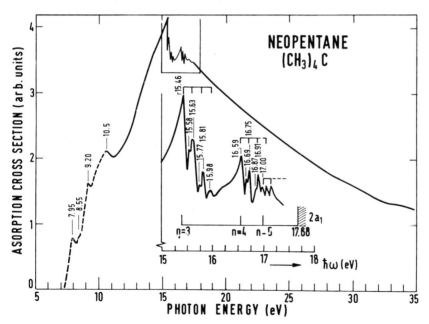

Fig. 6.26 Absorption spectrum of neopentane for photon energies between 5 eV and 35 eV. Below 10 eV the spectrum reported by SANDORFY [6.108] is shown [6.116]

term values and the resulting quantum defect (one obtains, for instance, from the
n = 4 member δ = 0.46) are very similar to those of the observed p-type series in
methane [6.110] (see above) and in benzene [6.119] and perdeuterated benzene [6.120]
in the 15-17 eV spectral range (C_6H_6: T_3 = 2.22 eV, T_4 = 1.03 eV, C_6D_6: T_3 = 2.21 eV).

The remaining bands in the neopentane spectrum could be assigned to vibrational
progressions based on the totally symmetric C-H deformation mode ν_2 (170 meV) of
symmetry species a_1 as indicated in Fig. 6.26. In addition, a further weaker progression has been attributed to the combination of $\nu_2 + \nu_3$ where ν_3 is the totally
symmetric C-CH_3 stretching mode.

Later in a careful study with a conventional light source PERSON and NICOLE
[6.113] obtained quantitative cross section data and were able to extend the Rydberg
series out to n = 9 with the same assignments, as given in [6.116]. They also fitted
the Fano-Beutler profiles for six levels and obtained from these fits an estimate
for the autoionization rates of about 10^{14} s^{-1} for n = 3 and 3 - 5 $\times$ 10^{13} s^{-1} for
n = 4 and 5. Certainly these studies as well as investigations of other molecules
can be improved with quantitative high resolution absorption spectroscopy in this
range made possible by the combination of storage rings sources with large focal
length instruments (see e.g. [6.114]).

b) *Molecules with bonding σ- and π-orbitals*

Ethylene and its derivatives as well as benzene and the aromatic hydrocarbons are
among the more frequently examined species falling into this category. Much of the
early work on these compounds has been summarized and discussed by HERZBERG [6.1]
and a detailed discussion of the more recent work including work done with synchrotron radiation has been given by ROBIN [6.3]. Both ethylene and benzene play a key
role in studies of the electronic states of organic compounds by virtue of their
typical but relatively simple structure. In spite of their importance the VUV-spectra
of these molecules and the decay mechanisms of highly excited states are not thoroughly understood. Some aspects of the ethylene spectrum will be touched upon in
Sect. 6.5 when we shall describe mass spectroscopic studies done with synchrotron
radiation. Absolute absorption cross section determinations for ethylene over a
large spectral range are still not complete in spite of the recent work by LEE and
collaborators [6.110] with synchrotron radiation. In this section we illustrate the
present state of our understanding of the VUV spectra for larger organic molecules
by discussing some aspects of the benzene spectrum and the spectra from various
fluorinated benzenes. These spectra will also serve to illustrate a certain routine
which has been developed in the past few years for making assignments much easier.

This routine which has been summarized by SANDORFY [1.108] consists of basically
four steps: (i) comparison with a highly resolved photoelectron spectrum, (ii) comparison with a good quantum-chemical calculation for the ground state. By these two
steps one gains knowledge of the orbitals from where transitions originate, (iii)

ordering of the observed absorption bands into ns, np, nd ... type Rydberg series, (iv) assignment to valence shell transitions by comparison with spectra from the solid phase [6.106]. Unfortunately the assignment of normal to valence transitions can be made with confidence only for excitation energies below ≈ 8 eV. Above this energy the various transitions overlap considerably. According to the many calculations now available higher σ-π* and σ-σ* transitions must fall in this energy region. Since these structures are broad only the Rydberg bands superimposed on the continuum like background can be assigned.

A glance at Fig. 6.27 showing a survey of the optical absorption spectrum of benzene obtained with synchrotron radiation, verifies this situation. Following the intense π-π* transition at about 6.9 eV, a broad and continuous absorption with a maximum at about 17.8 eV is observed. Here the cross-section reaches values of about 150 Mbarn. It decreases smoothly for higher energies. The main features of the absorption cross-section, except for additional details revealed by the synchrotron light experiment (see also Fig.6.29), are in good agreement with other measurements [6.121-123]. More recently the absorption cross section of benzene and substituted benzenes has been carefully determined ($\Delta\lambda \lesssim 0.15$ Å) with the high resolution instrument available at the DORIS SR-Laboratory [6.114].

Next we search for information about the occupied MO's of benzene. The presently accepted result is shown in Fig. 6.28. It is based on the interpretation and analysis of the many UPS and XPS photoemission studies for benzene (see e.g. [6.124-126] and in accord with most of the recent MO calculations e.g. [6.127-129]). We note in passing, that optical studies of the VUV spectrum [6.119,120] have been very helpful in establishing some aspects of this ordering, in particular the ordering of the second and third uppermost orbital, the $\sigma-3e_{2g}$ and the $\pi-1a_{2u}$. Now with the aid of group-theoretical arguments governing the selection rules, one starts to order the observed

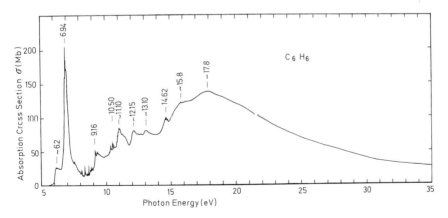

Fig. 6.27 Absorption cross section of benzene vapor from 5 to 35 eV [6.106]

314

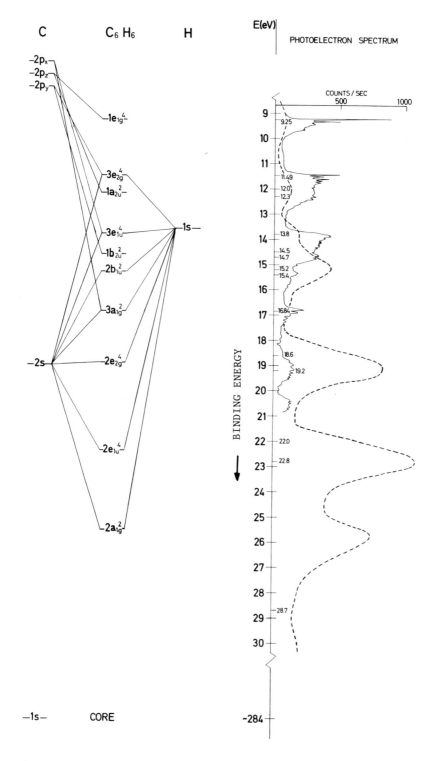

Fig. 6.28 Ordering of the molecular orbitals of benzene corresponding to the relative ionic energy levels (note that this is not an accurate molecular orbital energy level diagram owing to the deficiencies of Koopmans Theorem). Right part: UPS and XPS photoelectron spectra of benzene [6.9b] and [6.126]

315

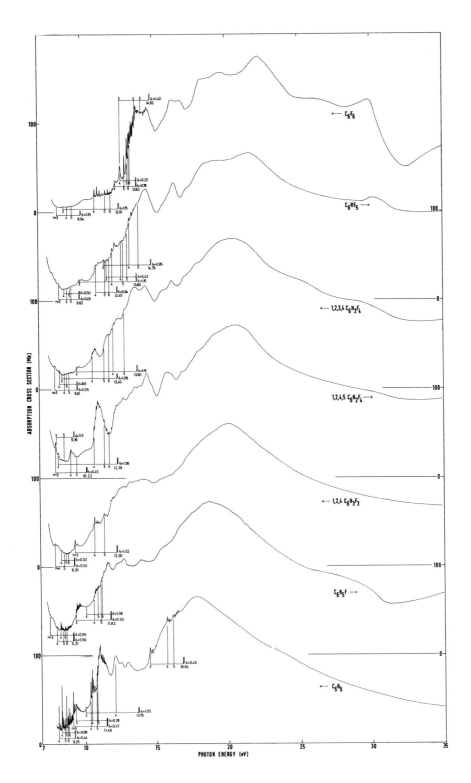

Fig. 6.29 Absorption cross section for benzene and various fluoroderivatives. A number of assignments for the Rydberg series are indicated [6.114]

316

absorption bands into the various Rydberg series. For higher IP's the complex
absorption pattern of overlapping bands usually complicates such an analysis. There
are, however, two additional pieces of information which can be exploited: (i) one
can generally assume, that the vibrational progression associated with a particular
Rydberg state closely resembles the progression observed for the respective ionic
state as observed in the photoelectron spectrum [6.119,130]. This is so, since the
potential curves for Rydberg states and for the corresponding ionic state are usually
quite similar. (ii) The observation of particular trends in a series of molecules
may serve as a yard stick in the assigment.

Concerning the first point the analysis of the high resolution spectra of benzene
and its fluor derivatives [6.114] yields good agreement with the recent calculations
of the vibrational structure of ionic states as performed by DUKE and coworkers
[6.131-133]. The latter point is illustrated in Fig. 6.29 where spectra for a number of fluorinated benzenes [6.114] are shown. In this particular case the perfluoro
effect [6.134] has been exploited: fluorine substitution stabilizes σ-orbitals more
than π-orbitals, and in this way benzene's overlapping absorption bands become
partly disentangled in the fluorinated molecules.

At this point we shall not go into the individual assignments. We rather refer to
a series of papers where these aspects have been discussed in detail for benzene
[6.119,120,135], naphthalene [6.135-136], anthracene [6.106,137], and benzene derivatives [6.114,120]. In addition to absorption experiments in the gas phase, spectra from solid polycrystalline films [6.135] and from single crystals [6.106,135,138]
have been obtained and recently absorption spectra from matrix isolated benzene
[6.139] and naphthalane [6.140] have been published.

6.2.11 Core spectra of organic compounds

There is an acute lack of data on energy levels and cross sections from K-shell
excited spectra for organic compounds containing C, N and O atoms. This is despite
of the growing interest for inner shell spectra from a theoretical point of view
(see e.g. [6.141,142]) and despite promising applications for chemical analysis,
e.g. determination of core electron binding energies and chemical shifts. Furthermore, the optical properties of large organic compounds, including polymers deserve
special attention in the soft X-ray region in order to understand the mechanisms of
radiation chemistry or to optimize the performance of organic compounds as resist
materials in soft X-ray microscopy and lithography [6.143]. As we have seen in a
previous section for the case of K-shell spectra from simple inorganic molecules
one expects transitions to occur between localized atomic like orbitals and empty
molecular orbitals, thus conceptually simplifying their analysis as compared to
the complex valence shell spectra. Consequently simple theoretical approximations
such as the equivalent core and one electron models become more acceptable.

The K-absorption spectrum of methane reported by CHUN [6.144] provided the first example of a K-shell spectrum from an organic compound obtained with synchrotron radiation. The X-ray absorption spectra of C_2H_2 and C_6H_6 had been investigated previously with the bremsstrahlen continuum [6.145], but the absorption structure in these spectra was too weak to allow for a reasonable interpretation. CHUN's spectrum of CH_4 [6.145] together with electron impact excitation spectra (e.g. [6.146,147]) has been discussed extensively in comparison with theoretical calculations [6.141, 142]. The strong absorption peak at 288.3 eV was identified as the first $1a_1 \rightarrow 2t*$ (3p) Rydberg transitions while the weak peak at 287.2 eV was interpreted as a transition to the $3a_1$, Rydberg orbital of essentially s-character made allowed by vibronic coupling [6.142].

Recently BROWN et al. [6.148] have reported soft X-ray absorption spectra obtained with synchrotron radiation for the K-edges of carbon in CH_4 and the fluoromethanes. These spectra appear in Fig. 6.30; the spectral information is summarized in Table 6.8 where the absolute energies observed and calculated [6.142-150] and some MO assignments are given. On the basis of these spectra previous assignments for the excited states of methane appear reasonable. The "forbidden" line at

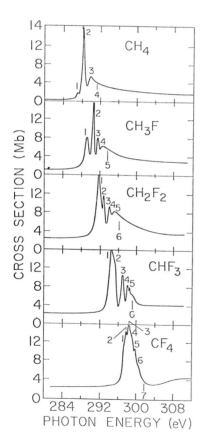

Fig. 6.30 Absorption spectra for methane vapor and the fluoro methanes in the range of the carbon K-edge. For the assignment of the various features see Table 6.8 [6.148]

Table 6.8 Spectral features, absolute energies observed and calculated, and final molecular orbital assignments [6.148]

Substance and point group spectral feature		Energy[a] E[eV]	ΔE[eV]	Calculated energy [eV]	Orbital assignment
CH_4	1	287.0	3.8	287.1,[c] 287,3[d]	$3a_1$ (3s)
	2	288.1	2.7	288.5[c] 288.4[d]	$3t_2$ (3p)
(T_d)	3	289.4	1.4		nt_2, $n \geq 4$
	4	290.8[b]	0.0	291.0[c]	ionization
CH_3F	1	289.1	4.4		3s
	2	290.5	3.0		3p
(C_{3v})	3	291.6	1.9		3d
	4	292.0	1.5		np
	5	293.5[e]	0.0	293.9[c]	ionization
CH_2F_2	1	291.9	4.5		3s
	2	292.9	3.5		
(C_{2v})	3	294.0	2.4		3p
	4	294.5	1.9		3d
	5	295.5	0.9		np
	6	296.4[e]	0.0	297.1[c]	ionization
CHF_3	1	294.8	4.3		-
	2	295.3	3.8		-
(C_{3v})	3	297.2	1.9		3d ?
	4	298.0	1.2		
	5	298.6	0.5		
	6	299.1[e]	0.0	300.3[c]	ionization
CF_4	1	297.6	4.2		-
	2	298.0	3.8		-
(T_d)	3	298.5	3.3		-
	4	298.9	2.9		-
	5	299.9	1.9		3d ?
	6	301	0.8		np
	7	301.8[e]	0.0		ionization

[a] absolute energies ±0.05 eV from [6.148];
[b] core ionization energy, from [6.149];
[c] [6.150];
[d] [6.142];
[e] core ionization energy [6.9a,151]

threshold is only little affected by the substitution of deuterium for hydrogen in
the optical experiment. However, recent electron energy loss experiments by HITCHCOCK
and BRION on the substituted compounds corroborate the assumption that this transition is being allowed by vibronic coupling [6.255].

Inspection of Fig. 6.30 shows that with increasing fluorination the Rydberg like
threshold spectra evolve into a complex band in CF_4. It is interesting to note that
oscillatory EXAFS structure above threshold has been observed in CF_4 while it is
absent in CH_4. This observation parallels the situation we have discussed before
when comparing the Si-$L_{II/III}$ spectrum in SiH_4 and the spectrum from SiF_4.

In Fig. 6.31 some of the electron yield spectra obtained by EBERHARDT et al.
[6.152] in a study of core excitations in gaseous methane, ethane, ethylene, benzene and acetylene are reproduced. In these experiments the yield of low-energy
secondary electrons was measured as the photon energy was swept through the carbon
K-edge. At least below the ionization limit the number of low energy secondary
electrons is closely related to the absorption cross section provided the carbon

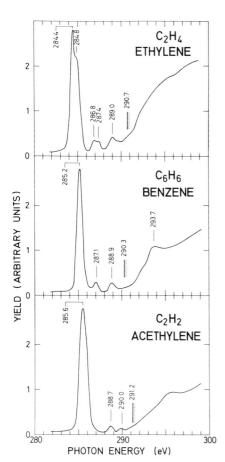

Fig. 6.31 Electron yield spectra of gaseous
ethylene, benzene and acetylene in the range
of the carbon K-edge. For the assignment of
the various features see Table 6.9 [6.152]

Table 6.9 Excitation energies, ionization potentials and term values for the carbon 1s shell for methane, ethane, ethylene, benzene, and acetylene. (All energies are in eV, sh denotes shoulders)

	excitation energies (a)	(b)	IP C1s (c)	Term Value (d)	Assignment (d)
methane CH_4	286.9			3.8	3s
	288.0			2.7	3p
	289.4			1.3	4p
			290.7	0	∞
ethane C_2H_6	286.8	286.9		3.9	3s
	288.0	287.9		2.7	3p
		289.3		1.1	4p
	289.6	289.7			
			290.7	0	∞
ethylene C_2H_4	284.4	284.68		6.2	π^*
		285.04			
	284.8sh	285.50(?)		5.8	π^*
		285.90(?)			
	286.8	287.4		3.8	3s
	287.4sh	287.8		3.2	3p
		288.3sh			
	289.0	289.3		1.6	4p
			290.6	0	∞
		292.6			shake up[b]
		295.2			
benzene C_6H_6	285.2	285.2		5.1	π^*
	287.1	287.2		3.1	3p,3s
		288.0			
		288.6sh			
	288.9	288.9		1.4	4p
			290.3	0	∞
		290.4			
		291.3			shake up[b]
	293.7	293.5			
acethylene C_2H_2	285.6	285.9		5.6	π^*
		288.1			
	288.7	289.0		2.5	3p
		290.0			
			291.2	0	∞
		291.4			shake up
		292.4			and
	295.6	295.6			shake-off[b]
		300.6			

(a) From SR electron yield measurements by EBERHARDT et al. [6.152]; (b) from electron energy loss spectra by HITCHCOCK and BRION [6.154]; (c) from XPS work [6.9a,126,155]; (d) according to [6.152]; only the final orbital is listed

K-shell hole decays predominantly by an Auger process and fluorescence is negligible. This technique was employed rather than conventional absorption spectroscopy since optical experiments in the 280 eV $\lesssim h\nu \lesssim$ 300 eV region are severely hampered by carbon contamination on the surface of the optical elements [6.153].

In spite of the limitations imposed by the yield technique and the limited resolution these results allowed for an interpretation and assignment of the observed structures in terms of core to valence (core → π*) and core to Rydberg (core → R) transitions. These assignments of the spectra together with the very recent results of electron impact experiments [6.154] which generally arrived at the same assignments, appear in Table 6.9. As expected the spectra are fairly simple compared to the valence shell absorption and are characterized by a small number of maxima below the ionization threshold and additional strong resonances for the molecules containing π-electrons. The assignments given in Table 6.9 are largely based on a simple one electron picture, although the strong resonances could perhaps also be interpreted as shape resonances. Note, that these prominent maxima are absent in the spectra from the saturated hydrocarbons. The assignment of the Rydberg transitions rests essentially on a consideration of their term values with an expected energy ordering or Rydberg orbitals for molecules containing first row elements of 3s < 3p < 3d ~ 4s < higher members [6.3].

The electron impact data [6.154] although they agree qualitatively with the SR-yield experiment, show quantitative deviations, which are most probably due to problems associated with the yield method applied in the SR experiment. With programs of absorption studies at the carbon K-edge in progress at several SR-Laboratories we may look forward to an increased resolution and more precise data which may justify the hopes cherished for this particularly interesting field of molecular absorption spectroscopy with SR.

6.3 Photoelectron spectroscopy

Within the last decade it has become clear that photoelectron spectroscopy is one of the most direct and successful ways of investigating the electronic structure of molecular systems [6.9]. Usually a small number of resonance lines from rare gas discharge lamps are used to eject electrons from their orbitals. The kinetic energies of the photoemitted electrons and other properties such as their intensity, and angular distribution are recorded. Synchrotron radiation offers a number of outstanding properties which allow to go beyond the conventional limits and to develop important new methods for photoelectron spectroscopy. Although some beautiful applications of new methods are already available, it is obvious that much remains to be done in order to fully develop these experimental methods for molecular spectroscopy. The main reason for a time lag as compared, for instance, to solid-state photoelectron spectroscopy with synchrotron radiation [6.156] is the fact that far higher intensities of the incident monochromatic light are required because of the low target density and the usually higher resolution necessary for the electron analysis. Further problems associated with the confinement of the target gas and contamination problems of the optical elements of the monochromators play a role. Consequently the new possibilities opened up by a tunable and polarized photon source have only been developed recently with the availability of high intensity storage ring sources.

6.3.1 Intensities of photoelectron spectra and partial photoionization cross sections

Apart from the quantitative data concerning orbital energies to be obtained from a photoelectron spectrum, the intensities of the bands obviously contain much information about the nature of the MO's [6.9d,157]. The measurement of the relative intensity of each energy group as a function of the exciting photon energy yields direct information about the branching ratios that is the partitioning of excited electrons amongst the various available excitation modes. If the absolute photoionization cross section is known the values of the branching ratio can be used to give absolute values of the partial photoionization cross section [6.9e,158], which in turn lend themselves for a direct comparison with theoretical predictions [6.159, 160].

For a number of molecules including N_2, CO, CO_2 and SF_6 PLUMMER et al. [6.40, 161,162] have performed such measurements. Fig. 6.32 shows photoelectron energy distribution measurements for N_2 and CO at several photon energies. The measured signal I in these spectra is proportional to

$$I \propto \frac{\sigma_i}{4\pi} \int_\Omega \left[(1 + \frac{\beta_i}{2})(3 \cos^2\theta - 1) \right] d\Omega$$

where σ_i is the partial photoionization cross section for a given ionic state i, θ the angle of collection with respect to the direction of polarization, β_i the asymmetry parameter characterizing the angular distribution of the ith ionic state at a given photon energy [6.163], and Ω the solid angle accepted by the analyzer. One way to eliminate the unknown asymmetry parameters β_i is a measurement under the magic angle of $54°44'$ [6.163]. The branching ratios $B^i = A_i/\Sigma A_i$ are obtained by recording an energy distribution at a fixed photon energy and then measuring the area A_i of the ith peak and taking the sum over all observed states of the ion.

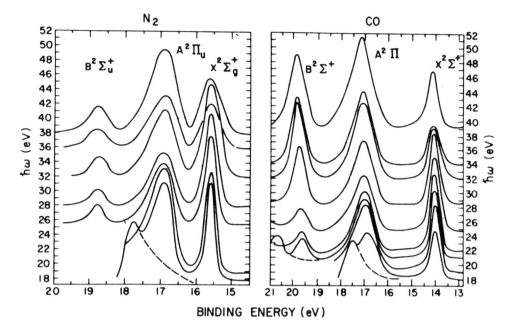

Fig. 6.32 Photoelectron energy distributions for N_2 and CO at selected photon energies $\hbar\omega$ plotted on a binding energy scale. The curves are displaced relative to each other with amounts proportional to the differences in photon energy (scales on the left and right hand side). The dashed lines indicate the background [6.40]

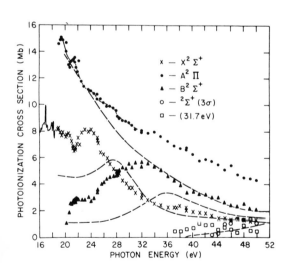

Fig. 6.33 Partial photoionization cross sections (still containing the angular parameters β_i) for five ionic states of CO (the four lowest energy one-electron states and one multi-electron excited state). The solid curves have been obtained from CIS-spectra. The dashed lines are the results of DAVENPORT's [6.164] calculation [6.40]

For CO the partial photoionization cross sections obtained in this manner are shown in Fig. 6.33. In order to calculate the partial photoionization cross sections which still contain the angular dependence as given by the above equation total absorption cross section data by LEE et al. [6.25] (Fig.6.5) had been used. The four one-electron excited states of the CO ion $X^2\Sigma^+, A^2\Pi, B^2\Sigma^+$ and $^2\Sigma^+(3\sigma)$ correspond to the removal of one electron from the 5σ, 1π, 4σ and 3σ molecular orbitals of CO, respectively. At approximately 10 eV kinetic energy one observes for two of the Σ states peaks in their cross sections due to scattering resonances in the final states. (For the $^2\Sigma^+(3\sigma)$ state this resonance is not within the range of the measurement.) These shape resonances have been discussed by DEHMER and DILL [6.20] for N_2 photoionization from the N 1s level (see Sect.6.24). DAVENPORT [6.164] has applied the same methods, namely the SCF-SWXα calculation, to determine the valence levels of N_2 and CO. His calculated results are also shown in Fig. 6.33 (dashed lines) and a qualitative agreement with the data can be noted.

Recently, WOODRUF and MARR [6.165,166] reported on similar experiments on N_2 in which they could partially resolve individual vibrational states and follow the intensity changes within a vibrational progression over the region covered by the Hopfield bands.

One important aspect of these quantitative partial photoionization cross section data is their usefulness for the discussion of photoemission intensities and assignments in spectra from adsorbed molecules [6.167,168]. There is now evidence from a number of experiments that the hν-dependence of the photoionization cross section plays also a predominant role in determining the surface sensitivity in these photoemission experiments (see e.g. [6.169]).

Near threshold the partial photoionization cross sections will exhibit sharp structure due to autoionization, which is difficult to observe using the above described method of integrating areas in energy distributions. There is, however, a very elegant mode of photoemission which takes particularly advantage of the continuous nature of the synchrotron radiation spectrum. In this mode the photon energy is swept while fixing the ionic state which is being observed. This can be done by sweeping the transmitted kinetic energy of the electron analyzer synchronously with and at the same rate as the photon energy. For example, if hν - E_{kin} is fixed at 15.7 eV for N_2 the ionic state is the $X^2\Sigma_g^+$ state. Therefore PLUMMER et al. [6.40] have suggested to interpret the abbreviation CIS (for constant initial state - energy spectra) as "constant ionic state" spectra. As an example the CIS curves obtained by PLUMMER et al. [6.40] for the X, A and B ionic state of N_2 as a function of the wavelength of the incidence light are shown in panel B of Fig. 6.34 together with the absorption spectrum as obtained by GÜRTLER et al. [6.36] at a resolution of 0.03 Å (panel A). It is obvious that this mode is a very convenient way to obtain the position and relative intensities in the rapidly varying partial photoionization cross section near the ionization threshold. We note, however, that these curves are not exactly proportional to the partial photoionization cross section due to the energy dependent efficiency of the detecting system [6.40]. Furthermore a higher resolution would certainly reveal even more structure in these spectra.

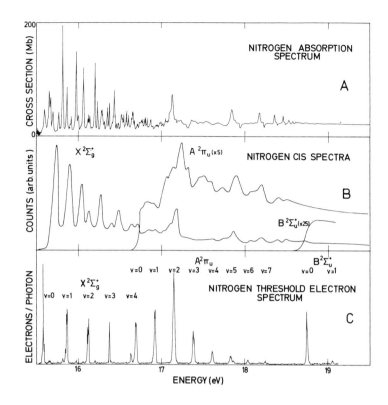

Fig. 6.34 *Upper part:* Nitrogen absorption spectrum in the range $15.5 \le \nu \le 19.5$ eV [6.36]. *Middle part:* Constant ionic state curves for the X $^2\Sigma_g^+$, A $^2\Pi_u$, and B $^2\Sigma_u^+$ ionic states of N_2 as a function of the energy of the incident light. The values $h\nu - E_{Kin}$ used correspond to the measured binding energies. The energy analyzer pass energy was 40 eV. There are some slight differences in the energy location of structure between this panel and the upper and lower, which are believed to be instrumental in origin [6.40]. *Lower part:* Threshold electron spectrum of N_2 [6.177]

6.3.2 Photoionization resonance spectroscopy and coincidence measurements

Still another very useful mode of photoelectron spectroscopy can be performed advantageously with synchrotron radiation: photoionization resonance spectroscopy (PIR). In this mode $h\nu$ is scanned and photoionization yield spectra are recorded with the additional constraint of detecting only threshold (zero kinetic energy) electrons [6.170-172]. Thus photoionization currents appear only for those photon energies where resonances of the ion occur. It is essential for these methods to have an accurate and efficient means of analyzing and detecting threshold electrons and rejecting those with higher kinetic energies. These methods have been developed previously [6.170,171] and are constantly improved [6.172,173].

One of the main advantages of the method is the possibility to determine very accurately ionization potentials and to achieve an energy resolution basically determined by the wavelength resolution of the monochromator for the primary excitation. This technique has been exploited by JOCHIMS et al. [6.174] for the study of small organic molecules with moderate resolution and more recently by FREY et al. [6.175,176] and PEATMAN et al. [6.177], using highly dispersed synchrotron radiation. Thus the accurate determination of photoionization thresholds, vibrational levels and recently, rotational levels [6.178] of the corresponding molecular ions have been reported for a number of molecules.

As an example we show in panel C of Fig. 6.34 the results for molecular N_2 as obtained by PEATMAN et al. [6.177]. It is important to note that since the threshold electron analyzer is a passive device registering the presence of zero kinetic energy electrons as the photon energy is scanned, no corrections of the transmission function of the analyzer are necessary. From the data shown in Fig. 6.34 the relative transition probabilities at threshold have been determined. From a comparison of these results with data obtained by absorption spectroscopy, conventional photoelectron spectroscopy [6.9] and SF_6-electron trapping experiments [6.179] it was concluded, that the interaction of ionizing Rydberg states with ionic states plays a significant role on the intensities observed at the photoionization threshold. In their study PEATMAN et al. [6.177] found that although the threshold spectra observed differ extensively in peak form, peak intensity and in the length of the vibrational progressions from those in photoelectron spectroscopy. The relative integrated intensities over the entire $X^2\Sigma_g^+$, $A^2\Pi_u$ and $B^2\Sigma_u^+$ state, respectively, are comparable to the non-threshold results.

Recently SCHLAG and collaborators [6.180-182] have extended the PIR spectroscopy by performing threshold electron-photon coincidence measurements. This method permits e.g. a direct measurement of the radiative lifetimes of individual vibronic states, since any measurements carried out in coincidence with the resonance zero kinetic energy electrons only pertain to the state prepared by the primary photons. Results obtained by this coincidence technique have so far been reported for individual vibronic states of the A-state and the B-state of CO_2^+ [6.180] and of the two different vibrational states of the $A^2\Sigma^+$ state of N_2O^+ [6.181,182].

6.4 Fluorescence

6.4.1 Fluorescence- and Excitation-spectra

The absorption of a VUV photon can lead to photoionization, photoionization excitation, photodissociation and dissociative excitation. For a simple molecule like N_2 these processes are:

$$N_2 + h\nu \rightarrow N_2^+ + e^- \qquad \text{photoionization}$$

$$N_2 + h\nu \rightarrow N_2^{+*} + e^- \qquad \text{photoionization excitation}$$

$$N_2 + h\nu \rightarrow N^* + N \qquad \text{dissociative excitation}$$

$$N_2 + h\nu \rightarrow N^+ + N^* + e^- \qquad \text{dissociative ionization and excitation.}$$

In general, absorption and photoemission measurements are insufficient to completely characterize the final state of the absorption process. Much information on the final state is contained in the fluorescence radiation emitted by the excited molecules, molecular ions and dissociation fragments. The analysis of the excitation- and fluorescence-spectra can contribute considerably to our knowledge on the geometry and the potential curves of the excited states. Fluorescence measurements are of immense importance for answering the fundamental question "how is the excitation energy distributed". This problem is intimately related to the investigation of decay and energy transfer processes. The relevance of these processes for, e.g., laser physics [6.183] and the physics of stellar atmospheres is obvious [6.12,184]. In interpreting fluorescence data one always should keep in mind that the emission of a fluorescence photon in most cases is one step in a sequence of decay processes. Often there are competing decay channels which make the determination of absolute production cross sections for well defined final states very difficult. Energy transfer processes e.g. can result in a complete quenching of the fluorescence.

The advantages synchrotron radiation offers for fluorescence measurements have only been exploited within the last years. CARLSON and coworkers [6.185-191] have studied the fluorescence from excited states of molecules, molecular ions and photodissociation fragments produced by the interaction of CO, CO_2, N_2O, N_2, O_2 molecules with synchrotron radiation in the photon energy range from 15 eV to 70 eV. The samples were contained in a gas cell attached to the exit slit of a VUV monochromator providing a band width of ~ 1 Å for the exciting radiation. A solar blind photomultiplier was used to detect the fluorescence light emitted perpendicular to the polarization vector of the exciting light and the direction of the incident photon beam. The fluorescence was viewed through both, gaseous (O_2), and solid (LiF, CaF_2, BaF_2, Al_2O_3, SiO_2) filters. In order to exclude secondary excitation processes by energetic photoelectrons the pressure dependence of the fluorescence radiation was carefully studied. Only pressures ($p < 10^{-2}$ Torr) low enough to guarantee a linear dependence of the fluorescence yield on the pressure were used for the determination of relative fluorescence production cross sections. Absolute fluorescence cross sections were obtained by normalizing to known cross sections [6.192]. The cross sections for the production of N_2 fluorescence [6.186] in the spectral range of various photomultiplier filter combinations as a function of the incident photon energy are presented in Figs. 6.35 and 36. The dominant components of the fluorescence are the NI 1200 Å, 1493 Å and 1743 Å lines which originate from the $N(2p^2 3s\ ^4P) \rightarrow N(2p^3\ ^4S^0)$;

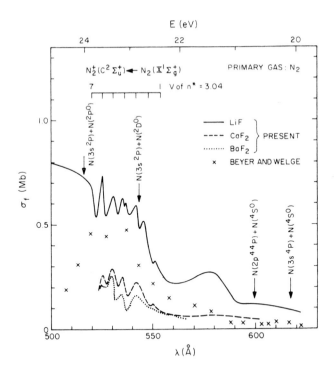

Fig. 6.35 Cross sections σ_f (Mb) for the production of fluorescence in the spectral regions of the photomultiplier itself (1050-1800 Å), and in the CaF$_2$ (1240-1800 Å) and BaF$_2$ (1350-1800 Å) filtered regions, as a function of the incident photon wavelength for N$_2$. The onsets of the processes that are possible responsible for the fluorescence are shown. The positions of the Rydberg absorption series converging to the N$^+$(C $^2\Sigma_u{}^+$) state are indicated. The data given by BEYER and WELGE [6.248] for the production cross sections of the NI 1200 Å line are also shown [6.186]

N($2p^2 3s$ ^2P) → N($2p^3$ $^2D^0$) and N($2p^2 3s$ ^2P) → N($2p^3$ $^2P^0$) transitions. The onsets of processes possibly contributing to the fluorescence in the energy range under consideration can be determined from spectroscopic data. Those considered most important are included in Figs. 6.35 and 36. Discrete structures show up for excitation energie between 22 eV and 24 eV. These structures are related to the Rydberg series converging to the N$_2{}^+$(C $^2\Sigma_u{}^+$) state. The presence of these structures in the excitation spectrum is due to the interaction of these Rydberg states with the dissociation continuum.

The interaction of Rydberg states with the ionization continuum gives rise to the structures found in the excitation spectrum of CO (Fig.6.37) [6.189]. The fluorescen between 3000 Å and 5200 Å is essentially the CO$^+$(A$^2\Pi$ → X$^2\Sigma^+$) emission system. At the onset the fluorescence shows a step like increase at the thresholds of the vibrational levels of the CO$^+$ (A $^2\Pi$) state. The step increase of 2.1; 1.9 and 0.7 Mb at the

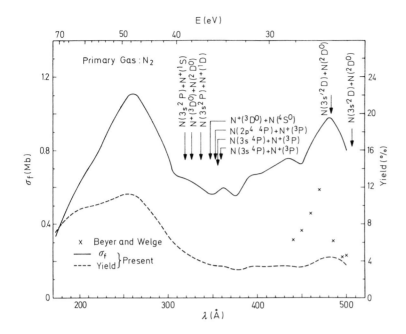

Fig. 6.36 Cross sections σ_f(Mb) for the production of fluorescence in the wavelength region 1050-1800 Å, as a function of the incident photon wavelength from 175-500 Å for N_2. The onsets of the processes that are possibly responsible for the fluorescence are indicated. The data given by BEYER and WELGE [6.248] for the production cross sections of the NI 1200 Å line are also shown [6.186]

v' = 0,1 and 2 thresholds are consistent with the Franck-Condon factors of 0.215; 0.176 and 0.079 [6.193]. For photon energies higher than the v = 2 threshold, the photoionization continuum is perturbed by the III-V, sharp and diffuse Rydberg series converging to the v' = 0 and 1 vibrational levels of the CO^+ (A $^2\Pi$) state.

In addition to using filters, a second monochromator to analyze the fluorescence of NO, N_2O and CO_2 was also employed [6.194-201]. The N_2O^+ fluorescence around 3500 Å excited by 17.3 eV photons is presented in Fig. 6.38. There are several well separated vibrational bands. The dominant process is

$$N_2O[\tilde{X}\ ^1\Sigma^+\ (0,0,0)] + h\nu_p \rightarrow N_2O^+[\tilde{A}\ ^2\Sigma^+\ (0,0,0)]$$

$$+ e^- \rightarrow N_2O^+[\tilde{X}\ ^2\Pi_i(0,0,0)] + h\nu_s + e^-.$$

The excitation spectrum for the (0,0,0) → (0,0,0) vibrational band is given in Fig. 6.39. There is a steep increase at threshold followed by prominent structures above

330

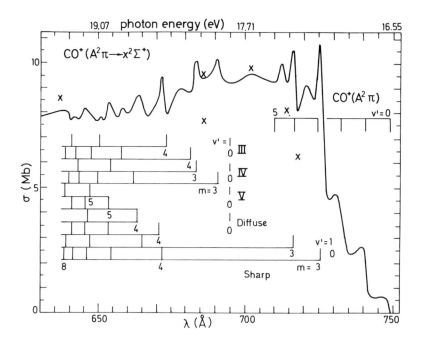

Fig. 6.37 The cross sections for the production of the CO^+ (A $^2\Pi \rightarrow$ X $^2\Pi^+$) fluorescence at the primary photon wavelengths between 630 and 750Å. The positions of the III - V, diffuse, and sharp Rydberg series converging to the CO^+ (A $^2\Pi$) state given by OGAWA and OGAWA [6.249] are shown. The number, m, is originally defined by OGAWA and OGAWA [6.249] as $\mu = m - n*$, where n* is the effective quantum number and μ is a minimum positive value. The cross sections given by JUDGE and LEE [6.250] are used to normalize the present data and are also indicated in the figure by "X" [6.189]

17 eV. These structures are due to the absorption Rydberg series (RVI_{abs}, $RVII_{abs}$) and the apparent emission Rydberg series (RIX_{ae}) converging to the $N_2O^+(\tilde{C}$ $^2\Sigma^+$) state. The interference of these Rydberg states with the continuum of the $\tilde{A}$ $^2\Sigma^+$ state results in the window type character of the RIX_{ae} series. The influence of the $RVIII_{ae}$ apparent emission series in the formation of $N_2O^+[\tilde{A}$ $^2\Sigma^+(0,0,0)]$ seems to be very small.

6.4.2 Time resolved fluorescence spectroscopy

The time dependence of the fluorescence also serves as a fingerprint of the excited states. Furthermore fluorescence decay-time measurements provide accurate data on rate constants both of radiative and nonradiative deactivation processes. Radiationless transitions due to unimolecular processes (e.g. predissociation, auto-ionization) or to the interaction of the excited molecule with the environment (e.g. vibrational-relaxation, photo-chemical reactions) play an important part in the deactivation and energy transfer processes and therefore have attracted much interest in recent years [6.202]. The short duration (FWHM: 0.2-2 ns) and the extreme stabilit

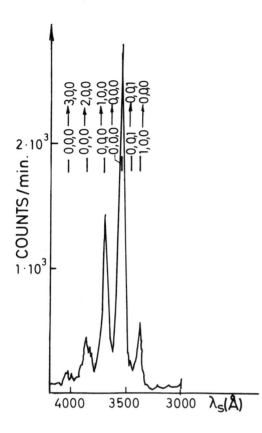

Fig. 6.38 $N_2O^+(\tilde{A}\ ^2\Sigma^+ \rightarrow X\ ^2\pi i)$ fluorescence spectrum excited by 17.3 eV photons. The vibrational transitions are indicated [6.194]

of the repetition frequency (1-100 MHz) of the light pulses emitted by electron storage rings open the field of subnanosecond spectroscopy in the VUV and X-ray region. The pilot experiments [6.203-207] performed up to now have given a flavor of the wealth of information that can be obtained by this technique.

Fig. 6.40 shows an experimental decay curve I(t) of the fluorescence of the first singlet exciton of tetracene single crystals centered at 5000 Å [6.205]. The pulse shape $I_0(t)$ of the exciting synchrotron radiation pulse is also given. I(t) is the convolution of the fluorescence decay function f(t) by the illumination function $I_0(t)$. The best fit to the data found by LOPEZ-DELGADO et al. [6.205] is based on a fluorescence decay function of the form

$$f(t) = \frac{1}{\tau_1}\ e^{-t/\tau_1} + \frac{m}{\tau_2}\ e^{-t/\tau_2} \tag{6.7}$$

where

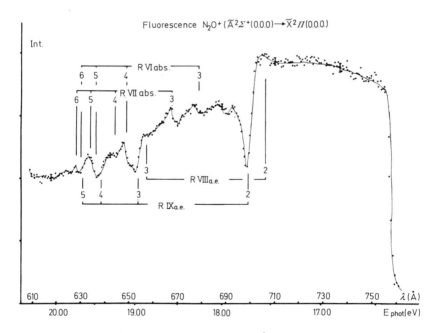

<u>Fig. 6.39</u> Fluorescence intensity at 3550 Å against incident photon energy for $N_2O^+[\tilde{A}\ ^2\Sigma^+\ (0,0,0) \rightarrow \tilde{X}\ ^2_\Pi\ (0,0,0)]$ [6.201]

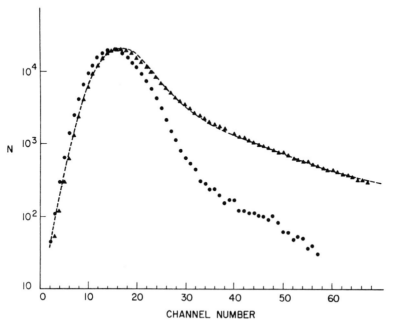

<u>Fig. 6.40</u> Tetracene single crystal fluorescence decay; λ_{exc} = 4080 Å, λ_{obs} = 5000 Å number of counts (N) per channel (56ps/chan.). ● Spear SR pulse ($I_0(t)$) at λ = 5000 Å; ▲experimental decay; -- calculated decay: $I(t) = I_0(t) \cdot f(t)$ [6.203]

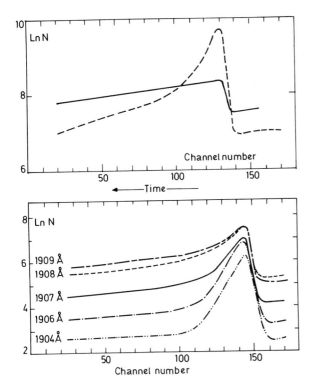

Fig. 6.41a Typical decay curve from the A (solid line) and the C^0 (dashed line) level. NO pressure ~ 0.2 torr; excitation bandwidth $\simeq 1.5$ Å (100 ns/200 chan.) [6.203]

Fig. 6.41b C^0 level decay shape as a function of the excitation wavelength. NO pressure ~ 0.2 torr; excitation bandwidth $\simeq 1.5$ Å (100 ns/200 chan.) [6.203]

$$\tau_1 = 0.200 \pm 0.025 \text{ ns}; \quad \tau_2 = 1.70 \pm 0.20 \text{ ns}; \quad m = 0.1$$

In contradiction to the results obtained by excitation with powerful lasers [6.208], LOPEZ-DELGADO et al. did not find a slower decay channel. They attributed this discrepancy to excitation density effects.

When the exciting pulse repetition rate $(1/T)$ and the fluorescence decay rate $(1/\tau)$ are of the same order of magnitude, the fluorescence intensity $I(t)$ becomes a periodic function in time

$$I(t) \approx \sum_{n=0}^{\infty} I_0 e^{-(t+nT)/\tau} \quad . \tag{6.8}$$

This holds as long as the amplitude I_0 and the period T are constant, which is a reasonable assumption. In Fig. 6.41a typical decay curve of fluorescence from the A $^2\Sigma^+(v = 0)$ level of NO, reported by BENOIST D'AZY et al. [6.203], is shown. The lifetime is directly given by the slope of the straight part of the decay curve. All vibrational levels belonging to the A $^2\Sigma^+$ state of NO decay exponentially. In contrast to this the decay curves of fluorescence from the vibrational levels of the C $^2\Pi$ states of NO are clearly non-exponential (Fig.6.41a). There are always a short-lived and a long-lived component. They attributed the long-lived component to the A $^2\Sigma^+ \to$ X $^2\Sigma^+$ fluorescence. The A $^2\Sigma^+$ state being populated from the C $^2\Pi$ state via infrared emission.

A detailed analysis of the C$^2\Pi$(v = 0) level decay curve shows that there are three components: (i) a long-lived component corresponding to the C $^2\Pi \to$ A$^2\Sigma^+ \to$ X $^2\Sigma$ cascade process, (ii) an intermediate component, collision free value of $\tau \sim 20$ nsec and (iii) a short-lived component, $\tau \sim 3$ ns.

Narrow band excitation across the rotational envelope of the C $^2\Pi$(v = 0) level reveals a drastic change of the decay curves when changing the excitation energy (Fig.6.41b). According to BENOIST D'AZY et al. this is not due to a wavelengths dependence of the rate constants but to a strong wavelengths dependence of the relative intensities of the components. Decreasing the excitation wavelength from 1909 Å to 1904 Å the intensity of the long-lived component goes to zero. For excitation wavelengths $\lambda < 1906$ Å, where the long-lived component is practically absent, high rotational levels (I > 7) are populated. Thus it was concluded [6.203] that a weak predissociation of all or most of the rotational levels above the dissociation limit, situated at I $\approx$ 5, is responsible for the disappearance of the long-lived component.

Before leaving this subject we want to mention the important part molecular fluorescence plays in deactivation processes of excited states in solids, and atomic gases at higher pressures (e.g. rare gases, solid rare gases [6.209-211]). Many interesting results recently obtained with SR have been reported in these areas which we have hardly touched upon.

6.5 Mass spectrometry

The identification of the charged fragments produced is one of the principal means to characterize the final state and to gain insight in the interaction between photons and molecules. Mass spectra of positive and negative ions obtained with variable energy of the exciting photon contain information on photoionization, photofragmentation, ion pair formation autoionization and predissociation processes. From these spectra appearance potentials, heats of formation, ionization- and bond-dissociation energies can be determined. Like photoabsorption, photoelectron and fluorescence spectroscopy mass spectrometry helps to establish the multidimensional potential energy diagram which enables an exact description of molecular processes. Furthermore these measurements form critical tests for theoretical models like the quasi-equilibrium theory [6.212].

Using a quadrupole mass spectrometer BAUMGÄRTEL and coworkers [6.174,213-220] determined the photoion spectra of small organic molecules for photon energies between 9 eV and 27 eV. In Fig. 6.42 the relative photoion yield Y_i of propylene is presented as an example

$$Y_i = \frac{\text{number of ions produced}}{\text{number of transmitted photons}}$$

$$= \frac{\sigma_i \; (1-e^{-\sigma_t n \cdot \ell})}{\sigma_t \cdot e^{-\sigma_t n \ell}}$$

(6.9)

where σ_i is the ion production cross section, σ_t the total absorption cross section, and $n\ell$ the molecular density times length of the ionization cell.

For $\sigma_t n\ell \ll 1$ the realation $Y_i \approx \sigma_i$ holds. The ionization potentials obtained by photoelectron spectroscopy $IP_1 ... IP_6$ are included in Fig. 6.42 [6.215].

The first ionization potential IP_1 is in good agreement with the appearance potential $AP(C_3H_6^+)$ of the $C_3H_6^+$ ion. There is also close correspondence between IP_2 and $AP(C_3H_5^+)$, $AP(C_3H_4^+)$ and between IP_3 and $AP(C_3H_3^+)$, $AP(C_2H_3^+)$. The underlying processes are

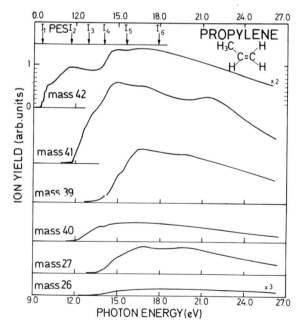

Fig. 6.42 Photoion spectra of propylene. The ionization potentials are marked by arrows [6.215]

$$C_3H_6 + h\nu \rightarrow C_3H_6^+ + e^- \qquad AP(C_3H_6^+) = 9.37 \text{ eV}$$
$$C_3H_5^+ + H + e^- \qquad AP(C_3H_5^+) = 11.88 \text{ eV}$$
$$C_3H_4^+ + H_2 + e^- \qquad AP(C_3H_4^+) = 11.91 \text{ eV}$$
$$C_3H_3^+ + H_2 + H + e^- \qquad AP(C_3H_3^+) = 13.19 \text{ eV}$$
$$C_2H_3^+ + CH_3 + e^- \qquad AP(C_2H_3^+) = 13.20 \text{ eV}$$
$$C_2H_2^+ + CH_4 + e^- \qquad AP(C_2H_2^+) = 12.92 \text{ eV}$$

For the isomeric cyclopropane there are additional processes contributing at threshold, namely

$$C_3H_6 + h\nu \rightarrow C_3H_6^+ + e^- \qquad AP(C_3H_6^+) = 9.91 \text{ eV}$$
$$C_3H_5^+ + H + e^- \qquad AP(C_3H_5^+) = 11.44 \text{ eV}$$
$$C_3H_5^+ + H^- \qquad AP(C_3H_5^+) = 10.74 \text{ eV}$$
$$C_3H_4^+ + H_2 + e^- \qquad AP(C_3H_4^+) = 11.64 \text{ eV}$$
$$C_3H_3^+ + H_2 + H + e^- \qquad AP(C_3H_3^+) = 12.86 \text{ eV}$$
$$C_3H_3^+ + H_2 + H^- \qquad AP(C_3H_3^+) = 12.1 \text{ eV}$$
$$C_2H_3^+ + CH_3 + e^- \qquad AP(C_2H_3^+) = 12.64 \text{ eV}$$
$$C_2H_2^+ + CH_4 + e^- \qquad AP(C_2H_2^+) = 12.71 \text{ eV}$$

For the photoreaction $AB + h\nu \rightarrow A^+ + B + e^-$ the following relation holds

$$AP(A^+) = \Delta H_f(A^+) + \Delta H_f(B) - \Delta H_f(AB) + E = D(A - B) + IP(A) + E$$

where ΔH_f denotes the standard heat of formation; E the excess energy; $D(A - B)$ the dissociation energy of bond A - B, and IP(A) the ionization energy of A.

The main problem in the evaluation of this equation arises from the excess energy E the determination of which requires the measurement of the internal and the kinetic energies of the outgoing fragments. Since for many fragmentation processes the excess energy is known to be negligible, BAUMGÄRTEL and coworkers [6.174,213-220] based their evaluation on the assumption E = 0. With the help of standard heats of formation given in the literature they were thus able to calculate heats of formation, bond dissociation and ionization energies. In many cases it was possible to deduce the fragmentation mechanism and the structure of the fragments from these energetic considerations. The photoion spectrum of $C_3H_6^+$ from propylen shows a well resolved vibrational structure at threshold, whereas there is no structure in the corresponding spectrum of cyclopropane (Fig.6.43). According to these investigators this might be due to the opening of the ring structure.

Valence and Rydberg transitions are responsible for the photoabsorption below the first ionization potential. The lowest ionization potentials can be determined from photoion spectra. The overlap of different ionization continua and the structures due to resonant transitions, e.g. Rydberg transitions, render the determination of higher ionization potentials very difficult. Photoelectron spectroscopy, especially resonance photoelectron spectroscopy with a continuous light source offers the best

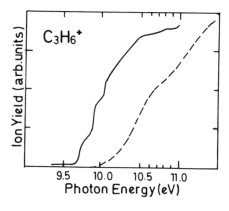

Fig. 6.43 $C_3H_6^+$ threshold in the photo-ion spectra of propylene ⸻ and cyclo-propane --- [6.215]

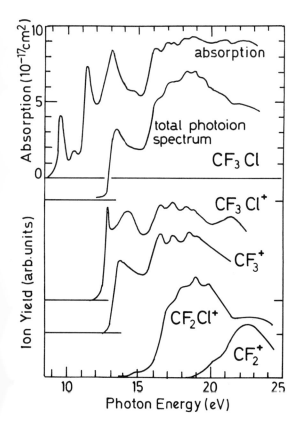

Fig. 6.44 Absorption spectrum and total and mass selected photoion spectra of CF₃Cl [6.174]

remedy to this problem, as has been discussed in Sect. 6.3. The coupling of excited molecular states with different dissociation channels manifests itself in the mass selected photoion spectra. Towards higher energies the total photoion spectra become very similar to the photoabsorption spectra since almost every absorption process results in ionization. This is borne out by the similarity of the photoabsorption and total photoion spectrum of CF_3Cl above the first IP as shown in Fig. 6.44, [6.174, 219]. Part of the weak structures superimposed on the continuum persist in the mass selected photoion spectra. The appearance potentials and the underlying reactions are

Reaction	AP	[eV]
$CF_3Cl + h\nu \rightarrow CF_3Cl^+ + e^-$	$AP(CF_3Cl^+)$	12.45
$CF_3^+ + Cl + e^-$	$AP(CF_3^+)$	12.55
$CF_2Cl^+ + F + e^-$	$AP(CF_2Cl^+)$	14.25
$CF_2^+ + F + Cl + e^-$	$AP(CF_2^+)$	18.85

The large relative intensity of CF_3^+ implies that the most probable fragmentation process involves the cleavage of the C-Cl bond. The instability of the molecular ion CF_3Cl^+ is documented by the low intensity. JOCHIMS et al. [6.174,219] conclude that the CF_3^+ ion is produced by predissociation of molecular ions. The onset regions of both the CF_2Cl^+ and CF_2^+ spectrum are typical for direct fragmentation processes. New investigations performed on the smaller fragments CF^+, Cl^+, Cl^- and F^- helped to elucidate further dissociation processes [6.254a,b]. The new results emphasize the importance of ion pair formation for the understanding of dynamical aspects of molecular fragmentation.

6.6 Acknowledgements

We would like to thank our colleagues from the synchrotron radiation group at DESY for their stimulating interest and support. We are especially indebted to all colleagues from other SR Laboratories and from the DESY group who generously made available to us results of experiments and calculations prior to publication. Valuable comments by W.H.E. Schwarz are gratefully acknowledged. Special thanks go to Erika Thumann for her patience and skill in carefully typing the manuscript and to V. Fischer, W. Knaut, J. Schmidt and M. Sörensen for preparing the figures.

6.7 Appendix

In the following table some information is collected for various molecules obtained in experiments exploiting synchrotron radiation. Most abbreviations are obvious; PED, photoelectron energy distribution

Molecule	Energy Range[eV]	Method	Remarks	Reference
H_2	16-68	absorption	photoelectric, absolute cross section	[6.226]
	14.5-16.5 16.5	fluorescence from fragments	H Lyα excitation spectrum, absorption spectrum	[6.221]
	14.5-18.3	fluorescence from fragments	H Balmer, α, β and γ excitation spectra	[6.222]
	14.4-18.3	fluorescence from fragments	H Lyα, Hα, Hβ, Hγ excitation spectra, absorption spectra predissociation, preionization	[6.223]
	14.6- 15.40	absorption and fluorescence	observation of dissociation products, detection of the predissociation of the $D'^1\Pi_u^+, v' > 1$ state	[6.225]
	14.70- 14.80	fluorescence from fragments	H Lyα, excitation spectra with $\Delta\lambda = 0.12$ Å, branching ratio $\sigma(2s)/[\sigma(2s) + \sigma(2p)]$	[6.224]
D_2	16-18	absorption	photoelectric, absolute cross section	[6.226]
	14.5-18.3	fluorescence from fragments	D Balmer α, β and γ excitation spectra	[6.222]
	14.4-18.3	fluorescence from fragments	D Lyα, Hα, Hβ, Hγ excitation spectra, preionization, pre-dissociation	[6.223]
N_2	20-120	absorption	photographic, $\Delta\lambda = 0.06$ Å, Rydberg series to $C^2\Sigma_u^+$ at 23.6 eV	[6.26]
	18-35	absorption photoioniza-tion	photoelectric, absolute cross section, line width	[6.227]
	18-68	absorption	photoelectric, absolute cross section	[6.25]
	10-35	absorption	photoelectric, $\Delta\lambda = 0.03$ Å, absolute cross section, Rydberg-valence interaction, line shape analysis	[6.36]
	~ 400	absorption	photographic, $\Delta\lambda = 0.03$ Å, fine structure at the K-edge	[6.51]
	~ 400	absorption	photographic, $\Delta\lambda = 0.1$ eV, line width	[6.58]
	390-450	absorption	photoelectric, absolute cross section f-type shape resonance 9.2 eV above 1s-ionization potential	[6.57]
	14-50	photoemission	energy dependence of PED's, branching ratios, partial cross sections	[6.40]
	10-30	reflection, solid	comparison to gas phase, de-perturbed vibrational progression	[6.39]

Molecule	Energy Range[eV]	Method	Remarks	Reference
N_2	10-30	photoemission	PED's, comparison to PED's gas phase	[6.228]
	excitation 30-70	fluorescence from fragments in the range 6.8-11.8	relative fluorescence intensity, excitation mechanisms	[6.185] [6.186b]
	excitation 16-70	fluorescence from fragments in the range 6.8-11.8	production cross sections for fluorescence, excitation mechanisms	[6.186]
O_2	18-68	absorption	photoelectric, absolute cross sections	[6.25]
	18-35	absorption	photoelectric, absolute cross sections, photoionization efficiencies	[6.227]
	20-25	absorption	photographic, $\Delta\lambda = 0.06$ Å Rydberg series	[6.27]
	14.6-15.5	photoemission	$h\nu$-dependence of the PED, intensity of vibrational progressions	[6.230]
	530-550	absorption	photographic, structure at O K-edge	[6.86]
	480-680	absorption	photographic recording	[6.53]
	10-22	reflection solid	comparison to gas phase	[6.39]
	10-30	photoemission solid	PED's comparison to PED's gas phase	[6.228]
	excitation 16-70	fluorescence from fragments in the range 6.8-11.8	production cross sections for fluorescence	[6.186]
	20.5-21.7	photoionization electron ion ionization time of flight	total photoionization cross section, partial cross section O^+, O_2^+	[6.231]
CO	18-35	absorption	photoelectric, cross sections photoionization efficiencies, fine structure	[6.227] [6.256] [6.257]
	18-68	absorption	photoelectric, absolute cross section	[6.25]
	281-298, 525-545	absorption	photographic, C K-edge, O K-edge	[6.86]
	14-50	photoemission	energy dependence of PED's, branching ratios, partial cross sections	[6.40]
	10-30	photoemission, solid	PED's, comparison to PED's gas phase	[6.232]

Molecule	Energy Range[eV]	Method	Remarks	Reference
CO	excitation 15.5-73	fluorescence from fragments in the range 6.8-11.8	fluorescence production cross section excitation mechanisms	[6.187]
	excitation 19.5-71	fluorescence	cross section for $CO^+(B^2\Sigma^+ \to X^2\Sigma^+)$ first negative bands	[6.188]
	15.5-71	fluorescence	cross section for $CO^+(A^2\Pi \to X^2\Sigma^+)$ fluorescence excitation $h\nu < 19.6$ eV	[6.189]
NO	18-68	absorption	photoelectric, absolute cross sections	[6.25]
	18-25	absorption	photoelectric, Rydberg series to $C^3\Pi$ state at 21.72 eV	[6.234]
	~ 10	absorption	photographic, Rydberg series to $C^3\Pi$ state at 21.72 eV	[6.235]
	395-415	absorption	photoelectric, $\Delta\lambda \approx 0.05$ Å, N K-edge	[6.55,54]
	excitation 15.5-21	fluorescence of fragments in the range 8.8-11.8	$NO + h\nu \to O^*(2p^33s\,^3S^0) + N(^4S^0)$ $NO + h\nu \to N^*(2p^22s\,^4P) + O(^3P)$ $NO + h\nu \to N^*(2s2s^4\,^4P) + O(^3P)$	[6.19b,197]
	excitation 19-22.5	fluorescence of fragments in the range 9.2-11.1, mass spectrometry	relative photoionization yield of NO^+, O^+, N^+	[6.198, 199]
	$h\nu < 7.5$ excitation	time resolved fluorescence in the range $h\nu < 6.5$	$A^2\Sigma^+(v = 0,1,2,3)$, $B^2\Pi(v = 5)$, $C^2\Pi(v = 0,1)$, $D^2\Sigma^+(v = 0,1,2)$, collision free lifetimes, self-quenching rate, narrow band excitation of $C^2\Pi(v = 0)$, J-dependence of lifetime	[6.203]
CO_2	16-68	absorption	photoelectric, absolute cross section	[6.25]
	10-30	reflection, solid	ε_2 spectra, comparison to gas phase	[6.233]
	10-30	photoemission solid	PED's, comparison to PED's gas phase	[6.232]
	18.1-73	fluorescence	production cross section for $CO_2^+(\tilde{B}^2\Sigma_u^+ \to \tilde{X}^2\Pi_g)$	[6.190]
	excitation 15.5-73	fluorescence from fragments in the range 6.8-11.8	production cross section, excitation mechanisms	[6.187]
	15.5-71	fluorescence	cross section for $CO_2^+(A^2\Pi_u \to X^2\Pi_g)$, excitation mechanisms	[6.189]

Molecule	Energy Range[eV]	Method	Remarks	Reference
CO_2	excitation 12.4-31	fluorescence in the range 3.1-11.8	$\tilde{B}^2\Sigma^+_{u2} \rightarrow \tilde{X}^2\Pi_g 3/2,1/2\ CO_2^+$ $A^2\Pi_u \rightarrow X^2\Pi_g$ CO_2^+ comparison with CO_2 of Martian atmosphere	[6.195] [6.194]
	25-50	photoionization, time of flight, electron-ion coincidence	total photoionization cross section	[6.231]
	13.6-13.9	photoionization	photon ion yield of CO_2^+	[6.224]
	12-20	photoionization resonance spectroscopy	ionization potentials, vibrational and spin orbit structure of $\tilde{X}$, $\tilde{A}$, $\tilde{B}$, $\tilde{C}$ states of CO_2^+; coincidence measurements between threshold electrons and ions	[6.175]
	17-18	threshold electron-photon coincidence	lifetimes and fluorescence quantum yields of selected vibronic states of the $\tilde{A}$ and $\tilde{B}$ state of CO_2^+	[6.180]
CS_2	10-20	photoionization resonance spectroscopy	ionization potentials for the $\tilde{X}$, $\tilde{A}$, $\tilde{B}$ and $\tilde{C}$ state of CS_2^+, spin orbit splittings, vibrational frequencies; Fermi resonances in the X and A state	[6.176]
COS	10-20	photoionization resonance spectroscopy	ionization potentials for the $\tilde{X}$, $\tilde{A}$, $\tilde{B}$ and $\tilde{C}$ state of COS^+, spin orbit splittings, vibrational frequencies	[6.176]
N_2O	18-68	absorption	photoelectric, absolute cross section	[6.25]
	390-450	absorption	photoelectric, absolute cross section	[6.57]
	excitation 16.5-73	fluorescence of fragments in the range 6.8-11.8	photoproduction cross section	[6.191] [6.188]
	excitation 14-31	fluorescence in the range 3-11.8	$N_2O + h\nu_p \rightarrow N_2O^{+*} + e^- \rightarrow N_2O^+ + h\nu_s + e^-$ $N_2O + h\nu_p \rightarrow NO + N^* \rightarrow NO + N + h\nu_s$ $N_2O^+(\tilde{A}^2\Sigma^+ \rightarrow \tilde{X}^2\Pi_i)$ emission, excitation spectrum of $N_2O^+[\tilde{A}^2\Sigma^+(0,0,0) \rightarrow \tilde{X}^2\Pi_i(0,0,0)]$	[6.200] [6.197] [6.199]
	excitation energy 16.3-20.4	fluorescence	relative cross section of $N_2O[\tilde{X}^1\Sigma^+(0,0,0)] + h\nu \rightarrow N_2O^+[\tilde{A}^2\Sigma^+(0,0,0)] + e^-$ via fluorescence of $N_2O^+[\tilde{A}^2\Sigma^+(0,0,0) \rightarrow \tilde{X}^2\Pi(0,0,0)]$	[6.197, 201]

Molecule	Energy Range[eV]	Method	Remarks	Reference
NO_2	400-412 530-541	absorption	photographic, structure in the N 1s and O 1s region, low lying double excitations	[6.56]
H_2O D_2O	10-20	absorption	photoelectric $\Delta\lambda = 0.03$ Å, absolute cross section, vibrational and rotational analysis, Rydberg transitions	[6.46]
	11-25	absorption	photographic, $\Delta\lambda = 0.3$ Å	[6.50]
	16-68	absorption	photoelectric, absolute cross section	[6.236]
H_2S	160-175	absorption	photoelectric, absolute cross section, Rydberg series converging at 171.1, 172.2 eV	[6.90]
HCl	200-210	absorption	Cl 2p excitation, spin-orbit split series converging at 207.1 and 208.7 eV	[6.90]
HBr	13-16.5	fluorescence	$HBr + h\nu \rightarrow HBr^+ + e$	[6.199]
	10-28	photoionization fluorescence (2-11.8 eV)	excitation measurements, decay processes	[6.237]
Se_2	50-150	absorption	photographic Se 3d excitation, influence of intraatomic interaction	[6.63]
Te_2	37-130	absorption	photographic, Te 4d excitation	[6.65]
I_2	45-160	absorption	absolute cross section, I 4d excitation, comparison solid-gas	[6.67]
Li_2	56-63	absorption	photographic, pressure variation, partly resolved vibrational structure theoretical analysis, CI calculations	[6.239]
LiF	55-70	absorption	photographic, Li 1s excitation, comparison to crystalline LiF, ionic model, Z + 1 core analogy model, analogous ionic core virtual orbital model	[6.72] [6.73]
LiCl (Li_2Cl_2)	55-70	absorption	photographic, Li 1s excitation, comparison to crystalline LiF, ionic model, Z + 1 core analogy model, analogous ionic core virtual orbital model	[6.72] [6.73]
NaCl	30-70	absorption	photographic Na 2s, 2p excitation ionic model, comparison to crystalline NaCl, resonances above threshold	[6.73] [6.238]
CsF	12-23	absorption	photographic Cs 5p excitation	[6.77]
	70-180	absorption	photographic Cs 4d excitation, comparison to crystalline CsF	[6.77] [6.78]

Molecule	Energy Range[eV]	Method	Remarks	Reference
CsCl	12-21	absorption	photographic Cs 5p excitation	[6.75]
	70-180	absorption	photographic Cs 4p,4d excitation, comparison to crystalline CsCl	[6.73] [6.79]
CsBr	12-21	absorption	photographic Cs 5p	[6.77]
	70-180	absorption	photographic Cs 4d excitation, comparison with crystalline CsBr	[6.77] [6.78]
CsI	12-21	absorption	photographic Cs 5p excitation	[6.77]
	160-175	absorption	photographic Cs 4p excitation, ionic model	[6.73]
XeF_2	6-35	absorption	valence and Rydberg transitions, vibrational structure	[6.82] [6.80]
	50-160	absorption	absolute cross section Xe 4d,4p excitation, Ligand field splitting of core level, comparison gas-solid	[6.81] [6.82]
XeF_4	6-35	absorption	valence and Rydberg transitions, vibrational structure	[6.80] [6.82]
	50-160	absorption	absolute cross sections, Xe 4d, 4p excitation, Ligand field splitting of core level, dynamic Jahn-Teller-effect, comparison gas-solid	[6.81] [6.82]
XeF_6	6-36	absorption	valence and Rydberg transitions	[6.80,82]
	50-170	absorption	Xe 4d,4p excitation nearly 0_h geometry	[6.83]
BF_3	192-212	absorption	B K-excitation	[6.240]
BCl_3	192-212	absorption	B K-excitation	[6.240]
PH_3	120-200	absorption	absolute cross sections, P 2p, 2s excitation, Rydberg series converging at 137.3 eV, 138.2 eV	[6.90]
	130-140	absorption	absolute cross sections, P 2p, 2s excitation, comparison molecular and solid spectra	[6.91]
SiH_4	100-200	absorption	absolute absorption cross sections, Si 2p,2s excitation, Rydberg series converging at 107.2, 107.8 eV	[6.93] [6.90]
	100-170	absorption	absolute cross sections, Si 2s, 2p excitation, comparison molecular and solid spectra	[6.91]
SiF_4	100-120	absorption	absolute cross sections, Si 2p excitation, Rydberg series converging at 110.8, 111.4 eV	[6.90]
	100-170	absorption	absolute cross sections, Si 2p, 2s excitation, comparison molecular and solid spectra	[6.91]

Molecule	Energy Range[eV]	Method	Remarks	Reference
GeH$_4$	120-130	absorption	Ge 3p excitation, two lines at 124.7 eV and 129.7 eV	[6.90]
SF$_6$	19-24	absorption	photographic $\Delta\lambda$ = 0.06 Å vibrational progression between 20.6 and 22.5 eV Rydberg series converging at 26.83 eV	[6.241]
	15-41	absorption	absolute cross section, ionization efficiencies	[6.227]
	10-40	absorption	photographic, $\Delta\lambda$ = 0.3 Å, absolute cross section, detailed analysis	[6.242]
	10-30	absorption (reflection solid)	absolute cross section, comparison gas-solid	[6.87]
	16-70	absorption	absolute cross section, $\Delta\lambda$ = 1 Å	[6.110]
	150-250	absorption	S 2p,2s excitation, comparison gas-solid, inner-well resonances	[6.87]
	170-250	absorption	photographic excitation, weak S 2p Rydberg series between 176 and 182 eV	[6.86]
CH$_4$ methane	8-35	absorption	absolute cross sections, comparison gas-solid	[6.109]
	18-68	absorption	absolute cross sections	[6.25]
	284-292	absorption	C K-edge	[6.144]
	280-300	e$^-$-yield	C K-edge	[6.152]
	16-70	absorption	absolute cross section, $\Delta\lambda$ = 1 Å	[6.110]
CF$_4$	16-70	absorption	absolute cross section, $\Delta\lambda$ = 1 Å	[6.110]
CF$_3$Cl	16-70	absorption	absolute cross section, $\Delta\lambda$ = 1 Å	[6.110]
C$_2$H$_6$ ethane	8-35	absorption	absolute cross section, comparison gas-solid	[6.109]
	18-68	absorption	absolute cross section	[6.25]
	80-300	e$^-$-yield	C K-edge	[6.152]
C$_2$F$_6$	16-70	absorption	absolute cross section, $\wedge\lambda$ = 1 Å	[6.110]
C$_3$H$_6$ propane	8-35	absorption	absolute cross section	[6.109]
C$_4$H$_{10}$ butane	8-35	absorption	absolute cross section	[6.109]
	18-68	absorption	absolute cross section	[6.25]
(CH$_3$)$_4$C neopentane	7-35	absorption	Rydberg series 2a$_1$ → np between 15 and 18 eV	[6.116]

Molecule	Energy Range[eV]	Method	Remarks	Reference
CF$_3$Cl CF$_2$Cl$_2$ CFCl$_3$	10-35	photoionization absorption resonance photo-electron spectra	absolute cross section photofragmentation	[6.174]
cis-2-decene CH$_3$(CH$_2$)$_3$CH: CH(CH$_2$)$_3$CH$_3$ trans-2-decene	13,15, 21.2	photoionization	comparison with electron impact parameters, masses 18-142	[6.243]
C$_2$H$_4$ ethylene	18-68	absorption	absolute cross section	[6.25]
	280-300	e$^-$-yield	C K-edge	[6.152]
	17.5-50	photoionization ion-electron coincidence	fragmentations of C$_2$H$_4$, C$_2$H$_3^+$, C$_2$H$_2^+$, C$_2$H$^+$, CH$_2^+$, total photoionization cross section	[6.231]
	17.5-50	photoionization ion-electron coincidence	fragmentation, CH$^+$, C$^+$, C$_2^+$, H$^+$, H$_2^+$, C$_2$H$^+$, C$_2$H$_2^+$, C$_2$H$_3^+$, CH$^+$	[6.231]
CH$_2$-CHF CH$_2$-CHCl C$_2$H$_2$F$_2$ vinylchloride vinylfluoride 1,1-difluor-ethylene	10-32	photoionization	appearance potentials, heats of formation, band energies, ionization potentials	[6.213] [6.214]
C$_2$H$_3$Br vinylbromide	9-25	photoionization, absorption, resonance photoelectron spectra	appearance potentials, heats of formation	[6.217]
CH$_2$CHF	9-25	photoionization	comparison to photoelectron spectra, binding energies, appearance potentials	[6.218]
C$_2$H$_2$ acetylene	11.2-11.7	photoionization	photoion yield, supersonic molecular beam	[6.224]
	280-300	e$^-$-yield	C K 1s edge	[6.152]
CH$_3$-CH-CH$_2$ propylene	9-27	photoionization	appearance potentials, ionization potentials, heats of formation	[6.215] [6.216]
CH$_2$———CH$_2$ ＼　／ CH$_2$ cyclopropane	9-27	photoionization	comparison of photoelectron spectra with differential photoion spectra, autoionization and predissoziation	[6.220]

Molecule	Energy Range[eV]	Method	Remarks	Reference
CH_3CHO acetaldehyde C_2H_2O ethylene oxide				
C_6H_6	6-35	absorption	Rydberg assignments, ordering of MO's	[6.119]
	5-35	absorption	Rydberg assignments, ordering of MO's	[6.120]
	10-30	absorption	absolute cross sections, comparison gas-solid, comparison to energy loss	[6.106]
	5-35	absorption	absolute cross sections, $\Delta\lambda = 0.03$ Å, assignments in comparison to fluorbenzenes	[6.114]
	280-300	e^--yield	C K-edge	[6.152]
	16.75	photoemission	PED curve, comparison to HeI spectrum discussion of the importance of the asymmetry parameter	[6.247]
C_6D_6	5-35	absorption	absolute cross sections, Rydberg assignments	[6.120] [6.114]
C_6H_5F	5-35	absorption	absolute cross sections, Rydberg assignments	[6.120] [6.114]
C_6F_6	5-35	absorption	absolute cross sections, Rydberg assignments	[6.120] [6.114]
C_5H_5N	5-35	absorption	absolute cross sections, Rydberg assignments	[6.120]
tetraphenyl tin	10-45	photoemission	PED's thin films, energy dependence of cross sections	[6.245]
C_6H_5-CH_3 toluene	12-21	photoionization	photoionization mass spectra, comparison to electron impact spectra	[6.246]
naphthalene	5-30	absorption	photoelectric, photographic Rydberg assignments	[6.136]
anthracene	5-8.5	absorption	photographic, valence and Rydberg assignments	[6.137]
	3-11.5	reflection	single crystals, ε_2 spectra sum rules, comparison gas-solid	[6.106] [6.138]
tetracene	fluorescence 2.4-3.2	time resolved fluorescence	two lifetimes 0.2 ± 0.02 ns 1.7 ± 0.2 ns	[6.205]
quinine sulfate fluorescein anion aniline	excitation $h\nu \lesssim 7.5$ visible fluorescence	time resolved fluorescence	discussion of method, lifetimes of 0.3 ns in principle distinguishable single vibronic level fluorescence	[6.204]

Molecule	Energy Range[eV]	Method	Remarks	Reference
pyrazine 2-naphtol fluorescein	excita- h$\nu \lesssim 7.5$ visible fluores- cence	time resolved fluorescence	0.5 ± 0.2 ns lifetime	[6.206]
carbazole	excita- tion near UV (3.8)	fluorescence time resolved	carbozole in solid ethascole in the presence of KI	[6.207]

References

6.1 G. Herzberg: *Molecular Spectra and Molecular Structure*, Vols. I, II and III (Van Nostrand Reinhold New York 1945, 1950, 1966)

6.2 G. Herzberg: *The Spectra and Structures of Simple Free Radicals, An Introduction to Molecular Spectroscopy* (Cornell Univ. Press, Ithaca 1971)

6.3 M.R. Robin: *Higher Excited States of Polyatomic Molecules*, Vols. I and II (Academic Press, New York 1974, 1975)

6.4 H.H. Jaffé, M. Orchin: *Theory and Applications of Ultraviolet Spectroscopy* (Wiley, New York 1962)

6.5 *Chemical Spectroscopy and Photochemistry in the Vacuum-Ultraviolet*, ed. by C. Sandorfy, P.J. Ausloos and M.B. Robin (Reidel Publ. Company, Dordrecht-Holland 1974)

6.6 *Some Aspects of Vacuum Ultraviolet Radiation Physics*, ed. by N. Damany, J. Romand and B. Vodar (Pergamon Press, Oxford 1974)

6.7 *Vacuum Ultraviolet Radiation Physics*, ed. by E.E. Koch, R. Haensel and C. Kunz (Pergamon/Vieweg, Braunschweig 1974)

6.8a Henry F. Schaefer III: *The Electronic Structure of Atoms and Molecules*, (Addison-Wesley, Reading, MA 1972) R. McWeeny, B

6.8b R. McWeeny, B.T. Sutcliffe: *Methods of Molecular Quantum Mechanics* (Academic Press, London 1969)

6.8c J.C. Slater: *Quantum Theory of Molecules and Solids*, Vol. I (McGraw Hill, New York 1963)

6.8d A.B.F. Duncan: *Rydberg Series in Atoms and Molecules* (Academic Press, New York 1971)

6.8e *Modern Quantum Chemistry*, Vol. I, II, III, Ed. by O. Sinanoğlu (Academic Press, New York 1965)

6.8f O. Sinanoğlu, K.B. Wiberg: *Sigma Molecular Orbital Theory* (Yale University Press, New Haven 1970)

6.8g J.A. Pople, D.L. Beveridge: *Approximate Molecular Orbital Theory* (McGraw Hill, New York 1970)

6.9a K. Siegbahn, C. Nordling, G. Johansson, J. Hedman, F. Hedan, K. Hamrin, U. Gelius, T. Bergmark, L. Werme, R. Manne, Y. Baer: *ESCA Applied to Free Molecules* (North Holland, Amsterdam 1969)

6.9b D.W. Turner, C. Baker, A.D. Baker, C.R. Brundle: *Molecular Photoelectron Spectroscopy* (Wiley, London 1970)

6.9c J.H.D. Eland: *Photoelectron Spectroscopy* (Butterworths, London 1974)

6.9d J.W. Rablais: *Principles of Ultraviolet Photoelectron Spectroscopy* (Wiley, New York 1977)

6.9e J.A.R. Samson: (Phys. Reps. <u>28</u>, 303 1976)

6.9f G.V. Marr: *Photoionization Processes in Gases* (Academic Press, New York 1967)

6.10a E.N. Lassettre: in [Ref. 6.5, p. 43] and references therein
6.10b C. Backx, M.J. van der Wiel: in [Ref. 6.7, p. 137] and references therein
6.11a K. Watanabe, M. Zelikoff, C.E.Y. Inn: *Absorption Coefficients of Several Atmosphereic Gases*, Geophys. Res. Report AFCRL-TR-53-23 (1953)
6.11b J.O. Sullivan, A.C. Holland: *A Congeries of Absorption Cross Sections for Wavelengths Less than 3000 Å*, NASA Report CR 371 (1966)
6.11c R.D. Hudson: *Critical Review of Ultraviolet Photoabsorption Cross Sections for Molecules of Astrophysical and Aeronomical Interest*, Rev. Geophys. and Space Phys. 9, Aug. 71 (1971)
6.12 F.S. Rowland, M.J. Molian: "Chlorofluormethanes and the Environment", Rev. Geophys. and Space Phys. $\underline{13}$, 1 (1975) and references therein
6.13 *Excited States of Biological Molecules*, ed. by J.B. Birks (Wiley, London 1976)
6.14a B. Sonntag: In *Rare Gas Solids*, Vol. 2, ed. by M.L. Klein and J.A. Venables (Academic Press, London 1977) p. 1021
6.14b H.C. Wolff: In *Festkörperprobleme* Bd. $\underline{4}$, ed. by F. Sauter, (Vieweg, Braunschweig 1965)
6.14c *Electronic Structure of Polymers and Molecular Crystals*, ed. by J.M. Andre, J. Ladik and J. Delhalle (Plenum Press, New York 1974)
6.15a B. Meyer: *Low Temperature Spectroscopy* (Elsvier Publ. Company, New York 1971)
6.15b *Cryochemistry*, ed. by M. Moskowits and G.A. Ozin (Wiley, New York 1976)
6.16a *Molecular Physics*, ed. by D. Williams, Methods of Experimental Physics, Vol. 3 (Academic Press, New York 1962)
6.16b J.A.R. Samson: *Techniques of Vacuum Ultraviolet Spectroscopy* (Wiley, New York 1967)
6.16c *Spectroscopy*, Part A, B, ed. by D. Williams, Methods of Experimental Physics, Vol. 13 (Academic Press, New York 1976)
6.17 See e.g. *Excited States*, Vols. I, II and III, ed. by E.C. Lim (Academic Press, New York 1974, 1978)
6.18 W.H.E. Schwarz: Angew, Chemie, Intern. Ed. $\underline{13}$, 454 (197)
6.19a C.J. Ballhausen: *Introduction to Ligand Field Theory* (McGraw Hill, New York 1962)
6.19b H.K. Schläfer, G. Gliemann: *Einführung in die Ligandenfeldtheorie* (Akademische Verlagsges., Frankfurt 1967)
6.20a V.A. Foymichev, R.L. Barinskii: J. Struc. Chem. $\underline{11}$, 810 (1960)
6.20b R.L. Barinskii, I.M. Kulikova: J. Struct. Chem. $\underline{14}$, 335 (1973)
6.20c J.L. Dehmer: Phys. Fennica $\underline{9b}$, Supplement $\underline{S1}$, 60 (1974) and references therein
6.20d J.L. Dehmer, D. Dill: J. Chem. Phys. $\underline{65}$, 5327 (1976) and references therein
6.21 K.H. Johnson: In *Advances in Quantum Chemistry*, Vol. 7 Ed. by P.O. Löwdin (Academic Press, New York 1973) p. 143
6.22 *Photoionization and other Probes of Many-Electron Interactions*, ed. by F.J. Wuilleumier (Plenum Press, New York 1976)
6.23 See, e.g. J. Jortner and S. Mukamel, In: *The World of Quantum Chemistry*, eds. R. Daudel and B. Pullman, Reidel, Dordrecht, 1975 p. 145 and references therein
6.24a Conf. Digest III, Intern. Conf. on VUV Radiation Physics, ed. by Y. Nakai, Tokyo (1971)
6.24b Extended Abstracts, V. Intern. Conf on VUV Radiation Physics, Vol. I, II, III, ed. by M.C. Castex, M. Pouey and N. Pouey, Meudon (1977)
6.25 L.C. Lee, R.W. Carlson, D.L. Judge, M. Ogawa: J. Quant. Spectr. Radiat. Transfer 13, 1023 (1973)
6.26 K. Codling: Astrophys. J. $\underline{143}$, 552 (1966)
6.27 K. Codling, R.P. Madden: J. Chem. Phys. $\underline{42}$, 3935 (1965)
6.28 T. Lyman: Astrophys. J. $\underline{57}$, 161 (1911)
6.29 J.J. Hopfield: Phys. Rev. $\underline{35}$, 1133 (1930); 36, 789 (1930)
6.30 P.K. Carroll, C.P. Collins: Can. J. Phys. $\underline{47}$, 563 (1969)
6.31 K. Dressler: Can. J. Phys. $\underline{47}$, 547 (1969)
6.32 A. Lofthus, P.H. Krupenie: J. Phys. Chem. Ref. Data $\underline{6}$, 113 (1977)
6.33 C. Duzy, R.S. Berry: J. Chem. Phys. $\underline{64}$, 2421; 2431 (1976)
6.34 S.K. Shih, W. Butscher, R.J. Buenker, S.D. Peyerimhoff: Chem. Phys. $\underline{29}$, 241 (1978)

6.35 T.N. Rescigno, C.F. Bender, B.V. McKoy, P.W. Langhoff: J. Chem. Phys. 68, 970 (1978)
6.36 P. Gürtler, V. Saile, E.E. Koch: Chem. Phys. Lett. 48, 245 (1977)
6.37 V. Saile, P. Gürtler, E.E. Koch, A. Kozevnikov, M. Skibowski, W. Steinmann: Appl. Opt. 15, 2559 (1976)
6.38 W. Leoni: Dissertation, Eidg. Tech. Hochschule, Zürich (1972)
6.39 R. Haensel, E.E. Koch, N. Kosuch, U. Nielsen, M. Skibowski: Chem. Phys. Lett. 9, 548 (1971)
6.40 E.W. Plummer, T. Gustafsson, W. Gudat, D.E. Eastman: Phys. Rev. A15, 2339 (1977)
6.41 U. Fano: Phys. Rev. 124, 1866 (1961); U. Fano, J.W. Cooper: Phys. Rev. 137A, 1364 (1965); Rev. Mod. Phys. 40, 441 (1968)
6.42 W.A. Goddard III, W.J. Hunt: Chem. Phys. Lett. 24, 464 (1974)
6.43 R.J. Buenker, S.D. Peyerimhoff: Chem. Phys. Lett. 29, 253 (1974)
6.44 D. Yaeger, U. McKoy, G.A. Segal: J. Chem. Phys. 61, 755 (1974)
6.45 N.W. Winter, W.A. Goddard III, F.W. Bobrowicz: J. Chem. Phys. 62, 4325 (1975)
6.46 P. Gürtler, V. Saile, E.E. Koch: Chem. Phys. Lett. 51, 386 (1977)
6.47 R.S. Mulliken: J. Chem. Phys. 3, 506 (1935)
6.48 A.D. Liehr: Z. Naturforsch. 11A, 752 (1956)
6.49 J.A. Pople, H.C. Longuet-Higgins: Mol. Phys. 1, 372 (1958)
6.50 E. Ishiguro, M. Sasanuma, H. Masuko, Y. Morioka, M. Nakamura: J. Phys. B11, 993 (1978)
6.51 M. Nakamura, M. Sasanuma, S. Sato, M. Watanabe, H. Yamashita, Y. Iguchi, A. Ejiri, S. Nakai, S. Samaguchi, T. Sagawa, Y. Nakai, T. Oshio: Phys. Rev. 178, 80 (1969)
6.52 M. Nakamura, Y. Morioka, T. Hayaishi, E. Ishiguro, M. Sasanuma: In [Ref. 6.24a]
6.53 S. Bodeur, C. Senemaud, C. Bonnelle, J.P. Connerade: In [Ref. 6.7, p. 94]
6.54 Y. Morioka, M. Nakamura, E. Ishiguro, M. Sasanuma: In [Ref. 6.7, p. 92]
6.55 Y. Morioka, M. Nakamura, E. Ishiguro, M. Sasanuma: J. Chem. Phys. 61, 1426 (1974)
6.56 W.H.E. Schwarz, T.C. Chang, J.P. Connerade: Chem. Phys. Lett. 49, 207 (1977)
6.57 A. Bianconi, H. Petersen, F.C. Brown, R.Z. Bachrach: Phys. Rev. A17, 1907 (1978)
6.58 J.P. Connerade, M.W. D. Mansfield, K. Thimm: Chem. Phys. 1, 256 (1973)
6.59 B.R. Kay, Ph.E. Van der Leeuw, M.J. Van der Wiel: J. Phys. B10, 2513 (1977)
6.60 G.C. King, F.H. Read, M. Tronc: Chem. Phys. Lett. 52, 50 (1977)
6.61 G.R. Wigth, C.E. Brion, M.J. van der Wiel: J. Electr. Spectr. Relat. Phenom. 1, 457 (1973)
6.62 G.J. Schulz: Rev. Mod. Phys. 45, 378 (1973)
6.63 J.P. Connerade, M.W.D. Mansfield: Proc. Roy. Soc. Lond. A356, 135 (1977)
6.64 K. Radler et al. to be published
6.65 B. Sonntag, T. Tuomi, G. Zimmerer: Phys. Stat. Sol (b) 58, 101 (1973)
6.66 B. Rosen: *Spectroscopic data relative to diatomic molecules* (Pergamon Press, London 1970)
6.67 F.J. Comes, U. Nielsen, W.H.E. Schwarz: J. Chem. Phys. 58, 2230 (1973)
6.68 M.Ya. Amusia, N.A. Cherepkov, L.V. Chernysheva: Sov. Phys. JETP 33, 90 (1971)
6.69 M.Ya. Amusia: In [Ref. 6.7, p. 205]
6.70 G. Wendin: In [Ref. 6.7, p. 225]
6.71 W. Brandt, S. Lundqvist: Phys. Rev. 132, 2135 (1963)
6.72 K. Radler, B. Sonntag, T.C. Chang, W.H.E. Schwarz: Chem. Phys. 13, 363 (1976)
6.73 K. Radler: Dissertation, University of Hamburg (1974); and DESY, internal report F41-74/9
6.74 W.H.E. Schwarz: In *Int. Conf. on the physics of X-ray Spectra, Extended Abstracts*, Washington (1970) p. 49
6.75 G. Sprüssel: Diplomarbeit, University of Hamburg (1976); and DESY, internal report F41-76/02
6.76 C.H.E. Moore: *Atomic Energy Levels*, Vols. 1,2,3, NBS Circular 467, Wasnington (1949)
6.77 H.W. Wolff: private communication (1977)
6.78 H.W. Wolff, B.F. Sonntag: In *Second Intern. Conf. on Inner Shell Ionization Phenomena*, Abstracts of contributed papers, Freiburg (1976) p. 78

6.79   K. Radler, B. Sonntag: Chem. Phys. Lett. 39, 371 (1976)
6.80   U. Nielsen, W.H.E. Schwarz: Chem. Phys. 13, 195 (1976)
6.81   F.J. Comes, R. Haensel, U. Nielsen, W.H.E. Schwarz: J. Chem. Phys. 58, 516 (1973)
6.82   U. Nielsen: Dissertation, University of Hamburg (1973); and DESY internal report F41-73/9
6.83   U. Nielsen, R. Haensel, W.H.E. Schwarz: J. Chem. Phys. 61, 3581 (1974)
6.84   R.E. LaVilla, R.D. Deslattes: J. Chem. Phys. 44, 4399 (1966)
6.85a T.M. Zimkina, V.A. Fomichev: Sov. Phys. Dokl. 11, 726 (1966)
6.85b T.M. Zimkina, A.C. Vinogradov: J. Phys. Paris 32 (Colloque C4) 3 (1971)
6.86   M. Nakamura, Y. Morioka, T. Hayaishi, E. Ishiguro, M. Sasanuma: In [Ref. 6.24a]
6.87   D. Blechschmidt, R. Haensel, E.E. Koch, U. Nielsen, T. Sagawa: Chem. Phys. Lett. 14, 33 (1972)
6.88   E.S. Gluskin, A.A. Krasnoperova, Y.A. Mazalov: J. Structural Chemistry, USSR 1, 185 (1976)
6.89a J.L. Dehmer: J. Chem. Phys. 56, 4496 (1972)
6.89b F.A. Gianturco, C. Guidotti, U. Lamma: J. Chem. Phys. 57, 840 (1972)
6.89c V.P. Sachenko, E.V. Polozhentsev, A.P. Kovtun, Yu.F. Migal, R.V. Vedrinski, V.V. Kolesnikov: Phys. Lett. 48A, 169 (1974)
6.90   W. Hayes, F.C. Brown: Phys. Rev. A6, 21 (1972)
6.91   H. Friedrich: Diplomarbeit, University of Hamburg (1976); W. Butscher, H. Friedrich, P. Rabe, W.H.E. Schwarz, B. Sonntag: to be published
6.92   W. Hayes, F.C. Brown, A.B. Kunz: Phys. Rev. Lett. 27, 774 (1971)
6.93   M.B. Robin: Chem. Phys. Lett. 31, 140 (1975)
6.93b W.H.E. Schwarz: Chem. Phys. 9, 157 (1975), 11, 217 (1975)
6.94   D. Schröder, B. Sonntag: to be published
6.95   B.M. Kincaid, P. Eisenberger: Phys. Rev. Lett. 34, 1361 (1975)
6.96   R. de I. Kronig: Z. Physik 70, 317 (1931), 75, 191, 468 (1932)
6.97   I. Azaroff: Rev. Mod. Phys. 35, 1012 (1963)
6.98   E.A. Stern: Phys. Rev. B10, 3027 (1974); F.W. Lytle, P.E. Sayers, E.A. Stern: Phys. Rev. B15, 2426 (1977)
6.99   P.E. Sayers, F.W. Lytle, E.A. Stern: In *Advances in X-ray Analysis*, Vol. 13, ed. by B.L. Henke, J.B. Newkirk and G.R. Mallett (Plenum Press, New York 1970) p. 248
6.100a C.A. Ashley, S. Donaich: Phys. Rev. B11, 1279 (1975)
6.100b P.A. Lee, J.B. Pendry: Phys. Rev. 11, 2795 (1975)
6.101 G. Martens, P. Rabe, N. Schwentner, A. Werner: Phys. Rev. Lett. 39, 1411 (1977)
6.102 U. Fano: Phys. Rev. Lett. 31, 234 (1973); U. Fano: In [Ref. 6.7 p. 84]; G. Strinati: Phys. Rev. B, to be published
6.103 See e.g. F. Gutmann, L.E. Lyons: *Organic Semiconductors* (Wiley, New York 1967); H. Meier: *Organic Semiconductors* (Verlag Chemie, Weinheim 1974)
6.104 *Computational Methods for Large Molecules and Localized States in Solids*, ed. by F. Herman, A.D. McLean and R.K. Nesbet (Plenum Press, New York 1973)
6.105 See also J.J. Ritsko, N.O. Lipari, P.C. Gibbons, S.E. Schnatterly, J.R. Fields, R. Devaty: Phys. Rev. Lett. 36, 210 (1976)
6.106 E.E. Koch, A. Otto: Int. J. Radiat. Phys. Chem. 8, 113 (1976)
6.107 R.S. Mulliken: J. Chem. Phys. 3, 517 (1935)
6.108 C. Sandorfy: J. Molec. Structure 19, 183 (1973) and references therein
6.109 E.E. Koch, M. Skibowski: Chem. Phys. Lett. 9, 429 (1971)
6.110 L.C. Lee, E. Phillips, D.L. Judge: J. Chem. Phys. 67, 1237 (1977)
6.111 A. Richartz, R.J. Buenker, S.D. Peyerimhoff: Chem. Phys. 31, 187 (1978)
6.112 J. Berkowitz: In [Ref. 6.7, p. 107]
6.113 J.C. Person, P.P. Nicole: In [Ref. 6.24b, Vol. I, p. 111]
6.114 K.E. Ockenga, P. Gürtler, S.S. Hasnain, V. Saile, E.E. Koch: to be published
6.115 J.C. Person, P.P. Nicole: Argonne National Lab. Report ANL-8060 (1974)
6.116 E.E. Koch, V. Saile, N. Schwentner: Chem. Phys. Lett. 33, 322 (1975)
6.117 C. Sandorfy: In *Sigma Molecular Orbital Theory*, ed. by O. Sinanoglu and K.B. Wiberg (Yale Univ. Press. New Haven 1970)
6.118 S. Evans, J.C. Green, P.J. Joachim. A.F. Orchard, D.W. Turner, J.P. Maier: J. Chem. Soc. Farad. Trans. II 68, 905 (1972)

6.119   E.E. Koch, A. Otto: Chem. Phys. Lett. 12, 476 (1972)
6.120   E.E. Koch, A. Otto, V. Saile, N. Schwentner: In [Ref. 6.7, p. 77]
6.121   M. Yoshino, J. Takeuchi, H. Suzuki: J. Phys. Soc. Japan 34, 1039 (1973)
6.122   S.M. Bunch, G.R. Cook, M. Ogawa, A.W. Ehler: J. Chem. Phys. 28, 740 (1958)
6.123   J.C. Person: J. Chem. Phys. 43, 2553 (1965)
6.124   L. Åsbrink, O. Edquist, E. Lindholm, L.E. Selin: Chem. Phys. Lett. 5, 192
        (1970)
6.125   L. Karlsson, L. Mattsson, R. Jadrny, T. Bergmark, K. Siegbahn: Phys.
        Scripta 14, 230 (1976)
6.126   U. Gelius: J. Electr. Spectr. 51, 985 (1974)
6.127   S.D. Peyerimhoff, R.J. Buenker: Theoret. Chim. Acta 19, 1 (1970)
6.128   W.C. Ermler, C.W. Kern: J. Chem. Phys. 58, 3458 (1973)
6.129   W. v. Niessen, L.S. Cederbaum, W.P. Kraemer: J. Chem. Phys. 65, 1378 (1976)
6.130   See e.g. E. Lindholm: Arkiv Fysik 40, 97 (1968)
6.131a C.B. Duke, N.O. Lipari, L. Pietronero: Chem. Phys. Lett. 30, 415 (1975)
6.131b C.B. Duke, N.O. Lipari: Chem. Phys. Lett. 36, 51 (1975)
6.131c N.O. Lipari, C.B. Duke, L. Pietronero: J. Chem. Phys. 65, 1165 (1976)
6.132   C.B. Duke, K.L. Yip. G.P. Ceasar, A.W. Potts, D.G. Streets: J. Chem. Phys.
        66, 256 (1977)
6.133   See also L.S. Cederbaum, W. Domcke: J. Chem. Phys. 64, 603 (1976)
6.134   C.R. Brundle, M.B. Robin, N.A. Kuebler: J. Am. Chem. Soc. 94, 1466 (1972)
6.135   E.E. Koch: Dissertation, University of Munich (1972), and DESY internal
        report F41-72/2
6.136   E.E. Koch, A. Otto, K. Radler: Chem. Phys. Lett. 10, 131 (1972)
6.137   E.E. Koch, A. Otto, K. Radler: Chem. Phys. Lett. 21, 501 (1973)
6.138   E.E. Koch, A. Otto, Phys. Stat. Sol. (b) 51, 69 (1972); Chem. Phys. 3,
        370 (1974)
6.139   S.S. Hasnain, T.D.S. Hamilton, I.H. Munro, E. Pantos: J. Molec. Spectr.
        72, (1978) in press
6.140   S.S. Hasnain, P. Brint, T.D.S. Hamilton, I.H. Munro: J. Mol. Spectro.
        72, (1978) in press
6.141   P.W. Deutsch, A.B. Kunz: J. Chem. Phys. 59, 1155 (1973)
6.142   P.S. Bagus, M. Krauss, R.E. LaVilla: Chem. Phys. Lett. 23, 13 (1973)
6.143   W. Gudat: Nuc. Instrum. Methods 152, 279 (1978)
6.144   H.U. Chun: Phys. Lett. 30A, 445 (1969)
6.145   H.U. Chun, T.K. Ha, R. Mann: Z. Phys. Chemie, NF 47, 313 (1965); H.U. Chun,
        H. Gebelein: Z. Naturforsch. 22a, 1813 (1967)
6.146   C.R. Wight, C.E. Brion: J. Electron Spectr. 4, 25 (1974)
6.147   A.P. Hitchcock, M. Pocock, C.E. Brion: Chem. Phys. Lett. 49, 125 (1977)
6.148   F.C. Brown, R.Z. Bachrach, A. Bianconi: Chem. Phys. Lett. 54, 425 (1978)
6.149   T.D. Thomas: J. Am. Chem. Soc. 92, 4184 (1970)
6.150   L.A. Curtis, P.W. Deutsch: J. Electron Spectr. 10, 193 (1977); see also
        P.W. Deutsch, L.A. Curtis: Chem. Phys. Lett. 34, 588 (1976)
6.151   D.W. Davis, D.A. Shirley, T.D. Thomas: J. Chem. Phys. 56, 671 (1972); also
        52, 3295 (1970)
6.152   W. Eberhardt, R.-P. Haelbich, M. Iwan, E.E. Koch, C. Kunz: Chem. Phys.
        Lett. 40, 180 (1976)
6.153   F.C. Brown, R.Z. Bachrach, N. Lien: Nucl. Instrum. Methods 152, 73 (1978)
6.154   A.P. Hitchcock, C.E. Brion: J. Electron Spectr. 10, 317 (1977)
6.155   T.D. Thomas: J. Chem. Phys. 52, 1373 (1970)
6.156   E.g., D.E. Eastman: In [Ref. 6.7, p. 417]
6.157   E.g., W.C. Price: In Advances in Atomic and Molecular Physics, Vol. 10
        (Academic Press, New York 1974) p. 131
6.158   J.A.R. Samson, J.L. Gardner: J. Electron Spectr. 8, 35 (1976)
6.159   I.H. Scofield: Theoretical Photoionization Cross Sections from 1 to 1500 keV,
        Lawrence Livermore Lab. Report TID-4500 UC-34 (1973)
6.160   E.W. Plummer: In Interactions on Metal Surfaces, ed. by R. Gomer, Topics
        in Applied Physics, Vol. 4 (Springer, Berlin, Heidelberg, New York 1975)
        p. 143
6.161   T. Gustafsson, E.W. Plummer, D.E. Eastman, W. Gudat: Phys. Rev. A 17, 175
        (1978)
6.162   T. Gustafsson: Phys. Rev. A to be published

6.163   J.C. Tully, R.S. Berry, B.J. Dalton: Phys. Rev. 176, 95 (1968); J. Cooper, R.N. Zare: J. Chem. Phys. 48, 942 (1968)
6.164   J.W. Davenport: Ph.D. Thesis, University of Pennsylvania (1976); J.W. Davenport: Phys. Rev. Lett. 36, 945 (1976)
6.165   G.V. Marr, P.R. Woodruff: J. Phys. B9, L377 (1976)
6.166   P.R. Woodruff, G.V. Marr: Proc. Roy. Soc. A 358, 87 (1977)
6.167   E.g., T. Gustafsson, E.W. Plummer, D.E. Eastman, J.L. Freeouf: Sol. State. Comm. 17, 391 (1975)
6.168   T. Gustafsson, E.W. Plummer: In *Photoemission from Surfaces*, ed. by F. Feuerbacher, B. Fitton and R.W. Willis (Wiley, London 1978) p. 353
6.169   See, e.g. C.L. Allyn, T. Gustafsson, E.W. Plummer: Chem. Phys. Lett. 47, 127 (1977)
6.170a W.B. Peatman, T.B. Borne, E.W. Schlag: Chem. Phys. Lett. 3, 492 (1969)
6.170b T. Baer, W.B. Peatman, E.W. Schlag: Chem. Phys. Lett. 4, 2431 (1969)
6.171   R. Sphor, P.M. Guyon, W.A. Chupka, J. Berkowitz: Rev. Sci. Instr. 42, 1872 (1971)
6.172   E.g. W.B. Peatman, G.B. Kasting, D.J. Wilson: J. Electr. Spectr. 7, 233 (1975)
6.173   W.B. Peatman: J. Chem. Phys. 64, 4093, 4368 (1976) and references therein
6.174   H.W. Jochims, W. Lohr, H. Baumgärtel: Ber. Bunsenges. phys. Chemie 80, 130 (1976)
6.175   R. Frey, B. Gotchev, O.F. Kalman, W.B. Peatman, H. Pollak, E.W. Schlag: Chem. Phys. 21, 89 (1977)
6.176   R. Frey, B. Gotchev, W.B. Peatman, H. Pollak, E.W. Schlag: Int. J. Mass Spectrom. Ion Phys. 26, 137 (1978)
6.177   W.B. Peatman, B. Gotchev, P. Gürtler, E.E. Koch, V. Saile: J. Chem. Phys. 69, 2089 (1978)
6.178   W.B. Peatman, P. Gürtler, V. Saile: To be published, see also Ref. 6.173
6.179   A. Chutjian, J.M. Ajello: J. Chem. Phys. 66, 4544 (1977)
6.180   E.W. Schlag, R. Frey, B. Gotchev, W.B. Peatman, H. Pollak: Chem. Phys. Lett. 51, 406 (1977)
6.181   R. Frey, B. Gotchev, W.B. Peatman, H. Pollak, E.W. Schlag: Chem. Phys. Lett. 54, 411 (1978)
6.182   B. Gotchev: Dissertation, Technical University of Munich (1977)
6.183   R.W. Waynant, R.C. Elton: Proc. IEEE, 64, 1059 (1976) and references therein
6.184   See, e.g. L.F. Stief: In [Ref. 6.5 p. 571]
6.185   L.C. Lee, R.W. Carlson, D.L. Judge, M. Ogawa: Chem. Phys. Lett. 19, 183 (1973)
6.186   L.C. Lee, R.W. Carlson, D.L. Judge, M. Ogawa: J. Chem. Phys. 61, 3261 (1974)
6.187   L.C. Lee, R.W. Carlson, D.L. Judge, M. Ogawa: J. Chem. Phys. 63, 3987 (1975)
6.188   L.C. Lee, R.W. Carlson, D.L. Judge, M. Ogawa: J. Geophys. Res. 79, 5286 (1974); L.C. Lee: J. Phys. B10, 3033 (1977)
6.189   L.C. Lee, R.W. Carlson, D.L. Judge: J. Phys. B9, 855 (1976)
6.190   R.W. Carlson, D.L. Judge, M. Ogawa: J. Geophys. Res. 78, 3194 (1973)
6.191   L.C. Lee, R.W. Carlson, D.L. Judge, M. Ogawa: J. Phys. B8, 977 (1975)
6.192   R.W. Carlson: J. Chem. Phys. 60, 2350 (1974)
6.193   M.E. Wacks: J. Chem. Phys. 41, 930 (1964)
6.194   R. Zietz: Diplomarbeit, University of Hamburg (1974); and DESY internal report F41-75/02
6.195   W. Sroka, R. Zietz: Phys. Lett. 43A, 493 (1973)
6.196   M. Hertz, H.W. Jochims, W. Sroka: Phys. Lett. 46A, 365 (1974)
6.197   H.W. Jochims: Diplomarabeit, University of Hamburg (1974); and DESY internal report F41-75/01
6.198   H. Hertz, H.W. Jochims, H. Schenk, W. Sroka: Chem. Phys. Lett. 29, 572 (1974)
6.199   H. Hertz, H.W. Jochims, H. Schenk, W. Sroka: In [Ref. 6.7, p. 68]
6.200   W. Sroka, R. Zietz: Z. Naturforsch. 28a, 794 (1973)
6.201   H. Hertz, H.W. Jochims, W. Sroka: J. Phys. B7, L548 (1974)
6.202   *Radiationless Processes in Molecules and Condensed Phases*, ed. by F.K. Fong, Topics in Applied Physics, Vol. 15 (Springer, Berlin, Heidelberg, New York 1976)
6.203   O. Benoist d'azy, R. Lopez-Delgado, A. Tramer: Chem. Phys. 9, 327 (1975)
6.204   R. Lopez-Delgado, A. Tramer, I.H. Munro: Chem. Phys. 5, 72 (1974)
6.205   R. Lopez-Delgado, J.A. Miehê, B. Sipp: SSRP Report No. 76/04 (1976)

6.206   L. Lindqvist, R. Lopez-Delgado, M. Martin, A. Tramer: Opt. Commun. 10, 283 (1974)
6.207   J. Najbar, I.H. Munro: Daresbury Report, DL/SRF/P50 (1976), J. Luminescence (1977)
6.208   A.W. Smith, C. Weiss: Chem. Phys. Lett. 14, 507 (1972)
6.209a G. Zimmerer: In *Proc. Intern. Summer School on Synchrotron Radiation Research*, Vol. 1, ed. by A.N. Mancini, I.F. Quercia, Intern. College on Appl. Phys. INFN, Catania (1976) p. 453
6.209b G. Zimmerer: In *Luminescence of Inorganic Solids*, ed. by B. Di Bartolo, (Plenum Press, New York 1978) 627
6.210   N. Schwentner: In *Luminescence of Inorganic Solids*, ed. by B. Di Bartolo, (Plenum Press, New York 1978) 645
6.211   See, e.g. S.S. Hasnain, T.D.S. Hamilton, I.H. Munro, E. Pantos, I.T. Steinberger: Phil. Mag. 35, 1299 (1977) and references therein
6.212   W. Forst: *Theory of Unimolecular Reactions* (Academic Press, New York 1973)
6.213   D. Reinke, R. Krässig, H. Baumgärtel: Z. Naturforsch. 28a, 1021 (1973)
6.214   D. Reinke: Dissertation, University of Freiburg (1973); and DESY internal report F41-73/6
6.215   R. Krässig, D. Reinke, H. Baumgärtel: Ber. Bunsenges. Phys. Chemie 78, 425 (1974)
6.116   R. Krässig: Dissertation, University of Freiburg (1974); and DESY internal report F41-74/8
6.217   W. Lohr, H.-W. Jochims, H. Baumgärtel: Ber. Bunsenges. Phys. Chemie 79, 901 (1975)
6.218   D. Reinke, H. Baumgärtel, T. Cvitas, L. Klasinc, H. Güsten: Ber. Bunsenges. Phys. Chemie 78, 1145 (1974)
6.219   H.-W. Jochims: Dissertation, University of Freiburg (1976); and DESY internal report F41-76/07
6.220   R. Krässig, D. Reinke, H. Baumgärtel: Ber. Bunsenges. Phys. Chemie 78, 154 (1974)
6.221   P.M. Guyon, J.E. Mentall: In *Electronic and Atomic Collisions*, Vol. 1, ed. by J.S. Risley, R. Geballe (Univ. of Washington Press, Seattle 1975) p. 10
6.222   P. Borrell, P.M. Guyon, M. Glass-Maujean: In [Ref. 6.7, p. 54]
6.223   P. Borrell, P.M. Guyon, M. Glass-Maujean: J. Chem. Phys. 66, 818 (1977)
6.224   J.M. Mentall, P.M. Guyon: J. Chem. Phys. 67, 3845 (1977)
6.225   M. Glass-Maujean, J. Breton, P.M. Guyon: Phys. Rev. Lett. 40, 181 (1978)
6.226   L.C. Lee, R.W. Carlson, D.L. Judge: J. Quant. Spectrosc. Radiat. Transfer 16, 873 (1976)
6.227   M. Sasanuma, E. Ishiguro, Y. Morioka, M. Nakamura: In [Ref. 6.24a]
6.228   F.-J. Himpsel, N. Schwentner, E.E. Koch: Phys. Stat. Sol. (b) 71, 615 (1975)
6.229   W.S. Watson, J. Lang, D.T. Stewart: J. Phys. B6, L148 (1973)
6.230   J.A. Kinsinger, J.W. Taylor: Intern. J. Mass Spectrom. and Ion Phys. 11, 461 (1973)
6.231   P.M. Guyon, J. Mentall, I. Nenner, N. Moulin, R. Botter: Adv. Mass Spectr. 7, (1977)
6.232   R. Nürnberger: Diplomarbeit, University of Hamburg (1977)
6.233   E.E. Koch, M. Skibowski: Chem. Phys. Lett. 14, 37 (1972)
6.234   M. Sasanuma, Y. Morioka, E. Ishiguro, M. Nakamura: J. Chem. Phys. 60, 327 (1974)
6.235   E. Miescher, Y.T. Lee, P. Gürtel: J. Chem. Phys. 68, 2753 (1978)
6.236   E. Phillips, L.C. Lee and D.L. Judge: J. Quant. Spectr. Radiat. Transfer 88, 309 (1977)
6.237   H. Schenk: Diplomarbeit, University of Hamburg (1976)
6.238   K. Radler, B. Sonntag, W.H. Wolff: Proc. Intern. Conf. Physics of X-ray Spectra, Washington (1976) p. 54
6.239   W.H.E. Schwarz, W. Butscher, D.L. Ederer, T.B. Lucatorto, B. Ziegenbein, W. Mehlhorn, H. Prömpeler: J. Phys. B11, 591 (1977)
6.240   W. Hayes, F.C. Brown: J. Phys. B4, L85 (1971)
6.241   K. Codling: J. Chem. Phys. 44, 4401 (1966)
6.242   M. Sasanuma, E. Ishiguro, H. Masuko, Y. Morioka, M. Nakamura: J. Phys. B: Atom. Molec. Phys. 11, 3655 (1978)

355

6.243 B.M. Johnson, J.W. Taylor: Intern. J. Mass Spectr. Ion Phys. $\underline{10}$, 1 (1972/73)
6.244 G.R. Parr, J.W. Taylor: Rev. Sci. Instr. $\underline{44}$, 1578 (1973)
6.245 G.M. Bancroft, K.T. Sham, D.E. Eastman, W. Gudat: J. Am. Chem. Soc. $\underline{99}$, 1752 (1977)
6.246 W.L. Stebbings, J.W. Taylor: Int. J. Mass Spectro. Ion Phys. $\underline{9}$, 471 (1972)
6.247 J.A. Kinsinger, J.W. Taylor: Int. J. Mass Spectro. Ion Phys. $\underline{10}$, 445 (1972/73)
6.248 K. Beyer, K.H. Welge: J. Chem. Phys. $\underline{51}$, 5323 (1969)
6.249 M. Ogawa, S. Ogawa: J. Mol. Spectr. $\underline{41}$, 393 (1972)
6.250 D. L. Judge, L.C. Lee: J. Chem. Phys. $\underline{57}$, 455 (1972)
6.251 W.C. Price, T.R. Passmore, D.M. Roessler: Discussions Faraday Soc. $\underline{35}$, 201 (1963)
6.252 D.L. Hildenbrabd, E. Murad: J. Chem. Phys. $\underline{44}$, 1524 (1966)
6.253 J.W. Hashi, J.E. Margrave: Fluorine Chem. Rev. $\underline{2}$, 77 (1968)
6.254a H. Schenk: Dissertation, University of Freiburg (1978)
6.254b H. Oertel, H. Schenk, H. Baumgärtel: to be published
6.255 A.P. Hitchcock, C.E. Brion: In [Ref. 6.24b, Vol. I, p. 43]
6.256 K. Codling, W.A. Potts: J. Phys. B$\underline{7}$, 163 (1974)
6.257 L.C. Lee, R.W. Carlson, D.L. Judge: Mol. Phys. $\underline{30}$, 1941 (1975)

7. Solid-State Spectroscopy

D. W. Lynch

With 23 Figures

A complete survey of the vacuum ultraviolet and soft X-ray spectroscopy of solids should include not only reflection and transmission spectroscopy, but X-ray and ultraviolet photoemission spectroscopy, electron energy loss spectroscopy, several other spectroscopies, and the theoretical bases for their interpretation, for all deal with the same microscopic phenomena. This chapter will be limited, of necessity, to absorption, reflection, luminescence, and modulation spectroscopy of solids in the 6-300 eV region. (The dividing line between soft X-ray and vacuum ultraviolet is not well-defined and will shift several times during this chapter). There is an enormous literature to cover, for conventional discharge sources have been in use for many years, and improved versions are still widely used. Synchrotron radiation has a number of advantages over them, and its use has led to the discovery of new kinds of spectral features. For example, the high intensity and temporal stability of synchrotron radiation permit high-resolution and modulation spectroscopy to be carried out almost routinely, revealing previously unresolved structure, and the polarization allows work on anisotropic materials. Because synchrotron radiation has a continuous spectrum over several decades of energy the search for spectral features can be carried out thoroughly. This chapter describes work done almost exclusively with synchrotron radiation, although some of it could have been carried out with conventional sources. In fact, some has been so carried out, often earlier, but space does not permit adequate referencing of much of such work.

The vacuum ultraviolet region will be emphasized because there are several recent reviews of the soft X-ray spectroscopy of solids [7.1-4]. The data from reflectance and transmittance measurements have technological value, being useful in the design of filters, mirrors, and other devices, but the aim of this chapter really is to describe the spectra and their interpretation in terms of the electronic structure of the solid as a whole or of its constituent atoms or molecules. The interpretation then will be in terms of one-electron transitions, nominal one electron transitions with many-body corrections, and collective excitations. These are reviewed briefly in Sect. 7.1.2, and more extensively in Chap. 4. Sect. 7.2-4 review the spectra of metals, semiconductors and insulators, respectively. In these three sections, most of the discussion is devoted to only a few solids, illustrative of a class of solids, with emphasis placed on those spectral features which have led to an increased understanding of a class of solids, or to those which are perplexing, the understanding of which awaits further work. Many other solids are disposed of with just a reference, so treated not because of lack of merit, but for lack of space. Results of other measurements will be introduced occasionally if they help clarify a particular problem. Because of the emphasis on the microscopic interpretation of the spectral features, little attention will be paid either to measurement methods or to the accuracy of the magnitudes of the resultant dielectric functions or absorption coefficients, although the magnitudes can sometimes be important in the interpretation. These considerations are discussed elsewhere [7.5,6].

7.1 Quantitative Description of Optical Properties

7.1.1 Macroscopic Optical Properties

The linear optical properties of matter are described by Maxwell's equations and a constitutive parameter [7.7,8], usually taken to be the frequency-dependent complex dielectric function,

$$\tilde{\varepsilon}(\omega) = \varepsilon_1(\omega) + i\varepsilon_2(\omega) \quad , \tag{7.1}$$

or the frequency-dependent complex conductivity,

$$\tilde{\sigma}(\omega) = \sigma_1(\omega) + i\sigma_2(\omega) \quad , \tag{7.2}$$

related by

$$\tilde{\varepsilon}(\omega) = -(4\pi i/\omega)\tilde{\sigma}(\omega) \quad . \tag{7.3}$$

The absorption of electromagnetic radiation arises from ε_2 or σ_1. Several generalizations of these functions are sometimes necessary. As written, these are local equations, relating, for example, the current density $\underline{J}$ at a point $\underline{r}$ to the electric field at that point,

$$\underline{J}(\underline{r}) = \tilde{\sigma}\underline{E}(\underline{r}) \quad . \tag{7.4}$$

If the current density depends on the field at other points, one must write

$$\underline{J}(\underline{r}) = \int \tilde{\sigma}(\underline{r},\underline{r}')\underline{E}(\underline{r}')d^3\underline{r}' \qquad (7.5)$$

and $\tilde{\sigma}(\underline{r},\underline{r}')$ is a non-local conductivity. Examples of such a situation are the anomalous skin effect in metal single crystals at low temperature, which occurs in the infrared and microwave regions, optical activity, and some cases of exciton absorption in the ultraviolet. In such cases in homogeneous solids a wave-vector dependent conductivity $\tilde{\sigma}(\omega,\underline{k})$, the Fourier transform of $\tilde{\sigma}(\omega,\underline{r}-\underline{r}')$, can be introduced. The other extension is to non-cubic crystals. With suitable axes, the dielectric function becomes a diagonal tensor, with either two (hexagonal, trigonal, and tetragonal systems) or three (orthorhombic, monoclinic, and triclinic systems) independent complex components, although the latter (biaxial crystals) require the equivalent of a 6-component non-diagonal tensor to specify the principal axes with respect to crystallographic axes [7.9]. For uniaxial crystals, one principal axis is the c-axis of the crystal. The application of a static stress or an electric field along a symmetry direction may lower the symmetry of a crystal, sometimes increasing the number of independent dielectric function components. The application of a static magnetic field along a symmetry direction will produce off-diagonal components, even for a cubic crystal, and the resultant tensor is antisymmetric [7.10]. It is in the study of non-cubic crystals that the polarized nature of synchrotron radiation is invaluable.

Other quantities often used are the complex refractive index,

$$\tilde{N} = n + ik \equiv \tilde{\varepsilon}^{1/2} \quad , \qquad (7.6)$$

from which

$$\varepsilon_1 = n^2 - k^2 \qquad \varepsilon_2 = 2nk \qquad (7.7)$$

$$2n^2 = (\varepsilon_1^2 + \varepsilon_2^2)^{1/2} + \varepsilon_1 \qquad 2k^2 = (\varepsilon_1^2 + \varepsilon_2^2)^{1/2} - \varepsilon_1 \quad , \qquad (7.8)$$

and the absorption coefficient,

$$\alpha = 4\pi k/\lambda_0 \quad , \qquad (7.9)$$

where λ_0 is the wavelength in vacuum. (We have dropped the explicit frequency dependence.) The atomic or molecular absorption cross section σ is related to the absorption coefficient and ε_2 of a medium containing N molecules per unit volume as

$$\sigma = \alpha/N = 2\pi\varepsilon_2/nN\lambda_0 \quad . \qquad (7.10)$$

At normal incidence from vacuum, the reflectance of an opaque sample is

$$R = \frac{(n-1)^2 + k^2}{(n+1)^2 + k^2} \quad , \tag{7.11}$$

and the transmission of a thin sample of thickness d, neglecting coherent interference effects, is

$$T = \frac{(1-R)^2 e^{-\alpha d}}{1-R^2 e^{-2\alpha d}} \approx \frac{1-R}{1-R} e^{-\alpha d} \quad , \tag{7.12}$$

where the last approximate equality is valid for $\alpha d \gg 1$. Generalizations of these for non-normal incidence may be found [7.11,12], as can the analog of (7.11) for anisotropic crystals [7.13].

As is evident in the equations above, the quantities one measures, e.g., R or T, involve both real and imaginary parts of the functions whose spectra we wish to interpret, $\tilde{\varepsilon}$ or $\tilde{\sigma}$. Thus two measurements at each frequency are generally required. A very large number of pairs of measured quantities have been analyzed for the accuracy with which they give, e.g., n and k, when experimental errors in the measured quantities are considered [7.6,14-16]. In the vacuum ultraviolet region, the only one of these used more than a few times is the measurement of the reflectance at several angles, usually using p-polarized radiation. The experimental technique most frequently employed is to measure R over as wide a frequency range as possible and use the Kramers-Kronig integral (below) to obtain the phase of the reflected electric field, from which $\tilde{N}$ can be obtained.

In the soft X-ray region the above quantities become simpler. Above an energy in the 30 to 100 eV range $n \approx 1$ and $k \ll 1$. Then (7.7), (7.11) and (7.12) simplify to

$$\tilde{\varepsilon}_1 \approx 1$$

$$\varepsilon_2 \approx 2k \ll 1 \tag{7.13}$$

$$R \ll 1 \tag{7.14}$$

$$T \approx \exp(-\alpha d) \tag{7.15}$$

and one measurement, usually that of $\dot{T}$, at each frequency is sufficient.

A useful technique for obtaining the absorption spectrum is that of photoyield spectroscopy, documented elsewhere [7.17-20]. The photoyield spectrum for photon energies above about 30-50 eV is very similar to the absorption spectrum, especially in the spectral features, but these may lie on a smoothly varying background that cannot yet be subtracted. The only other disadvantage of photoyield spectroscopy is that it does not give the absolute magnitude of the absorption coefficient. An important advantage is that it can be used on samples of arbitrary shape and on liquids, obviating preparing thin films of highly reactive materials. In discussing results, we will not specifically label spectra obtained by photoyield methods.

Another material function is the so-called electron energy loss function,

$$\text{Im}\{-1/\tilde{\varepsilon}\} = \varepsilon_2/(\varepsilon_1^{\,2} + \varepsilon_2^{\,2}) \quad . \tag{7.16}$$

This may be shown to be proportional to the probability per unit path length that a fast electron loses energy $\hbar\omega$ in passing through a medium [7.21,22]. Some peaks in the loss function occur near the zeroes of $\tilde{\varepsilon}$. Microscopically, the zeroes in $\tilde{\varepsilon}$ are caused by the occurrence of longitudinal excitations while peaks in ε_2 are the result of transverse excitations. (We assume that the longitudinal and transverse dielectric functions, in reality not the same [7.8], are equal. This is, to date, a good approximation). If $\varepsilon_2 \ll 1$, the loss function peaks at zeroes of ε_1. For a free electron gas, these zeroes occur with $d\varepsilon_1/d\omega > 0$ and $d\varepsilon_2/d\omega < 0$ at $\omega_p = (4\pi Ne^2/m)^{1/2}$ (N: electron density), the frequency of collective excitations of the electron gas, "plasmons", when quantized. In other solids, plasmons in the "gas" of valence electrons, or perhaps in other groups of electrons, give a zero in ε_1 with $d\varepsilon_1/d\omega > 0$ and $d\varepsilon_2/d\omega < 0$. Plasmons localized at the interface between media A and B occur at those frequencies for which $\text{Im}\{-1/(\tilde{\varepsilon}_A + \tilde{\varepsilon}_B)\}$ peaks, i.e., near the zeroes of $\varepsilon_{A1} + \varepsilon_{B1}$ if $\varepsilon_{A2}, \varepsilon_{B2} \ll 1$. For the surface of a solid in vacuum, surface plasmons occur at the peaks of $\text{Im}\{-1/(\tilde{\varepsilon} + 1)\}$. Other peaks in the loss function occur near peaks in ε_2 when $|\varepsilon_1| \gg \varepsilon_2$. These are caused by interband transitions. In the soft X-ray region

$$\text{Im}\{-1/\tilde{\varepsilon}\} \approx \varepsilon_2 \quad . \tag{7.17}$$

The loss function can be generalized to take account of non-locality and anisotropy [7.23.24].

Most of the previously mentioned material parameters have real and imaginary parts related by Kramers-Kronig integrals. For example,

$$\varepsilon_2(\omega) = -\frac{2\omega}{\pi}\, p \int_0^\infty \frac{\varepsilon_1(\omega')d\omega'}{\omega'^2 - \omega^2} \quad , \tag{7.18}$$

$\hbar = h/2\pi$ (normalized Planck's constant)

and

$$\varepsilon_1(\omega) - 1 = \frac{2}{\pi} \int\limits_0^\infty \frac{\varepsilon_2(\omega')\omega' d\omega'}{\omega'^2 - \omega^2} \quad ; \tag{7.19}$$

for others, see [7.7,8.25]. These form a useful check on experimentally derived data and give physical insight into optical properties. They also lead to sum rules and superconvergence sum rules [7.26-29], the most useful of which are

$$\int\limits_0^\infty \varepsilon_2(\omega)\omega d\omega = \frac{\pi}{2} \omega_p^2 \quad , \tag{7.20}$$

and

$$\int\limits_0^\infty \frac{\varepsilon_2(\omega) d\omega}{\omega} = \frac{\pi}{2}[\varepsilon_1(0) - 1] \quad , \tag{7.21}$$

and to partial sum rules,

$$\frac{m}{2\pi^2 e^2} \int\limits_0^\omega \omega' \varepsilon_2(\omega') d\omega' = N_{eff}(\omega) \quad , \tag{7.22}$$

and

$$\int\limits_0^\omega \frac{\varepsilon_2(\omega') d\omega'}{\omega'} = \frac{\pi}{2}[\varepsilon_{eff}(\omega) - 1] \quad , \tag{7.23}$$

in which N_{eff} is the density of electrons contributing to absorption below frequency ω, and $\varepsilon_{eff}(\omega)$ is the contribution to the static dielectric constant of those electrons absorbing below ω.

Kramers-Kronig integrals connect $Im\{-1/\tilde{\varepsilon}\}$ and $Re\{-1/\tilde{\varepsilon}\}$, permitting a determination of $\tilde{\varepsilon}$ from electron energy loss measurements. Conversely, the energy loss spectrum can be computed from $\tilde{\varepsilon}$ determined by optical measurements.

7.1.2 Microscopic Description

The dielectric function can be calculated by several techniques from any microscopic model. We quote here an often-used expression, suitable for a one-electron model of a solid:

$$\varepsilon_2 = \frac{\hbar^2 e^2}{\pi m^2 \omega^2} \sum_{i,f} |\hat{e} \cdot \underline{p}_{if}|^2 \delta(E_f - E_i - \hbar\omega) f_{FD}(E_i)[1 - f_{FD}(E_f)] \quad , \tag{7.24}$$

in which $\hat{e}$ is a unit polarization vector for the incident field, f_{FD} is the Fermi-Dirac distribution function and

$$\underline{p}_{if} = \int \psi_f^* \underline{p} \, \psi_i \, d\tau \tag{7.25}$$

is the momentum matrix element between initial state i and final state f. This is the electric dipole approximation, valid even in the soft X-ray region for most known optical transitions in solids [7.30]. ε_1 can be found from the calculated ε_2 by (7.19), with suitable changes for conducting materials [7.7].

In deriving (7.24) is was assumed that the electric field causing the transition was the average field of the electromagnetic wave in the crystal. This neglects local field effects [7.31-39]. If the initial state is very well localized, a local field correction, the Lorentz-Lorenz correction, must be applied, while for a less well localized initial state, a local field correction smaller than the Lorentz-Lorenz correction must be used. On a band picture, no correction is needed for wide bands, but one is needed for a narrow band, the quantitative definition of wide not being well known in actual cases. The local field correction is not important for most core levels, for at the energy of their excitations, ε_1 is close to unity and ε_2 is small, making any corrections small. For valence and some core electron spectra, however, these corrections are very large and cause spectral peak positions and strengths to shift [7.40,71], depending on how large a local field correction is used. This is illustrated in Fig. 7.1. The interpretation of spectra is invariably done without the inclusion of local field effects, however.

In solid-state spectroscopy it is traditional to think in terms of energy bands and Bloch functions. The quantum numbers are the wave vector $\underline{k}$, a band index, spin, and a symmetry label. In the reduced zone scheme, wave vector is conserved upon optical excitation. For symmetry points and lines in k-space, one can use group theory to identify pairs of states, i, f, for which $\underline{p}_{ef} = 0$. For non-vanishing $\underline{p}_{if}$, (7.24) is often simplified by assuming the matrix element to be constant, or slowly varying with energy. Then

$$\varepsilon_2(E) \propto \int \frac{dS_{E_f - E_i}}{|\nabla_k(E_f(\underline{k}) - E_i(\underline{k}))|} \tag{7.26}$$

and the structures in the spectra are attributed to the singularities in the intergrand, i.e., to the critical points. The integrand is expanded in a series about each critical point and characteristic spectral shapes result [7.42-45].

The introduction of the attraction between the excited electron and the hole causes the one-electron picture to break down. New bound states, excitons, can form [7.46,47], and not only is the ϵ_2 spectrum below the lowest M_0 critical point altered by the appearance of exciton "lines" or bands, but the regions below all critical points and above all critical points except M_3 are altered, usually reduced, but certainly changed in shape, by the electron–hole interaction for an energy range of the order of the exciton Rydberg [7.48-50]. For a large exciton Rydberg, the entire spectrum is distorted. In fact, it makes little sense then to talk about critical points. Tightly bound, highly localized excited states are composed of states from large volumes of k-space, i.e., large regions of one or more bands. The spectral shape expected at a critical point becomes mixed with that of another type of critical point [7.48-50].

The wave vector is not a good quantum number for well-localized initial states. Excitation of core electrons to the conduction band occurs only to states which overlap the core level. One can think of a localized density of conduction band states, possibly only for a particular projection of the angular momentum. In such cases, the absorption spectrum should resemble that of an atom, but there may be solid state effects near threshold, due to the electron-hole attraction which may be modified in the solid, just above threshold where some density of states structure may occur (although this is a rare occurrence), and in the delayed absorption continuum where there may be interference effects from near neighbors of the excited atom. These will be discussed subsequently. It is still difficult to say *a priori* whether a localized or delocalized picture will be better for the near-threshold absorption of a particular core level [7.51].

All solids exhibit collective excitations, plasmons, at energies usually corresponding to vacuum ultraviolet wavelengths, although in normal geometries they are

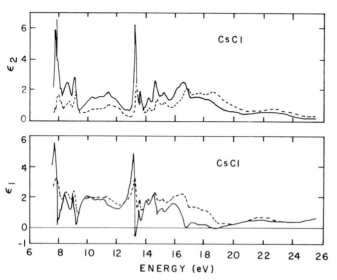

Fig. 7.1 Real and imaginary parts of the dielectric function of CsCl from the Kramers-Kronig analysis of the reflectance spectrum measured at 90 K. The solid lines are the usual dielectric functions with no local field correction and the dashed lines are the results with the full Lorentz-Lorenz local field correction [7.40]

not directly excited by photons [7.52-54]. Since they are longitudinal excitations, they are readily excited by fast electrons [7.55,56]. They may be excited with photons in special geometries, with p-polarized light, and on rough surfaces. Standard optical measurements which give the dielectric function can, however, give the zero-wave-vector plasmon spectrum, for it is proportional to $Im\{-1/\tilde{\varepsilon}\}$ and $Im\{-1/(\tilde{\varepsilon} + 1)\}$ for volume and surface plasmons, respectively. Plasmons can be directly studied by electron energy loss measurements, for the volume and surface plasmons may be separated by different relative intensities as a function of angle and sample thickness, but in the event that the requisite measurements cannot be made, optical determinations of the loss function are very useful.

A free electron gas will have a volume plasmon energy of $\hbar\omega_p = \hbar(4\pi Ne^2/m)^{1/2}$ at zero wave vector, and the corresponding surface plasmon energy is $\hbar\omega_p/\sqrt{2}$. Here N is the number density of electrons. For a simple system with an energy gap, the same Coulomb interaction giving rise to plasmons can give a longitudinal exciton, the latter below the band gap E_G, and the former at $[(\hbar\omega_p)^2 + E_G^2]^{1/2}$. Early theories [7.57,58] concluded that one or the other type of excitation, but not both, occurred, but later work shows that both can occur in one system [7.59,60], as is known experimentally (see Sects.7.4.1 and 2). For real systems, the interband transitions may have a drastic effect on the plasmons, introducing partial screening on the Coulomb interaction by the electrons responsible for the interband absorption. Generally the plasmons then have energies other than $\hbar\omega_p$ (the above expression being an example) and the surface plasmon energy is not $\hbar\omega_p/\sqrt{2}$. Such complicated systems may be treated phenomenologically [7.61], sometimes resulting in more than one volume and more than one surface plasmon. (Often one volume plasmon is near $\hbar\omega_p$, however.) Microscopic treatments of such complicated systems are not yet available.

7.1.3 Modulation Spectroscopy

Modulation spectroscopy [7.62-64] consists of the application of an oscillating perturbation to a sample, often at an audio frequency, and the measurement of the component of the transmitted or reflected light at the modulation frequency, or a harmonic. Typical perturbations are uniaxial strain, an electric or magnetic field, or a temperature change. Each of these produces a change in the dielectric function, $\Delta\tilde{\varepsilon}$, which causes a change in the measured quantity, e.g., ΔR in the reflectance R. By measuring ΔR or $\Delta R/R$ over a wide frequency range, one can get the associated change in the phase shift $\Delta\theta$ by a Kramers-Kronig integral [7.65,66]. These, and the dielectric function, allow one to obtain $\Delta\tilde{\varepsilon}$ as

$$\Delta\varepsilon_1 = [n(\varepsilon_1 - 1) - k\varepsilon_2]\frac{\Delta R}{2R} + [k(\varepsilon_1 - 1) + n\varepsilon_2]\Delta\theta \tag{7.27}$$

$$\Delta\varepsilon_2 = [k(\varepsilon_1 - 1) + n\varepsilon_2]\frac{\Delta R}{2R} - [n(\varepsilon_1 - 1) - k\varepsilon_2]\Delta\theta \quad , \tag{7.28}$$

where the prefactors of $\Delta R/R$ and $\Delta\theta$ are known as Seraphin coefficients [7.67]. (Normal incidence at a vacuum-solid interface is assumed.) This method can also be applied to a three-material system such as an oxide-coated sample [7.68,69].

The advantages of modulation spectroscopy can be seen by expanding the dielectric function near a critical point in a series in the perturbing parameter ΔF, here taken to be a scalar. The change in the imaginary part, (7.24), can be written as

$$\Delta\varepsilon_2(E) = \frac{\partial\varepsilon_2}{\partial|p_{if}|^2} \frac{\partial|p_{if}|^2}{\partial F}\Delta F + \frac{\partial\varepsilon_2}{\partial(E_{if}-E)}\frac{\partial E_{if}}{\partial F}\Delta F + \cdots \tag{7.29}$$

The first term varies in energy as $\varepsilon_2(E)$, and represents the change in matrix element due to the perturbation-induced wave function changes. This is negligible in most cases. The second term represents a shift of the energies of the initial and final states under the perturbation. Near a critical point all states are expected to shift together. More terms in this series may be needed to allow for the influence of the perturbation on the broadening parameters [not shown in (7.29)] or on the occupation factors (if states i or f are near the Fermi energy), or for cases in which the second term vanishes by symmetry. Vector perturbations lift degeneracies, a mechanism not included in (7.29). The first factor in the second term in (7.29) shows that the $\Delta\varepsilon_2$ spectrum will be proportional to $d\varepsilon_2/dE$; hence structure in $\varepsilon_2(E)$ will be emphasized, especially near critical points. Just as different types of critical points give characteristic shapes in ε_2, they give characteristic shapes in $\Delta\varepsilon_1$ and $\Delta\varepsilon_2$. The other factor in the second term of (7.29) is a measure of the response of the states associated with the interband critical point to the perturbation, and this is a powerful interpretative tool when combined with theoretical expectations of how high-symmetry states respond to perturbations. This factor emphasizes or diminishes the roles of certain state pairs in $\Delta\varepsilon_2$.

One can, of course, measure the spectra with and without the application of a static perturbation, but if equal amounts of time are spent collecting the data, the difference spectra will have the same signal to noise ratio [7.70]. However drift in the system often introduces errors, so modulation spectroscopy is usually preferable. One can also take the energy derivatives of measured spectra numerically, both first and second derivatives being useful and possible, if the signal to noise ratio is high enough, or, by various means, one can sinusoidally modulate the wavelength emerging from the monochromator so that $dR/d\lambda$ is measurable directly [7.71]. These methods, energy derivative spectroscopy and wavelength modulation spectroscopy, are both useful, affording the "resolution" of structure difficult to see in the normal spectra, but they lack the additional feature of modulation spectroscopy which gives the response of a group of levels to the perturbation.

Thermomodulation is perhaps the easiest to carry out, for the requirements on the sample are easy to achieve, but it is the most difficult to interpret, for so many mechanisms can contribute to the response [7.72-72]. Scalar perturbations, e.g., a temperature change, gives no symmetry information. The vector perturbations, uniaxial stress [7.75] and electric field [7.76-80], can lift the degeneracy of equivalent points in reciprocal space, depending on the orientation, and this can give additional information, the direction in reciprocal space of the critical point transitions. Moreover, electroreflectance on semiconductors can give a third derivative spectrum, with its attendant increase in resolution [7.81]. Magnetic field perturbations have been applied with most success to the lowest interband edges of semiconductors and insulators [7.82] (but see Sect.7.2.1).

Two samples of slightly different composition can be alternated in the beam of measuring radiation, giving so-called composition modulation spectroscopy, for the detector yields an alternating signal proportional to the difference in reflectance or transmission and a DC signal proportional to the average reflectance or transmission. This is especially useful in the study of alloys [7.83].

A radiation source for modulation spectroscopy should be continuous, intense, and stable in both position and in time. Synchrotron radiation, especially from storage rings, meets all these requirements, and, moreover, its polarization is necessary in some measurements. The signal-to-noise ratio in modulation spectroscopy at best varies as the square root of the number of photons counted, and in this respect synchrotron radiation from storage rings is superior to all other conventional sources for photon energies above about 5 eV.

7.1.4 Summary

Qualitatively the gross optical properties of most solids in the visible and vacuum ultraviolet, arising from transitions from valence to conduction bands, can be described as follows. For semiconductors, most structure in ε_2 arises from interband critical points, and there is considerable oscillator strength from many such transitions. Exciton effects are detectable, but not prominent. Insulators have very strong exciton effects which may dominate the ε_2 spectra. Metals, however, show few signs of critical points. Transitions throughout large volumes of the Brillouin zone in metals are important, though they may contribute little structure to ε_2. The effect of the Fermi level is to cause some additional structure in ε_2 as transitions suddenly become allowed to or from a flat band.

This picture is expected to be valid further in the vacuum ultraviolet with three exceptions. Electron and hole life-times may shorten, broadening the spectra, the final states may become more like plane waves, and the localization of the hole will increase. In the latter case, and for the excitation of core electrons, a more atomic approach will have to be used. Exciton effects then are best described as multiplet structure in the transitions. Away from absorption edges, absorption structure arises from the radial part of the dipole matrix element. These subjects are discussed in Chap. 4.

7.2 Metals and Alloys

7.2.1 Vacuum Ultraviolet

a) Simple Metals

The dielectric functions of many simple metals have been determined below 6, 10 or 20 eV with conventional sources [7.84,85]. Because of the lack of structure above 6 eV, there has not been much interest in repeating the measurements with synchrotron radiation, and interest has focused on the sum rules and on plasmons, comparing optical data with data from electron energy loss measurements. An unsettled question in interpreting the interband absorption is the role of empty d-bands as final states [7.86-94].

Exceptions to this lack of interest are the simple metals which have d core electrons, of which the 3d and 4d electrons of Zn and Cd are the best examples. (We defer the s and p core levels to Sect.7.2.2.) Both are about 10 eV below the Fermi level and very close to the bottom of the conduction band. These deep states are not completely atomic-like, however, for photoemission studies have shown them to be somewhat broadened by the crystal potential, and their splitting in metallic Zn and Cd is about 50% larger than the spin-orbit splitting of the free atom, another effect of the crystal potential [7.95-99]. Measurements on Cd have shown that transitions from the uppermost 4d level begin at 9.35 eV and persist to very high energy, only about a third of the oscillator strength having been used by 30 eV [7.100]. There is considerable structure in the absorption just above threshold, but band calculations do not extend to high enough energy to allow even a rudimentary interpretation of the spectrum. The 3d, 4d and 5d transitions of other simple metals, Ga, In, Sn, Hg, Tl, and Pb have not been measured with synchrotron radiation, although the In 4d edge does appear in data on InBi [7.101]. The properties of these core levels themselves are best studied by X-ray induced photoemission spectroscopy (XPS). Then the absorption from these levels can be used to study the region of the conduction band above the Fermi level, although it has not been clearly established that the weak structures seen are primarily due to band structure and dipole matrix element effects.

Although not a simple metal, polysulfur nitride lacks d-electrons and is mentioned here. The reflectance of a film of $(SN)_x$ was measured to 27 eV [7.102].

The resultant dielectric function agreed qualitatively with that calculated from several different band structures, affording no discrimination among them.

b) Noble Metals

The noble metals have been studied frequently in the visible and vacuum ultraviolet. The absorption features below 10 eV arise primarily from transitions from the filled d bands to the conduction bands, with some contribution from interconduction band transitions. All three noble metals have peaks in their reflectances above 20 eV. the peak in Au being especially prominent. (All heavy transition metals such as Pt and Os have similar peaks.) These peaks contain weak structure which has been brought out in thermomodulation measurements on Au [7.103]. The width of these structures is as small as 0.1 eV in thin film samples. Fig. 7.2 shows the thermoreflectance spectrum and the differential dielectric function spectra obtained by Kramers-Kronig analysis. No core electrons can contribute to spectra at these energies, so the increase in absorption must arise from the excitation of conduction or 5d electrons into a region of high density of states far above the Fermi level. The energy bands of Au show such a region, arising from bands derived from the Au 5f levels, which become f- and p-like at some symmetry points in the Brillouin zone.

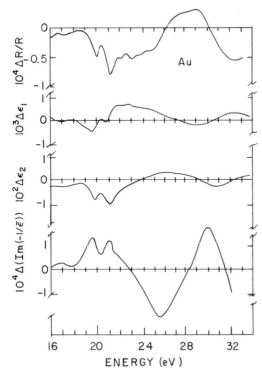

Fig. 7.2 The measured thermoreflectance spectrum of gold at about 200 K and the differential real and imaginary parts of the dielectric function, $\Delta\varepsilon_1$ and $\Delta\varepsilon_2$, derived from it by Kramers-Kronig analysis. $\Delta\mathrm{Im}\{-1/\tilde{\varepsilon}\}$ is the differential energy loss function [7.104]

The three sharpest peaks in ΔR/R (19.9-22.7 eV in Fig.7.2) were assigned originally to transitions from the three levels in the 5d band at the zone center to a common
final state, f-like at the zone center. The 5d states were spin-orbit and crystal-
field split. Since the bands involved are so flat, these transitions may occur in
other parts of the Brillouin zone as well (or instead). Several alternative in-
terpretations were suggested [7.105,106] and two transitions were observed near
these energies in photoemission [7.105]. A definitive assignment of these transitions localized in the Brillouin zone and from extended regions of k-space.
These transitions are unusual in that the lifetime of the final state must be fairly
long, possibly due to its f-like symmetry. The states just below it in energy have
a low density of states, which reduces the scattering rate. Additional evidence for
or against any proposed model should come from moving some of the initial states
above the Fermi level, as in Pt or Ir, and preliminary spectra for Pt show structures
corresponding to those of Au [7.107].

To study the degree of localization of these transitions, BEAGLEHOLE and
THIEBLEMONT [7.108] measured the reflectances of Cu, Au, and Cu-Au alloys. In the
alloys it is known that the Cu 3d and Au 5d states merge into a single band. If
the final states are localized on a single site, one expects the alloy spectra
above 19 eV to be a weighted sum of the pure metal spectra. (The spectral features
presumably arising from the 3d-"f" transitions in Cu begin about 5 eV higher in Cu
than those of Au and they are weaker.) A preliminary analysis of the data indicates
that on the Au-rich side, the final states do not mix, i.e., they are localized
[7.108].

The thermoreflectance spectra of Au also show other structures [7.103,109]. Below 19 eV there are some interband features which cannot be assigned with certainty
to transitions between the 5d and the conduction bands. Above 25 eV there are two
structures which have the shape expected for plasma edges, shifted in position by
the change in electron density caused by thermal expansion. These occur at regions
of peaks in $\mathrm{Im}\{-1/\tilde{\varepsilon}\}$ [7.110,111].

Thermomodulation is one of the few modulation techniques applicable to metals in
the vacuum ultraviolet. Electromodulation in metals requires the use of liquid elec-
trolytes to date, and piezomodulation gives rather small signals, especially on
metal single crystals which have very low elastic strain limits. Thermomodulation
and wavelength or energy derivative spectroscopy seem likely to remain the best
methods for studying band structure effects in optical spectra. Line shapes for
both core to conduction band and conduction band to non-free-electron-like conduction
band transitions should be calculated for a few systems.

c) Transition Metals

Transition metals have been studied extensively by optical and electron energy loss techniques, and a general picture is emerging [7.112-120], despite the fact that very often the data have been taken on samples of these reactive metals with oxygen on the surface or in the bulk. This affects the magnitudes of ε_1 and ε_2 but not their spectral features, provided the contamination is slight, but errors in the magnitude alter the shape of the electron energy loss function calculated from $\bar{\varepsilon}$. Surface contamination can also distort the measured loss function.

Interband transitions begin at energies as low as 0.1 eV because of the incompletely filled d-bands. They persist, with some structure, throughout the visible and near ultraviolet, then a minimum in ε_2 and in the reflectance occurs at an energy which depends primarily on the crystal structure, roughly at 10, 6 and 10 ev for bcc, hcp, and fcc structures, respectively. There is then another rise in ε_2 and the reflectance at higher energies. Fig. 7.3 shows examples of this minimum in the conductivities of the bcc transition metals Nb and Mo. This behavior can be understood on the basis of the density of states alone. Transitions from filled to empty d-states are forbidden in free atoms, but allowed in the solid because of the hybridization of the d bands with s-p bands. The matrix elements for such transitions are not often known, but limited calculations show that they can depend strongly on wave vector [7.122]. The interband transitions begin weakly at an arbitrarily low energy, and suddenly become strong at higher energies if the Fermi level lies in a minimum of the density of states, i.e., in a "gap" in the density of states.

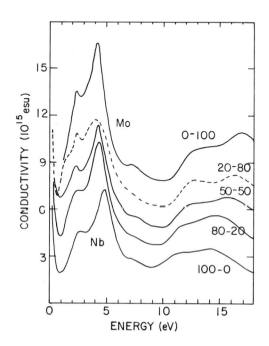

Fig. 7.3 Optical conductivity of Nb, Mo and three Nb-Mo alloys. The alloy concentrations are given in atomic percent. The scale is correct for Nb. Each other curve has been displaced upward 1.5×10^{15} esu from the curve below [7.121]

There is an absorption edge at the gap energy (hcp and bcc, but not fcc because
for the latter the Fermi energy is not at a minimum in the density of states.)
Strong absorption then persists till the energy of the full width of the d-band,
above which energy the density of either initial or final states is the small
density of states of just the s-p bands. Since there are no core levels in the
10-20 eV region, the subsequent rise in absorption must occur from a rise in the
density of final states. Band calculations on a few 4d and 5d bcc transition metals
show again that there is a flat band partly derived from f and p levels somewhere
above the Fermi level and that this band has a fairly large density of states [7.122].
The dipole matrix elements have not often been evaluated. The optical data show that
the position of this band above the Fermi level depends primarily upon crystal structure, being lowest for hcp transition metals and highest for fcc transition metals
(and noble metals, for this corresponds to the transitions in Au discussed in the
preceeding section.) Transitions to these states were studied by thermoreflection in
Mo films [7.123], but no sharp structures like those in Au were seen.

Several alkali metal (M) tungsten bronzes, M_xWO_3 with x < 1, have been measured
to 38 eV [7.124]. The interband absorption below 8 eV has been ascribed to transitions between bands based on hybridized O 2p and W 5d functions. There is a prominent
increase in the conductivity at 8 eV which cannot arise from any core level. It, too,
must be ascribed to excitations from the Fermi level to states with a high density
of states beginning 8 eV above the Fermi level, states based on f and p levels of
W. In this case the W atoms form a simple cubic lattice (or a distorted one) in
the perovskite-, or distorted perovskite-, structured bronzes, but the W-W distance
is sufficiently large that a localized final state description may be more appropriate than a band picture.

The transition metal alloys Nb_xMo_{1-x} are expected to show rigid band behavior
[7.125]. Their band structures are nearly identical, so the principal effect of adding Mo to Nb is an increase in the Fermi level. The edge in absorption at about
10 eV should then shift to lower energy as Mo is added to Nb, but as Fig. 7.3 clearly
shows, the shift is in the other direction. This may be an effect of the crystal
potential on the position of the "f-band", for, on a scale of 1 or 2 eV, it may
depend on more than just the crystal structure. The shift is not the obvious one of
a drop in the 4f level as the atomic number increases, for it is in the opposite
direction. From the shapes of the edges in Fig. 7.3 it is clear that the alloy spectra above 10 eV are not weighted sums of the spectra of the constituent elements.
The final states must be somewhat delocalized. These features of the transitions
from the d bands to "f bands" are not universal, however, for in Fe-Ni alloys there
is virtually no effect of composition on the corresponding threshold (although the
strength of the absorption grows with increasing Ni concentration), despite a change
in crystal structure [7.126]. Moreover, in Au-Cu alloys, discussed in the previous
section, the spectra imply considerable localization of the final states. It is

clear that more measurements of this edge in other alloy systems are needed before
a detailed interpretation can be made.

Almost all transition metals have electron energy loss spectra which exhibit at
least two peaks arising from volume collective excitations, here to be called plasmons, although, as will be explained, they are unlike those in simpler materials such
as alkali metals [7.127]. (Alkaline earth metals also exhibit two volume plasmons
[7.128].) Early electron energy loss measurements on transition metals often were
made in reflection on samples prepared by continuous evaporation to avoid oxidation.
The measurements of the spectra as a function of scattering angle and sample thickness, needed to separate volume from surface losses [7.55], could not be made, although some later work in which a surface was deliberately oxidized led to the recognition of surface excitations [7.129]. The early spectra had four peaks which
were interpreted in order of increasing energy as the surface and volume plasmons
and double excitations of each, although the plasmon energy expected for a free
electron gas often coincided with the energy of the third or fourth loss peak.
Vacuum ultraviolet optical data (Fig.7.4) showed, however, that there were two peaks
in Im$\{-1/\tilde{\varepsilon}\}$ and two in Im$\{-1/(\tilde{\varepsilon} + 1)\}$ (W has three in each spectrum) with the lower
peak in each spectrum nearly coincident, making them unresolved in electron energy
loss spectra, especially those using backscattered electrons; see, however [7.130].
The lower peak in each spectrum falls in the region of the minimum of ε_2. Thus after
the d- to d- transitions are exhausted (The sum rule indicates that, e.g., only
about 3 of the 5 electrons/atom of Ta and 3 of the 6 of Mo have contributed to ε_2
below 10 eV [7.113]) there is a collective oscillation of each type, and after the
d-f transitions are exhausted (the sum rule shows nearly all valence electron oscillator strength is exhausted) there is another. The higher energy volume plasmon is

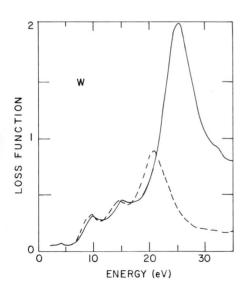

Fig. 7.4 Volume and surface electron
energy loss functions, solid and dashed,
respectively, calculated from the optically measured dielectric function for
W [7.116]

very close in energy to that expected for the free electron gas model treating all the s-p-d-valence electrons as free. The higher energy surface plasmon is, however, much higher in energy that the free electron gas result of $\hbar\omega_p/\sqrt{2}$. In the case of the Nb-Mo alloys, the high energy plasmon energy shifts with composition as the expected square root of the electron density [7.121]. It is tempting to ascribe the lower frequency plasmon to a group of electrons, the d-electrons or the s-p electrons, but it is clear that it does not shift with an increase in d-electron concentration.

It is well-known that interband transitions play a role in collective excitations. Phenomenologically, their contribution to ε_2 can damp plasmons out of existence as elementary excitations, and their contribution to ε_1 can shift the position of the resonances in $Im\{-1/\tilde{\varepsilon}\}$ [7.61]. The normal-mode problem for such complicated systems has not often been studied, especially in systems as complicated as transition metals. CAZAUX [7.131,132] has modelled the two-plasmon behavior of layered semiconductors with a pair of Lorentzian oscillators, one for the lone-pair electrons and one for the bonding electrons. These two sets of electrons have interband transitions to a common final band, the antibonding band, which are well localized and separated in energy. The composite dielectric function yields two peaks in $Im\{-1/\tilde{\varepsilon}\}$, one near the free electron gas result, and one much lower, although, in the former case, only one set of electrons is "directly" involved in the excitation. For transition metals (and other simpler metals which also show two plasmon behavior) the two groups of electrons are really the same electrons, but with transitions to two sets of final states. The overlap of these transitions and their spread in energy is so large, however, that modeling them with two oscillators is not very realistic.

d) Rare Earths

The rare earths as a group have not been thoroughly studied in the vacuum ultraviolet [7.133]. They are very reactive, and even if handled in ultra-high vacuum, may contain considerable oxygen from their previous chemical and metallurgical history, oxygen which may be incorporated in films prepared in good vacuum. This oxygen does not affect all measurements with equal severity, however. Magneto-optic Kerr effect measurements [7.134,135] seem not to be so severely affected. In these measurements, first carried out in the visible [7.136,137], an applied alternating magnetic field causes the magnetization $\underline{M}$ to reverse periodically. In cubic ferromagnetic materials magnetized along the z-axis the conductivity tensor has the form

$$\begin{pmatrix} \tilde{\sigma}_{xx} & \tilde{\sigma}_{xy} & 0 \\ -\tilde{\sigma}_{xy} & \tilde{\sigma}_{xx} & 0 \\ 0 & 0 & \tilde{\sigma}_{zz} \end{pmatrix} \quad ,$$

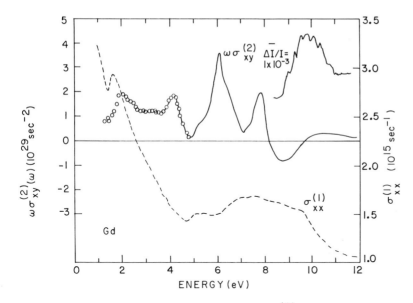

Fig. 7.5 Magneto-optical spectrum of Gd. $\sigma_{xx}^{(1)}$ is the optical conductivity [7.140] and $\sigma_{xy}^{(2)}$ is the absorptive part of the off-diagonal component of the conductivity tensor of a magnetized crystal [7.138]. The inset shows fine structure in the 8 eV peak. The dots are data obtained by an ellipsometric technique [7.137]

in which $\tilde{\sigma}_{xy}$ is proportional to M and is small, making modulation spectroscopic techniques necessary for good signal-to-noise ratios. The imaginary part of $\tilde{\sigma}_{xy}$ represents the difference in absorption of left- and right-hand circularly polarized light, and, as such, may be positive or negative. Spectra of $\mathrm{Im}\tilde{\sigma}_{xy}$ have been obtained in the vacuum ultraviolet by ERSKINE for Fe, Ni, Co, and Gd [7.138,139]. In ferromagnetics the magneto-optic Kerr effect arises from the spin-orbit interaction, but its interpretation may vary with the nature of the absorbing electrons [7.140]. For Gd the effect is important for locating the optical excitation of the 4f electrons, which have weak, but not negligible, dipole matrix elements near threshold. They contribute strongly to $\mathrm{Im}\{\tilde{\sigma}_{xy}\}$, which is proportional to the spin-polarization of the initial (4f) states and to their spin-orbit splitting, both of which are large compared with those of other electrons being excited at the same energy. The sign of $\mathrm{Im}\{\tilde{\sigma}_{xy}\}$ at threshold is governed by the sign of the spin polarization of the initial states. Fig. 7.5 shows both the conductivity ($\mathrm{Re}\{\tilde{\sigma}_{xx}\}$) and the absorptive part of the off-diagonal component of the conductivity ($\mathrm{Im}\{\tilde{\sigma}_{xy}\}$) for polycrystalline Gd. Sum rule considerations placed the onset of absorption of the 4f electrons at a rise in $\mathrm{Re}\{\tilde{\sigma}_{xx}\}$ at 6.1 eV [7.141]. The peak in $\mathrm{Im}\{\tilde{\sigma}_{xy}\}$ at this energy is the corresponding structure, and it can be seen to be the largest feature in the spectrum. The inset shows fine structure seen around 8 eV, attributed to multiplet structure in the 4f shell. The onset of 4f excitation was 2 eV lower than expected from the positioning of the 4f levels by XPS data, a shift attributed to the binding of a conduction band electron to the 4f hole in optical, but not X-ray, excitation.

For Ni, the magneto-optic Kerr effect shows that the absorption edge at 7 eV [7.126] arises from minority-spin electron, i.e., from the excitation of d-electrons from states just below the Fermi energy to p-like states 7 eV above the Fermi energy [7.139].

7.2.2 Soft X-ray

a) Simple Metals

The core electron excitations in simple metals have attracted considerable attention in recent years because they have been expected to exhibit unusual threshold edge shapes, attributed to many-body effects, when observed in absorption and emission, and characteristic asymmetric line shapes in X-ray photoelectron spectra. Measurement of the absorption edges of these metals with synchrotron radiation has played an important role in the study of this "edge problem" [7.142-155]. The history of the efforts to reconcile the theory of these edge shapes with experimental data on the K-edge in Li and the $L_{2,3}$ edges in Na, Mg, and Al was given by KUNZ [7.2] and KOCH et al. [7.4]. We give only a brief outline of the problem, and describe the most recent work in this area. According to the theory of MAHAN, NOZIERES and DE DOMENICIS (MND) [7.156-158], the K edges (s → p) are expected to be rounded and the $L_{2,3}$ edges (p → s) peaked, both in absorption and emission, from what is expected for a one-electron model, usually a free-electron gas model. If these edges are measured in electron energy loss spectra, the rounded K edges should become less rounded as the momentum transfer increases, eventually becoming peaked, while the $L_{2,3}$ edges should become less peaked, although these expectations are model-dependent [7.159]. The asymmetric XPS line shape for a core excitation should also be described by the many-body theory using a parameter that must also be related to a parameter, the threshold exponent, used in fitting the edge shapes for transitions from the same core level [7.160,161]. Several innconsistencies existed in fitting all the data in 1975, and the temperature dependence of the Li K edge was not understood.

In the past two years there have been several new experimental results, and several additional microscopic effects introduced, further complicating the fitting of edge data. BAER et al. have measured the Li K level XPS line shape as a function of temperature [7.162]. They were able to produce a fit at all temperatures by convolving the MND shape with a Lorentzian function to represent the broadening due to the core hole lifetime, the instrument function, and a Gaussian function representing the effect of phonons. Only the latter had any temperature dependence. Not only was a good fit achieved, but the same parameters could be used to obtain a satisfactory fit of the absorption edge in Li, especially when using a density of final states weighted for p-character. The largest single contribution to the edge broadening or rounding was the phonon term, even at low temperature. CITRIN et al. [7.163] subsequently made similar measurements on the K levels of Na, Mg and Al with the result

that the hole lifetime is the dominant broadening mechanism, and on the $L_{2,3}$ levels
in Na, Mg and Al. They calculated the expected one-electron edge shapes for the
latter and used the parameters obtained from XPS plus the MND theory to fit measured
edge shapes, finding the many-body contribution to be significant. In Na, exchange
coupling is needed as well. The inclusion of dipole matrix elements and a realistic density of final states was not found to be important in Na and Mg.

The latter conclusion is at variance with the work of GUPTA and FREEMAN [7.164-
166], who calculated the $L_{2,3}$ edge shapes for Na and Mg and the K edge shape for Li
using only a one-electron picture, incorporating dipole matrix elements and a very
fine mesh for the Brillouin zone integration. They find that the one electron edge
shapes by themselves are in reasonable agreement with the measured edges, and no
large additional effect is needed. Adding to the confusion are the calculated results
that exchange effects and broadening alone can account for the Na $L_{2,3}$ edge shape
[7.167], and that exchange effects should make a significant contribution to the
rounding of the Li K edge, and make consistent fits with the Li K XPS data easier
to achieve [7.168a].

The K $M_{2,3}$, Rb $N_{2,3}$, and Cs $O_{2,3}$ edges have been measured recently by ISHII et
al. [7.168b]. All exhibit the spike phenomenon, but in Rb and Cs, only at the lower
energy component. The sharpness of the spike decreases with increasing principal
quantum number and the threshold exponents do not follow the electron density dependence of the MND theory.

There have been additional measurements which have been evaluated on the basis of
the MND model, but not always with the use of realistic final states and dipole matrix elements. Electron energy loss measurements have been made on the Li K edge
[7.169] and the $L_{2,3}$ edges of Na and Mg and Al [7.170,171]. At large momentum trans-
fers, no significant change in edge shape is seen. New soft X-ray emission spectra
of the Li K edge have been made at several temperatures [7.172]. These spectra are
mirror images of the absorption spectra and broaden considerably with increasing
temperature. The mirroring is not expected if the density of final states enters
into the interpretation of the absorption spectrum above the edge. The $N_{2,3}$ edge in
Cs has been measured, as well as the $L_{2,3}$ edge in liquid Na. The latter revealed a
spike, but whether there is enough local order in liquid Na to give a one electron
"band structure" feature is not known. Finally, it should be pointed out that all
of the experimental data used may not be self-consistent, as emphasized by GUPTA et
al. [7.166]. Emission measurements are plagued by self-absorption, photoyield measurements may have some self-absorption effects, and the preparation and retention of
clean films of these reactive metals for transmission measurements is difficult.

Several eV above the edges, the soft X-ray absorption spectra of simple metals
exhibit undulatory structure, originally ascribed to a variety of effects, but now
apparently arising from interference between the final state wave function components
representing outgoing electrons and electrons backscattered from the neighboring

atoms, as in extended X-ray absorption fine structure (EXAFS) [7.173]. The analysis
of this structure is more difficult than in the harder X-ray region because of the
limited energy range available before the structure is overlapped by that from other
transitions. There is a small region above the edge from which limited band structure information may be obtained [7.174] (but see [7.175] and [Ref. 7.2, Sect.6.1.2])
perhaps better in alloys than in pure metals. The edge and this region have been
observed to shift and change upon alloying but a quantitative interpretation is not
yet available [7.2,176-178].

Core transitions in Cu, Ag, Sn, Au, and Bi have been observed and the sum rule
tested along with that for Al [7.179-181].

b) Transition metals

The nd transition metals have np core electrons with excitation thresholds in the
30-80 eV region. It was hoped that the empty part of the nd band would appear in the
spectra, but measurements on 3d and 5d transition metals clearly showed that this
was not the case [7.182-185]. The observed structures occupy far too large an energy
region. Similar structures occur in the 4d transition metals as well [7.186]. The
spin-orbit splitting of the 3p levels is barely resolved while that of the 4p and 5p
levels is easily resolved. The structure was originally assigned to the interaction
of the excited nd electron with the np hole [7.187,188], as occurs dramatically with
the 4d-4f excitations of the rare earths, but with a smaller interaction energy and
fewer (unresolved) multiplet components. Later calculations on Sc, Ti, V, and Co
showed this could account for the observed structures [7.189]. This explanation
cannot hold for Ni, Pd, and Pt, for the excited electron completes a d-shell and
there can be no multiplet splitting except the spin-orbit splitting of the hole.
DIETZ et al. [7.190] proposed a model for Ni involving the interference of two excited configurations, the discrete $3p^53d^{10}$ and the continuum $3p^63d^8$ εf which led to
the observed Fano line shape [7.191]. (The former is broadened by the $3p^53d^{10} \rightarrow$
$3p^63d^8$ εf Auger transition.) DIETZ and McRAE [7.192] later discussed applying this
model to Pd and Pt. DAVIS and FELDKAMP [7.193] extended this model, including multiplet structure, to the 3p absorption spectra of Cr, Mn, Fe, and Co with success,
offering an improved explanation for the observed line shape. The multiplet structure contributes to the width of the edge structure in all four metals, but additional broadening, ascribed to a small local magnetic moment (S = 1/2), is needed
to fit the edge of Cr.

The corresponding transitions have been observed in several transition metal
alloys with Al [7.176-178], and in Cu-Ni alloys [7.194], but these were discussed
elsewhere [7.2]. The results on Cu-Ni alloys show clearly that the absorption is
localized on a single site, as the alloy absorption spectrum is a weighted sum of
the spectra of its constituents. There is no effect on the Ni spectrum of filling

the 3d band. The extreme localization of the 3p transitions of Ni in NiAl and NiAl$_3$ is not so evident, however.

The 3p excitation of Ni is unusual in that the material is ferromagnetic and the final states are exchange-split. The magneto-optic Kerr effect has been calculated to be quite large and to be capable of distinguishing between two models for the ferromagnetism of Ni [7.141]. Measurements are in progress [7.195].

c) Rare Earths

The 4d-4f excitations have been described several times, along with their relation to the corresponding transitions in Ba, Cs, and Xe [7.196,197]. Recent measurements have been reported on Gd and Dy [7.198], the former being in poorer agreement with theory than those on several other rare earths. In all cases, the measured spectra for the metals agree fairly well with those calculated for free tripositive ions; for Ce, the measured spectra for metal and vapor are nearly identical [7.199]. The 5d-5f excitations in Th and U show features similar to the 4d-4f transitions in the corresponding rare earths [7.200].

7.2.3 Summary

The vacuum ultraviolet spectra of metals and alloys have aided in the understanding of the electronic structures and collective excitations. Additional theoretical work, especially on bands far above the Fermi level and on dipole matrix elements, is needed, along with calculations of the optical properties of alloys. The "edge problem" for light metals is still with us, but its significance to many-body theory seems to be fading as other effects are being seen as playing large roles in the edge shapes. The structures above the edges in some metals can be interpreted as EXAFS, but we cannot use the limited energy range of such spectra to gain much information about the positions of nearest neighbors of the absorbing atoms. In transition metals and rare earths, atomic effects are dominant at and above many core excitation edges, with solid-state effects playing only a minor role. The edges of both light metals and of transition metals in alloys, however, still require interpretation, for there are shifts and changes of shape upon alloying, as well as apparent effects of the density of states, none of which are well understood. There is a need for more data on well-chosen alloys, as well as for theoretical work

7.3 Semiconductors

7.3.1 Vacuum Ultraviolet

Group IV and III-V Semiconductors. The vacuum ultraviolet optical properties of many group IV and III-V semiconductors were measured before the era of synchrotron radiation sources because of the need to confirm calculated band structures of these important materials [7.42-44,201,202]. The spectra of a few of these semiconductors have been remeasured with synchrotron radiation, measurements of many have been made to higher energies, new semiconductors, including anisotropic ones, have been measured, and modulation spectroscopy has been applied perhaps with more success to semiconductors that to insulators and metals.

Optical transitions from the valence band to the conduction band begin in the infrared or visible and continue to at least 20 eV, although most of the oscillator strength is exhausted by 6 to 8 eV. Transitions from the 4d core levels of In in several In-group V compounds were identified in early work, as were systematic trends in the spectra of some IV, III-V, and II-VI semiconductors, trends based on ionicity, crystal symmetry, and relativistic effects. The 3d core level excitations in Ge were first found with the aid of synchrotron radiation, although they cause very little structure in the reflectance and are very difficult to detect [7.203]. (This paper also discussed the difficulty of extracting the dielectric function from reflectance measurements when $n \approx 1$, $k \ll 1$.) An extensive survey of transitions from d core levels to the conduction band in III-V semiconductors was carried out by CARDONA et al. [7.204-206], who measured both absorption and reflectance spectra. The Ga 3d and In 4d absorption edges all showed the spin-orbit splitting of the core level. Several peaks above the edges were attributed to peaks in the conduction band density of states, e.g., those in Fig. 7.6. In this spectral region, $n \approx 1$ and $k \ll 1$, so the reflectance and absorption coefficient spectra are similar, although there are small peak shifts between the two spectra. The optical constants of 6 III-V compounds were obtained in the 15-40 eV region. The sum rule on ε_2 showed that in the first 25 eV above the onset of the 3d or 4d core transitions only about 3 (Ga) or 4 (In) of the expected 10 d-electrons had contributed to the absorption. The positions of the excitation thresholds of the d-levels were compared with those expected from XPS and bandgap measurements. There was agreement within the errors of several tenths of an eV estimated for the comparison, indicating no exciton effects larger than this.

Electroreflectance (ER) measurements on GaP, GaAs, and GaSb have been carried out in the 5-30 eV region [7.207-209,69]. Since the ER spectra in the limit of small fields are essentially third derivative spectra [7.210,211], there is a great deal of structure to be found in them, provided there is not excessive broadening due,

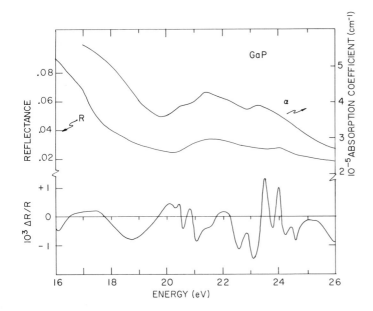

Fig. 7.6 Absorption coefficient α, reflectance R, and electroreflectance $\Delta R/R$ spectra of GaP. The former two spectra [7.206] were taken at 300 K, the latter spectrum [7.208] at 110 K. The structures above 20 eV probably all arise from excitations of Ga 3d electrons, but not all have been interpreted

e.g., to short electron or hole lifetimes. Fig. 7.6 shows a comparison of the reflectance, absorption, and ER spectra of GaP in the region of the excitation of the Ga 3d electrons, all obtained with synchrotron radiation. The enhanced structure in the ER measurements is obvious.

The electroreflectance spectra in the 5-20 eV region arise from transitions between valence and conduction bands, and the critical points involved are not necessarily at symmetry points, nor along symmetry lines. Band calculations alone are not enough to allow an assignment of the structures. Nonetheless most of the structures found in the ER spectra of GaP, GaAs, and GaSb between 6 and 20 eV have been interpreted in terms of particular interband critical points.

The excitations from the flat 3d levels of Ga (bandwidth estimated to be less than 0.1 meV [7.212]) can be interpreted in terms of critical points in just the conduction band, if the electron-hole interaction can be neglected. Possible final states are critical points at Γ, X, and L, and the lowest minimum is known for each of the three compounds studied. Estimates of matrix elements for transitions to each of these minima were sufficiently different that they allow identification of up to three spin-orbit split pairs of structures. The transitions to the Γ_6 minimum are too weak to be seen unless Γ_6 is the lowest conduction band minimum.

In GaP the lowest conduction band minima are the X_6 minima which give a strong ER spectral feature. The ER spectrum for these transitions was Kramers-Kronig analyzed

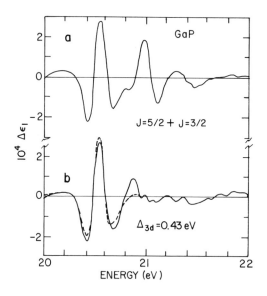

Fig. 7.7 $\Delta\varepsilon_1$ spectrum of GaP obtained by Kramers-Kronig analysis
of the electroreflectance spectrum
(a) as obtained. (b) The contribution from only the j = 5/2 component
(solid) and the spectrum calculated
for an M_0 critical point with exciton effects and broadening included
(dashed) [7.218]

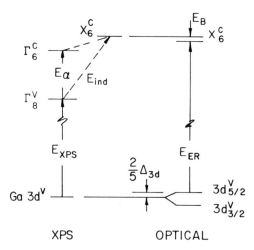

Fig. 7.8 Energy levels and gaps in a
Ga-V semiconductor, e.g., GaP. E_{XPS} is
obtained from X-ray induced photoemission. E_α is the direct band gap, obtained from infrared or visible absorption measurements. E_{ind} can be found
from optical band gap measurements at
high pressure, and from pressure coefficients of this gap. E_{ER} is the gap
observed for Ga3d excitation E_B =
E_{xps} + E_{ind} - E_{ER} is interpreted as
an exciton binding energy. Δ is the Ga
3d spin-orbit splitting

to give a $\Delta\bar{\varepsilon}$ spectrum, shown in Fig. 7.7, which was separated into two spin-orbit
component spectra [7.213]. The Ga 3d level has a spin-orbit splitting of 0.43 ± 0.02
eV and the transitions from the two sub-levels to the X_6 minima occur with the expected statistical weights, indicating that the dipole matrix elements are spin-
independent and that the effects of exchange are small. Each of the spin-orbit components can then be fit to BLOSSEY'S theory [7.214-216] for weak field electroabsorption, which included exciton effects, yielding an exciton binding energy of 160
meV and a broadening parameter of 160 meV, the latter setting a lower limit on the
lifetime of the excited state. This fit, shown in Fig. 7.7, implies that the ER spectra from the 3d levels of Ga arise from the same mechanism as those from valence
band levels.

By comparing the energy to excite an electron from the upper 3d sublevel to a conduction band minimum, as observed in the ER spectrum, with the energy expected for this excitation on the basis of XPS data, fundamental band gap data, and if needed, data from high pressure measurements, one finds a discrepancy, shown as E_B in Fig. 7.8, of 170 meV for GaP, in good agreement with the exciton binding energy obtained by fitting the line shape of the $\Delta\bar{\varepsilon}$ spectrum for the transition to X_6. The transitions to the X_6 minima in GaAs and GaSb have exciton binding energies of 90 meV each, according to an analysis like that of Fig. 7.8 [7.217]. To the extent that the exciton binding energy is the same for all conduction band minima in one compound, the absolute energies of all the conduction band minima which give features in the ER spectra can be determined, even those whose final states have not yet been identified. Perhaps the most striking feature of the results on GaAs is the fact that the L-point minima in the conduction band are lower than the minima at X, a reversal from the ordering accepted for over ten years [7.218,219].

Binding energies of the order of 100 meV for excitons with the hole in the Ga 3d level are somewhat larger than the values expected on the effective mass model with the hole mass set equal to infinity, which are 16 to 38 meV for these compounds, but the discrepancy is not so large that the effective mass model is not a good zero-order model for the excitons. The BLOSSEY model for the electroabsorption is still expected to be valid. The largest experimental error in determining the binding energies by the energy cycle shown in Fig. 7.8 is in determining the onset of valence band transitions from the XPS data, which usually have a spectral bandpass larger than the resultant exciton binding energies.

Recently the matrix elements used to identify the transitions were recalculated, as were the $\bar{\varepsilon}$ spectra for transitions to the conduction band from the Ga 3d levels in GaP and GaAs [7.220]. The matrix elements to X_6 and L_6 were found to be comparable in magnitude, making the identification of ER structures less certain. The $\bar{\varepsilon}$ spectra were calculated without exciton effects, and the threshold energies were placed to agree with the XPS spectra. The calculated reflectance spectra for GaP and GaAs had the same shapes as the measured spectra, but the largest peak in each calculated spectrum was 1.5 and 1.0 eV too high, respectively. The meaning of this shift is not clear at present, for exciton effects do not normally give a rigid shift, which seems to occur, and the shift is surprisingly large. Similar effects have been found in Pb-chalcogenides to be discussed shortly. However, the recent recalculation and reinterpretation of Ga 3d excitations does not help explain the very good fit of the ER line shape to BLOSSEY's theory with 100-200 meV binding energies, nor the systematics leading to the level ordering in GaAs. A possible error in the use of XPS spectra for predicting thresholds for optical spectra is the neglect of relaxation processes during the X-ray excitation [7.221-223] . For work on the Ga 3d levels and valence bands, the relaxation energy of the hole in the valence band must be well under 100 meV to be ignorable.

Independently of the detailed interpretation in terms of excitons, the shifts of the ER spectra with temperature can be understood on the basis of the temperature dependence of the final state conduction band minima [7.224]. For GaP, on which the analysis has been carried out, the shift of the X_6 minima is all that is needed to produce a fit to the data. No temperature dependence of the 3d core level, such as could arise from a temperature dependence of the ionicity, is required.

The ER spectra from the Ga 3d levels showed no anisotropy in a series of measurements on a (110) face of GaP [7.213]. When the modulating field is normal to the surface and the electric field of the external photon beam is along [110] or [100], respectively, the photon polarization inside the sample is approximately along [100] or [111], respectively, an effect of the refractive index of less that unity and of non-normal incidence. Since the transitions to equivalent points become inequivalent when the modulating field is present, anisotropy might be expected [7.225-227]. This effect, and that of matrix element modulation, are not found in the ER spectra of the Ga 3d levels, but they are found in the ER spectra of valence to conduction band transitions. Such effects, unexplored to date in the 6-20 eV range, should be used to confirm the assignments of ER structures. There are also a number of ER structures in the 22-30 eV region in GaP, GaAs, and GaSb which have not been interpreted.

Attempts to obtain ER spectra at the As 3d edge in GaAs were unsuccessful [7.228] for several possible reasons. In addition to the very low reflectance in this region, the expected structures may be very broad, because the excitation to the lowest X conduction band minima, X_6, has little oscillator strength due to the small As p-character at X_6, while the stronger transitions at higher energy to X_4 may have a shorter electron lifetime. The wavefunctions of both X_6 and X_4 states have more Ga 3p character than As 3p, making the expected oscillator strength smaller for As 3d excitations than for Ga 3d. Moreover, As 3d hole lifetimes may be considerably shorter than Ga 3d.

Other group IV and III-V compounds have been studied with synchrotron radiation. Graphite, unlike the other semiconductors just discussed, lacks tetrahedral bonds, and occurs in a highly anisotropic hexagonal layered structure. Single crystals may be obtained, but only basal plane surfaces are suitable for optical study. In measurements of reflectance with p-polarized radiation at angles of incidence far from normal, the component of $\tilde{\varepsilon}$ parallel to the c-axis makes a contribution. Such measurements were made between 3 and 40 eV, and both components of the dielectric function tensor obtained [7.229]. Very strong optical anisotropy was found to above 20 eV. Comparison with electron energy loss data was good, except for the region near a peak at 11 eV in the component of ε_2 parallel to the.c-axis, which is particularly sensitive to measurement accuracy. This peak was stronger in the spectrum of this component of ε_2 obtained from the energy loss data. Band structure calculations and matrix elements calculations should be extended to higher energies to resolve the discrepancy and to interpret the transitions.

SiC, a IV-IV compound, like graphite, has no core excitations in the vacuum ultra-violet so valence to conduction band transitions alone account for its optical pro-perties. The reflectance of a 6H polytype sample was measured to 25 eV at several angles of incidence, and the dielectric function obtained [7.230]. As expected, it is very simple, with a broad peak in ε_2 near 15 eV. A strong peak in the loss func-tion appears at 22 eV arising from a volume plasmon.

A brief report on the large-band-gap material AlN has appeared, giving reflectance and luminescence excitation spectra on basal plane samples from 3 to 40 eV [7.231]. The reflectance is similar to that of other III-V compounds.

a) II-VI Compounds

Many II-VI compounds were studied before the advent of synchrotron radiation. How-ever many have been remeasured for a systematic comparison with band models. The first to be studied was ZnO, because no data existed above the fundamental absorption edge. Single crystal reflectance measurements were analyzed to give both components of the dielectric function tensor to 25 eV [7.232]. It proved to be unlike those of previously measured II-VI compounds, but in good agreement with expectations from band structure calculations.

FREEOUF [7.233] carried out extensive reflectance measurements on ten II-VI com-pounds in the 0.6-30 eV range, using oriented single crystals for studying the aniso-tropy of those with the wurtzite structure. (These II-VI compounds include some, e.g., MgO and BeO, with such large band gaps that they are classed as insulators). Fig. 7.9 shows the reflectance of CdS at several temperatures for $\underline{E}$ parallel and perpendicular to the c-axis, as an example of many of the spectra FREEOUF obtained. Such spectra were Kramers-Kronig analyzed to give dielectric functions, electron energy loss functions and partial sum rules. In many cases calculated ε_2 spectra were available for comparison. The region of strong peaks below about 12 eV arises from valence to conduction band transitions and the spectra could be explained quite well by the band structure results. The partial sum rule on ε_2, illustrated in Fig. 7.10, shows that for CdO, the curve is low, probably indicating too low a meas-ured reflectance value. It also shows that the other Cd chalcogenides all have between 4 and 8 electrons per molecule contributing to absorption below about 12 eV, while 6 should be the limiting number. Above about 10 eV the Cd 4d core levels begin to con-tribute structure to ε_2 and R. Although these transitions appear very weak in the reflectance spectrum (Fig.7.9) about half of the 10 electrons in the 4d levels have contributed to ε_2 in the first 20 eV above threshold (Fig.7.10). FREEOUF notes a number of discrepancies between his measured spectra and those of others, but the origin of these is not clear. The valence band to conduction band spectra in these compounds scaled linearly with the Penn gap. This study, although time consuming, was far less so than if it had been carried out with conventional sources. Synchro-tron radiation sources make such systematic studies on a large set of related

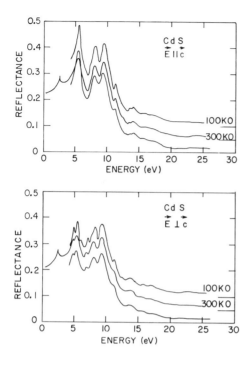

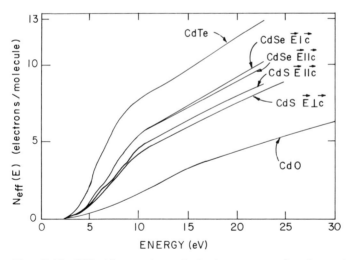

Fig. 7.9 Reflectance spectra of CdS for $\underline{E} \perp \underline{c}$ and $\underline{E} \parallel \underline{c}$ at 400, 300, and 100 K. The scale is correct for the 400 K data. The other curves have been displaced upward by 0.05 and 0.10 [7.233]

Fig. 7.10 Effective number of electrons per molecule contributing to absorption below the energy on the abscissa axis for several cadmium compounds. This is the result of the partial sum rule on ε_2 [7.233]

materials over a wide spectral region considerable easier to carry out. Reflectance and luminescence measurements on other II-VI compounds, e.g., CaS, SrS, CaO, have been made in the Soviet Union [7.234,235].

b) Pb-chalcogenides

Pb-chalcogenides have been studied in reflectance by CARDONA et al. [7.236] and by
MARTINEZ et al. [7.237-240]. The former noted that the transitions from the Pb 5d
levels at 19 eV did not give the expected pairs of spin-orbit doublets. They showed
with a model calculation that this resulted from the selection rule for total angular momentum on the dipole matrix element. The upper Pb 5d level has j = 5/2 and the
conduction band minimum, based largely on Pb 6p functions, has j = 1/2. Thus transitions to the conduction band minimum and the concomitant exciton effects are not
allowed, but transitions become allowed at slightly larger energy as other angular
momentum components are mixed into the lower conduction band states away from the
critical point. The 5d (j = 3/2) to 6s (j = 1/2) edge is allowed at threshold. This
illustrates dramatically the importance of selection rules. In this work XPS spectra
were invaluable in showing that the Pd 5d levels did indeed have the expected spin-
orbit splitting, though it did not appear in the optical spectra.

PINCHAUX et al. [7.237] and MARTINEZ et al. [7.238] repeated such measurements,
taking the second derivative of the reflectance with respect to wavelength as well.
MARTINEZ et al. [7.239,240] computed the dielectric function from their calculated
bands, placing the threshold for excitation of Pb 5d electrons to agree with XPS

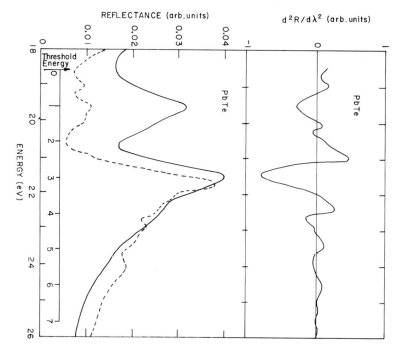

Fig. 11 Reflectance and its second derivative with respect to wavelength for PbTe.
The dashed line is a calculated spectrum with the threshold set to agree with
$E_{xps} + E_\alpha$ in Fig. 7.8 [7.238]

data, and compared the measured reflectance with that calculated. The agreement of the spectral shapes was good, as can be seen in Fig. 7.11, but the experimental thresholds were too low by 0.8 and 0.9 ($\pm$ 0.1) eV for PbTe and PbSe, respectively. This shift could be accounted for by core exciton effects, but it is extraordinarily large for these materials with such high dielectric constants. Moreover, as MARTINEZ et al. point out, a nearly rigid shift is not expected from electron-hole attraction. (The situation is similar to the comparison of the calculated and measured spectra reported by this group for the Ga 3d excitations in III-V compounds [7.220]). At this time, a good explanation for this rigid shift is not at hand, but it may arise in relaxation effects about the 5d hole during X-ray excitation, making the placement of the threshold of the calculated spectrum the uncertain point.

c) Other semiconductors

A host of non-cubic semiconductors has been studied by synchrotron radiation spectro-scopy, although in many cases, the energy bands are so complicated that it is diffi-cult to interpret the spectra. Trigonal Se and Te were investigated, and both com-ponents of the dielectric functions obtained [7.241]. Transitions from several val-ence bands to the conduction bands are strong below about 18 eV, exhibiting consider-able anisotropy. The first core levels are the 3d levels of Se at 55 eV and the 4d of Te at about 42 eV. The latter have been studied on thin films, both crystalline and amorphous, with little difference in the spectra due to crystallinity [7.242,243]. The core excitation spectra above threshold did not agree well with the density of conduction band states, and an atomic calculation suggested that multiplet structure was large enough to contribute significantly to the observed spectrum. The Se 3d spectrum, however, fit the expected density of states more closely, except for a weak tail toward lower energy. Again, when comparing the energy of the absorption edge for these core transitions with the expected energy from XPS data, there is a shift of 1 to 2 eV for the Se 3d level.

The Se 3d and Te 4d spectra were measured in the alloy series Se_xTe_{1-x} [7.244]. The shapes of the spectra were the same in all samples, but there was an increase in the energy of the absorption peak above threshold with increasing x. The shift was linear in x for the Te 4d peak and, at low Te concentration, for the Se 3d peak, but for large Te concentration, the Se 3d peak shifted more rapidly than linearly. The linear shifts were the same for both core levels, and were attributed to a shift in the mean energy of the lower conduction band, but the additional nonlinear shift is unexplained.

Amorphous As, a semiconductor, and crystalline As, a semi-metal have been studied to 30 eV, with very little difference in the spectra appearing in the vacuum ultra-violet [7.245]. The Se 3d and As 4d excitations, the latter at about 41 eV, have been measured in Se, As, and As_2Se_3 amorphous films [7.246]. The spin-orbit splitting

shows clearly, and several weak shoulders appear in the first few eV above threshold. The amorphous Se spectrum is similar to that of crystalline Se, and the same
is true for As. Both spectra show shifts in the positions of the thresholds from
those expected from XPS and band gap data, 0.3 eV for As and 2.0 eV for Se, both in
the excitonic direction. The shift is not a rigid one for the As 4d spectrum, however, for the density of states needed to explain the valence band absorption spectrum of amorphous As rises slowly from the band minimum, while a sharp 4d edge is
observed, indicative of excitonic effects. In amorphous As_2Se_3 both edges appear as
they do in the spectra of the elements, indicating that the absorption is localized
near one atom site. The origin of the weak shoulders above the edges may be the same
as that for the 3p excitations of 3d transition metals, multiplet splitting, but this
remains to be demonstrated.

The reflectance of single crystals of orthohombic GeS has been measured, using
all three polarization orientations necessary for the determination of the three
principal components of the dielectric function [7.247]. The valence to conduction
band transitions extended to about 20 eV, showing considerable anisotropy. About half
of the structures could be identified from band calculations.

Several layered semiconducting compounds have been measured, but usually with the
electric field in the basal plane. As the refractive index rises, it becomes more
difficult to obtain meaningful data leading to $\varepsilon_\parallel$ by the technique applied to graphite. Amorphous and orthohombic Sb_2Se_3 [7.248] and hexagonal Bi_2Ti_3 [7.249] have
been measured in reflectance to 30 eV. The band structures of these compounds are
very complicated, and interpretation was based on sum rule considerations, separating
the transitions of the non-bonding valence bands from those of the bonding valence
bands. The Bi 5d levels at 24.5 eV give weak, but sharp, structures in the reflectance, but, like the 5d levels of Pb in Pb-chalcogenides, the structures cannot be
separated into spin-orbit components because of the spin-dependent electric dipole
matrix elements.

The reflectance of GaSe has been measured and the second derivative with respect
to wavelength taken [7.250]. The transitions from the Ga 3d states give three pairs
of spin-orbit split structures, arising from transitions to the lowest conduction
band minima and to two higher sets of minima, in agreement with the calculated bands.

The layer-structured transition metal chalcogenides $HfSe_2$, $NiTe_2$, and $TiTe_2$ have
been measured in reflectance between 12 and 42 eV, a region in which interband
transitions, but not all of them, occur, as well as core transitions [7.251]. The
core levels of Hf (4f and 5p), Te (4d), and Ti (3p) all provide reflectance edges
which exhibit spin-orbit splitting, but little structure above the reflectance edges.

ZIVITZ and STEVENSON [7.252] measured the reflectances of the tetragonal semiconductors Zn_3As_2 and Cd_3As_2, finding evidence of the Cd 4d and Zn 3d thresholds
at about 11 eV. These transitions push the plasmon peak in the loss function to
about 9.5 eV from the free electron gas value of 13.13 eV for Cd_3As_2. SURZANARAZANAN

et al. [7.253] measured the reflectance and photoemission spectra of TmTe, a semi-conductor in which the valence of Tm was a question. Much of the absorption spectrum from below 1 to 20 eV arises from the valence to conduction band transitions. They calculated the $4f^{13} \rightarrow 4f^{12}5d$ spectrum for Tm^{2+} and the $4f^{12} \rightarrow 4f^{11}5d$ spectrum for Tm^{3+}, including the cubic crystal field splitting of the final 5d state. The transitions of Tm^{2+} could account for the measured absorption and electron energy distribution curves. There was no evidence that Tm^{3+} excitations were occurring, except in the photoemission spectra, where the ratio of Tm^{3+} to Tm^{2+} grew with time after cleavage from an initial value of about 0.06, clearly a surface phenomenon.

7.3.2 Soft X-ray

The soft X-ray spectra of semiconductors have been discussed by BROWN [7.1], but some additional work has appeared. The $L_{2,3}$ edge of Si has been measured by BROWN et al. [7.254-256] on very thin single crystals and amorphous films while FUJITA et al. [7.257] have obtained a reflectance spectrum in this region, where the reflectance should, and does, resemble the absorption spectrum. Fig. 7.12 shows the absorption spectrum at this edge and a spectrum calculated from the conduction band density of states, using two such spectra, weighted in the ratio of 2:1, separated by the spin-orbit splitting of the Si 2p levels. The structures above the edge resemble

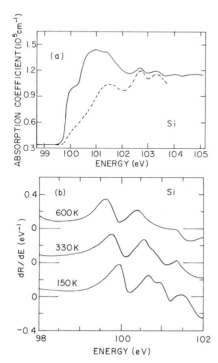

Fig. 7.12 Absorption coefficient [7.256] (a) and derivative of the reflectance at 85° angle of incidence and at several temperatures [7.260] (b) for crystalline Si. The dashed line is from the calculated conduction band density of states but includes the Si 2p spin-orbit splitting and statistical weights

structure in the density of conduction band states, but the relative weights of the structures are not in agreement. The edge rises very steeply, indicating a long hole lifetime, and the steep edge region is far stronger than the density of states picture suggests. ALTARELLI and DEXTER [7.258] have shown that this can be understood on the basis of the electron-hole Coulomb interaction, which leads to an exciton binding energy of 40 meV, too small to give a resolvable structure, and an enhanced dipole matrix element which can account for the shape of the edge. PANTELIDES [7.259] later pointed out the necessity of considering valley-orbit splitting for core excitons, and of a more detailed treatment of the spatial extent of the core hole wave function. He calculated 37 meV for the binding energy of the deepest of the valley-orbit split $n = 1$ Si $L_{2,3}$ excitons, but a detailed fit to the edge shape was not carried out.

Electroreflectance measurements on this edge gave no signal larger than $\pm 1 \times 10^{-4}$ in $\Delta R/R$ [7.260], but thermoreflectance measurements were successful, both by modulating the temperature, and by measuring R at several temperatures and taking derivatives numerically [7.260,261]. Fig. 7.12 shows the energy derivative spectrum of this edge at several temperatures. The principal effect of temperature is a shift of the edge by $(5 \pm 1) \times 10^{-4}$ eV/K, which should arise largely from the shift of the final state. These spectra also show some structure at 0.9 eV above threshold which is attributed to final states in the conduction band minima at L, while the threshold final states are at the absolute minima along Δ. A comparison of the Si $L_{2,3}$ threshold observed in reflection with the threshold expected from XPS data indicates an excitonic effect of 0.30 eV, consistent with the failure to detect an ER signal, but much larger than the 40 meV needed to fit the shape of this edge, and expected on the basis of the effective mass approximation. In this case, the shift is not a rigid shift of the expected one-electron ε_2 or reflectance, for the edge is steeper than the calculated density of states. It is not clear whether the inclusion of dipole matrix elements alone is sufficient to sharpen the calculated edge. Studies on heavily-doped Si samples [7.261] show no change in the edge shape for carrier concentrations as high as 10^{20} cm^{-3} although earlier measurements on heavily-doped n-type samples showed a decrease in absorption at and above the edge as though the free electrons were screening the electron-hole Coulomb interaction [7.262]. PANTELIDES [7.263] has calculated that in n-type Si an excited core electron can bind to a donor, producing structure in absorption up to 1 eV below the corresponding edge. Such structures have not yet been observed, but the requisite low temperatures and proper narrow range of donor concentrations were not achieved in previous measurements.

CARDONA and HAENSEL [7.264] measured the absorption spectra in the 30-150 eV region of a number of II-IV and II-VI compounds, and of Ge, Se, PbS, and PbSe. A few of the constituent elements gave sharp spin-orbit split thresholds for some transitions but many expected transitions did not produce discrete structures. In general the overall absorption spectra resembled those of the atoms. The S $L_{2,3}$ absorption in CdS and ZnS, however, showed richly-structured spectra in the first 10 eV above the 163 eV threshold [7.265,266]. Many of the peaks in CdS were assigned to critical points in the conduction band. The spectra were similar to the $L_{2,3}$ spectra of Cl in alkali chlorides. WATANABE et al. [7.267] made transmission measurements on the S $L_{2,3}$ region of single crystals of CdS, observing considerable anisotropy in the first four peaks above threshold. Cooling to liquid nitrogen temperature sharpened the structures, but did not shift them. A comparison of one spin-orbit component spectrum for each polarization with the corresponding valence to conduction band spectrum showed close similarities for the first two peaks above threshold.

The absorption above threshold for 4d excitations in I_2 is about the same for solid and vapor phases, but the sharp structures at threshold in the vapor give only a few peaks in the spectrum of the solid [7.268]. These could be n = 1 and 2 members of two spin-orbit split exciton series, but a charge transfer model may be as satisfactory.

SONNTAG and BROWN [7.269] measured the 30-200 eV absorption spectra of several transition metal chalcogenides, using very thin crystals of these layered materials, some of which are metallic. The S $L_{2,3}$ excitation spectrum in MoS_2 was measured carefully at high resolution and separated into spin-orbit components, each of which showed several peaks in the first 6 eV above threshold. These peaks agreed with several in the density of states. The edge was so steep that exciton enhancement was believed to be present. In $NbSe_2$, the Se $M_{2,3}$ absorption showed only spin-orbit structure, but the Se $M_{4,5}$ exhibited some additional structure which resembled structure in the conduction band density of states. The edge was less sharp, presumably due to the lack of exciton effects in this metallic crystal. The S $L_{2,3}$ edge in TaS_2 was simple, but there was a shift in the threshold between the 1T and 2H polytypes. The S $L_{2,3}$ absorption in TiS_2 has a structured region in the first 6 eV above threshold, then a distinct minimum before the rise into the delayed continuum. The final states here seem to be unoccupied molecular orbitals, or a separate, unoccupied pair of d-bands.

Hexagonal BN has been studied at the K edge of B, near 190 eV [7.270]. This is perhaps the highest energy at which anisotropy has been observed, and it is considerable. Fig. 7.13 shows the absorption spectrum for $\underline{E} \perp c$ and for $\underline{E}$ with a component parallel to c. The marked anisotropy has been explained in terms of selection rules for transitions to two-dimensional conduction bands.

7.3.3 Summary

The vacuum ultraviolet optical properties of semiconductors have contributed significantly to our understanding of the electronic structure of these materials, and will continue to do so. A new development is that additional information on the

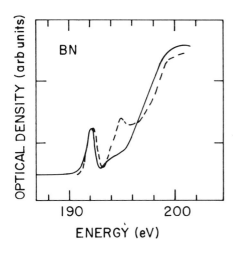

Fig. 7.13 Optical density of a thin single crystal of BN. The solid curve is for
$\underline{E} \perp c$ and normal incidence. The dashed
curve is for 23° angle of incidence, giv-
ing a component of $\underline{E} \parallel c$ [7.270]

conduction bands can be obtained from core electron spectra, at least from the d-
levels with excitation energies in the 10-30 eV region. Such spectra give information on the conduction band itself, a band not accessible to other spectroscopies
with such high resolution. An unsolved problem is the role of exciton effects,
which clearly are present. The use of XPS and band gap data always leads to larger
exciton effects than expected from fitting ER line shapes, or from other considerations, and the spectra sometimes, but not always, seem to be shifted rigidly. This
is not understood at present, but it is reasonable to believe that there may be a
problem in interpreting the energy difference between the valence band edge and the
core level in XPS data. Deeper core levels have not given much information on the
electronic structure, except perhaps for the S 2p level in CdS, but they·have not
yet been studied extensively enough in, e.g., mixed crystals, for us to know whether
they may be useful for such purposes.

7.4 Insulators

7.4.1 Rare gas solids

We discuss the spectra of these large-gap solids only briefly, because recent reviews
are available [7.271,272]. We take solid Ar as our example, but equally thorough data
are available for the other rare gas solids, except He, which solidifies only under
pressure. SONNTAG [7.271] has pointed out the difficulty of getting accurate data on
these materials because of sample preparation problems, and reviews the large body
of vacuum ultraviolet work on rare gas solids carried out with conventional light
sources.

394

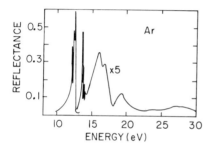

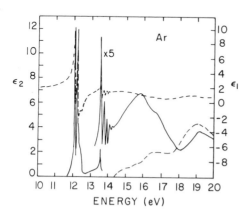

Fig. 7.14 Reflectance of solid Ar
[7.273] and the dielectric function derived therefrom by Kramers-Kronig analysis [7.274]. The solid line is ε_2, the
dashed, ε_1

Fig. 7.14 shows the vacuum ultraviolet reflectance of solid Ar [7.273] and the
imaginary part of the dielectric function derived from it by Kramers-Kronig analysis [7.274]. The valence band is composed of the 3p states which have a spin-orbit
splitting of 0.20 eV for the free atom. Two exciton series can be recognized in the
ε_2 spectrum, split by about this energy. A better fit has been made, taking exchange
into account, for it can alter the observed splitting and the relative strengths of
the two "spin-orbit" components [7.275], and a Wannier series of excitons was obtained for each such component. As expected, the n = 1 exciton energy is not that
predicted by the Wannier model because the electron and hole are too localized for
the model to be completely valid. A total of four excitons in the Wannier series
for one spin-orbit partner have been identified, the higher terms of which lead to
an exciton Rydberg of 2.32 eV and a band gap of 14.15 eV at about 20 K. The binding
energy of the n = 1 exciton is 2.08 eV, making the central cell correction to its
energy -0.24 eV. This arises from the breakdown of the validity of a dielectric continuum model for a strongly localized exciton, as well as from the need to use states
from parts of the conduction band other than its lowest minimum. There are also exchange contributions to this correction. (The n = 1 Wannier exciton has a radius
of only 1.8 Å compared to an interatomic spacing of 2.76 Å). From the exciton Rydberg
an exciton reduced effective mass of 0.44 is obtained. Although the two n = 1 exciton

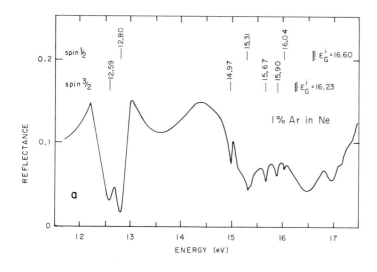

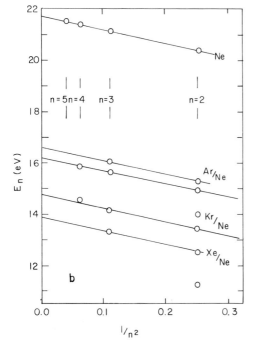

Fig. 7.15 (a) Reflectance spectrum of an 1800 Å film of 1% Ar in Ne on a gold substrate. The exciton energies and series limits are shown for the two spin-orbit split series. (b) Exciton peak positions for Ne, and Ar, Kr, and Xe in Ne vs. n^{-2}. n is the exciton principal quantum number. Double points result from resolved spin-orbit splittings [7.278]

peaks fall very close to the expected positions for the atomic excitations $3p^6 \rightarrow 3p^5 4s$, it is difficult to describe these as Frenkel excitons [7.276,277].

PUDEWILL et al. [7.278] have doped solid Ne with other rare gases and measured the reflectance of a gold mirror covered with the condensed gases. They measured a series of up to four exciton-like absorption peaks below the absorption edge of the host Ne, shown in Fig. 7.15. The peaks formed Wannier series, except for the lowest member of each series (see Fig.7.15). The exciton Rydberg was independent

396

of the nature of the impurity and was the same as that for excitons in Ne, indicating
that the hole was essentially localized and the electron not strongly coupled to the
impurity. The series limits, spin-orbit splittings, and central cell corrections did,
however, depend on the nature of the impurity.

SAILE et al. [7.279] have discovered several weak absorption peaks below the
n = 1 and 2 exciton peaks in Ar, Xe, and Kr. These peaks are found only on carefully-
prepared very thin samples. They can be removed by a thin overlayer of another rare
gas. In the case of Xe, not only does an Ar overlayer destroy this absorption, but
the evaporation of the overlayer restores the original absorption. SAILE et al.
attribute the absorption to surface excitons. The shift to lower energy with respect
to volume excitons can be visualized by using an average dielectric constant in the
Wannier exciton model, the average of the dielectric constant of the rare gas and
that of the vacuum, but a more detailed calculation of the excitation of a valence
electron on a surface atom has been carried out [7.280]. Polarization of neighboring
atoms replaces the average dielectric constant of the two media in causing the surface exciton to have a greater binding energy than its volume counterpart.

Above the exciton series limit there is continuum absorption, peaking at about
16 eV (Fig.7.14). Structures in this region may be connected with structure in the
conduction band, but the effect of the dipole matrix elements and the electron-hole
Coulomb interaction are expected to be large. The latter should distort the expected
single-particle ε_2 for a few exciton Rydbergs above the continuum threshold. A qualitative interpretation of the structures is not yet available. The loss function
yields a plasmon peak at 19 eV, compared with a value of 20.6 eV for
$[(\hbar\omega_p)^2 + E_G^2]^{1/2}$, using the valence electron density and the band gap energy.

The onset of absorption by 3s electrons is expected near 30 eV. The absorption
spectrum (Fig.7.16) shows features there, but they are of the window type, indicating

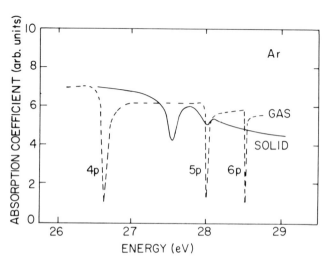

Fig. 7.16 Absorption co-
efficient of solid Ar
(solid curve) and Ar gas
(dashed) in the region of
excitation of the 3s e-
lectrons. The final states
are indicated for the gas
phase spectrum [7.281]

an interference between the transitions to discrete final states and continuum transitions [7.281]. The continuous absorption goes through a minimum at 45 eV, then rises slowly till the onset of transitions from the 2p electrons at 245 eV [7.282]. The minimum is a "Cooper minimum" due to the vanishing of the radial part of the dipole matrix element for the $3p \rightarrow \epsilon d$ transitions [7.283]. Except for the threshold regions, all absorption above about 30 eV is about the same for the solid as for the vapor. The spectrum of the 2p excitations can be understood surprisingly well in terms of the density of conduction band states, plus one or two pairs of spin-orbit split Frenkel excitons at threshold, despite the failure of such a model for the valence band excitations [7.284,285].

The other rare gas solids are similar, except that Kr and Xe have core electrons of d-symmetry as well, and Ne is practically devoid of core electrons [7.286-289]. These spectra have been discussed elsewhere [7.271,272]. The interpretation of the threshold regions of the spectra has been facilitated by studies on mixed rare gas solids [7.290].

The rare gas solids form an excellent laboratory in which to study the dynamics of the excited states produced by absorption. Measurements of the yield of photoelectrons as a function of the energy of the absorbed photons and of luminescence yield as a function of exciting photon energy [7.291-293] together account for those excitations leading to emission of an electron or a photon. The luminescence emission spectrum and lifetime give information on the emission process itself. These measurements can be augmented by measurements on doped rare gases, which give information on migration of the excitations. Such measurements, and others, have been reviewed recently by ZIMMERER [7.272] and earlier by JORTNER [7.294]. We outline briefly the situation for just one rare gas, Ar, space not permitting the more extensive treatment the subject deserves.

The principal feature of the luminescence emission spectrum is a simple peak, 0.55 eV wide, at 9.8 eV, about 2.2 eV below the exciton absorption peak [7.291]. This is attributed to the decay of a self-trapped exciton, an electronically excited Ar_2^+ molecule-ion in the crystal, from its vibrational ground state to its anti-bonding electronic ground state. The Stokes shift of the luminescence arises from vibrational relaxation both before and after emission. Lifetime measurements are not yet reported for Ar, but some interesting lifetime measurements have been made on solid Xe [7.295]. Lifetimes often are studied by using energetic ionizing radiation, but the short pulses of ultraviolet from large storage rings allow selective excitation in just one absorption peak. This has been done for solid Xe at 80 K. Below 50 K the principal emission is a peak at 7.1 eV, but at higher temperatures, a peak at 7.6 eV grows, becoming the only peak at 80 K. At 80 K its lifetime is 58 ns, and it can be excited only be absorption in the n = 2, j = 3/2 exciton peak. The authors conclude that this emission is the result of a self-trapped exciton in the n = 2 state, while the 7.1 eV emission is from an n = 1 self-trapped exciton. It is possible to excite an n = 2 exciton which can relax to the n = 1 self-trapped state.

The photoluminescence yield spectrum of solid Ar is shown in Fig. 7.17. At low energies there are minima at the positions of the exciton absorption peaks which are interpreted as the result of radiationless decay at the sample surface of excitons created near the surface. These minima become more pronounced as the surface becomes more contaminated. Both the diffusive motion of excitons to the surface and the dipole-dipole transfer of the excitation energy of "stationary" excitons [7.296] have been proposed to explain the shapes of the yield curves, and both can be made to fit with reasonable values of the unknown parameters [7.297-301]. A choice between the two mechanisms can be made from photoelectron yield spectra. See Fig. 7.17b. In pure thin film samples the photons travel to the substrate (Au) and liberate photoelectrons which pass back through the sample, escape, and are detected. The use of various sample thicknesses allows the determination of the diffusion length of the excitation and the scattering lengths of the photoelectrons. (The dipole-dipole transfer

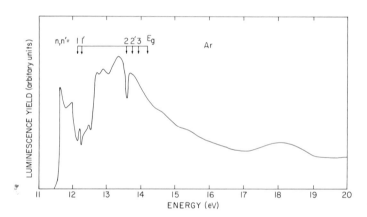

Fig. 7.17(a) Excitation spectrum (photoluminescence yield spectrum) for 9.8 eV emission in solid Ar corrected for the reflectance. The positions at the exciton peaks are marked [7.291]

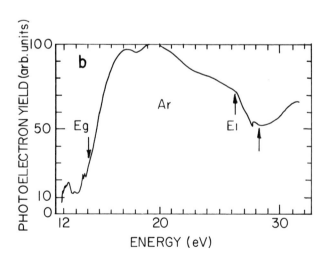

Fig. 7.17(b) Photoelectron yield spectrum for solid Ar [7.302]

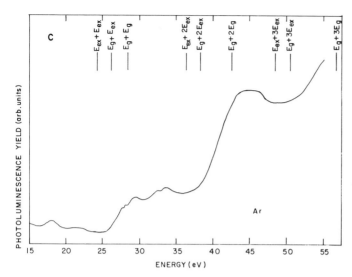

Fig. 7.17(c) Photo-
luminescence yield for
solid Ar at higher ener-
gies [7.298]. The letters
along the top indicate
the energy in units of
band gap and n = 1 exci-
ton energies

model does not fit the data well in pure rare gas solids, but it does fit for some
doped systems [7.302-313].) Free exciton diffusion lengths are in the 50-300 Å range,
except for much larger values in Ne, and electron scattering lengths are 1000 Å and
more, due to the lack of inelastic scattering mechanisms for electrons with kinetic
energies of less than about a band gap. Photoelectron yield spectra for doped rare
gas solids have been equally informative. The motion of host excitons to those impurities which can capture them with the emission of an energetic electron can be
studied as a function of impurity concentration and sample thickness, leading to
similar parameters as in the study of pure samples. Even more information can be obtained if the energy distribution of the emitted electrons is measured. Doubly-doped
solid rare gases have also been studied [7.310].

At higher photon energies, the photoyield spectrum, Fig. 7.17c, has a series of
steps, marking the energies at which the primary excitation, a fast electron, can
create a second excitation, an exciton, both of which can lead to luminescence. At
these energies the photoelectron yield spectrum drops, a result of the suddenly-
shortened mean free path of the primary electron, which can fall to 1-5 Å, and of
the fact that the secondary excitations cannot produce emitted electrons. Above this
drop, the primary electron can scatter, creating an exciton, and still have enough
energy to escape. From Fig. 7.17c it is clear that at 50 eV, three excitons are produced by the primary electron and its resultant excitations.

The absorption spectra of XeF_2 and XeF_4 were measured between 50 and 160 eV, the region of excitation of the Xe 4d and 4p levels [7.314]. The spectra of both phases were nearly identical. The sharper structures were interpreted as transitions into unoccupied molecular orbitals and Rydberg states. An unusual feature is the appearance of a ligand field splitting of the Xe 4d levels.

7.4.2 Alkali Halides

Alkali halides have been studied more than any other class of insulators, although the detailed study of excited states has not been carried as far as in the rare gas solids. The simplest such compound is LiF, [7.315,316] the reflectance, ε_2 and loss function spectra of which are shown in Fig. 7.18. The first large peak is not immediately followed by a series of other peaks, although in other alkali halides another peak or two can be seen, exclusive of spin-orbit pairs. Model parameters for excitons and the band gap have not been well known for LiF because there is no obvious structure from which one can read off energies. Ab-initio band calculations cannot be relied on yet as the sole means of supplying a missing link in a series of energies leading to the gap energy. Recently PIACENTINI [7.317] applied the Wannier exciton model to the valence band spectrum of LiF, even though it was known that the n = 1 Wannier exciton, and perhaps others, would have an appreciable central cell correction. He obtained a good fit to the conductivity spectrum by assuming a Wannier model for the positions of all excitons except the n = 1, for the relative intensities of all excitons, and for the effect of the electron-hole Coulomb interaction on the continuum. The fit was good, and a band gap of 14.2 eV resulted, placing the gap at a structureless region of the spectrum. The exciton Rydberg was 2.1 eV and the n = 1 exciton had a central cell correction of -0.5 eV. The broadening for all excitons was obtained from that of the n = 1 peak, and it led to the merging of all higher excitons into a continuum.

The region between the first peak and about 20 eV is rather structureless, presumably arising from transitions between the valence and conduction bands, with

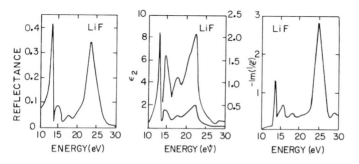

Fig. 7.18 Reflectance, ε_2, and electron energy loss function of LiF at 100 K, the latter two obtained from a Kramers-Kronig analysis of the former [7.316]

critical points and excitons not much in evidence. There is a prominent structure
above 20 eV whose origin is controversial, a peak in ε_2 and a more prominent one in
the loss function at higher energy. These were at first attributed to the simultaneous excitation of two excitons by one photon, but calculated oscillator strengths
were too small [7.318,319]. A later assignment was to a double excitation process,
but of an exciton and an electronic polaron, the latter possibly bound to the exciton [7.230]. When unbound, this should begin to contribute to ε_2 at the sum of the
band gap and exciton energies (26.8 eV) and peak at somewhat higher energy. If bound,
the peak should occur at 25.2 eV, somewhat above the observed peak. The loss function
peak in the latter case should be at twice the energy of the longitudinal valence
band exciton, a higher energy than observed. The calculated oscillator strength is
still to low, but the calculated lineshape, somewhat triangular, is close to the
observed shape.

Synchrotron radiation, especially from storage rings, enables modulation spectra
to be obtained in the vacuum ultraviolet to as high an energy as lineshapes permit.
Temperature modulation measurements on LiF were carried out [7.321,322]. They proved
to be of value in interpreting the spectrum of LiF over the entire range discussed
so far. Fig. 7.19 shows the measured thermoreflectance spectrum and the differential
dielectric and loss function spectra derived from it by Kramers-Kronig analysis.

The first exciton peak has a line shape in the differential dielectric function
that arises from the shift of the exciton energy with temperature, with only a small
contribution from the temperature dependence of the width. The region above the first
peak can be fit qualitatively with the Wannier model of PIACENTINI [7.317], using the
temperature dependence of the n = 1 peak. Thus the Wannier model can fit both the
spectral shape and its temperature dependence. Between 15 and 20 eV several small
features are seen better in the differential spectra than in the ε_2 spectrum. These

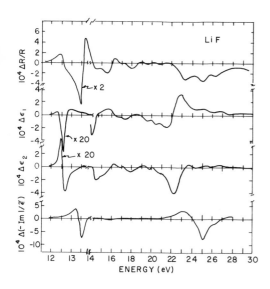

Fig. 7.19 Thermoreflectance spectrum
($\Delta R/R$) of single crystal LiF at 110 K
and differential dielectric function
and energy loss function derived from
it by Kramers-Kronig analysis [7.322]

were assigned tentatively to critical point transitions modified by exciton effects, but genuine proof for these assignments is lacking. The peak in ϵ_2 at 21.7 eV has a corresponding structure at 22.2 eV in $\Delta\epsilon_2$, but its line shape in $\Delta\epsilon_2$ is not that of the exciton structure at 13.6 eV. It arises almost entirely from an increase in broadening with temperature, not from a peak shift. A longitudinal counterpart appears at 25.0 eV in the differential loss function, but it is overlapped by a second structure. The other structure, centered at 25.4 eV, has no transverse counterpart in the $\Delta\epsilon_2$ spectrum, and fits the line shape expected for the valence electron plasmon whose energy shifts from thermal expansion. Until further calculations of the temperature dependence of the electronic polaron can account qualitatively for the difference in the temperature dependence of the valence band exciton and of the electronic polaron, the latter does not seem as likely a source for the structures under discussion as does a simpler model, an energetic exciton (or at least a transition highly localized in energy) and a valence plasmon. Additional evidence for the overlap of longitudinal exciton and a plasmon is found in the characteristic electron energy loss measurements of GOUT and PRADAL [7.323], in which the peak at 25 eV was slightly split. Later measurements with sufficient resolution to see this splitting did not find it, but when a large wave vector transfer was effected in measurements on single crystals, two loss function peaks emerged [7.324]. The nature of the energetic exciton mentioned above is not clear. Calculated band structures show no gap in the density of states,but there are gaps at several symmetry points on the Brillouin zone boundary. A gap at X is in the right energy region for both longitudinal and transverse excitons and for the plasmon. A higher conduction band along Λ is degenerate with this gap, and both excitons and plasmons can decay into single particle excitations, but if the above interpretation of the spectra is correct, the decay rate must be small. We shall discuss other alkali halides later, but it is appropriate to point out here that there are similar structures in the thermomodulation spectra and energy loss functions of several other alkali halides, structures which are more complicated than in LiF [7.315]. There are three transverse "excitons" in $\Delta\epsilon_2$ and there are three longitudinal "excitons" overlapping the plasmon in the differential loss function.

The Li 1s level in LiF gives rise to an absorption spectrum of considerable controversy. The states at the conduction band minimum have s-like symmetry about the Li site, so excitons with s-like envelopes and low energy conduction band states cannot be reached by electric dipole transitions. The p-like parts of the conduction band are some 10 eV above the minumum. The absorption spectrum consists of a prominent peak at 62 eV preceded by a long threshold beginning at 53.5 eV. There are weak structures in the threshold region and above the peak. Absorption spectra by SONNTAG [7.325] and photoyield spectra by GUDAT et al. [7.326] show the region below the peak more extensively than earlier spectra [7.327,328]. Calculated interband absorption spectra [7.329,330] have a long threshold region due to s-like final states and a peaked structure from the p-like final states, as observed, but a comparison with XPS [7.331] and band gap data led to the placing of the calculated interband threshold 11 eV above the 53.5 eV threshold, and the calculated peak 10 eV above the measured peak [7.326,332-334]. Assuming this shift arises from the electron-hole Coulomb interaction leads to an exciton binding energy of about 10 eV, a reasonable value, considering that both initial and final states are localized on one site at which there can be little screening from the valence electrons and from other electrons on the Li^+ ion [7.332]. The problem then is to interpret the threshold region, which would have an equally large exciton binding energy but no prominent structure due to an exciton. The agreement of the shape of the measured peak with that in the interband spectrum calculated without exciton effects also is not clear, as a rigid shift is not expected.

Additional work has clarified the situation. Electron energy loss measurements show that the only part of the spectrum that changes as the momentum transfer increases is at 61 eV, at the foot of the 62 eV peak [7.324]. The only dipole-forbidden transitions are in this region; there are none in the long threshold region. Use of the XPS spectrum and the band gap to place the interband threshold leads to exciton binding energies of 3 eV for the forbidden exciton, involving a 2s electron, and 11 eV for the allowed exciton with a 2p electron. The larger binding energy of the latter is attributed to the pushing up of the 2p-like states in the crystal upon overlap with the F^- 2p valence states, i.e., a pushing up of the extended band states rather than a pulling down of the exciton. This interpretation receives support from recent calculations on clusters of LiF in which relaxed final states are used [7.335]. These give exciton binding energies of 2.0 and 9.2 eV for the 2s and 2p core excitons, respectively, placing them close to each other in the spectrum. The calculations also suggest that the long threshold arises from the F^- 2s electrons. Although these should have a threshold at about 38 eV, there are not F^- p-like states in the first 15 eV above the bottom of the conduction band, so the strong threshold is expected near 53 eV, as observed. The structure above the 62 eV peak then arises from interband transitions strongly distorted by the Coulomb interaction of electron and hole, perhaps so strongly distorted that the band picture may not be a good starting point.

Other alkali halides have valence band exciton spectra complicated by larger spin-orbit effects. In some cases the lowest peak from one spin-orbit component of the valence band may overlap the continuum from the other component, leading to interference effects [7.336]. There is an extensive literature on this subject, most of the measurements having been obtained with conventional sources.

A feature common to all alkali halides except Li-halides is the occurrence of excitations of p-like core electrons of the alkalies [7.337-351,71]. These occur roughly at 33, 21, 16, and 13 eV for the Na 2p, K 3p, Rb 4p, and Cs 5p, respectively. The latter three seriously overlap the valence band spectra. They have been studied extensively by SKIBOWSKI et al. Fig. 7.20 shows a typical example, the 4p spectrum of

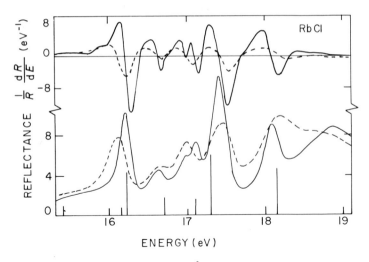

Fig. 7.20 Reflectance R and $R^{-1}dR/dE$, measured directly as a wavelength derivative, for RbCl at 8 K (solid) and 300 K (dashed). The lines at the bottom are the positions and strengths of the Rb 4p exciton spectrum calculated with a localized excitation model [7.71]

Rb in RbCl. The reflectance and its wavelength derivative were measured at several temperatures. The interpretation of these, and similar, spectra is by no means yet settled. The most recent use of the conventional conduction-band-exciton model is by PANTELIDES [7.332], who uses XPS and band-gap data to determine the threshold for interband transitions from the core level to the conduction band. Peaks below this energy are due to excitons, and peaks above, to interband transitions, possibly strongly modified by the Coulomb interaction. ÅBERG and DEHMER [7.352] consider these transitions to be basically atomic transitions, with the final states perturbed by the crystal field, and for d-like states, split by it. They do not attempt to account for every one of the structures. BALZAROTTI et al. [7.353] viewed the corresponding region in KCl near threshold as possibly excitonic, although they did not try to interpret it, and the region more than a few eV above threshold was assumed to be interband in nature. They computed the spectral shape by using a model appropriate to EXAFS, allowing backscattering from both cation and anion neighbors. The agreement with experiment was encouraging, but qualitative. Finally, SATOKO and SUGANO [7.254] took the view that all of the spectrum was excitonic. They placed the electron and hole on the same lattice site and allowed the final states to split in the cubic crystal field. The field strength parameter, 10 Dq, was used as a fitting parameter. A single reasonable value produced a good fit to the positions and strengths of most of the lines, but a few that were calculated were not observed, and the temperature dependence of the oscillator strength of two lines was not explained. Recent high-resolution measurements of the region of K 3p excitation in KI at low temperatures did not bring out any previously unresolved structure [7.255]

At higher energies, the less-pronounced structure in ϵ_2 of the Rb-halides could be explained qualitatively with calculated densities of states for the conduction bands, but the measured peaks were larger, probably due to matrix element and exciton effects. The four models described above are not all mutually exclusive, but a demonstration of the equivalence of several of them is still forthcoming.

The deeper core levels of alkali halides have been discussed before [7.1]. The spectra of both d and p core levels on cations and anions have been measured [7.356-364]. All give a series of sharp peaks followed by broader lines and a continuum with a peak, the latter nearly identical to that found in atomic spectra. The solid state features near the threshold still lack a quantitative interpretation. Spin-orbit partners can be identified, and the expected threshold obtained from XPA and band-gap data, but the nature of the excitons and interband transitions is not yet clear. PANTELIDES [7.333] and KUNZ [7.334] have shown, using expected interband thresholds, that alkali ion core excitation spectra exhibit large exciton effects because the valence electrons cannot provide much screening for the electron-hole attraction. The interband absorption is sgrongly distorted by this interaction. Exciton effects also occur for halogen core electron excitations, but they are smaller, not small enough, however, to permit the interband absorption above threshold to be interpreted in terms of critical points.

Alkali halides exhibit intrinsic luminescence like that of the solid rare gases. It also arises from an excited dimer, for in the alkali halides at low temperatures, holes can be trapped in the lattice, forming (halogen)$_2^-$ molecule-ions which can attract and bind electrons, forming self-trapped excitons. The emission spectra are well known, and are the same for several modes of excitation. Photoluminescence yield curves for alkali halides have been obtained recently to high photon energies, Fig. 7.21 being an example [7.365,366a]. The yield begins to rise at E_1, the energy of the first exciton, then falls slightly as the probability of forming a self-trapped exciton from an excited electron-hole pair falls. (The fine structure is from structure in the reflectance and absorption coefficient). At $E_1 + E_G$, point B in the figure, the excited electron has enough energy to create an exciton in a scattering event and still be trapped as an exciton itself, so that two luminescent photons may appear. Were it not for other excitation processes, one would expect a step in the yield at every increase in energy of E_1. Such a step occurs at point E, i.e., at $F_G + 2F_1$. In this region the incident photons can excite a Cl$^-$ 3s electron with enough kinetic energy to produce a valence band exciton upon inelastic scattering. This should occur near point E, but since the Cl$^-$ 3s threshold is very weak, no prominent structure is expected. The step at G arises from a similar process, with the Na 2p electrons being excited to an energy E_1 above their interband threshold. The absorption threshold for Na 2p transitions produces the dips at F. Similar spectra have been measured for NaCl, RbCl, and RbBr. These photoluminescence yield spectra may be compared profitably with photoelectron yield spectra, as was done

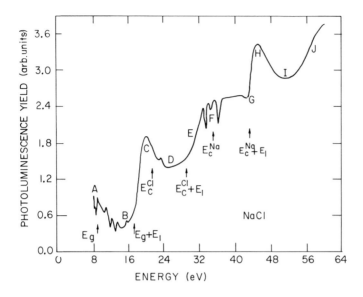

Fig. 7.21 Photoluminescence yield spectrum of NaCl at 4 K. Energies indicated are E_G, the gap energy, E_1, the energy of the exciton, $E_c{}^{Cl}$ and $E_c{}^{Na}$, the excitation energies of core levels on each ion [7.365]

for rare gas solids. The peaks in one spectrum correlate well with minima in the other. The detailed studies of exciton dynamics carried out on the rare gas solids have not yet been carried out on alkali halides.

Synchrotron radiation has been used in a study of the U-center (substitutional H^- or D^-) in LiF, the absorption band of which is at 9.9 eV [7.366b].

7.4.3 Other Metal Halides

Many other insulators have been studied in the valence band region with conventional sources. The Cl 2p spectra in AgCl and TlCl have been measured and discussed, but not as recently as the corresponding spectra of alkali chlorides, so that XPS data have not been used to estimate the threshold [7.367,368].

Alkaline earth halides have been measured with conventional sources and synchrotron radiation but the lack of band structures precludes detailed interpretation [7.349,369]. RUBLOFF [7.349] measured the reflectance of CaF_2, SrF_2, and BaF_2 to 36 eV at several temperatures. The valence band excitation spectra appear to have strong excitonic peaks arising from transitions from the F^- 2p valence band to s- and d-like regions of the conduction band. The Ca 3p, Sr 4p, and Ba 5p states give exciton-like structures beginning at 24, 22, and 16 eV, respectively, but with far less structure than found for spectra of the corresponding alkali p-levels in alkali halides. The luminescence yield spectrum of CaF_2 is similar to that of alkali halides although the interpretation cannot be carried as far because of uncertainties in the assignment of transitions in the absorption spectra [7.370].

Continuing with other metal halides, we expect to encounter new effects. The crystal symmetry often will be lower than cubic, cation electron states could mix into the valence band, or constitute it nearly exclusively, there may be increased covalent binding, and there may be partially filled d- or f-states in, or near, the band gap. In recent years crystals exhibiting all of these effects have been studied, but the interpretation is necessarily more primitive than that for alkali halides and rare gas solids. The valence band absorption spectra of CuCl and CuBr were measured to 30 eV [7.271] but the effects of the Cu 3d levels in the valence band primarily appear near threshold, which had been previously studied.

$SrCl_2$ and CdF_2, both with the cubic fluorite structure, have been measured to 56 eV [7.372]. The former is far more covalently bonded than CaF_2 or BaF_2 and has less prominent excitonic structure at the threshold of the Sr 4p excitations than is found in corresponding spectra for other alkaline earth halide fluorides. Valence band excitons seem to be evident, but the interband transitions persist to higher energy as a result of the wide valence bands in this material. The Cd 4d levels in CdF_2 give no excitons at the expected energy because the valence band minimum is believed to be s-like. The valence band spectrum resembles that of CaF_2, which is unexpected because the Cd 4d levels should be very close to the valence bands.

PbI_2 and PbF_2 have also been measured [7.373]. The former is a layer compound with considerable anisotropy, while both of these crystals differ from previously discussed insulators in that the top of the valence band is composed of 6s states on the Pb^{+2} ions, while lower parts of these bands come from halogen p-states. Transitions from the Pb 5d states at about 20 eV have large exciton effects, as determined by a comparison of XPS and optical data.

BORDAS et al. [7.374] have measured the reflectance of BiI_3 and BiOI to 40 eV. The upper part of the BiI_3 valence band should be composed of Bi^{3+} 6s states, with I^- 5p states somewhat lower in energy. This was strongly suggested by the optical data, especially the sum rule. That the expected prominent Bi^{3+} states are not found in XPS measurements is attributed to the small excitation probability for X-rays and to the enhancement of band gap absorption by exciton effects. The Bi^{3+} 5d levels can be excited to Bi^{3+} 6p states at the bottom of the conduction band. There is one expected transition that is forbidden by the selection rule on total angular momentum, as in the Pb-chalcogenides, in agreement with the observed three sharp lines. According to the XPS and band gap data, these should occur 1.5 eV higher than observed, an effect of Coulomb interaction on these highly localized transitions. Further evidence for extreme localization of these essentially atomic transitions is the lack of any interband structure in the region above the three sharp peaks.

The Mg 2p excitations of Mg halides have been measured in absorption [7.375]. There are sharp structures at the 53 eV threshold and weaker structures on the rising continuum at higher energies. The continuum is like that of the metal, and arises from delayed 2p → εd transitions, an atomic effect. The sharp structures were interpreted first as atomic excitations on the Mg^{2+} ions, but the relative strengths and splitting of the lines were different from the atomic Mg^{2+} spectrum, indicating that the exchange interaction differs in the solid. The exchange contribution to the splitting was the same for three Mg halides, implying localization of the excited state on the Mg site.

SATO et al. [7.376] measured the Cl 2p spectra of 12 metal chlorides. Most of the spectra had sharp structures at threshold attributed to excitons, the exceptions being $CaCl_2$, $SrCl_2$, and $BaCl_2$, but not $MgCl_2$.

The valence band spectra of the transition metal chlorides and bromides are very strong and nearly devoid of structure [7.377,378]. No band structures exist, but a model based on a transition metal ion in an octahedron of halogens has been used to interpret the spectra. Ignoring the forbidden d-d transitions, the first absorption is from the halogen p-like valence band to empty metal 3d states. These transitions, all below about 10 eV, shift as expected with the atomic number of the metal increases. The next transitions, not clearly resolved from the others, are attributed to interband transitions, although the admixture of cation and anion states in the conduction band is not known.

Much more complicated spectra are found in the perovskite-structured compounds KMF_3, where M is a transition metal [7.379,380]. M can also be Mg, Zn, or Cu for comparison spectra, and K may be replaced by Rb. The 3p electrons of the K give one or more exciton-like structures around 20 eV, with the number of structures decreasing as the transition metal atomic number increases. This seems to be due to the lowering of the conduction band minimum which contains 3d states, as the 3d levels drop in energy with increasing atomic number. The first strong transitions, again excluding d-d and d-s transitions on the metal ion, are from the F^- 2p valence band to the 3d and 4s levels on the transition metal, or the two bands formed from them, and at higher energy, to bands arising from the K 4s levels. The first peak is very strong and temperature-dependent, indicative of its excitonic origin. Transitions analogous to the 3d → 4p on the transition metal ions are not manifest in the spectra.

Transition metal chloride and bromide absorption spectra have been measured at the metal 3p threshold [7.381]. Fig. 7.22 shows an example, Cr and two of its dihalides. The shape of the large step in Cr has been discussed in Sect. 7.2.2 in terms of multiplet structure and interference effects. There are marked differences between the metal and halide spectra, and even between those of the two halides. Exciton-like peaks appear below threshold, and structures appear in the continuum, presumably a band structure effect. Final states in or below the conduction band of the halide may overlap poorly with valence band states, making matrix elements leading to Auger broadening and interference with a continuum much smaller than for

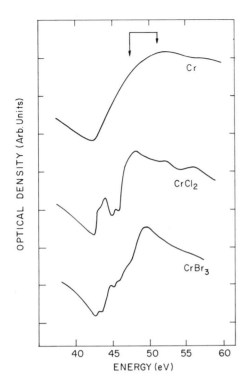

Fig. 7.22 Absorption of the Cr 3p excitations in Cr, $CrCl_2$, and $CrBr_3$. The arrows mark the expected spin-orbit splitting [7.381]

metallic Cr. Such spectra occur for a number of transition metal halides, but not for Ni halides, for which all structures appear above the threshold for the metal.

Rare earth halides form a long series of insulators which can be studied effectively with synchrotron radiation. Single crystals of uniaxial LaF_3, CeF_3, PrF_3, and NdF_3 have been measured to 30 eV, along with an unoriented DyF_3 crystal and polycrystalline GdF_3 [7.382-384]. The forbidden 4f → 4f transitions occur in the visible and ultraviolet. Four or five peaks arising from 4f → 5d transitions localized on the rare earth ions appear in the 5-8 eV region. These are similar to transitions on the free rare earth ions themselves, and to the corresponding transitions observed in rare-earth-doped LaF_3, whose absorption spectra, luminescence excitation spectra, and luminescence lifetimes have been measured [7.385-387]. (There is a shift in the energy of these transitions in absorption for the different ion environments, however and, for CeF_3 and Ce^{+3} in LaF_3, a Stokes shift in emission). The higher-energy transitions begin around 10 eV for all the salts measured, and give reflectance peaks at 11-14 eV, as in Fig. 7.23. These arise from the excitation of F^- 2p valence electrons into the conduction bands, presumably composed of lanthanide 6s and 5d states. There seem to be no valence band excitons at all. It is possible that the rare earth 4f-5d transitions play this role, except for LaF_3, for which the empty 4f states may shorten the lifetime of any La 5d final state. Luminescence studies should prove helpful. (It is known from XPS data [7.388] that

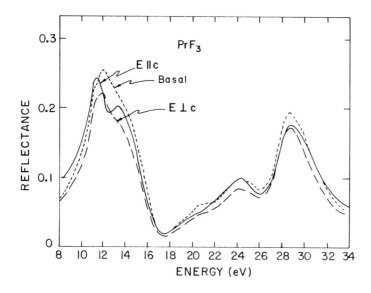

Fig. 7.23 Reflectance of single crystal PrF_3 for $\underline{E}$ parallel and perpendicular to the optic axis. The basal plane reflectance is also for $\underline{E}$ perpendicular to the optic axis [7.383]

the filled 4f levels overlap the F^- 2p bands only in DyF_3 of the fluorides measured, but there is no effect of this in the spectra). The weak structures beginning at about 18 eV in all the salts are due to valence band excitations to higher parts of the conduction band (the structureless part) and to excitations of the lanthanide 5p levels which give the sharper structures, structures which move to higher energy with increasing atomic number. These may be the excitons associated with the excitation of the 5p levels, and the continuum transitions of these levels form the broad peak which moves to higher energy with increasing rare earth atomic number. The "exciton" peaks show some anisotropy.

Transmission spectra of $LaCl_3$, $CeCl_3$, $LaBr_3$, and $CeBr_3$ have been measured in the vacuum ultraviolet, and, for the former two, in the region of the Cl 2p absorption [7.389]. The Cl 2p absorption edge region has several peaks, but none as sharp as those found in alkali chlorides.

The rare earth 4d excitations occur at higher energies [7.390]. These give spectra much like those of the rare earth metals, extensively discussed elsewhere. The chief differences are in the continuum region above the threshold, and the differences are small. There is a bit more structure in the spectra of the trifluorides [7.390] than in those of the metals. This region of the spectrum can be sensitive to the environment of the rare earth either by changes in the conduction band states into which the localized excitations decay or through EXAFS effects.

7.4.4 Other Inorganic Insulators

Several other simple insulators have been measured. The vacuum ultraviolet reflectances of $CaWO_4$ and $CaMoO_4$ have been measured, but the lack of a band structure precluded interpretation except for a molecular orbital assignment for the lower energy transitions [7.391].

NiO was investigated near the 3p edge of Ni, and the spectrum, appearing unlike that of Ni, was interpreted in terms of transitions to empty Ni^{2+} multiplets for the $3p^5 3d^9$ and $3p^5 3d^8 4s$ configurations on the free ion [7.392]. The Al 2p edge structure was investigated in Al_2O_3 in both crystalline (γ, not α) and amorphous phases [7.393,394]. A shift of 0.8 eV for the threshold was found for both forms upon comparison with XPS and tunneling estimates of the threshold. Structures in the first 13 eV above threshold were attributed to interband transitions, but a detailed assignment was not possible.

The condensed gases N_2, O_2, and CO_2 are all large-band-gap insulators, Their spectra have been measured and compared with those of the gases [7.395,396]. Only solid N_2 retains any vibrational structure (see also Ch.6).

The reflectance of solid SF_6 has been measured in the 10-30 eV region, and the absorption coefficient in the 170-300 eV range [7.397]. The latter data were almost identical with those for the vapor phase, accounted for by the explanation of DEHMER [7.398,399], that the S 2p excitations, even well above threshold, are to final states localized on the S site by a potential well formed by the neighboring F^- ions, through the electrostatic potential, by exchange effects, or both. The valence band region has a spectrum far less structured that that of the vapor, and one that differs qualitatively in the first 6 eV above threshold, an effect presumably due to the strong interaction of the least-bound electrons, the F^- 2p, between neighboring molecules in the solid.

7.4.5 Organic Insulators

The first, and simplest, organic insulators to be measured were solid methane and ethane [7.400]. The structures in the reflectance spectra below 30 eV were similar to those in the absorption spectra of the vapors and at nearly the same positions.

Single crystals of anthracene have been studied extensively by KOCH and OTTO [7.401-404]. These monoclinic crystals have only one component of the dielectric tensor parallel to a crystallographic axis, making measurement of two of the three components difficult. Nevertheless, meaningful information can be extracted from the observed anisotropies by determining dielectric functions for polarization directions parallel, or nearly parallel, to the axes of the molecules within the crystal. This allows identification of the transitions in the solid for comparison with vapor and solution phase (where possible) spectra.

Thin films of cytosine, a nucleic acid base, were studied by reflectance and electron energy loss measurements [7.405]. The reflectance calculated from the energy loss measurements agreed well with the measured spectrum, supporting the $\tilde{\epsilon}$ spectra derived from the latter.

7.4.6 Summary

The study of excited states in rare gas solids by many optical techniques has been very productive, and in the future we may expect similar studies to be carried out on other insulators. Indeed a start has been made. The vacuum ultraviolet optical properties of insulators give less information on the electronic structure than was the case for semiconductors because of the relatively larger exciton effects, which may be so large that the band picture may not be the best starting place. Many classes of insulator have not yet been studied. The core electron spectra seem to show exciton-like features at threshold, but these have not yet been interpreted unambiguously.

References

7.1 F.C. Brown: Solid State Phys. **29**, 1 (1974).
7.2 C. Kunz: In *Optical Properties of Solids – New Developments*, ed. by B.O. Seraphin (North-Holland, Amsterdam 1975) p.473.
7.3 R. Haensel: In *Festkörperprobleme*, Vol. **15** (Vieweg, Braunschweig 1975) p. 203.
7.4 E.-E. Koch, C. Kunz, B. Sonntag: Phys. Reports **28C**, 154 (1977).
7.5 D.W. Lynch: In *Proc. Quebec Summer Workshop on Synchrotron Radiation Facilities*, ed. by J.W. McGowan and E.M. Rowe (June, 1976).
7.6 D.E. Aspnes, D.W. Lynch: In *I.C.A.P.-I.N.F.N. Course on Synchrotron Radiation Research* ed. by A.N. Mancini and I.F. Quercia, Alghero, Italy (Sept. 1976) Vol. 1, p. 230.
7.7 F. Stern: Solid State Phys. **15**, 300 (1963).
7.8 F. Wooten: *Optical Properties of Solids* (Academic Press, New York 1972).
7.9 J.F. Nye: *Physical Properties of Crystals* (Oxford University Press, London 1957) Sect. 5.1.
7.10 S.D. Smith: *Handbuch der Physik*, Vol. **25/2a**, ed. by S. Flügge (Springer, Berlin, Heidelberg, New York 1967) p. 234.
7.11 M. Born, E. Wolf: *Principles of Optics* (Pergamon Press, London 1959).
7.12 H. Wolter: Handbuch der Physik, Vol. **24**, ed. by S. Flügge (Springer, Berlin, Heidelberg, New York 1956) p.461.
7.13 E.E. Koch, A. Otto, K.L. Kliewer: Chem. Phys. **3**, 362 (1974).
7.14 S.P.F. Humphreys-Owen: Proc. Phys. Soc. (London) **77**, 949 (1961).
7.15 W.R. Hunter: J. Opt. Soc. Am. **55**, 1197 (1965).
7.16 W.R. Hunter: Appl. Opt. **6**, 2140 (1967).
7.17 M.A. Rumsh, V.N. Shchemelev, Kh. Prois: Fiz. Tverd Tela **4**, 69 (1962) [Sov. Phys. Solid State **4**, 49 (1962)].
7.18 A.P. Lukirskii, O.A. Ershov, T.M. Zimkina, E.P. Savinov: Fiz. Tverd. Tela **8**, 1787 (1966) [Sov. Phys. Solid State **8**, 1422 (1966)].
7.19 W. Gudat, C. Kunz: Phys. Rev. Lett. **29**, 169 (1972).
7.20 H. Petersen, C. Kunz: Phys. Rev. Lett. **35**, 863 (1975).
7.21 J. Hubbard: Proc. Phys. Soc. (London) **68A**, 976 (1955).
7.22 R.H. Ritchie: Phys. Rev. **106**, 874 (1957).
7.23 E. Tosatti: Nuovo Cimento **63B**, 54 (1964).

7.24 E. Tosatti: Nuovo Cimento 65B, 280 (1970).
7.25 M. Cardona: In *Optical Properties of Solids*, ed. by M. Cardona, S. Nudelman and S.S. Mitra (Academic Press, New York 1972).
7.26 M. Altarelli, D.L. Dexter, H.M. Nussenzveig, D.Y. Smith: Phys. Rev. B 6, 4502 (1972).
7.27 A. Villani, A.H. Zimmerman: Phys. Rev. B 8, 2914 (1973).
7.28 M. Altarelli, D.Y. Smith: Phys. Rev. B 9, 1290 (1974); B 12, 3511 (1975).
7.29 D.Y. Smith: Phys. Rev. B 13, 5303 (1976).
7.30 M. Piacentini: Nuovo Cimento 63B, 458 (1969).
7.31 S. L. Adler: Phys. Rev. 126, 413 (1972).
7.32 N. Wiser: Phys. Rev. 129, 62 (1963).
7.33 M. Kastner: Phys. Rev. B 6, 2273 (1972).
7.34 M. Kastner: Phys. Rev. B 7, 5237 (1973).
7.35 D.L. Johnson: Phys. Rev. B 9, 4475 (1974).
7.36 D.L. Johnson: Phys. Rev. B 12, 3428 (1975).
7.37 W. Hanke, L.J. Sham: Phys. Rev. B 12, 4501 (1975).
7.38 S.G. Louie, J.R. Chelikowski, M.L. Cohen: Phys. Rev. Lett. 34, 155 (1975).
7.39 R.D. Turner, J.C. Inkson: J. Phys. C 9, 3583 (1976).
7.40 T.K. Bergstresser, G.W. Rubloff: Phys. Rev. Lett. 30, 974 (1973).
7.41 S.R. Nagel, T.A. Witten: Phys. Rev. B 11, 1623 (1975).
7.42 J.C. Phillips: Solid State Phys. 18, 56 (1966).
7.43 D.L. Greenaway, G. Harbeke: *Optical Properties and Band Structure of Semiconductors* (Pergamon, New York 1968).
7.44 H.R. Philipp, H. Ehrenreich: In *Semiconductors and Semimetals*, Vol 3, ed. by R.K. Willardson and A.C. Beer, (Academic, New York 1967) p. 93.
7.45 F. Bassani, G. Pastori-Parravicini: *Electronic States and Optical Transitions in Solids* (Pergamon Press, Oxford 1975).
7.46 D.L. Dexter, R.S. Knox: *Excitons*, (Interscience, New York 1965).
7.47 R.S. Knox: *Theory of Excitons* (Academic Press, New York 1963).
7.48 B. Velický, J. Suk: Phys. Stat. Sol. 16, 147 (1966).
7.49 Y. Toyozawa, J. Inoue, T. Inui, M. Okazaki, E. Hanamura: J. Phys. Soc. Japan 22, 1337 (1967).
7.50 M. Okazuki, M. Inoue, Y. Toyozawa, T. Inui, E. Hanamura: J. Phys. Soc. Japan 22, 1349 (1967).
7.51 C. Kunz: Comments Solid State Phys. 5, 31 (1973).
7.52 D. Pines: Solid State Phys. 1, 362 (1955).
7.53 S. Raimes: Repts. Prog. Phys. 20, 1 (1957).
7.54 D. Pines: *Elementary Excitations in Solids* (Benjamin, New York 1963).
7.55 H. Raether: In *Springer Tracts Mod. Phys.*, Vol. 38, ed by G. Höhler (Springer, Berlin, Heidelberg, New York 1965) p. 84.
7.56 J. Daniels, C.L. Festenberg, H. Raether, K. Zeppenfeld: In *Springer Tracts Mod. Phys.*, Vol. 54, ed. by G. Höhler (Springer, Berlin, Heidelberg, New York 1970) p. 77.
7.57 C. Horie: Prog. Theor. Phys. 21, 113 (1959).
7.58 T. Miyakawa: J. Phys. Soc. Japan 24, 768 (1968).
7.59 P.V. Giaquinta, E. Tosattti, M.P. Tosi: Solid State Comm. 19, 123 (1976).
7.60 P.V. Giaquinta, M. Parrinello, R. Tosatti, M.P. Tosi: J. Phys. C 9, 2031 (1976).
7.61 C.B. Wilson: Proc. Phys. Soc. (London) 80, 481 (1960).
7.62 M. Cardona: *Modulation Spectroscopy* (Academic Press, New York 1967).
7.63 R.K. Willardson, A.C. Beer (eds.). *Semiconductors and Semimetals*, Vol. 9, *Modulation Techniques*, (Academic Press, New York 1972).
7.64 Surf. Sci. 37, (1973).
7.65 A. Balzarotti, E. Colavita, S. Gentile, R. Rosei: Appl. Opt. 14, 2412 (1975).
7.66 R.E. Prange, H.D. Drew, J.B. Restorff, unpublished.
7.67 B.O. Seraphin, N. Bottka: Phys. Rev. 145, 628 (1966).
7.68 D.E. Aspnes: J. Opt. Soc. Am. 63, 1380 (1973).
7.69 D.E. Aspnes, C.G. Olson, D.W. Lynch: J. Appl. Phys. 47, 602 (1976).
7.70 D.D. Sell: Appl. Opt. 9, 1926 (1970).
7.71 W. Zierau, M. Skibowski: J. Phys. C 8, 1671 (1975).
7.72 B. Batz: In *Semiconductors and Semimetals*, Vol. 9, ed. by R.K. Willardson

414

and A.C. Beer, (Academic Press, New York 1972) p. 316.
7.73 R. Rosei, D.W. Lynch: Phys. Rev. B 5. 3883 (1972).
7.74 I. Balslev: In *Semiconductors and Semimetals*, Vol. 9, ed. by R.K. Willardson and A.C. Beer, (Academic Press, New York 1972) p. 403.
7.75 E.O. Kane: Phys. Rev. 178, 1368 (1969).
7.76 N. Bottka, J.E. Fischer: Phys. Rev. B 3, 2514 (1971).
7.77 B.O. Seraphin: In *Semiconductors and Semimetals*, Vol. 9, ed. by R.K. Willardson and A.C. Beer, (Academic Press, New York 1972) p. 1.
7.78 D.F. Blossey, P. Handler: In *Semiconductors and Semimetals*, Vol. 9, ed. by R.K. Willardson and A.C. Beer, (Academic Press, New York 1972) p. 257.
7.79 D.E. Aspnes and N. Bottka: In *Semiconductors and Semimetals*, Vol. 9, ed. by R.K. Willardson and A.C. Beer, (Academic Press, New York 1972) p.457.
7.80 V. Rehn: Surf. Sci. 37, 443 (1973).
7.81 D.E. Aspnes: Surf. Sci. 37, 418 (1973).
7.82 R.C. Aggarwal: In *Semiconductors and Semimetals*, Vol. 9, ed. by R.K. Willardson and A.C. Beer, (Academic Press, New York 1972) p.151.
7.83 W. Gudat, C. Kunz, J. Karlau: Appl. Opt. 13, 1412 (1974).
7.84 F. Abelés: F. Abelés (ed.): *Optical Properties of Solids*, (North-Holland, Amsterdam 1972) p. 93.
7.85 P.O. Nilsson: Solid State Phys. 29, 139 (1974).
7.86 J.C. Sutherland, R.N. Hamm, E.T. Arakawa: J. Opt. Soc. Am. 59, 1581 (1969).
7.87 J.C. Sutherland, E.T. Arakawa: J. Opt. Soc. Am. 57, 645 (1967); 58, 1080 (1968).
7.88 U.S. Whang, E.T. Arakawa, T.A. Callcott: J. Opt. Soc. Am. 61, 740 (1971).
7.89 U.S. Whang, E.T. Arakawa, T.A. Callcott: Phys. Rev. B 5, 2118 (1972).
7.90 U.S. Whang, E.T. Arakawa, T.A. Callcott: Phys. Rev. B 6, 2109 (1972).
7.91 B. Feuerbacher, M. Skibowski, R.P. Godwin: Phys. Lett. 26A, 595 (1968).
7.92 M. Skibowski, B. Feuerbacher, W. Steinmann, R.P. Godwin: Z. Physik 211, 329 (1968).
7.93 M.R. Potter, G.W. Green: J. Phys. F 5, 1426 (1975).
7.94 P.O. Nilsson, G. Forssell: J. Phys. F 5, L159 (1975).
7.95 R.A. Pollak, S.P. Kowalczyk, L. Ley, D.A. Shirley: Phys. Rev. Lett. 29, 274 (1972).
7.96 R.T. Poole, R.C.G. Leckey, J.G. Jenkin, J.G. Liesegang: Phys. Rev. B 8, 1401 (1973).
7.97 L.A. Hisscott, P.T. Andrews: J. Phys. F 5, 1077 (1975).
7.98 A.D. McLachlan, J.G. Jenkin, J. Liesegang, R.C.G. Leckey: J. Electron. Spectr. 3, 207 (1974).
7.99 L. Ley, S.P. Kowalczyk, F.R. McFeely, D.A. Shirley: Phys. Rev. B 10, 4881 (1974).
7.100 C.G. Olson, D.W. Lynch: Phys. Rev. B 9, 3159 (1974).
7.101 R.L. Kroes, R.C. Linton, F.E. Martin: J. Opt. Soc. Am. 66, 999 (1976).
7.102 J. Bordas, A.J. Brant, H.P. Hughes, J. Jakobssen, H. Kamimura, F.A. Levy, K. Nakao, Y. Natsume, A.D. Yoffe: J. Phys. C 9, 2277 (1976).
7.103 C.G. Olson, M. Piacentini, D.W. Lynch: Phys. Rev. Lett. 33, 644 (1974).
7.104 M. Piacentini, Il Nuovo Cim. 39B, 682 (1977).
7.105 J. Hermanson, J. Anderson, G.J. Lapeyre: Phys. Rev. B 12, 5410 (1975).
7.106 N.E. Christensen: Phys. Rev. B 13, 2698 (1976).
7.107 C.G. Olson and D.W. Lynch, to be published.
7.108 D. Beaglehole and B. Thieblemont, Il Nuovo Cim. 39B, 477 (1977).
7.109 C.G. Olson and D.W. Lynch (unpublished).
7.110 J.L. Robins: Proc. Phys. Soc. London 78, 1177 (1961).
7.111 M. Creuzburg: Z. Physik 196, 433 (1966).
7.112 J.H. Weaver, D.W. Lynch, C.G. Olson: Phys. Rev. B 7, 4311 (1973).
7.113 J.H. Weaver, D.W. Lynch, C.G. Olson: Phys. Rev. B 10, 501 (1974).
7.114 J.H. Weaver: Phys. Rev. B 11, 1416 (1975).
7.115 D.W. Lynch, C.G. Olson, J.H. Weaver: Phys. Rev. B 11, 3617 (1975).
7.116 J.H. Weaver, C.G. Olson, D.W. Lynch: Phys. Rev. B 12, 1293 (1975).
7.117 J.H. Weaver, C.G. Olson: Phys. Rev. B 15, 590 (1977).
7.118 J.H. Weaver, C.G. Olson, D.W. Lynch: Phys. Rev. B 15, (1977).
7.119 J.H. Weaver, C.G. Olson: Phys. Rev. B (1977).
7.120 J.H. Weaver, C.G. Olson: Phys. Rev. B (1977).

7.121 E.S. Black, D.W. Lynch, and C.G. Olson, Phys. Rev. B 16, 2337 (1977).
7.122 I. Petroff, C.R. Viswanathan: Phys. Rev. B 4, 799 (1971).
7.123 J.H. Weaver, C.G. Olson, D.W. Lynch, M. Piacentini: Solid State Comm. 16, 163 (1975).
7.124 D.W. Lynch, R. Rosei, J.H. Weaver, C.G. Olson: J. Solid State Chem. 8, 242 (1973).
7.125 W.E. Pickett, P.B. Allen: Phys. Rev. B 11, 3599 (1975).
7.126 T.J. Moravec, J.C. Rife, R.N. Dexter: Phys. Rev. B 13, 3247 (1976).
7.127 C. Kunz: Z. Physik 196, 311 (1966).
7.128 J.G. Endriz, W.E. Spicer: Phys. Rev. B 2, 1466 (1970).
7.129 G.W. Simmons, E.J. Scheibner: J. Appl. Phys. 43, 693 (1972).
7.130 V.V. Zashkvara, K.Sh. Chokin: Fiz, Tverd. Tela 17, 1935 (1975) [Sov. Phys.-Solid State 17, 1268 (1975)].
7.131 J. Cazaux: Solid State Comm. 8, 545 (1970).
7.132 J. Cazaux: Opt. Commun. 3, 221 (1971).
7.133 S.H. Liu: In Handbook on the Physics and Chemistry of the Rare Earths, ed. by K.A. Gschneidner and L. Eyring, (North-Holland, Amsterdam 1977), Ch. 3.
7.134 P.N. Argyres: Phys. Rev. 97, 334 (1955).
7.135 H.S. Bennett, E.A. Stern: Phys. Rev. 137, A448 (1965).
7.136 J.L. Erskine, E.A. Stern: Phys. Rev. Lett. 30, 1329 (1973).
7.137 J.L. Erskine, E.A. Stern: Phys. Rev. B 8, 1239 (1973).
7.138 J.L. Erskine: Phys. Rev. Lett. 37, 157 (1976).
7.139 J.L. Erskine, Physica B 89, 83 (1977).
7.140 J.L. Erskine, C.P. Flynn: Phys. Rev. B 14, 2197 (1976).
7.141 J.L. Erskine, E.A. Stern: Phys. Rev. B 12, 5016 (1975).
7.142 T. Sagawa, Y. Iguchi, M. Sasanuma, A. Ejiri, S. Fujiwara, M. Yokota, S. Yamaguchi, M. Nakamura, T. Sasaki, T. Oshio: J. Phys. Soc. Japan 21, 2602 (1966).
7.143 N. Swanson, K. Codling: J. Opt. Soc. Am. 58, 1192 (1968).
7.144 R. Haensel, G. Keitel, B. Sonntag, C. Kunz, P. Schreiber: Phys. Stat. Sol. (a) 2, 85 (1970).
7.145 A. Ejiri, S. Yamaguchi, M. Saruwatari, M. Yokuta, K. Inayoshi, G. Matsuoka: Opt. Comm. 1, 349 (1970).
7.146 C. Gähwiller, F.C. Brown: Phys. Rev. B 2, 1918 (1970).
7.147 C. Kunz, R. Haensel, G. Keitel, P. Schreiber, B. Sonntag: NBS Special Publication 323, ed. by L.H. Bennett, (Washington, 1971) p.275.
7.148 J.H. Slowik, F.C. Brown: Phys. Rev. Lett. 29, 934 (1972).
7.149 B.F. Sonntag: J. Phys. F 3, L255 (1973).
7.150 J.H. Slowik: Phys. Rev. B 10, 416 (1974).
7.151 N. Kosuch, G. Wiech, A. Faessler: In Proc. 4th Intern. Conf. on Vacuum Ultraviolet Radiation Physics, Hamburg, 1974, ed. by E.E. Koch, R. Haensel and C. Kunz, (Pergamon/Vieweg, Braunschweig 1974) p. 398.
7.152 C. Kunz, H. Petersen, D.W. Lynch: Phys. Rev. Lett. 33, 1556 (1974).
7.153 H. Petersen: Phys. Rev. Lett. 35, 1365 (1975).
7.154 H. Petersen, C. Kunz: Phys. Rev. Lett. 35, 863 (1975).
7.155 H. Petersen: Phys. Stat. Sol. (b) 72, 591 (1975).
7.156 G.D. Mahan: Phys. Rev. 163, 612 (1967).
7.157 P. Nozieres, C.T. De Domenicis: Phys. Rev. 176, 1097 (1969).
7.158 G.D. Mahan: Solid State Phys. 29, 75 (1974).
7.159 S. Doniach, P. Platzman, J.T. Yue: Phys. Rev. B 4, 3345 (1971).
7.160 S. Doniach, M. Sunjić: J. Phys. C 3, 285 (1970).
7.161 J.W. Gadzuk, M. Sunjić: Phys. Rev. B 12, 524 (1975).
7.162 Y. Baer, P.H. Citrin, G.K. Wertheim: Phys. Rev. Lett. 37, 49 (1976).
7.163 P.H. Citrin, G.K. Wertheim, M. Schlüter, Y. Baer: Intern. Conf. Physics of X-Ray Spectra, NBS (Aug. 30-Sept. 2, 1976) extended abstracts, p. 29.
7.164 R.P. Gupta, A.J. Freeman: Phys. Rev. Lett. 36, 1194 (1976).
7.165 R.P. Gupta, A.J. Freeman: Phys. Lett. 59A, 223 (1976).
7.166 R.P. Gupta, A.J. Freeman, J.D. Dow: Phys. Lett. 59A 226 (1976).
7.167 Y. Onodera: J. Phys. Soc. Japan 39, 1482 (1975).
7.168a S.M. Girvin, J.J. Hopfield: Phys. Rev. Lett. 37, 1091 (1976).

416

7.168b T. Ishii, Y. Sakisaka, S. Yamaguchi, T. Hanyu, H. Ishii: J. Phys. Soc. Japan 42, 876 (1977).
7.169 J.J. Ritsko, S.E. Schnatterly, P.C. Gibbons: Phys. Rev. B 10, 5017 (1974).
7.170 P.C. Gibbons, T.H. Moog, S.G. Slusky, S.E. Schnatterly, J.J. Ritsko, J.R. Field Bull. Am. Phys. Soc. (Ser. 11.) 20, 434 (1975).
7.171 J.J. Ritsko, S.E. Schnatterly, J.R. Fields: Phys. Rev. Lett. 36, 326 (1976).
7.172 T.A. Callcott, E.T. Arakawa: Phys. Rev. Lett. 38, 442 (1977).
7.173 J.J. Ritsko, S.E. Schnatterly, P.C. Gibbons: Phys. Rev. Lett. 32, 671 (1974).
7.174 A. Balzarotti, A. Bianconi, E. Burattini: Phys. Rev. B 9, 5003 (1974).
7.175 W. Gudat and C. Kunz, unpublished.
7.176 W. Gudat, C. Kunz, J. Karlau: In *Proc. Intern. Symp. X-Ray Spectra Electronic Structure of Matter* (1975) ed. by A. Faessler and G. Wiech, Vol. 1, p. 295.
7.177 H.J. Hagemann, W. Gudat, C. Kunz: Solid State Comm. 15, 655 (1974).
7.178 H.J. Hagemann, W. Gudat, C. Kunz: Phys. Stat. Sol. (b) 74, 507 (1976).
7.179 R. Haensel, C. Kunz, B. Sonntag: Phys. Lett. 25A, 205 (1967).
7.180 R. Haensel, C. Kunz, T. Sasaki, B. Sonntag: Appl. Opt. 7, 301 (1968).
7.181 R. Haensel, C. Kunz, T. Sasaki, B. Sonntag: J. Appl. Phys. 40, 2046 (1969).
7.182 B. Sonntag, R. Haensel, C. Kunz: Solid State Comm. 7, 597 (1969).
7.183 R. Haensel, R. Radler, B. Sonntag, C. Kunz: Solid State Comm. 7, 1495 (1969).
7.184 F.C. Brown, C. Gähwiller, A.B. Kunz: Solid State Comm. 7, 597 (1971).
7.185 M. Cukier, P. Dhez, F. Wuilleumier, P. Jaeglé, M. Lamoureaux, F. Combet Farnot: In *Proc. 4th Intern. Conf. on Vacuum Ultraviolet Radiation Physics, Hamburg (1974)*, ed. by E.E. Koch, R. Haensel, and C. Kunz, (Pergamon/Vieweg, Braunschweig 1974) p. 102.
7.186 J.H. Weaver, C.G. Olson: Phys. Rev. B 14, 3251 (1976).
1.187 F. Combet Farnoux, M. Lamoureaux: In *Proc. 4th Intern. Conf. on Vacuum Ultraviolet Radiation Physics. Hamburg (1974)*, ed. by E.E. Koch, R. Haensel and C. Kunz, (Pergamon/Vieweg, Braunschweig 1974) p. 89.
7.188 F. Combet Farnoux: T. Åberg, E. Källne, R. Manne (eds.): *Proc. Intern. Conf. on X-Ray Processes in Matter, Otaniemi, Finland*, Physica Fennica 9 (Suppl. s1), 80 (1974).
7.189 F. Combet Farnoux, F. Keller: Intern. Conf. Physics of X-Ray Spectra, NBS (Aug. 30-Sept. 2, 1976) extended abstracts, p. 310.
7.190 R.E. Dietz, E.G. Mcrae, Y. Yafet, C.W. Caldwell: Phys. Rev. Lett. 38, 1372 (1974).
7.191 U. Fano: Phys. Rev. 124, 1866 (1961).
7.192 R.E. Dietz, E.G. Mcrae: Intern. Conf. Physics of X-Ray Spectra, NBS (Aug. 30-Sept. 2, 1976) extended abstracts, p.217.
7.193 L.C. Davis, L.A. Feldkamp: Solid State Comm. 19, 413 (1976).
7.194 W. Gudat, C. Kunz: Phys. Stat. Sol. (b) 52, 433 (1972).
7.195 J.L. Erskine: private communication.
7.196 R. Haensel, P.Rabe, B. Sonntag: Solid State Comm. 8, 1845 (1970).
7.197 P. Rabe, K. Radler, H.W. Wolff: In *Proc. 4th Intern. Conf. on Vacuum Ultraviolet Radiation Physics, Hamburg (1974)*, ed. by E.E. Koch, R. Haensel and C. Kunz, (Pergamon/Vieweg, Braunschweig 1974) p. 242.
7.198 M. Cukier, P. Dhez, P. Jaeglé: Intern. Conf. Physics of X-Ray Spectra, NBS (Aug. 30-Sept. 2, 1976) extended abstracts, p.321.
7.199 H.W. Wolff, R. Bruhn, K. Radler, B. Sonntag: Phys. Lett. 59A, 67 (1976).
7.200 J.H. Weaver, unpublished
7.201 M. Cardona: In *Semiconductors and Semimetals*, Vol. 3, ed. by R.K. Willardson and A.C. Beer, (Academic, New York 1967) p. 125.
7.202 D.L. Greenaway, G. Harbeke: *Optical Properties and Band Structures of Semiconductors* (Pergamon Press, Oxford 1968).
7.203 B. Feuerbacher, R.P. Godwin, T. Sasaki, M. Skibowski: J. Opt. Soc. Am. 58, 1434 (1968).
7.204 M. Cardona, W. Gudat, B. Sonntag, P.Y. Yu: In: *Proc. 10th Intern. Conf. on the Physics of Semiconductors*, Cambridge, Mass. (1970), ed. by S.P. Keller, J.C. Hensel, F. Stern, (US AEC Div. of Technical Information, Springfield, VA 1970) p. 209.
7.205 M. Cardona, W. Gudat, E.E. Koch, M. Skibowski, B. Sonntag, P.Y. Yu: Phys. Rev. Lett. 25, 659 (1970).

7.206 W. Gudat, E.E. Koch, P.Y. Yu, M. Cardona, C.M. Penchina: Phys. Stat. Sol. (b) 53, 327 (1972).
7.207 D.E. Aspnes, C.G. Olson: Phys. Rev. Lett. 33, 1605 (1974).
7.208 D.E. Aspnes, C.G. Olson, D.W. Lynch: Phys. Rev. B 12, 2527 (1975).
7.209 D.E. Aspnes, C.G. Olson, D.W. Lynch: Phys. Rev. B 15, 5331 (1976).
7.210 D.E. Aspnes, J.E. Rowe: Phys. Rev. B 5, 4022 (1972).
7.211 D.E. Aspnes: Surf. Sci. 37, 418 (1973).
7.212 J.C. Phillips: Phys. Rev. Lett. 22, 285 (1969).
7.213 D.E. Aspnes, C.G. Olson, D.W. Lynch: Phys. Rev. B 14, 2534 (1976).
7.214 D.F. Blossey: Phys. Rev. B 2, 2976 (1970).
7.215 F.C. Weinstein, J.D. Dow, B.Y. Lao: Phys. Rev. B 4, 3502 (1971).
7.216 D.F. Blossey: Phys. Rev. B 3, 1382 (1971).
7.217 D.E. Aspnes, C.G. Olson, D.W. Lynch: In Proc. 13th Intern. Conf. Physics of Semiconductors, Rome (1976), ed. by F.G. Fumi (Tipografia Marves, Rome 1977) p. 1000.
7.218 D.E. Aspnes, C.G. Olson, D.W. Lynch: Phys. Rev. Lett. 37, 766 (1976).
7.219 D.E. Aspnes: Phys. Rev. B 14, 5331 (1976).
7.220 P. Thiry, Y. Petroff, R. Pinchaux, J. Chelikowsky, M.L. Cohen: Solid State Comm. 20, 1107 (1976).
7.221 D.A. Shirley: Chem. Phys. Lett. 16, 220 (1972).
7.222 S.P. Kowalczyk, L. Ley, F.R. Mcfeely, R.A. Pollak, D.A. Shirley: Phys. Rev. B 9, 381 (1974).
7.223 A.B. Kunz: Phys. Rev. B 12, 5890 (1975).
7.224 D.E. Aspnes, C.G. Olson, D.W. Lynch: Phys. Rev. Lett. 36, 1563 (1976).
7.225 J.C. Phillips: Phys. Rev. 146, 584 (1966).
7.226 N. Bottka, J.E. Fischer: Phys. Rev. B 3, 2514 (1971).
7.227 V. Rehn: Surf. Sci. 37, 443 (1973).
7.228 C.G. Olson, D.E. Aspnes: unpublished.
7.229 R. Klucker, M. Skibowski, W. Steinmann: Phys. Stat. Sol. (b) 65, 703 (1974).
7.230 V. Rehn, J.L. Stanford, V.O. Jones, W.J. Choyke: In Proc. 13th Int. Conf. on the Physics of Semiconductors, Rome (1976), ed. by F.G. Fumi (Tipografia Marves, Rome 1977) p. 985.
7.231 V.V. Mikhailin, W.E. Oranovskii, S. Pacesová, J. Pastrňák, A.S. Salamatov: Phys. Stat. Sol. (b) 55, K51 (1973).
7.232 R. Klucker, H. Nelkowski, Y.S. Park, M. Skibowski, T.S. Wagner: Phys. Stat. Sol. (b) 45, 265 (1971).
7.233 J.L. Freeouf: Phys. Rev. B 7, 3810 (1973).
7.234 V. Mikhailin: In Luminescence of Crystals, Molecules, and Solutions, ed. by F.Williams, (Plenum Press, New York 1973) p. 269.
7.235 V.N. Meleshkin, V.V. Mikhailin, V.E. Oranovskii, P.A. Orekhnov, J. Pastrňák, S. Pacesová, A.S. Salamatov, M.V. Fok, A.S. Yarov: Trudy Fiz. Inst. Akad. Nauk. SSSR 80, 140 (1975), in English: The Lebedev Institute Physics Series, Vol. 80, N.G. Basov Ed. (Consultants Bureau, New York, 1976) p. 139.
7.236 M. Cardona, C.M. Penchina, E.E. Koch, P.Y. Yu: Phys. Stat. Sol. (b) 58, 127 (1973).
7.237 R.Pinchaux, P. Thiry, Y. Petroff, M. Balkanski: In Proc. of the 4th Intern. Conf. on Vacuum Ultraviolet Radiation Physics, Hamburg (1974), ed. by E.E. Koch, R. Haensel and C. Kunz, (Pergamon/Vieweg, Braunschweig 1974) p. 489.
7.238 G. Martinez, M. Schlüter, M.L. Cohen, R. Pinchaux, P. Thiry, D. Dagneaux, Y. Petroff: Solid State Comm. 17, 5 (1975).
7.239 G. Martinez, M Schlüter, M.L. Cohen: Phys. Rev. B 11, 660 (1975).
7.240 G. Martinez, M. Schlüter, M.L. Cohen: Phys. Rev. B 11, 651 (1975).
7.241 P. Bammes, R. Klucker, E.E. Koch, T. Tuomi: Phys. Stat. Sol. (b) 49, 561 (1972).
7.242 N.J. Shevchik, M. Cardona, J. Tejeda: Phys. Rev. B 8, 2833 (1973).
7.243 B. Sonntag, T. Tuomi, G. Zimmerer: Phys. Stat. Sol. (b) 58, 101 (1973).
7.244 B. Sonntag, G. Zimmerer, T. Tuomi: In Proc. 11th Int. Conf. on the Physics of Semiconductors, Warsaw (1972)(PWN-Polish Scientific Publishers, Warsaw 1972) p.813.
7.245 G.N. Greaves, E.A. Davis, J. Bordas: Phil. Mag. 34, 265 (1976).
7.246 J. Bordas, J.B. West: Phil. Mag. 34, 501 (1976).

418

7.247 J.D. Wiley, W.J. Buckel, W. Braun, G.W. Fehrenbach, F.J. Himpsel, E.E. Koch: Phys. Rev. B $\underline{14}$, 697 (1976).
7.248 J.C. Shaffer, B. Vanpelt, C. Wood, J. Freeauf, K. Murase, J.W. Osmun: Phys. Stat. Sol. $\underline{54}$, 511 (1973).
7.249 Z. Hurych, J.C. Shaffer, D.L. Davis, T.A. Knecht, G.J. Lapeyre, P.L. Gobby, J.A. Knapp, C.G. Olson: Phys. Rev. Lett. $\underline{33}$, 830 (1974).
7.250 P. Thiry, R. Pincheaux, D. Dagneaux, Y. Petroff: In *Proc. 12th Intern. Conf on the Physics of Semiconductors*, ed. by M.H. Pilkuhn (B.G. Teubner, Stuttgart (1974)) p. 1324.
7.251 R. Mamy, B. Thiêblemont, O. Cerclier: J. Physique Lett. $\underline{37}$, L85 (1976).
7.252 M. Zivitz, J.R. Stevenson: Phys. Rev. B $\underline{10}$, 2457 (1974).
7.253 R. Suryanarayanan, G. Güntherodt, J.L. Freeouf, F. Holtzberg: Phys. Rev. B $\underline{12}$, 4215 (1975).
7.254 C. Gähwiller, F.C. Brown: Phys. Rev. B $\underline{2}$, 1918 (1970).
7.255 C. Gähwiller, F.C. Brown: In *Proc. 10th Intern. Conf. Physics of Semiconductors Cambridge, Mass (1970)*, ed. by S.P. Keller, J.C. Hensel and F. Stern, (US AEC Div. of Technical Information, Springfield, VA 1970) p. 213.
7.256 F.C. Brown, O.P. Rustgi: Phys. Rev. Lett. $\underline{28}$, 497 (1972).
7.257 H. Fujita, Y. Iguchi, Y. Okada, T. Sasaki: J. Phys. Soc. Japan $\underline{33}$, 1494 (1972).
7.258 M. Altarelli, D.L. Dexter: Phys. Rev. Lett. $\underline{29}$, 110 (1972).
7.259 S. Pantelides: Solid State Comm. $\underline{16}$, 217 (1975).
7.260 R.S. Bauer, R.Z. Bachrach, D.E. Aspnes, and J.C. McMenamin, Il Nuovo Cim. $\underline{39B}$, 409 (1977).
7.261 F.C. Brown, R.Z. Bachrach, and M. Skibowski, Phys. Rev. B $\underline{15}$, 4781 (1977).
7.262 H. Fujita, Y.Iguchi: Jap. J. Appl. Phys. $\underline{14}$, 220 (1975).
7.263 S.T. Pantelides: Solid State Comm. $\underline{16}$, 1151 (1975).
7.624 M. Cardona, R. Haensel: Phys. Rev. B $\underline{1}$, 2605 (1970).
7.265 C. Sugiura, Y. Hayasi, H. Konuma, S. Sato, M. Watanabe: J. Phys. Soc. Japan $\underline{29}$, 1645 (1970).
7.266 C. Sugiura, Y. Hayasi, H. Konuma, S. Kiyono: J. Phys. Soc. Japan $\underline{31}$, 1784 (1971).
7.267 M. Watanabe, H. Yamashita, Y. Nakai, S. Sato, S. Onari: Phys. Stat. Sol. $\underline{43}$, 631 (1971).
7.268 F.J. Comes, U. Nielsen, W.H.E. Schwarz: J. Chem. Phys. $\underline{58}$, 2230 (1973).
7.269 B. Sonntag, F.C. Brown: Phys. Rev. B $\underline{10}$, 2300 (1974).
7.270 F.C. Brown, R.Z. Bachrach, M. Skibowski: Phys. Rev. B $\underline{13}$, 2633 (1976).
7.271 B. Sonntag, In *Rare Gas Solids*, Vol. $\underline{11}$, ed. by M.L. Klein and J.A. Venables (Academic, New York 1976), Chap. 17.
7,272 G. Zimmerer: In *I.C.A.P.-I.N.F.N. Course on Synchrotron Radiation Research*, ed. by A.N. Mancini and I. F. Quercia, Alghero, Italy (Sept. 1976) Vol. $\underline{1}$, p. 409.
7.273 R. Haensel, G. Keitel, E.E. Koch, M. Skibowski, P. Schreiber: Phys. Rev. Lett. $\underline{23}$, 1160 (1969).
7.274 M. Skibowski: unpublished.
7.275 Y. Onodera, Y. Toyozawa: J. Phys. Soc. Japan $\underline{22}$, 833 (1967).
7.276 U. Rössler, O. Schütz: Phys. Stat. Sol. (b) $\underline{56}$, 483 (1973).
7.277 W. Andreoni, M. Altarelli, F. Bassani: Phys. Rev. B $\underline{11}$, 2352 (1975)..
7.278 D. Pudewill, F.-J. Himpsel, V. Saile, N. Schwentner, M. Skibowski, E.E. Koch: Phys. Stat. Sol (b) $\underline{74}$, 485 (1975).
7.279 V. Saile, M. Skibowski, W. Steinmann, P. Gürtler, E.E. Koch, A. Kozenvnikov: Phys. Rev. Lett. $\underline{37}$, 305 (1976).
7.280 H.W. Wolff, unpublished.
7.281 R. Haensel, G. Keitel, C. Kunz, P. Schreiber: Phys. Rev. Lett. $\underline{25}$, 208 (1970).
7.282 R. Haensel, G. Keitel, N. Kosuch, U. Nielsen, P. Schreiber: J. Physique $\underline{32}$, C4-326 (1971).
7.283 U. Fano, J.W. Cooper: Rev. Mod. Phys. $\underline{40}$, 441 (1968); $\underline{41}$, 724 (1969).
7.284 U. Rössler: Phys. Stat. Sol. (b) $\underline{45}$, 483 (1971).
7.285 A.B. Kunz, D.J. Mickish: Phys. Rev. D $\underline{8}$, 779 (1973).
7.286 R. Haensel, G. Keitel, E.E. Koch, N. Kosuch, M. Skibowski: Phys. Rev. Lett. $\underline{25}$, 1281 (1970).
7.287 R. Haensel, G. Keitel, E.E. Koch, M. Skibowski, P. Schreiber: Opt. Comm. $\underline{2}$, 59 (1970).

7.288 R. Haensel, G. Keitel, P. Schreiber, C. Kunz: Phys. Rev. Lett. 22. 398 (1969).
7.289 R. Haensel, G. Keitel, P. Schreiber, C. Kunz: Phys. Rev. 188, 1375 (1969).
7.290 R. Haensel, N. Kosuch, U. Nielsen, B. Sonntag, U. Rössler: Phys. Rev. B 7,
       1577 (1975).
7.291 R. Brodmann, U. Hahn, G. Zimmerer, R. Haensel, U. Nielsen: In *Proc. 4th Intern.*
       Conf. on Vacuum Ultraviolet Radiation Physics, Hamburg (1974), ed. by E.E.
       Koch, R. Haensel and C. Kunz, (Pergamon/Vieweg, Braunschweig 1974) p. 344,
       U. Gerick: Diplomarbeit, University of Hamburg (1977).
7.292 R. Brodmann, R. Haensel, U. Hahn, U. Nielsen, G. Zimmerer: Chem. Phys. Lett.
       29, 250 (1974).
7.293 Ch. Ackermann, R. Brodmann, U. Hahn, A. Suzuki, G. Zimmerer: Phys. Stat. Sol.
       (b) 74, 579 (1976).
7.294 J. Jortner: In *Proc. 4th Intern. Conf. on Vacuum Ultraviolet Radiation Physics,*
       Hamburg (1974), ed. by E.E. Koch, R. Haensel and C. Kunz, (Pergamon/Vieweg,
       Braunschweig 1974) p. 263.
7.295 K. Monahan, V. Rehn, E. Matthias, and E. Poliakoff, J. Chem. Phys. 67, 1784
       (1977).
7.296 Th. Förster: Ann. Physik 2, 55 (1948).
7.297 C. Ackermann, R. Brodmann, R. Haensel, V. Hahn, G. Tolkien, and G. Zimmerer,
       J. Luminesc. 12, 13, 315 (1976).
7.298 H. Möller, R. Brodman, G. Zimmerer, U. Hahn: Solid State Comm. 20, 401 (1976).
7.299 Ch. Ackermann, R. Brodmann, G. Tolkiehn, G. Zimmerer, R. Haensel, U. Hahn:
       J. Luminesc. 12/13, 315 (1976).
7.301 S.S. Hasnain, T.D.S. Hamilton, I.H. Munro, E. Pantos, and I.T. Steinberger,
       Phil. Mag. 35, 1299 (1977).
7.302 N. Schwentner, M. Skibowski, W. Steinmann: Phys. Rev. B 8, 2965 (1973).
7.303 E.E. Koch, B. Raz, V. Saile, N. Schwentner, M. Skibowski, W. Steinmann: Jap.
       J. Appl. Phys. Suppl. 2 Pt2, 775 (1975).
7.304 Z. Ophir, B. Raz, J. Jortner, V. Saile, N. Schwentner, E.E. Koch, M. Skibowski,
       W. Steinmann: J. Chem. Phys. 62, 650 (1975).
7.305 D. Pudewill, F.-J. Himpsel, V. Saile, N. Schwentner, M. Skibowski, E.-E. Koch,
       and J. Jortner, J. Chem. Phys. 65, 5226 (1976).
7.306 E.E. Koch, V. Saile, N. Schwentner, M. Skibowski: Chem. Phys. Lett. 28, 562
       (1976).
7.307 N. Schwentner, F.-J. Himpsel, V. Saile, M. Skibowski, W. Steinmann, E.E. Koch:
       Phys. Rev. Lett. 34, 528 (1975).
7.308 A.B. Kunz, D.J. Mikisch, S.K.V. Mirmira, T. Shima, F.-J. Himpsel, V. Saile,
       N. Schwentner, E.E. Koch: Solid State Comm. 17, 761 (1975).
7.309 N. Schwentner: Phys. Rev. B 14, 5490 (1976).
7.310′ E. Pantos, S.S. Hasnain, and I.T. Steinberger, Chem Phys. Lett. 46, 395
       (1977).
7.311 S.S. Hasnain, I.H. Munro, and T.D.S. Hamilton, J. Phys. C 10, 1097 (1977).
7.312 N. Schwentner, E.E. Koch: Phys. Rev. B 14, 4687 (1976).
7.313 S.S. Hasnain, I.H. Munro, and T.D.S. Hamilton, Il Nuovo Cim. 39B, 500 (1977).
7.314 F.J. Comes, R. Haensel, U. Nielsen, W.H.E. Schwarz: J. Chem. Phys. 58, 516
       (1973).
7.315 D.M. Roessler, W.C. Walker: J. Phys. Chem. Solids 28, 1507 (1967).
7.316 K.K. Rao, T.J. Moravec, J.C. Rife, R.N. Dexter: Phys. Rev. B 12, 5937 (1975).
7.317 M. Piacentini: Solid State Comm. 17, 697 (1975).
7.318 T. Miyakawa: J. Phys. Soc. Japan 17, 1898 (1962).
7.319 J. Hermanson: Phys. Rev. 177, 1234 (1960).
7.320 J.T. Devreese, A.B. Kunz, T.C. Collins: Solid State Comm. 11, 673 (1972).
7.321 D.W. Lynch, M. Piacentini, C.G. Olson: Phys. Rev. Lett. 35, 1658 (1975).
7.322 M. Piacentini, D.W. Lynch, C.G. Olson: Phys. Rev. B 13, 5530 (1976).
7.323 C. Gout, F. Pradal: J. Phys. Chem. Solids 29, 581 (1968).
7.324 J.R. Fields, P.C. Gibbons, S.E. Schnatterly: Phys. Rev. Lett. 28, 430 (1977).
7.325 B.F. Sonntag: Phys. Rev. B 9, 3601 (1974).
7.326 W. Gudat, C. Kunz, H. Petersen: Phys. Rev. Lett. 32, 1370 (1974).
7.327 R. Haensel, C. Kunz, B. Sonntag: Phys. Rev. Lett. 20, 262 (1968).
7.328 F.C. Brown, C. Gähwiller, A.B. Kunz, N.O. Lipari: Phys. Rev. Lett. 25, 927
       (1970).

420

7.329 W.P. Menzel, C.C. Lin, D.F. Fouquet, E.E. Lafon, R.C. Chaney: Phys. Rev. Lett. 30, 1313 (1973).
7.330 A.B. Kunz, D.J. Mickish, T.C. Collins: Phys. Rev. Lett. 31, 756 (1973).
7.331 S.P. Kowalczyk, F.R. Mcfeely, L.Ley, R.A. Pollak, D.A. Shirley: Phys. Rev. B 9, 3573 (1974).
7.332 S.T. Pantelides, F.C. Brown: Phys. Rev. 33, 298 (1974).
7.333 S.T. Pantelides, Phys. Rev. B 11, 2391 (1975).
7.334 A.B. Kunz: Phys. Rev. B 12, 5890 (1975).
7.335 A. Zunger and A.J. Freeman, Phys. Rev. B 16, 2901 (1977).
7.336 Y. Onodera: Phys. Rev. B 4, 2751 (1971).
7.337 R. Haensel, C. Kunz, T. Sasaki, B. Sonntag: Phys. Rev. Lett. 20, 1436 (1968).
7.338 D. Blechschmidt, R. Klucker, M. Skibowski: Phys. Stat. Sol. 36, 625 (1969).
7.339 H. Saito, M. Watanabe, A. Ejiri, S. Sato, H. Yamashita, T. Shibaguchi, H. Nighida, S. Yamaguchi: Solid State Comm. 8, 1861 (1970).
7.340 D. Blechschmidt, R. Haensel, E.E. Koch, U. Nielsen, M. Skibowski: Phys. Stat. Sol. (b) 44, 787 (1971).
7.341 D. Blechschmidt, V. Saile, M. Skibowski, W. Steinmann: Phys. Lett. 35A, 221 (1971).
7.342 F.A. Korolev, O.F. Kulikov: Opt. Spectr. 31, 822 (1971) [Opt. i Spekt 31, 422 (1971)].
7.343 Y. Iguchi, T. Sasaki, H. Sugiwara, S. Sato, T. Nasu, A. Ejiri, S. Onati, K. Kojima, T. Oya: Phys. Rev. Lett. 26, 82 (1971).
7.344 H. Saito: Sci. Light 20, 1 (1971).
7.345 M. Watanabe, H. Yamashita, A. Ejiri, H. Nishida, H. Sato, T. Shibaguchi, S. Sato: J. Phys. Soc. Japan 31, 1085 (1971).
7.346 G.W. Rubloff, J. Freeouf, H. Fritzsche, K. Murase: Phys. Rev. Lett. 26, 1317 (1971).
7.347 C.J. Peimann, M. Skibowski: Phys. Stat. Sol. 46, 655 (1971).
7.348 V. Saile, M. Skibowski: Phys. Stat. Sol. (b) 50, 661 (1972).
7.349 G.W. Rubloff: Phys. Rev. B 5, 662 (1972).
7.350 V. Saile, N. Schwentner, M. Skibowski, W. Steinmann, W. Zierau: Phys. Lett. 46A, 245 (1973).
7.351 M. Piacentini (unpublished).
7.352 T. Aberg, J.L. Dehmer: J. Phys. C 6, 1450 (1973).
7.353 A. Balzarotti, A. Bianconi, E. Burattini, G. Strinati: Solid State Comm. 15, 1431 (1974).
7.354 C. Satoko, S. Sugano: J. Phys. Soc. Japan 34, 701 (1973).
7.355 M. Skibowski: private communication.
7.356 T. Sagawa, Y Iguchi, M. Sasanuma, T. Nasu, S. Yamaguchi, S. Fujiwara, M. Nakamura, A. Ejiri, T. Masuoka, T. Sasaki, T. Oshio: J. Phys. Soc. Japan 21, 2584 (1966).
7.357 Y. Iguchi, T. Sagawa, S. Sato, M. Watanabe, H. Yamashita, A. Ejiri, M. Sasanuma, S. Nakai, M. Nakamura, S. Yamaguchi, Y. Nakai, T. Oshio: Solid State Comm. 6, 575 (1968).
7.358 H. Fujita, C. Gähwiller, F.C. Brown: Phys. Rev. Lett. 22, 1369 (1969).
7.359 F.C. Brown, C. Gähwiller, H. Fujita, A.B. Kunz, W. Scheifley, N.J. Carrera: Phys. Rev. B 2, 2126 (1970).
7.360 M. Cardona, R. Haensel, D.W. Lynch, B. Sonntag: Phys. Rev. B 2, 1117 (1970).
7.361 Y. Nakamura, N.I. Watanabe, S. Sato, Y. Nakai: Solid State Comm. 9, 2017 (1971).
7.362 O. Aita, I. Nagakura, T. Sagawa: J. Phys. Soc. Japan 30, 1414 (1971).
7.363 M. Watanabe: J. Phys. Soc. Japan 34, 755 (1973).
7.364 W. Scheifley, F.C. Brown, S.T. Pantelides: In *Proc. 4th Intern. Conf. on Vacuum Ultraviolet Physics, Hamburg (1974)*, ed. by E.E. Koch, R. Haensel and C. Kunz, (Pergamon/Vieweg, Braunschweig 1974) p. 396.
7.365 J.H. Beaumont, A.J. Bourdillon, M.N. Kabler: J. Phys. C9, 2961 (1976).
1.366a R. Onaka, H. Onuki: In *Proc. 3rd Intern. Conf. on Ultraviolet Radiation Physics Tokyo (1971)*, ed. by Y. Nakai, (Physical Society of Japan, Tokyo 1972) abstract, 30p A1.
7.366b J.H. Beaumont, J. Bordas, A.J. Bourdillon, and M.R. Hayns: J. Phys. C 7, L349 (1974).

7.367 S. Sato, M. Watanabe, Y. Iguchi, S. Nakai, Y. Nakamura, T. Sagawa: J. Phys. Soc. Japan 33, 1638 (1972).
7.368 N.J. Carrera, F.C. Brown: Phys. Rev. B 4, 3651 (1971).
7.369 W. Hayes, A.B. Kunz, E.E. Koch: J. Phys. C 4, L200 (1971).
7.370 A.J. Bourdillon, J.H. Beaumont: J. Phys. C 9, L479 (1976).
7.371 T. Ishii, S. Sato, T. Matsukawa, Y. Sakisaka, T. Sagawa: J. Phys. Soc. Japan 32, 1440 (1972).
7.372 A.J. Bourdillon, J.H. Beaumont: J. Phys. C 9, L473 (1976).
7.373 J.H. Beaumont and A.J. Bourdillon, J. Phys. C 10, 761 (1977).
7.374 J. Bordas, J. Robinson, and A. Jakobsson, J. Phys. C 11, 2607 (1978).
7.375 P. Rabe, B. Sonntag, T. Sagawa, R. Haensel: Phys. Stat. Sol. (b) 50, 559 (1972).
7.376 S. Sato, T. Ishii, I. Nagakura, O. Aita, S. Nakai, M. Yokota, K. Ichikawa, G. Matsuoka, S. Kono, T. Sagawa: J. Phys. Soc. Japan 30, 459 (1971).
7.377 T. Ishii, Y. Sakisaka, T. Matsukawa, S. Sato, T. Sagawa: Solid State Comm. 13, 281 (1973).
7.378 Y. Sakisaka, T. Ishii, T. Sagawa: J. Phys. Soc. Japan 36, 1365 (1974).
7.379 H. Onuki, F. Sugawara, Y. Nishihara, M. Hirano, Y. Yamaguchi, A. Ejiri, H. Takahashi, H. Abe: Solid State Comm. 20, 35 (1976).
7.380 J.H. Beaumont, A.J. Bourdillon, J. Bordas: J. Phys. C 10, 333 (1977).
7.381 S. Nakai, H. Nakamori, A. Tomita, K. Tsutsumi, H. Nakamura, C. Sugiura: Phys. Rev. B 9, 1870 (1974).
7.382 D.W. Lynch, C.G. Olson: Solid State Comm. 12, 661 (1973).
7.383 D.G. Olson, D.W. Lynch, and M. Piacentini, Phys. Rev. (1979) B18, 5740 (1978).
7.384 S. Sato: J. Phys. Soc. Japan 41, 913 (1976).
7.385 W.S. Heaps, D.S. Hamilton, W.M. Yen: Opt. Commun. 9, 304 (1973).
7.386 L.R. Elias, W.S. Heaps, W.M. Yen: Phys. Rev. B 8, 4989 (1973).
7.387 W.S. Heaps, L.R. Elias, W.M. Yen: Phys. Rev. B 13, 94 (1970).
7.388 G.K. Wertheim, A. Rosencwaig, R.L. Cohen, H.J. Guggenheim: Phys. Rev. Lett. 27, 505 (1971).
7.389 S. Suzuki, T. Ishii, T. Sagawa: J. Phys. Soc. Japan 38, 156 (1975).
7.390 C.G. Olson and D.W. Lynch, to be published.
7.391 R. Grasser, E. Pitt, A. Scharmann, G. Zimmerer: Phys. Stat. Sol. (b) 69, 359 (1975).
7.392 F.C. Brown, C. Gähwiller, A.B. Kunz: Solid State Comm. 9, 487 (1971).
7.393 K. Codling, R.P. Madden: Phys. Rev. 167, 587 (1968).
7.394 A. Balzarotti, A. Bianconi, E. Burattini, M. Grandolfo, R. Habel, M. Piacentini: Phys. Stat. Sol. (b) 63, 77 (1974).
7.395 R. Haensel, E.E. Koch, N. Kosuch, U. Nielsen, M. Skibowski: Chem. Phys. Lett. 9, 548 (1971).
7.396 E.E. Koch, M. Skibowski: Chem Phys. Lett. 14, 37 (1972).
7.397 D. Blechschmidt, R. Haensel, E.E. Koch, U. Nielsen, T. Sagawa: Chem Phys. Lett. 14, 33 (1972).
7.398 J.L. Dehmer: J. Chem. Phys. 56, 4496 (1972).
7.399 F.A. Gianturco: Chem. Phys. Lett. 17, 127 (1972).
7.400 E.E. Koch, M. Skibowski: Chem. Phys. Lett. 9, 429 (1971).
7.401 E.E. Koch, S. Kunstreich, A. Otto: Opt. Commun. 2, 365 (1971).
7.402 E.E. Koch, A. Otto: Phys. Stat. Sol. (b) 51, 67 (1972).
7.403 E.E. Koch, A. Otto: Chem Phys. 3, 370 (1974).
7.404 E.E. Koch, A. Otto: Int. J. Rad. Phys. Chem. 8, 113 (1976).
7.405 D.E. Johnson, M. Isaacson: Opt. Commun. 8, 406 (1973).

Additional References with Titles

Chapter 5

Y. Farge, M. Pouey (eds): Proc. 5th Intern. Conf. Vacuum Ultraviolet Radiation Physics, Montpellier, Sept. 5-9, 1977, J. Phys. (Paris) *39*, Colloq. C-4 (1978), especially:
F. Combet Farnoux: Atomic effects in solids, p. C4-1
B. Sonntag: Atomic and molecular effects in the VUV spectra of solids, p. C4-9
H.P. Kelly: Electron correlations in photoionization and vacancy decay processes, p. C4-16
P.G. Burke: R-matrix method - advantages and applications, p. C4-27
D.A. Shirley: Correlation effects in photoelectron spectrometry of atoms, p. C4-35
J.L. Dehmer: Angular distribution of photoelectrons and non-thermal photoions from atoms and molecules, P. C4-42
K. Codling, J.R. Hamley, J.B. West: The absolute photoabsorption cross section of atomic cadmium from the 4d threshold to 250 eV. J. Phys. B *11*, 1713 (1978)
D.M.P. Holland, K. Codling, J.B. West, G.V. Marr: Multiple photoionization in the rare gases from threshold to 280eV. Report Daresbury Laboratory DL/SRF/P195 (Oct. 1978) and submitted to J. Phys. B
J.B.West, J. Morton: Absolute photoionization cross-selection tables for xenon in the VUV and the soft x-ray regions. Report Daresbury Laboratory DL/SRF/P130 June 1978 and submitted to Atomic Data and Nuclear Data Tables
V. Schmidt, N. Sandner, W. Mehlhorn: Post-collision interaction in the xenon $N_{4,5}$-Oo Auger spectrum excited by photon impact. Phys. Rev. Lett. *38*, 63 (1977)
M.Y. Adam, F. Wuilleumier, N. Sandner, V. Schmidt, G. Wendin: Satellite lines in the 5s-5p photoelectron spectrum of xenon. J. Phys. (Paris) *39*, 129 (1978)
M.Y. Adam, F. Wuilleumier, S. Krummacher, V. Schmidt, W. Mehlhorn: Correlation satellites in the outer-shell photoelectron spectrum of argon. J. Phys. B *11*, L413 (1978)
F. Wuilleumier: Photonisation simple et multiple dans les atomes. J. Phys. (Paris) *39*, C1-71 (1978)
F. Wuilleumier, M.Y. Adam, P. Dhez, N. Sandner, V. Schmidt, W. Mehlhorn: Synchrotron radiation as a photon source for photoelectron spectroscopy in the soft x-ray range. Jpn. J. Appl. Phys. *17-2*, 281 (1978)
M.Y. Adam, F. Wuilleumier, N. Sandner, S. Krummacher, V. Schmidt, W. Mehlhorn: New determination of partial subshell photoionisation cross-section in argon and xenon. Jpn. J. Appl. Phys. *17-2*, 43 (1978)
R. Bruhn, B. Sonntag, H.W. Wolff: 3p-excitation of atomic Mn; experimental evidence for the super Coster-Kronig decay. Phys. Lett. *69A*, 9 (1978)
M.Y. Adam, F. Wuilleumier, S. Krummacher, N. Sandner, V. Schmidt, W, Mehlhorn: Recent progress in the study of photoionisation processes of atomic species by spectroscopy using synchrotron radiation. J. Electron Spectrose. *15*,211 (1979)
R. Bruhn, B. Sonntag, H.W. Wolff: 3p excitation of atomic and metallic Fe, Co, Ni and Cu. J. Phys. B *12*, 203 (1979)
W. Eberhardt, G. Kalkoffen, C. Kunz: Measurement of the Auger decay after resonance excitation of Xe 4d and Kr3d resonance lines. Phys. Rev. Lett. *41*, 156 (1978)
J.P. Connerade: Inter-subshell correlations and simultaneous ejection of two photoelectrons in the absorption spectrum of Ga I. Proc. Soc. Lond. *A 354*, 511 (1977)

J.P. Connerade, M.W.D. Mansfield: Molecular damping of centrifugal barrier effects in the 3d absorption spectrum of selenium vapour. Proc. Roy. Soc. Lond. A *356*, 135 (1977)

J.P. Connerade: On double photoionization. J. Phys. B *10*, L239 (1977)

J.P. Connerade, D.M. Tracy: On "collective excitation" in the 5p spectra of barium and the lanthanides. J. Phys. B *10*, L235 (1977)

J.P. Connerade, W.R.S. Garton, M.W.D. Mansfield, M.A.P. Martin: Interchannel interaction and series quenching in the 5d and 6s spectra of PbI. Proc. Roy. Soc. Lond. A *357*, 499 (1977)

J.P. Connerade, M.W.D. Mansfield: A correction to an apparent discrepancy between theory and experiment in 3d subshell absorption spectra. Proc. Roy. Soc. Lond. A *352*, 557 (1977)

J.P. Connerade, M.A.P. Martin: On the outermost d- subshell absorption spectra of GeI and SnI. Proc. Roy. Soc. Lond. A *357*, 103 (1977)

D.H. Tracy: Photoabsorption structure in lanthanides: 5p subshell spectra of SmI, EuI, DyI, HoI, EuI, TmI, and YbI. Proc. Roy. Soc. Lond. A *357*, 485 (1977)

M.W.D. Mansfield, M.P. Connerade: On the simultaneous excitation of two electrons in neutral atomic zinc. Proc. Roy. Soc. Lond. A *359*, 389 (1978)

M.A. Baig, J.P. Connerade: Extensions to the spectrum of doubly excited MgI in the vacuum ultraviolet. Proc. Roy. Soc. Lond. A *364*, 353 (1978)

J.P. Connerade: The controlled raising of discrete levels into the far continuum. J. Phys. B *11*, L409 (1978)

J.P. Connerade: Controlling the collapse of atomic wavefunctions. J. Phys. B *11*, L381 (1978)

J.P. Connerade, M. Baig, M.W.D. Mansfield, E. Radtke: The absorption spectrum of AgI in the vacuum ultraviolet. Proc. Roy. Soc. Lond. A *361*, 379 (1978)

M.W.D. Mansfield: The simultaneous excitation of two electrons in atomic cadmium. Proc. Roy. Soc. Lond. A *362*, 129 (1978)

J.P. Connerade, M.W.D. Mansfield, G.H. Newsom, D.H. Tracy, M.A. Baig, K. Thimm: A study of 5p excitation in atomic barium I. The 5p absorption spectra of BaI, CsI and related elements. Philos. Trans. Roy. Soc. London A *290*, 327 (1979)

T.N. Chang, U. Fano: Many-body theory of atomic transitions. Phys. Rev. *A13*, 263 (1975)

T.N. Chang, U. Fano: Transition matrices for the theory of spectra. Techniques for their construction and calculation. Phys. Rev. *A13*, 282 (1975)

L.C. Davis, L.A. Feldkamp: Interpretation of 3p-core-excitation spectra in Cr, Mn, Fe, Co and Ni. Solid State Commun. *19*, 413 (1976)

L.C. Davis, L.A. Feldkamp: Interaction of many discrete states with many continua. Phys. Rev. *B15*, 2961 (1977)

L.C. Davis, L.A. Feldkamp: $M_{2,3}$ spectrum of atomic Mn. Phys. Rev. *A17*, 2012 (1978)

J.R. Swanson, L. Armstrong, Jr.: Multiconfiguration Hartree-Fock calculation of photoionization cross sections of rare gases. Phys. Rev. *A15*, 661 (1977)

M.Ya. Amusia, N.A. Cherepkov, I. Pavlin, V. Radojevic, Dj. Zivanovič: Photoabsorption for sodium and magnesium atoms in the random-phase approximation with exchange. J. Phys. B *10*, 1413 (1977)

M.Ya. Amusia, V.K. Ivanov: On the existence of a collective level in the Xe atom. Phys. Lett. *65A*, 217 (1978)

M. Ohno, G. Wendin: Many-electron effects in the 3p x-ray photoelectron spectrum of Kr. J. Phys. B *11*, 1557 (1978)

Chapter 6

Y. Farge, M. Pouey (eds.): Proc. 5th Intern. Conf. Vacuum Ultraviolet Radiation Physics, Montpellier, Sept. 5-9, 1977, J. Phys. (Paris) *39*, Colloq. C-4 (1978) especially

B. Sonntag: Atomic and molecular effects in the VUV spectra of solids, p. C4-9

J.L. Dehmer: Angular distributions of photoelectrons and nonthermal photoions from atoms and molecules, p. C4-42

A. Tramer: Decay of excited molecular states, p. C4-51

P.M. Guyon: Recent advances in the dynamics of photoselected excited states of molecules and ions. LURE preprint, Nov. 1978

J.P. Connerade: Non-Rydberg spectroscopy. Contemp. Phys. *19*, 415 (1978).

H. Petersen, A. Bianconi, F.C. Brown, R.Z. Bachrach: The absolute N_2 K-photoabsorption cross section up to $h\nu = 450$ eV. Chem. Phys. Lett. *58*, 263 (1978)

M. Sasanuma, E. Ishiguro, T. Hayaishi, H. Masuko, Y. Morioka, T. Nakajima, M. Nakamura: Photoionization of SF_6 in the VUV region. Preprint, Dec. 1978

G.V. Marr: Photoionization studies using synchrotron radiation. Daresbury Lab. preprint DL/SRF/P 147 (July 1978)

D.G. McCoy, J.M. Morton, G.V. Marr: The angular distribution of photoelectrons as a function of photon energy for the ground state photoionization of molecular oxygen. J. Phys. B, in press and Daresbury Lab. preprint DL/SRF/P 139

G.V. Marr, J.M. Morton, R.M. Holmes, D.G. McCoy: The angular distribution of photoelectrons from free molecules of N_2 and CO as a function of photon energy. J. Phys. B, in press and Daresbury Lab. preprint DL/SRF/P 142

P.M. Guyon, T. Baer, L.F.A. Ferrerira, I. Nenner, A. Tabchê-Fouhaillê, R. Botter, T. Govers: Observation of dissociative states of O_2^+ by threshold photoelectron-photoion coincidence. J. Phys. B *11*, L 141 (1978)

T. Baer, P.-M. Guyon, I. Nenner, A. Tabchê-Fouhaillê, R. Botter, L.F.A. Ferreira, T.R. Govers: Non-Franck-Condon transitions in resonant autoionization of N_2O. J. Chem. Phys., in press

T. Baer, P.M. Guyon, I.Nenner, A. Tabche-Fouhaille, B. Botter, L.F. Ferreira, Franck-Condon transitions in resonant autoionization of N_2O. Submitted to J. Chem. Phys.

R. Zietz: VUV-Fluoreszenzspektroskopie von NO und H_2 bei selektiver Anregung mit Synchrotronstrahlung. Dissertation, Universität Kaiserslautern (1978) DESY internal report F41-78/05

U. Hahn: Energie- und zeitaufgelöste Lumineszenzuntersuchungen an festen Edelgasen. Dissertation, Universität Hamburg (1978)

B. Alpert, D. Jameson, R. Lopez-Delgado, R. Schooley: On triptophan fluorescence lifetimes. Submitted to Chem. Phys. Lett.

H. Schmoranzer, R. Zietz Observation of selectively excited continuous vacuum ultraviolet emission in molecular hydrogen. Phys. Rev. A *18*, 1472 (1978)

D. Grimbert, M. Lovallêe, A. Nitzan, A. Tramer: Mechanism of collision-induced intersystem crossing in CO. Chem. Phys. Lett *57*, 45 (1978)

E.D. Poliahoff, M.G. White, R.A. Rosenberg, G. Thornton, E. Matthias, D.A. Shirley: Lifetimes of Xe excimer levels. J. Chem. Phys., in press; LBL 7651 preprint (April 1978)

H.W. Jochims. W. Lohr, H. Baumgärtel: Photoionization mass spectrometry studies of deuterated acetalhydes CH_3CDO and CD_3CHO. Chem. Phys. Lett. *54*, 594 (1978)

H.W. Jochims. W. Lohr, H. Baumgärtel: Photoreactions of small organic molecules VI: Photoionization processes of difluorethylenes. Preprint Sept. 1978

T. Bally, H. Baumgärtel, U. Büchler, E. Haselbach, W. Lohr, J.P. Maier, J. Vogt: Tris (methylidene)-cyclopropane ("/3/Radialene") Part 1: enthalpy of formation and strain energy. Chim. Acta *61*, 741 (1978)

E. Boursey, V. Chandrasekharan, P. Gürtler, E.E. Koch, P. Kunsch, V. Saile: Phonon induced fine structure of excitons in solid nitrogen. Phys. Rev. Lett. *41*, 1516 (1978)

N. Schwentner, E.E. Koch, Z. Ophir, J. Jortner: Photoelectron energy distribution measurements from benzene in rare gas matrices. Chem. Phys. *34*, 281 (1978)

J. Stöhr: Surface EXAFS studies using electron yield spectroscopy. oxygen on Ni (100). Jpn. J. Appl. Phys., in press

J. Stöhr, D. Denley, P. Perfetti: Surface EXAFS in the soft x-ray region: A study of an oxydized Al-surface. Phys. Rev. *B18*, (1978) (SSRL Report 78/06)

Chapter 7

In September 1977, an international conference on vacuum ultraviolet radiation physics was held in Montpellier. The review papers from that conference appeared in a supplement to Journal de Physique [*39*, C-4 (1978)] while contributed papers will appear in the normal literature. We list below a bibliography of some papers which have appeared since the review was written, as well as pertinent contributed papers from the Montpellier conference; the latter show the abstract number. (Those known to have been published have been omitted.)

A. Ejiri, F. Sugawara, H. Onuki, M. Hirano: The absorption spectra of polyvalent metals (Ga, In, Sn, Pb and Bi) in the extreme ultraviolet. II 11

D. Beaglehole, M. deCrescenzi, B. Thiêblemont, N. Binh, M.L. Theye, G. Vuye: Alloy and surface layer studies by differential reflectivity measurements. II 28

D. Beaglehole, M. deCrescenzi, B. Thiêblemont, M.L. Theye, G. Vuye: d-electron excitation in Au, Cu and Cu-Au alloys between 18 and 30 eV. II 29

R. Wall, J.R. Stevenson, J. Larsen, M.W. Ribarsky, K. Legg: Optical reflectivity and Auger spectroscopy of titanium surfaces with controlled exposures to oxygen and nitrogen. II 39

J.H. Weaver, C.G. Olson: Soft x-ray absorption studies of thorium 5d-5f structures in thorium and thorium compounds. II 21

S. Yamaguchi, H. Hanyu, H. Koike: $M_{2,3}$ and $L_{2,3}$ absorption spectra of Al based 3d transition metal alloys. II 23

R. Mamy, R. Thieblemont, F. Pradal, A. Chevy: Reflectivite de InSe de 2 a 30 eV. II 34

M. Piacentini, C.G. Olson: Optical properties and band structure of the layer compounds InSe, GaSe, and GaS. II 35

J. Stöhr, R.S. Williams, G. Apai, P.S. Wehner, D.A. Shirley: Carbon K-edge absorption fine structure of thin films. II 17

K.M. Monahan, V. Rehn: Time resolved spectroscopy of doped rare-gas solids. I 70

E. Pantos, S.S. Hasnain, I.H. Munro, T.D.S. Hamilton: Benzene absorption and excitation spectra in rare-gas matrices. I 71

S.S. Hasnain, I.H. Munro, T.D.S. Hamilton, P. Brint: Exciton diffusion or surface quenching of the luminescence center in rare gas solids. I 72

V. Saile, W. Steinmann, E.E. Koch: Properties of surface and bulk excitons in rare gas solids. I 74

V. Hahn, B. Jordan, N. Schwentner: Decay and relaxation of excitons in rare gas solids investigated by pulsed photon excitation. I 76

H.W. Wolff: Electronic structure of surface and bulk excitons in rare gas solids. I 80

W. Andreoni, M. deCrescenzi, E. Tosatti: Molecular aspects in the optical properties of rare gas solids. I 81

M. Iwan, C. Kunz: Photoemission of autoionizing Na^+-2p core excitons in NaCl.

M. Watnabe, E. Nishioka: Fundamental absorption in RbF-CsF solid solutions. II 19

D. Bäuerle, W. Braun, V. Saile, G. Sprüssel, E.E. Koch: Vacuum ultraviolet reflectivity and band structure of $SrTiO_3$ and $BaTiO_3$

D.E. Aspnes: Modulation spectroscopy with synchrotron radiation. Festkörperprobleme *XVII*? 235 (1977)

G. Jezequel, J. Thomas, J.C. Lemonnier: Optical transitions from d core states in polycrystalline indium, tin and lead. Solid State Comm. *23*, 559 (1977)

H. Petersen, C. Kunz: Density of states in the K excitation spectra of solid and liquid lithium. J. Phys. F *7*, 2495 (1977)

P.H. Citrin, G.K. Wertheim, Y. Baer: Many-body processes in x-ray photoemission line shapes from Li, Na, Mg, and Al metals. Phys. Rev. B *16*, 4256 (1977)

T.A. Callcott, E.T. Arakawa, D.L. Ederer: Emission and absorption x-ray edges of Li. Phys. Rev. B *16*, 5185 (1977)

M. Cukier, P. Dhez, B. Gauthê, P. Jaegle, C. Wehenkel, M. Combet-Farnoux: The 5d excitations in thorium and uranium observed by both photoabsorption and fast electron energy loss spectroscopy. J. Phys. Lett. (Paris) *39*, L315 (1978)

D.E. Aspnes, R.S. Bauer, R.Z. Bachrach, J.C. McMenamin: Thermoreflectance and temperature dependence of the $L_{2,3}$ soft x-ray threshold in Si. Phys. Rev. B *16*, 5436 (1977)

R. Mamy, B. Thiêblemont, L. Martin, F. Pradal: Reflectivity of layer-type transition metal dichalcogenides from 6 eV to 40 eV. Il Nuovo Cimento *38B*, 196 (1977)

J.C. Rife, R.N. Dexter, P.M. Bridenbaugh, B.W. Veal: Optical properties of the chalcopyrite semiconductors $ZnGeP_2$, $AnGeAs_2$, $CuGaS_2$, $CuAlS_2$, $CuInSe_2$, and $AgInSe_2$. Phys. Rev. B *16*, 4491 (1977)

427

F.C. Brown, R.Z. Bachrach, M. Skibowski: $L_{2,3}$ threshold spectra of doped silicon and silicon compounds. Phys. Rev. B *15*, 4781 (1977)

W. Eberhardt, G. Kalkoffen, C. Kunz, D.E. Aspnes, M. Cardona: Photoemission studies of 2p core levels of pure and heavily doped silicon. Phys. Status Solidi (b) *88*, 135 (1978)

E. Tegeler, N. Kosuch, G. Wiech, A. Faessler: Anisotropic emission of the x-ray K-emission band of nitrogen in hexagonal boron nitride. Phys. Status Solidi (b) *84*, 561 (1977)

S.S. Hasnain, T.D.S. Hamilton, I.H. Munro: Absorption coefficient in the n = 1 [Γ(3/2, Γ(1/2)] exciton region of solid Kr. J. Phys. C *11*, L261 (1978)

I.P. Kaminow, B.G. Bagley, C.G. Olson: Measurements of the absorption edge in fused silica. Appl. Phys. Lett. *32*, 98 (1978)

E. Boursey, V. Chandrasekharan, P. Gürtler, E.E. Koch, P. Kunsch, V. Saile: Phonon induced fine structure of excitons in solid nitrogen. Phys. Rev. Lett. *26*, 1516 (1978)

A.J. Bourdillon, J. Bordas, F. Khumalo: The reflection spectrum of lead glass. Philos. Mag. B *37*, 371 (1978)

Subject Index

432

Springer Series in Chemical Physics

Editors: V. I. Goldanskii, R. Gomer, F. P. Schäfer, J. P. Toennies

This series is devoted to graduate-level single- and multi-author monographs in the cross-disciplinary fields of physics and chemistry. Typical areas covered are photochemistry, isotope separation, molecular beam techniques, chemisorption, catalysis, and surface sciences in general.

Springer-Verlag
Berlin
Heidelberg
New York

Topics
in Applied Physics

Founded by H. K. V. Lotsch

A Selection

Springer-Verlag
Berlin
Heidelberg
New York